8051 MICROCONTROLLER

With deepest respect

to

Dr Raj Kamal

as my humble homage

CONTENTS

26   Advanced Microcontrollers   369

PREFACE TO THE FIRST EDITION

This is a textbook on 8051 microcontroller with a little difference. A glance at its table of contents would indicate that its chapter organization is suitable for classroom teaching. Furthermore, it contains plenty of pedagogical features for understanding the subject matter. Certain additional chapters, such as Power Management, Case Study and Advanced Microcontrollers, may be helpful for students for a better understanding of 8051 Microcontroller. A detailed Index at the end would help interested readers.

This book is a gradual development of my lecture notes on 8051 used for the last 10 years or so. Initially it consisted of a few pages and started growing as students reacted to it in a positive way. My teaching experience tells me that a textbook, which is meant for leading a student, who is trying to understand the subject matter for the first time, must clarify all the basic features in the most systematic and lucid way. The objectives must be well defined for any section right in its beginning, basic points must be quickly and clearly explained, suitable well-planned examples should follow and finally, as far as possible, diagrams should replace elaborately written descriptions for a quicker grasp of all fundamentals and critical points.

The chapters of this book are prepared in a way that can be covered within an hour lecture in the classroom including the question–answer session. Only a few chapters would demand two hours to complete. Solved examples, figures, tables and exercises of this book are designed to help students in the best possible manner. As the laboratory classes are expected to start at the earliest, instruction details are covered at the beginning before discussing Interrupts and Interfacing Issues.

All comments and criticisms regarding this work would be acknowledged by the author at subrataghoshal52@yahoo.co.in.

Subrata Ghoshal

PREFACE

As compared with the time-tested and well-established titles on the subject matter, the present title, *8051 Microcontroller: Internals, Instructions, Programming and Interfacing, 2nd Edition*, might have been considered as *yet-another-one* type entry. However, the steady growth of yearly sales figure indicates that the effort of the author was well accepted and appreciated by the readers, which was further confirmed by the mails received from different parts of this country. This may be attributed to its simple and easy-to-understand approach, application-oriented examples and rich pedagogical features.

Considering demands and expectations of readers, several additional features are included in this new edition. For many example cases, C-versions of assembly language programs are incorporated in different chapters. Few more solved examples are added in chapters 15 and 16 for clarification of the subject matter and for easier understanding. One new chapter (chapter 27) dealing with Intel 8255 PPI interfacing is included, a new section (section 2.8) on Assembler & IDE is placed in chapter 2, and several typographical errors are corrected. It is expected that these additional features would make the present work more student as well as teacher friendly.

I express my gratitude to the Vice Chancellor of Sikkim Manipal University, Brig. Dr. Somnath Mishra (Retd.) and the Director of Sikkim Manipal Institute of Technology, Maj. Gen. Dr. S. S. Dasaka, SM, VSM, (Retd.), for theirconstant encouragement and help extended to me for the preparation of this new edition. I sincerely thank the whole team of Pearson India, and specially Thomas Mathew Rajesh, Tushar Mishra, and Vipin Kumar for their excellent effort to bring out this new edition in most professional manner. I must thank my colleague, Gangotri Chakraborty and my student, Rudra Rimal, for pointing out several typographical mistakes in the previous edition. I also thank Ghanashyam Sharma, Karma W. Tamang and Purna Bahadur Subba of IT department of SMIT for helping me in various ways. I must not forget to thank my ex-student Neeraj (now in ENTESLA) for his help in C-program developments and testing. Finally I thank my wife Rita and my daughter Gayatri for their constant support and encouragements to me for completing the present task.

All comments and criticisms related to this edition would be thankfully acknowledged by the author through subrataghoshal52@yahoo.co.in.

Subrata Ghoshal

ACKNOWLEDGEMENTS

My sincere gratitude to P. P. Chhabria, Chairman; Padmashree Dr Vijay Bhatkar, Director; Dr Atanu Rakshit, Deputy Director; and Professor Shashank Pujari, HOD of Embedded System Design Department of International Institute of Information Technology, Pune for their support. I must acknowledge Professor Sahana Bhosle, Dr Kallol Das and Professor Tathagata Bhattacharya for their continuous encouragement during the preparatory stage of this manuscript. Kumud and Mahesh, of the Systems Department, extended their whole-hearted cooperation. I must thank many of my students, specially Ishani Pandit, Toral Panchal, Mehul Asher, Narendra Rathore, Rahul Sangole and others, for their help at various phases of this manuscript preparation.

I must also thank my unknown reviewers for their constructive criticisms and comments, which has enriched this work. I sincerely thank the whole team of Pearson Education and, in particular, Ruchi Sachdev for completing this project within such a short period of time and in such an efficient manner. Special mention of Thomas Mathew Rajesh and Sachin Saxena of Pearson Education who practically forced me to complete this manuscript with their constant encouragement and enormous help in every respect. I sincerely thank Intel Corporation for permitting me to reprint the instruction set definitions of MCS-51.

I am obliged to Dr Raj Kamal, Senior Professor and vice chancellor of School of Computer Sciences and Electronics and Information Technology, Faculty of Engineering Sciences of Devi Ahilya Viswavidyalaya, Indore, for kindly allowing me to dedicate this work to him. I am grateful to him in many respects, not only for his critical comments for this manuscript but also for sharing his deep insight with me regarding various aspects of modern technology and education and continuously encouraging me in technical writing.

Finally, I must profoundly thank my wife, Rita, and my daughter, now a young lady, Gayatri, for their cooperation and support throughout the preparatory stage of this manuscript.

Subrata Ghoshal

1 INTRODUCTION

CHAPTER OBJECTIVES

In this chapter, the reader is introduced to microcontrollers, its evolution, applications and main features. After completion of the chapter, the reader should be able to understand:

- Functions of a microcontroller.
- Essential peripherals for a microcontroller.
- Difference between a microprocessor (like 8085) and a microcontroller (e.g. MCS-51).
- The difference between Princeton and Harvard architectures.
- Variations in MCS-51 family of microcontrollers.
- Importance of power management in microcontrollers.
- Different types of packaging used for microcontrollers.

1.1 | Introduction

In last 40 years or so, we have witnessed a silent but rapid revolution as far as electronic computing, communication and control are concerned. Proof of this statement would be available in mobile handsets, MP3 players, automatic washing machines, microwave ovens, portable digital blood pressure- or blood sugar-measuring medical instruments, billing system in STD phone booths, etc., only to mention a few. This list is enormous and growing exponentially. All these are possible because of the usage of microcontrollers, which evolved from microprocessors. A small system fabricated with a microcontroller and having a PC interface through serial port may be seen in Fig. 1.1. The 40-pin IC is the microcontroller.

Figure 1.1 A microcontroller-based system with PC interface

1.2 | Microprocessor

A microprocessor, or its successor, a microcontroller, is a device which is capable of continuously *fetching* and *executing* instruction after instruction, as long as it is not switched off (Fig. 1.2). These instructions are electronic instructions generally represented by a group of bits. Fetching denotes bringing these instructions from their storage area, which is designated as memory. The term 'execution' means performing arithmetic, logical or control operations.

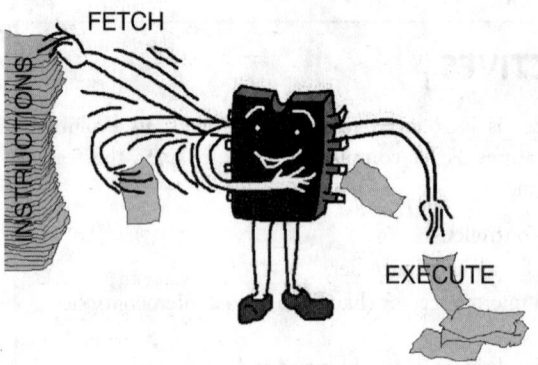

Figure 1.2 Function of a microprocessor or microcontroller

With respect to the human world, the speed of this execution is very fast. The fastest human may tap a computer key, say, five times a second. On the other hand, a processor is capable of executing a million of instructions within this second. To negotiate between this extremely fast and superbly slow operations of these two entities, some special mechanism is necessary. These mechanisms or devices are designated as input or output ports, depending upon their functions, or simply *ports*.

1.2.1 | Microprocessor-Based System

In case of microprocessors, the storage area for instructions, input and output ports are outside the micro-processor, in the form of additional devices. These are to be interfaced with the microprocessor through a system bus, composed of address bus, data bus and control bus (Fig. 1.3). Therefore, a microprocessor-based system must integrate a microprocessor and several other devices, which are externally interfaced through the system bus.

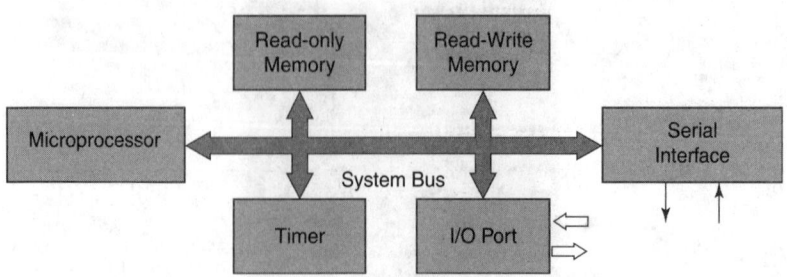

Figure 1.3 Schematic arrangement of a microprocessor-based system

 On an average about 250 to 300 soldering to be done for a small microprocessor based system. If a general purpose board is used, about 125 pieces of wires may be necessary for connections. This may provide the reader a rough idea about the complexity of fabricating a microprocessor based system.

1.3 | Microcontroller

A microcontroller is a device, which contains all the devices necessary to form a working system, in a single chip. Therefore, inside the microcontroller we find a central processing unit (CPU), program memory, data memory, I/O ports, timers, serial communication interface (USART), etc. (Fig. 1.4). In general, no external interfacing is necessary with memory or I/O devices (unless some expansion is desired) to make any microcontroller-based circuit working. In most of the cases, an external quartz crystal (for system clock), a reset circuit (made of a resistor and a capacitor) and a DC power supply (sometimes even from batteries) are sufficient to make such a unit operational.

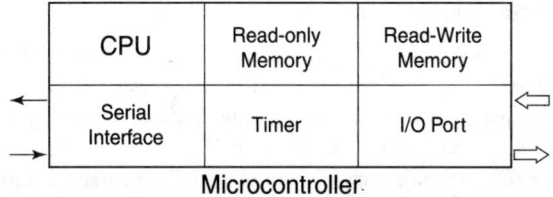

Figure 1.4 Schematic internal architecture of a microcontroller

Therefore, to summarize, a microcontroller is a self-sufficient standalone device (IC) but a microprocessor is not. Major advantages of microcontrollers over microprocessors are presented in Table 1.1.

Table 1.1 Advantage of microcontrollers over microprocessors

	Microprocessors	Microcontrollers
Support devices	External	Internal
Size of any working unit	Larger	Smaller
Power consumption	More	Less
Reliability of the system	Less	More
Time for constructing a circuit	More	Less
Throughput	Less	More
Software protection	Not possible	Possible

1.3.1 | General Architecture

In Fig. 1.4, a general architecture of the microcontroller is presented. However, the components of a microcontroller are not limited to these items only. As a matter of fact, it is quite common to find an analog comparator or an analog-to-digital converter (ADC) or a digital-to-analog converter (DAC) or a Watch Dog Timer (WDT), within a microcontroller package. The size and type of on-chip memory also varies substantially. However, a microcontroller never contains a DC motor or a stepper motor inside it.

Once I have really faced this question in my microcontroller class about the existence of any small DC motor built-in within any microcontroller. However, in some cases, drivers, like PWM control, are offered within a microcontroller. For example, MCS-96 offers three PWM outputs (refer chapter 26 of this book).

1.3.2 | Software Protection in Microcontrollers

Apart from making so many functions available in a single chip, microcontrollers offer another unique feature, which was not possible to implement in microprocessor-based systems. We are referring to the protection available in microcontrollers against unwanted software piracy. Most of the present date microcontrollers offer the feature of locking its on-chip program memory so that reading it from external circuit is not possible.

1.3.3 | Brief History of Intel Microcontrollers

First microcontroller, designated as MCS-48, from Intel was available in 1976. It was (and still being) extensively used in IBM PC keyboards. Its popularity resulted in its successor, MCS-51, also from Intel in 1980. Both of these devices are 8-bit, which means that at any time they can handle 8-bit of data. MCS-96 was the first 16-bit microcontroller from Intel made available in 1982.

Just like the success of 4004 microprocessor from Intel made many manufacturers to bring out their own microprocessors, similarly, the popularity of MCS-48 brought a few more microcontrollers in the market. Presently, different types of microcontrollers are manufactured by many semiconductor companies like Zilog, Motorola, Texas Instruments, Dallas Semiconductors, Phillips, Siemens, Atmel, Microchip, Analog Devices, etc.

1.4 | MCS-51 Family

Many of these manufacturers brought out their own version of microcontrollers around the MCS-51 *core*, but with different architecture. Here, the term 'core' indicates the instruction decoding part of a processor. Therefore, microcontrollers using MCS-51 core are capable of executing all MCS-51 instructions. An example of this is ADμC812 from Analog Devices which is fabricated around MCS-51 core but has an on-chip 12-bit ADC and two DACs, which are not available in the original MCS-51 architecture. Another example is Atmel's AT89C2051 which also uses MCS-51 core and has an on-chip analog comparator, not available in original MCS-51 architecture.

This wide variety is also observed in the size and type of memory offered in different devices. The available options are ROM, PROM, EPROM, EEPROM, FLASH, etc., and the size varies from 64 bytes onwards. Even some versions are available without any program memory (ROM-less version), which are expected to be externally interfaced. However, only two variations of the memory architecture are available, namely Princeton architecture and Harvard architecture.

Variety may also be observed in other microcontroller families, like: PIC or ARM. This is because of the fact that the present trend of design is to select the most appropriate microcontroller and use it to its maximum extent.

1.4.1 | Princeton and Harvard Architectures

Princeton architecture, also known as Von (pronounced as 'fon' and *not* as 'bhon') Neumann architecture, does not distinguish between program memory and data memory. The entire addressable memory area might have any combinations of program instructions and data [Fig. 1.5(a)]. Harvard architecture, on the other hand, treats program memory and data memory separately [Fig. 1.5(b)]. Intel MCS-51 is designed around this Harvard architecture. Those readers, who are familiar with Intel 8085 microprocessor, may be reminded that Intel 8085 was fabricated around Princeton architecture.

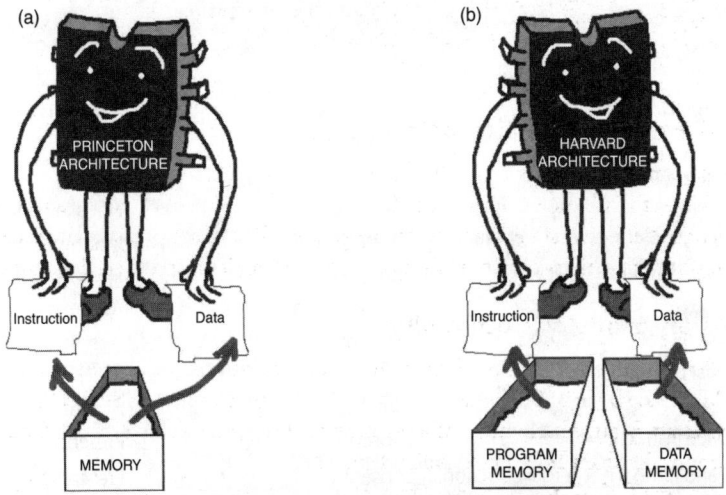

Figure 1.5 (a) Princeton and (b) Harvard architecture

 The reader's attention is drawn to the fact that Princeton architecture does not demand any additional address, data and control bus for memory communications, but Harvard architecture does. Each one is having its own advantage as well as limitations.

1.4.2 | Comparison Between 8085 and MCS-51

As some of the readers have already studied Intel 8085 microprocessor, therefore, a comparative study of 8085 and 8051 may not be totally irrelevant. Some of the similarities between these two devices are presented in Table 1.2.

Table 1.2 Similarities between 8085 and 8051

Parameters	8085	MCS-51
8-bit CPU	Yes	Yes
Multiplexed lower address and data bus	Yes	Yes
16 address lines	Yes	Yes
Eight data lines	Yes	Yes
ALE signal to de-multiplex address from data	Yes	Yes
Memory-mapped I/O	Yes	Yes
On-chip serial interface	Yes	Yes

On the other hand, the differences between these two devices are shown in Table 1.3. As we can observe, this table has more parameters than the table of similarities. It may be kept in mind that many of these features of MCS-51 have not been discussed yet and would be introduced in subsequent chapters.

Table 1.3 Differences between 8085 and 8051

Parameters	8085	MCS-51
System stack	Outside, unlimited	Inside, limited
I/O interfacing	Both memory-mapped I/O and I/O-mapped I/O	Only memory-mapped I/O
Reset input	Active low	Active high
Instruction execution time (minimum)	1.2 μsec	1 μsec
Max. size of memory (prog. + data)	64K	128K + 128 bytes
Architecture adopted	Princeton	Harvard
On-chip external interrupt pins	5	2
DMA request	Allowed	Not allowed
Serial communication	By polling	Interrupt driven
READY signal for slower peripheral devices	Exists	Does not exist
Pipeline architecture	Not adopted	Adopted
Status indicating signal (like S0, S1)	Present	Absent
Read signal for external program memory	$\overline{RD}$	$\overline{PSEN}$
Z flag in PSW	Present	Absent
8-bit multiply and divide instruction	Absent	Present
'Decrement register and jump if not zero' instruction	Absent	Present
Program counter (PC)	Available to the programmer	Not available to the programmer
No. of general purpose registers	Six (B, C, D, E, H, L)	Eight (R0, R1, R2, R3, R4, R5, R6, R7)
Swithing of register banks	Not possible	Possible
Size of SP	16-bit	8-bit
Power management features	Not provided	Provided

MCS-51 core is available in many variations. Table 1.4 presents some of the MCS-51 family members. The numbering convention adopted for MCS-51 is indicated in Fig. 1.6. If this numbering convention is known then the Table 1.4 may be generated from that. Note that last three rows of the table follow a different numbering convention not shown in Fig. 1.6.

Table 1.4 MCS-51 family members characteristics

Part no.	Pins in DIP	On-chip program memory	On-chip data memory (in bytes)	No. of I/O lines	No. of timers/ counters	Serial interface
8031	40	Not provided	128	32	2	1
8032	40	Not provided	256	32	3	1
8051	40	ROM 4K	128	32	2	1
8052	40	ROM 8K	256	32	3	1
8054	40	ROM 16K	256	32	3	1
8751	40	EPROM 4K	128	32	2	1

(Continued)

Part no.	Pins in DIP	On-chip program memory	On-chip data memory (in bytes)	No. of I/O lines	No. of timers/counters	Serial interface
8752	40	EPROM 8K	256	32	3	1
8754	40	EPROM 16K	256	32	3	1
89C51	40	FLASH 4K	128	32	2	1
89C52	40	FLASH 8K	256	32	3	1
89C54	40	FLASH 16K	256	32	3	1
89C1051	20	FLASH 1K	64	15	1	-
89C2051	20	FLASH 2K	128	15	2	1
89C4051	20	FLASH 4K	128	15	2	1

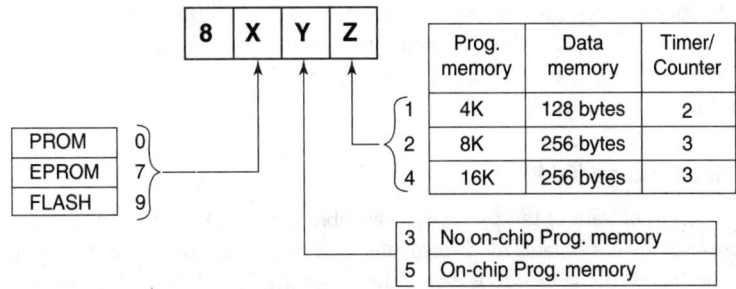

Figure 1.6 Numbering convention of MCS-51 microcontrollers

The second digit from left indicates the type of program memory (PROM/EPROM/FLASH) provided that the third digit from left is '5'. If the third digit from left is '3' then it should be taken as a version with no on-chip program memory. In that case external interfacing of program memory is essential. The last or fourth digit from left indicates the size of on-chip program memory, data memory and number of timers/counters. The standard speed of MCS-51 was set as 12 MHz. However, even a version with a speed of 100 MHz is available from SiLabs.

1.5 | Power Management

As most of the microcontroller-based systems are battery operated, one of the most important issues of microcontroller is its power management. Modern microcontrollers are designed with less power-hungry Complementary Metal-Oxide Semiconductor (CMOS) technology. Apart from that most microcontrollers offer power management modes, like Sleep Mode or Idle Mode. In these modes, microcontrollers consume less power than at the time when they are fully functional. Some microcontrollers (like AVR) allow to shut down their unused sectors (like ADC or DAC) so that less power could be consumed.

Power management is a unique feature of microcontrollers. As very few microprocessor based system were battery operated, therefore it was overlooked for microprocessors. As a matter of fact, a microprocessor based system consumes about 10 times more power than a microcontroller based system.

1.6 | Microcontroller Packaging

Modern microcontrollers are available in different packages. This packaging forms the outer shape of the IC with the silicon wafer encapsulated inside it. For certain devices, only some specific type of packaging is available. Therefore, in this section, we quickly take an overview of some common type of packaging for microcontroller ICs.

1.6.1 | Plastic Dual-Inline Package (PDIP)

This is one of the oldest and most widely adopted packaging. In general, in a rectangular package, two rows of pins (or leads) come out from two larger sides of the package. Spacing of these pins is 0.1 inch along the row. Across the rows, the spacing is 0.3 or 0.6 inches. These packages may be directly inserted through the holes of the printed circuit board (PCB) or it may be placed over an inexpensive socket (IC base).

This through-the-hole technique of component mounting prevailed for about three decades. In the meantime, another technology evolved which is known as surface mounting technology (SMT). In this case, instead of mounting the IC on the PCB through drilled holes, the IC is placed directly on solder pads without any holes. This technique reduced the device packing size. Moreover, components may be placed on both sides of a PCB. Nowadays, apart from ICs, all major electronic components, like resistor, capacitor, diode, transistor and light-emitting diode (LED), are available as surface mounting device (SMD). This generated a few new types of packaging as discussed below.

1.6.2 | Quad Flat Package (QFP)

As the name suggests, pins or leads of this package are available from all four sides of a square or rectangular package. However, these leads are not suitable for through-the-hole mounting. The device is to be placed over the PCB and soldering is to be done at the same side as that of the device. The size of the package is largely reduced and the spacing of the leads are also less (0.4–1 mm). Number of leads may vary from 32 to as much as 200 in some cases.

1.6.3 | Plastic-Leaded Chip Carrier (PLCC)

The difference between PLCC and QFP is basically in the shape of their leads. In case of QFP, the leads (projecting out of the package) are generally Gull-Wing type [Fig. 1.7(b)]. For PLCC, these leads are called J-leads (having the shape of the upper case 'J'), as shown in Fig. 1.7(a). As this J-lead reduces the footprint of the IC, therefore it is preffered and the PCB becomes more compact.

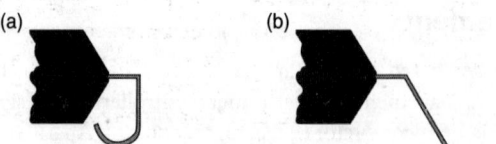

(a)   (b)

Figure 1.7 (a) J-lead and (b) Gull-Wing lead of ICs

1.6.4 | Plastic Quad Flat Pack (PQFP)

This packaging is similar to QFP as described above. Leads are provided at all four sides of the package. The shape of the leads might be gull-wing or bat-wing. Nowadays PQFPs are replaced by thin quad flat packs (TQFPs).

1.6.5 | Thin Quad Flat Pack (TQFP)

With respect to PQFP, this package, TQFP, is thinner in height. The maximum thickness ranges from 1 to 1.4 mm with a lead footprint of 2.0 mm (Fig. 1.8). The decreased thickness helps in increasing device density in circuit boards. The size of TQFP varies from 5 to 20 mm per side.

Figure 1.8 Top view of a TQFP

1.7 | Future Trend

With constant improvements in the throughputs, reduction of unit price and economy in power requirements, microcontrollers are foreseen to be used in more and more areas. Already these have percolated in various requirements of our daily life. In future, we expect to use a variety of compact and smart gadgets operated by microcontrollers.

SUMMARY

Microcontrollers are essentially microprocessors with on-chip program and data memory along with ports, timers, serial communication features, etc., all in a single-packaged IC. This integration makes the microcontrollers power efficient, compact and more reliable. MCS-48 was the first microcontroller designed by Intel, which was extensively used in IBM PC keyboards. MCS-51 was the next microcontroller designed by Intel and made available by 1980.

The MCS-51 family of microcontrollers was designed around Harvard architecture, which treats program memory and data memory separately. Presently, many manufacturers offer their own unique microcontrollers of 8-, 16- or 32-bit capabilities. These are available in various types of packages, like TQFP or dual-inline package (DIP), etc. Most microcontrollers offer some protection features for the on-chip program memory against unwanted software piracy. Power management also forms an important feature of microcontrollers, as, in general, these are powered from battery sources.

POINTS TO REMEMBER

- Once switched on, any microprocessor or microcontroller would keep on fetching and executing Instruction after instruction.

- Princeton architecture does not distinguish between program and data memory but Harvard architecture does.

REVIEW QUESTIONS

Evaluate Yourself

1. The basic job of a microcontroller is to
   (a) fetch and execute instructions
   (b) communicate with its peripheral devices
   (c) control external world
   (d) none of these

2. With respect to the human world, the speed of execution of microcontrollers is
   (a) very slow   (b) same
   (c) very fast   (d) none of these

3. Which feature of microcontroller is not available in a microprocessor-based system?
   (a) Registers   (b) Program memory lock
   (c) ADC   (d) None of these

4. How do you pronounce 'Von Neumann Architecture'?
   (a) Bone Human Architecture
   (b) Bhon Numan Architecture
   (c) Fon Newman Architecture
   (d) None of these

5. Which of the following architecture was adopted for Intel 8085 microprocessor?
   (a) Stack architecture
   (b) Register architecture
   (c) 8-bit architecture
   (d) None of these

6. MCS-51 microcontrollers were introduced by _____ in the year _____.
   (a) Fairchild, 1981   (b) Intel, 1976
   (c) Intel, 1980   (d) None of these

7. The type of program memory available in 87XX series of microcontrollers is
   (a) EPROM   (b) FLASH
   (c) OTP   (d) None of these

8. Which of the following leads are available in DIP ICs?
   (a)   (b)

   (c)    (d) None of these

9. 'Sleep' and 'idle' modes are offered in microcontrollers for
   (a) serial communications
   (b) memory management
   (c) I/O configuration
   (d) none of these

10. Multiplexed lower address and data bus are available both in
   (a) 8085 and MCS-51
   (b) 8086 and Z80
   (c) 6800 and AVR
   (d) none of these

Search for Answers

1. How the instructions are represented in a microcontroller? Where are they stored? What is meant by 'executing an instruction'?

2. Why some special mechanism is necessary to communicate between the human world and microcontrollers? What is the designation of this mechanism?

3. What is the difference between a microprocessor and a microcontroller?

4. What is the advantage of J-lead packages over Gull-Winged packages?

5. What is the meaning of 'Power Management' in case of microcontrollers?

6. Which was the first microcontroller introduced by Intel? Where was it extensively used?

7. A microcontroller contains CPU, RAM, ROM, I/O, Timers, ADC, DAC, UART, WDT, etc. in a single package but not a stepper motor or a servo motor. Why?

8. What is meant by MCS-51 core?

9. Name one 16-bit and one 32-bit microcontroller. Is there any 64-bit microcontroller available?

10. What is meant by CMOS technology?

Think and Solve

1. Why it is not possible to protect a microprocessor-based system from software piracy?

2. What are the advantages of a microcontroller over a microprocessor?

3. What are the advantages and disadvantages of Princeton (Von Neumann) Architecture against Harvard Architecture?

4. In spite of offering on-chip program and data memory, why additional external memory interfacing provision is available in microcontrollers?

5. What is the major disadvantage of modern SMD technology?

6. Make a list of the applications of microcontrollers around your own world.

7. Why microprocessors or microcontrollers are designed with their bit-wise width always a multiple of 8 (e.g. 8-bit, 16-bit, 32-bit, 64-bit)?

8. What are the differences between ALU, CPU, MPU and a microcontroller?

9. In which year MCS-51 was available in the market?

10. Which of the following two phrases is correct? 'RAM' or 'RAM memory'? Why?

2 GENERAL ARCHITECTURE

CHAPTER OBJECTIVES

In this chapter, we will learn about some external and internal features of MCS-51 microcontroller. After completion of the chapter, the reader should be able to

- Name major external signals and their types.
- Identify minimum input to make the system working.
- Understand alternate functions of some pins.
- Explain basic internal architecture of MCS-51.
- Know details of program memory and data memory.
- Understand registers, system clock characteristics and reset circuit.

2.1 | External Features

As mentioned in the previous chapter, MCS-51 core is available in many variations. For easier discussions, we select a standard version of MCS-51, the 8051 or XX51. This XX may be replaced by 80, 87 or even by 89C. However, the features discussed would remain identical for all these devices. For indicating any special cases, we would mention the version number of the concerned device, explicitly.

Externally, XX51 is a 40-pin IC in dual inline package (DIP). It is also available in other types of packaging, like plastic leaded chip carrier (PLCC) and thin quad flat pack (TQFP). However, for the present text we refer to the DIP. Schematically, major external features of XX51 are shown in Fig. 2.1.

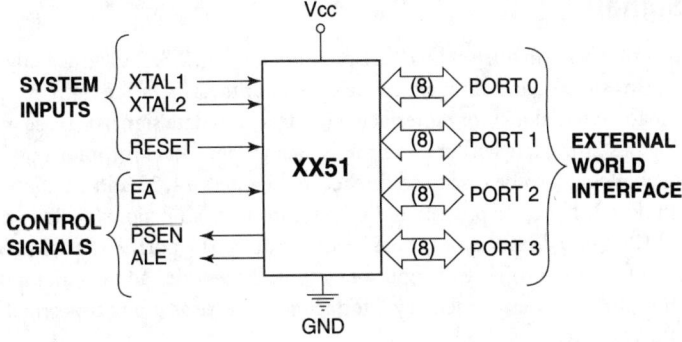

Figure 2.1 XX51 schematic inputs and outputs

The signals seen on the right-hand side of Fig. 2.1 are for external world interfacing. These are called input/output (I/O) ports, or simply ports. There are a total of four such ports designated 0, 1, 2 and 3. Each of these four ports has eight I/O lines, which are *bit-programmable*. Being bit-programmable means that individually each port bit may be programmed as either input or output, irrespective of the condition of any other bits of that port.

Along the remaining three sides of XX51 of Fig. 2.1, all system input signals, necessary for running the device, are shown. These are power input [common-collector voltage (Vcc) and system ground potential (GND)], external crystal input for clock circuit (XTAL1 and XTAL2) and the RESET input. Power, as we can easily understand, is the most important input, without which the device would not function at all. External crystal input is necessary to generate the *heartbeats* of the processor. The third input signal, RESET, is necessary to initialize various features of the processor to make a fresh start.

Three more signals are shown at the left-hand side of the XX51 in Fig. 2.1. These [External Access ($\overline{EA}$), Program Store Enable ($\overline{PSEN}$) and Address Latch Enable (ALE)] are control signals to be used if the system demands interfacing of external memory. Otherwise, these signals are not used. We will discuss in details about these three signals along with the external memory interfacing in Chapter 17. However, at present, we mention only that if the external memory interfacing is not necessary, then the $\overline{EA}$ input should be pulled high (connected with Vcc). Fig. 2.2 shows the schematic system inputs for the stand-alone operation of XX51. Note that a decoupling capacitor of 0.1 µF is necessary at Vcc to avoid unwanted high-frequency voltage oscillations at the power input.

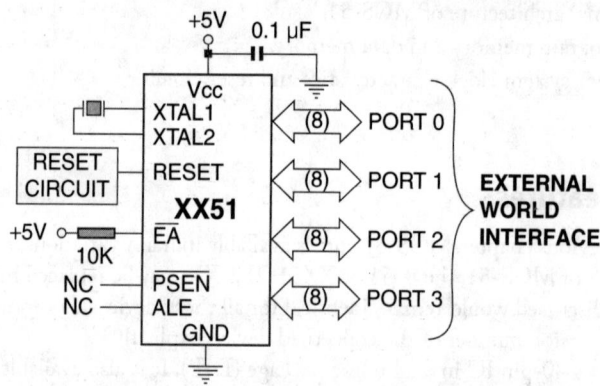

Figure 2.2 Schematic system inputs for stand-alone operation

2.2 | Pins and Signals

Detailed pin-wise signals of XX51 (in 40-pin DIP) are presented in Fig. 2.3. A careful study reveals that some pins are associated with two signals, one of which is within a parenthesis. On the right, for ports 0 and 2, the signals in parenthesis are for external memory interfacing (address and data signals). These will be discussed in detail in Chapter 17. Signals at the left within parenthesis against port 3 are for interrupts, counters, control and serial interface, which would also be discussed further in Chapters 14, 15 and 16. Note that no alternate signal is associated with port 1. Here, we present a brief description of all major signals of the device.

Pins 32-39: Port 0 (P0.0–P0.7) is the only true bi-directional I/O port. External pull-up registers are necessary for this port to function properly as input or output. Low-order address and data bus signals are also multiplexed with this port for external memory interfacing. To serve as input, concerned bit(s) of this port must be written with 1s.

Pins 1-8: Port 1 (P1.0–P1.7) is a quasi-bi-directional I/O port. It is internally pulled up. To serve as input, concerned bit(s) of this port must be written with 1s. This port does not offer any alternate functions.

```
                    ┌───────∪───────┐
              ┌──┐  │               │  ┌──┐
            1 │  │  P1.0       Vcc  │40│
              └──┘  │               │  └──┘
              ┌──┐  │               │  ┌──┐
            2 │  │  P1.1    [AD0]P0.0 │39│
              └──┘  │               │  └──┘
              ┌──┐  │               │  ┌──┐
            3 │  │  P1.2    [AD1]P0.1 │38│
              └──┘  │               │  └──┘
```

1	P1.0	Vcc	40
2	P1.1	[AD0]P0.0	39
3	P1.2	[AD1]P0.1	38
4	P1.3	[AD2]P0.2	37
5	P1.4	[AD3]P0.3	36
6	P1.5	[AD4]P0.4	35
7	P1.6	[AD5]P0.5	34
8	P1.7	[AD6]P0.6	33
9	RESET	[AD7]P0.7	32
10	P3.0[RxD]	[VPP]$\overline{EA}$	31
11	P3.1[TxD]	[PROG]ALE	30
12	P3.2[INT0]	$\overline{PSEN}$	29
13	P3.3[INT1]	[A15]P2.7	28
14	P3.4[T0]	[A14]P2.6	27
15	P3.5[T1]	[A13]P2.5	26
16	P3.6[$\overline{WR}$]	[A12]P2.4	25
17	P3.7[$\overline{RD}$]	[A11]P2.3	24
18	XTAL2	[A10]P2.2	23
19	XTAL1	[A9]P2.1	22
20	Vss	[A8]P2.0	21

(XX51)

Figure 2.3 Pin-wise signal assignment of XX51

Pins 21-28: Port 2 (P2.0–P2.7) is also a quasi-bi-directional I/O port with internal pull-ups. To serve as input, concerned bit(s) of this port must be written with 1s. High-order address bus is multiplexed with this port for external memory interfacing.

Pins 10-17: Port 3 (P3.0–P3.7), like the previous two ports, is also a quasi-bi-directional with internal pull-ups. To serve as input, concerned bit(s) of this port must be written with 1s. Some alternate functions like interrupt input, timer/counter input and serial communication signals are shared with this port pins. Control signals necessary for external data memory interface, namely $\overline{RD}$ and $\overline{WR}$, are also available from this port.

Pin 9: RESET is the input for initializing the device during initial power-on or thereafter. To initialize, this input must remain high for a minimum of two machine cycles when the power is on and the system clock is active.

Pins 18, 19: XTAL1 and XTAL2 are to be used for interfacing an external crystal to generate the system clock. External clock source may also be used through these pins as discussed later in this chapter.

Pin 40: Vcc is for main power source input. Generally, it is 5V DC but the tolerance range varies from device to device.

Pin 20: Vss is provided for system ground reference (0V) connection.

Pin 30: ALE outputs signal as long as the device is operational. This signal is to be used for de-multiplexing the multiplexed address-data signal of port 0 during external memory interfacing. In each machine cycle, two ALE pulses are available. This signal may also be used to test whether the device is operational or not.

Pin 31: $\overline{EA}$ input is provided to allow or disallow the internal memory. Generally, this input is to be pulled high if internal program memory is to be used.

Pin 29: $\overline{PSEN}$ stands for Program Store Enable and may be taken as the substitute of the *read* signal for external program memory, interfaced with the device.

Is there any reason for making $\overline{EA}$ and $\overline{PSEN}$ as active low? Yes, there is. Some times unconnected inputs may generate high signal as input (specially for TTL inputs). Making these signals active low ensures that in no such unwanted cases the signal would trigger a false action.

2.3 | Internal Architecture

Internally, XX51 is composed of an 8-bit data bus and a 16-bit address bus along with some control signals. All these form the system bus and important features like program memory (4K bytes), data memory (128 bytes), parallel ports (P0–P3), timers, serial interface, bus controller, interrupt controller, oscillator circuits along with the central processing unit are placed and interfaced with this system bus, as shown in Fig. 2.4. Note that serial port input and output, namely RxD and TxD, are interfaced with port 3 of the system, as alternate functions. We should now go for detailed discussions of program memory, data memory, system clock and reset features. Other internal parts of the microcontroller, like external interrupts, timers and serial interrupts, would be discussed later in Chapters 14, 15 and 16.

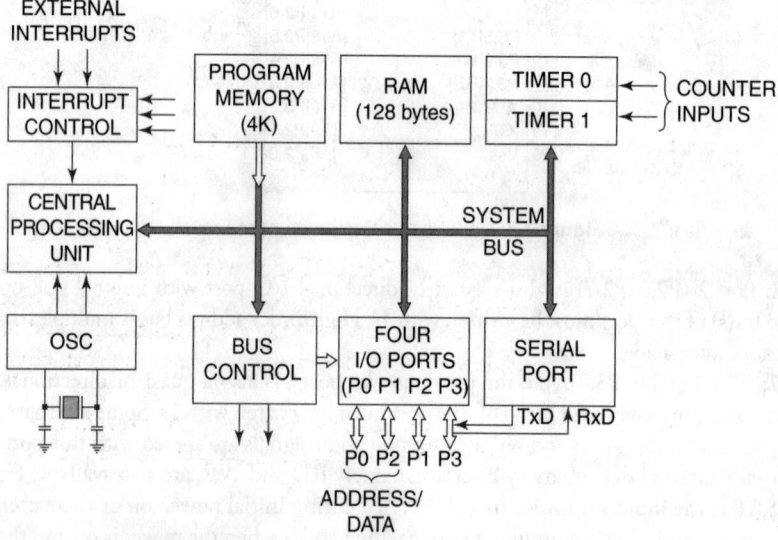

Figure 2.4 Simplified internal architecture of XX51

The reader may observe that in Fig. 2.4, the interfacing of program memory with the system bus is marginally different from other interfacing. We must remember that data bus is unidirectional from program memory while address and control bus directions are similar as other devices.

2.4 | Program Memory Organization

As indicated in Chapter 1, MCS-51 is designed around Harvard architecture, which distinguishes program memory from its data memory. Moreover, XX51 offers some amount of on-chip program memory and also a provision of expanding it by external interfacing of suitable non-volatile memory devices. Schematically, the organization of program memory of XX51 is explained through Fig. 2.5. Note that the $\overline{EA}$ input must be connected to GND if all instruction fetches must be performed from external program memory [Fig. 2.5(b)]. If $\overline{EA}$ is connected with Vcc, then initial instruction fetches are performed from internal program memory. However, once the boundary is crossed, the fetch would automatically be performed from external program memory [Fig. 2.5(a)]. The maximum size of total program memory, internal plus external, may be 64K bytes for any configuration.

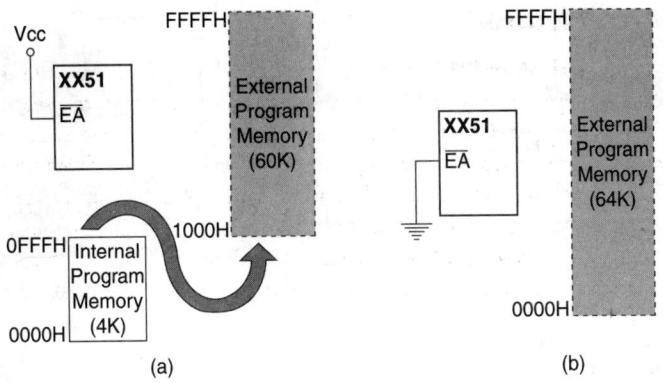

Figure 2.5 Arrangement of program memory in XX51 with (a) $\overline{EA}$ = 1 and (b) $\overline{EA}$ = 0

As discussed earlier, to access external program memory, XX51 uses P0 and P2 ports for address and data bus. The signal, $\overline{PSEN}$, is meant for program memory-read operations.

2.4.1 | Pipeline Architecture

In the previous chapter, in Table 1.3, it was indicated that 8051 adopted the pipeline architecture. A pipeline architecture allows the processor to fetch an instruction's code during the decoding or execution time of its previous instruction. Intel 8086 microprocessor is capable of storing six such advanced bytes. 8051 also uses this pipeline architecture. However, it fetches and stores only one byte of the next instruction's code in its pipeline.

Pipeline architecture makes a processor work faster. However, when the instructions are not in sequence, like program branching, etc., the pre-fetched code byte must be discarded or abandoned.

2.4.2 | Program Lock Bits

One important feature of program memory of MCS-51 is the program lock bits. These bits are provided to avoid unwanted software piracy. Once these lock bits are programmed, content of internal program memory cannot be read from the device by some outside circuit. However, if external program memory is also used, then these program lock bits cannot be utilized. A standard program memory-erase operation would unlock these lock bits but also would simultaneously erase the content of internal program memory completely.

 It is not possible to implement any such program memory locking arrangement in a microprocessor based system as the program memory must be externally interfaced with the microprocessor.

2.5 | Data Memory Organization

Just like the program memory, the maximum size of external data memory may be 64K bytes. However, this 64K of data memory limit is irrespective of the amount of internal (on-chip) data memory. All versions of XX51 offer 128 bytes of internal data memory, while others, like XX52 and XX54, offer 256 bytes of internal data memory. In either case, the maximum allowable external data memory is 64K. Therefore, the maximum size of data memory for MCS-51 devices may be taken as 64K plus size of available internal data memory (128 or 256 bytes). The organization of internal and external data memory of XX51/52 is shown in Fig. 2.6.

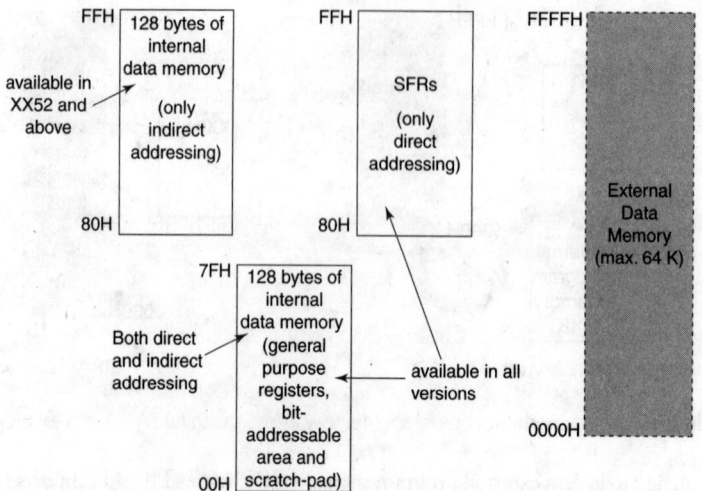

Figure 2.6 Arrangement of internal and external data memory of XX51/52

The internal data memory of 8051 is divided into three separate functional parts as

- register banks (address 00H to 1FH),
- bit-addressable area (address 20H to 2FH) and
- scratch-pad area (address 30H onwards).

This is shown in Fig. 2.7.

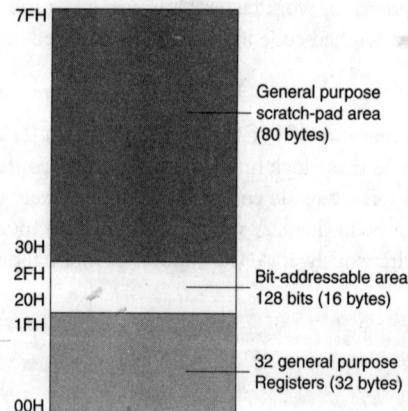

Figure 2.7 Major divisions of lower 128 bytes of internal RAM of XX51

2.5.1 | Register Banks

The lowest part of the internal RAM accommodates four general-purpose register banks designated as bank #0, #1, #2 and #3. Each bank contains eight general-purpose registers designated as R0, R1, R2 and so on, up to R7 as shown in Fig. 2.8. Multiple banks help in faster interrupt servicing. Out of these four banks, only one bank may be selected to be active at any time for operations. Whichever bank is selected, its registers may be addressed as R0, R1, R2, etc., by their respective nomenclature. This reduces the addressing time and generates faster response. However, the unselected bank's registers may also be used by their direct addresses. Initially, after system reset, bank #0 is selected as default. This bank selection is achieved through the Program Status Word (PSW) register, which is described in the next chapter.

These registers may be used to store any data or an operand during arithmetic or logical operations. They can also be used as counters to count down. Registers R0 and R1 of the active bank may also be used as data pointers for indirect addressing. This would be discussed in greater details in Chapter 3 under the section titled *Addressing Modes*. If not used as registers, then they may be used as general-purpose storage area.

Example 2.1

Purpose: To understand the correlation between individual register names and their direct addresses. Note that they might be addressed in both ways, if they are within the active register bank.

Problem

Assume that at present, register bank #1 is selected. Which of its registers would be addressed by the direct address 0AH, when some other bank is selected?

Solution

Referring to Fig. 2.8, we observe that the direct address 0AH belongs to register R2 of bank #1. Therefore, the correct answer is R2.

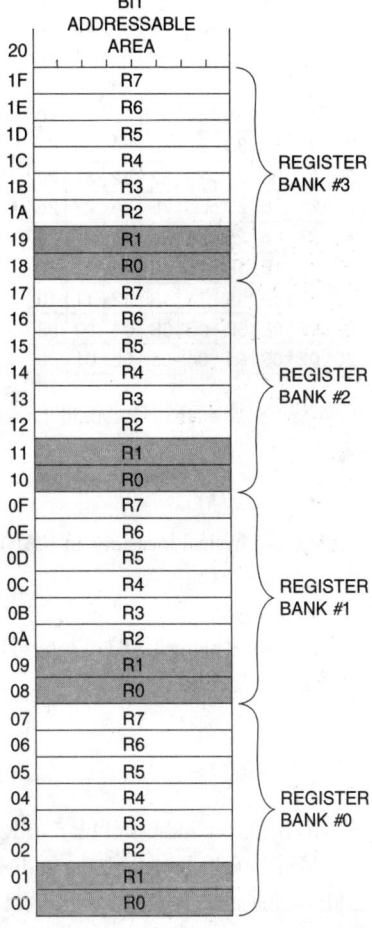

Figure 2.8 Organization of register banks and general-purpose registers

2.5.2 | Bit-Addressable Area

Next 16 bytes after four register banks, starting from address 20H to 2FH, contain 128 bits in total and are designated as bit-addressable area. Each one of these 128 bits has an individual bit address varying from 00H to 7FH. This area is expected to store bit variables for the application program.

In many control operations, the data might be of a single bit rather than a full byte. For example, the condition of a motor, running or not, may be represented through a single bit. Using a full byte for storage of this information is not necessary. Bit-addressable area helps the software developer to store these types of Boolean variables resulting in economical usage of the system memory. If not used for bit-variable storage, then such unused bytes of this area may also be used to store information in the form of bytes. Both byte and bit addresses within this area are shown in Fig. 2.9.

```
BYTE
ADDRESS  MSB                                    LSB
     2F  7F  7E  7D  7C  7B  7A  79  78
     2E  77  76  75  74  73  72  71  70
     2D  6F  6E  6D  6C  6B  6A  69  68
     2C  67  66  65  64  63  62  61  60
     2B  5F  5E  5D  5C  5B  5A  59  58
     2A  57  56  55  54  53  52  51  50
     29  4F  4E  4D  4C  4B  4A  49  48
     28  47  46  45  44  43  42  41  40  BIT
     27  3F  3E  3D  3C  3B  3A  39  38  ADDRESSES
     26  37  36  35  34  33  32  31  30
     25  2F  2E  2D  2C  2B  2A  29  28
     24  27  26  25  24  23  22  21  20
     23  1F  1E  1D  1C  1B  1A  19  18
     22  17  16  15  14  13  12  11  10
     21  0F  0E  0D  0C  0B  0A  09  08
     20  07  06  05  04  03  02  01  00
```

Figure 2.9 Bit-addressable area of RAM with bit addresses

Example 2.2

Purpose: To correlate directly addressable bits' physical locations in RAM and their individual addresses.

Problem

   (i)   What would be the address of bit 5 (bit 0 being LSB) of internal RAM location 2BH?
   (ii) Indicate the location of the bit whose bit address is 37H.
   (iii) Where is the bit E7H?

Solution

   (i)   Referring to Fig. 2.9, we find that the direct address of bit 5 of RAM location 2BH is 5DH.
   (ii) Referring to Fig. 2.9, it is bit 7 (MSB) of the byte with the address 26H.
   (iii) The bit addresses of the bit-addressable area vary from 00H to 7FH. Therefore, the bit is outside the bit-addressable area. (E7H is MSB of accumulator to be discussed in the next chapter).

2.5.3 | Scratch-Pad Area

The last part of the internal RAM area, from 30H to the rest, may be used for any general-purpose storage including the stack. It may be noted here that after system reset, the stack pointer (SP) is initialized at J7H so that the stack may start from 08H onwards. Any unused bytes of the bit-addressable area or unused register banks may also be used for general-purpose data storage.

For XX51 devices, the highest address of scratch-pad area is 7FH. However, for XX52, XX54, etc., the highest address is FFH as the size of on-chip RAM is 256 bytes. In this context, it may be noted that the higher 128 bytes if available on-chip, then it may only accessed by indirect addressing (through R0 and R1) and not by direct addressing. The lower 128 bytes of RAM, present in all systems, may be addressed both directly as well as indirectly (refer to Fig. 2.6).

2.6 | System Clock

The system clock in a processor is as important as the heartbeat of a human body. All duties performed by a processor are basically dependent on this system clock. As a normal practice, this clock is driven by an external quartz crystal attached with XTAL1 (pin 19) and XTAL2 (pin 18) of XX51, as shown in Fig. 2.10(a). It is a normal practice to use quartz crystals as they produce accurate oscillations.

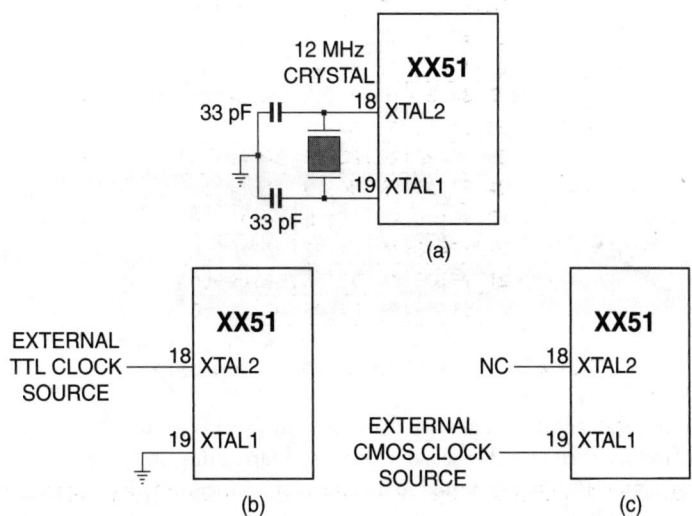

Figure 2.10 Oscillator circuit with (a) external crystal, (b) external TTL clock source for HMOS device and (c) external CMOS clock source for CHMOS device

However, if necessary, external clock sources may also be used to drive the system clock. For a HMOS device, this external clock signal may be applied to XTAL2 (pin 18) while XTAL1 (pin 19) must be grounded, as shown in Fig. 2.10(b). For a CHMOS device, this external clock signal must be supplied at XTAL1 (pin 19) while XTAL2 (pin 18) input must be left as it is [refer to Fig. 2.10(c)].

The allowable maximum and minimum clock frequency varies from device to device. A general practice is to use an external crystal of 12 MHz frequency unless serial communications are necessary. For the systems using serial communications facilities, a crystal of frequency 11.0592 MHz is more appropriate as the standard baud rates (9600, 4800, etc.) may easily be generated using this frequency (refer to Chapter 16 for further explanations).

Internally, multiple numbers of this system clock oscillation are used for various purposes. If we designate each complete oscillation of the external crystal or clock source as a *pulse*, then two such pulses produce a *state*. Six states generate one *machine cycle*. Instructions need one, two or four such machine cycles for fetching the opcode and its execution (Fig. 2.11).

We have previously been introduced to the ALE signal of XX51. Two pulses of this ALE signal are available for each machine cycle. A quick method to check whether an XX51 is functioning properly or not is to monitor this ALE signal through an oscilloscope. As a normal practice, every machine cycle would generate two ALE pulses as shown in Fig. 2.11.

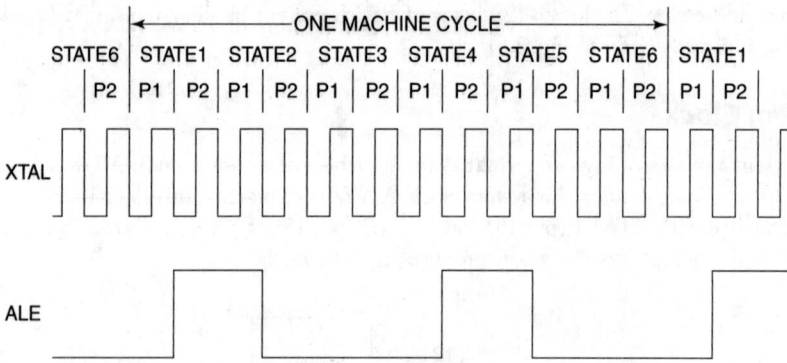

Figure 2.11 System clock for one machine cycle

Fig. 2.11 shows that 12 clock pulses are incorporated within a machine cycle. As MCS-51 depends on micro-programming or micro-coding, therefore so many pulses are necessary. Micro-programming reduces the hardware overhead for instruction decoding and its execution at the cost of the speed of execution.

2.7 | Reset

Reset, or power-on reset, is an another important feature of any processor. For XX51, this reset must be active high for at least two machine cycles (or 24 oscillator periods). During this time, many internal registers are initialized. The program counter (PC) is also cleared so that the first instruction, to be executed after reset, is fetched from address 0000H. Fig. 2.12(a) shows the standard power-on reset circuit, suitable for XX51. For manual reset operation without switching off and on the power input, a reset switch might be connected in parallel with the 10 µF capacitor [Fig. 2.12(b)].

Reset does not change the content of internal RAM of the system. However, after power-on reset, content of this internal RAM is unpredictable. In either case, following are the four major actions implemented by a system reset:

- PC is cleared,
- stack pointer (SP) is initialized by 07H (system stack starts from 08H),
- register bank #0 is selected and
- all ports output FFH.

Table 2.1 below indicates the conditions of all Special Function Registers (SFRs) after reset. These reset values are in binary. Conditions of the bits, marked as 'X', remain undefined after reset.

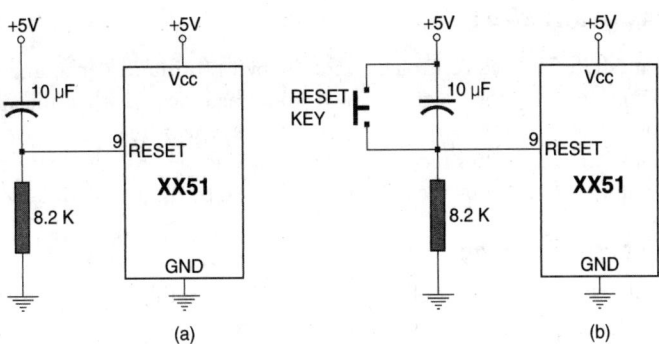

Figure 2.12 (a) Power-on reset circuit and (b) with manual reset option

Table 2.1 Reset values for XX51

Register	Reset value (B)	Remarks
A	00000000	
B	00000000	
DPTR	00000000 00000000	
IE	0XX00000	All interrupts disabled
IP	XXX00000	
P0	11111111	Output FFH, ready to read
P1	11111111	Output FFH, ready to read
P2	11111111	Output FFH, ready to read
P3	11111111	Output FFH, ready to read
PC	00000000 00000000	Address of first instruction To be executed after reset
PCON	0XXXXXXX	Timer 1 as baud rate gen.
		Power-saving modes unpredictable
PSW	00000000	Selects register bank #0
SBUF	XXXXXXXX	
SCON	00000000	
SP	00000111	Stack from 08H onwards
TCON	00000000	Stop both timers
TH0	00000000	
TH1	00000000	
TL0	00000000	
TL1	00000000	
TMOD	00000000	

To understand various operations performed by the processor during and after reset, we have to be familiar with some of the SFRs of XX51. These will be studied in the next chapter.

When the system or microcontroller is switched on, the reset generated is called power-on reset or *cold start*. Without switching off and on the power, if manual reset key is pressed, then it is known as *warm start*. We shall discuss more about it in the power management chapter (Chapter 24).

2.8 | How to Program 8051

We already know that MCS-51 microcontrollers offer 4K bytes (or more, depending upon the version) of on-chip program memory to store the program which would be executed when the system is switched on. Any beginner is curious to know that *how a program, written on a piece of paper, may be placed inside this program memory of the microcontroller chip*. In this section we discuss about the procedure regarding how to load the program memory of a microcontroller by the program, developed by the system designer.

2.8.1 | Overview of Programming

Any program, as we all know, is composed of several instructions. In this book, the complete instruction set of MCS-51 is presented in Appendix-B and explained in chapters 4 to 12. Several example programs are also presented throughout the book and we assume that presently we are ready with one such example program, composed of MCS-51 instructions, written on a piece of paper. The reader may note that a program written using this instruction set is known as *assembly language program*. The next step would be to transfer this assembly language program within the program memory inside the microcontroller chip (IC). We already know that this program memory may store several groups of bits and presently our program is not composed of bits but instructions. Therefore, these instructions of our program are to be *translated* into corresponding bits.

To implement this translation and putting translated bits inside microcontroller chip, we shall require the service of an ***universal programmer*** and few more ***software tools***. This universal programmer must be interfaced with any standard PC. Apart from offering necessary interfacing connections for PC (USB or similar) every universal programmer has a ZIF (Zero Insertion Force) socket to insert or remove any IC (to be programmed) easily, without the need of any external additional force. For this purpose all ZIF sockets are equipped with a mechanical lever which, when turned, grips or releases all IC pins simultaneously. The IC to be programmed must be inserted within this ZIF socket of the universal programmer.

The overall procedure for programming any microcontroller chip is as follows:

- Develop the program using an ***editor***.
- Translate the program to binary using an ***assembler***.
- Transfer the binary data within the microcontroller using ***universal programmer***.

We shall now discuss these steps with adequate details in following sections.

2.8.2 | Editor

To develop the assembly language program, we shall need the help of an editor. Editor is a software tool which allows us to freely type the program and save it as and when necessary. Apparently, this editor functions very much similar to any word-processing software. We may type our already developed software from paper to the editor as a new file. Alternately, we may use the editor to write necessary instructions one after another and finish our program development. Remember that the final outcome from the editor is always a file containing the soft copy of the program composed of MCS-51 instruction set. We should save this file before proceeding to the next step, where this file would be used as the *source file*.

2.8.3 | Assembler

Once software writing part is complete using the editor, help of an assembler is necessary to generate the binary code of each instruction of the program. Assembler is a software which accepts any program (source file) written in assembly language (using instruction set) and generates the executable machine code in binary. Depending upon the type of assembler, this conversion or translation may be completed in one-pass, two-pass, or multi-pass. The binary output file created by the assembler is to be saved for further use. The reader may note that assembler is processor dependent and for our purpose we shall need a MCS-51 assembler.

2.8.4 | Burning the Chip

The binary code, generated by the assembler, is to be fed within the microcontroller chip. This duty is carried out by the universal programmer and the process is known as *burning the chip*. To implement it, the microcontroller chip is to be plugged into the ZIF socket of the universal programmer. The 'program' command by clicking the mouse would transfer the bytes of binary output file (generated by the assembler) into the program memory within the microcontroller chip. After this, the microcontroller chip may be taken out of the ZIF socket of the universal programmer and may be fitted within the circuit board for functioning. Some microcontrollers offer in-circuit burning facility, means, they need not be fitted within the ZIF socket of the universal programmer, burnt, and then to be taken out and fitted within the circuit board. Rather the circuit, with the microcontroller to be programmed, is to be connected with the universal programmer through USB or other suitable interface and the *burning* operation may be implemented.

2.8.5 | IDE

Commercially, in most of the cases, the editor, assembler and other related software tools (like: debugger, simulator, etc.) are available as a single package which is known as IDE or integrated development environment. In remaining cases, they are available as separate software packages and to be individually executed.

The term IDE was originally introduced by Borland Corporation through their now-famous product, Turbo Pascal. Thereafter the concept of IDE became popular and was adopted by many software vendors. In case, of microcontrollers, some commercially available IDE packages offer data bases of different types of microcontrollers and an interactive, user-friendly package for solving all software development related problems.

SUMMARY

XX51 external pins and signals and internal architecture along with clock and reset operations are discussed in this chapter.

Externally, XX51 is available in a 40-pin DIP. Apart from four 8-bit bit-programmable I/O ports designated as P0, P1, P2 and P3, it also offers pins for power (5V DC) input, crystal oscillator connectors, active high reset input, external memory access allowing input ($\overline{EA}$) and two control signals (output) for external memory interface, namely $\overline{PSEN}$ and ALE. When interfaced with external memory, ports P0 and P2 are used for multiplexed address (16) and data (8) signals.

Internally, XX51 contains a 8-bit processor, 4K program memory, 128 bytes of data memory, two 16-bit timer/counters, interrupt handling circuits and a serial communication interface. Provision is there to lock this internal program memory using lock bits to avoid unwanted software piracy. Internal data memory of 128 bytes includes four general-purpose register banks of eight registers each, 16 bytes of bit-addressable area and 80 bytes of scratch-pad area. Out of four register banks, only one is active at any time and selectable through PSW register. Registers R0 and R1 of the selected bank may also be used as pointers.

The system clock may be generated by either using a quartz crystal or an external clock source. Every machine cycle is composed of 12 oscillator periods or six states. XX51 accepts an active high reset input, which should remain high for two consecutive machine cycles. Apart from initializing various internal registers, this reset clears PC, loads SP by 07H, selects register bank #0 and outputs FFH through all ports.

POINTS TO REMEMBER

- Only one of four register banks would be active any time, providing eight general-purpose registers from R0 to R7.

- Contents of bits 3 and 4 of PSW SFR select the active register bank at any time of operation.

REVIEW QUESTIONS

Evaluate Yourself

1. The number of pins for a XX51 in DIP is
   (a) 20   (b) 40
   (c) 48   (d) none of these

2. Which of the following statements is false?
   (a) All XX51 ports are bi-directional
   (b) All XX51 ports are bit-programmable
   (c) All XX51 ports output FFH after reset
   (d) All XX51 ports need external pull-ups

3. $\overline{PSEN}$ may be taken as
   (a) READ signal for external memory
   (b) WRITE signal for external memory
   (c) Either of the above two
   (d) none of these

4. The size of on-chip program memory for XX51 is
   (a) 4K bytes   (b) 8K bytes
   (c) 16K bytes   (d) none of these

5. The size of on-chip data memory for XX51 is
   (a) 64 bytes   (b) 128 bytes
   (c) 256 bytes   (d) none of these

6. One machine cycle of XX51 takes
   (a) 12 clock pulses   (b) 6 states
   (c) Both of these   (d) none of these

7. Reset input of XX51 should remain high for
   (a) 1 machine cycle   (b) 2 states
   (c) 3 clock periods   (d) none of these

8. Reset input for XX51
   (a) clears all ports   (b) loads SP by 00H
   (c) selects bank #1   (d) none of these

9. For a power-on reset circuit of XX51, the capacitor should be of
   (a) 0.001µF   (b) 0.01 µF
   (c) 10 µF   (d) none of these

10. After reset, PCON register is initialized as
    (a) 0xxxxxxx   (b) x0000000
    (c) xxxxxxxx   (d) none of these

Search for Answers

1. Try to annotate the following functional block diagram of XX51.

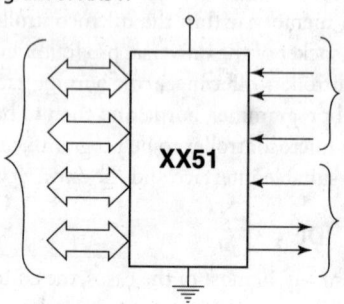

2. What is the major difference between port 0 and the other three ports of XX51?

3. What is the purpose of the ALE signal?

4. Try to annotate the following simplified schematic diagram indicating internal architecture of XX51.

5. What is locked by program lock bits? What is the purpose of providing this?

6. How should the following on-chip data memory locations be addressed?
   (a) Lower 128 bytes from 00H to 7FH.
   (b) Upper 128 bytes from 80H to FFH.
   (c) Special function registers located within 80H to FFH.

7. At any time how many general-purpose registers located between 00H and 1FH of XX51 may be used? Why R0 and R1 are different from the remaining registers?

8. If bit 0 is taken as LSB and bit 7 as the MSB, then what would be the bit-address of bit 3 of internal data memory location 29H?

9. In general, a 12 MHz crystal is used for XX51-based systems. However, in certain cases, a frequency of 11.0592 is preferred. What is the reason behind it?

10. What is the relation between machine cycle, state and pulse for any XX51 microcontroller?

Think and Solve

1. Why is it important to study the reset initialization details of any microcontroller? Are the internal RAM locations cleared by a system reset?

2. What is the maximum total size of data memory, which may be directly addressed by the 8752 microcontroller? What type of address bus makes it possible?

3. What purpose does the pair of 33 pF capacitor serve in the external crystal circuit? What would happen if these two capacitors are removed from an otherwise perfectly working circuit?

4. Would the reset circuit function properly if its 8.2K resistor is replaced by a 10K resistor? Justify your answer.

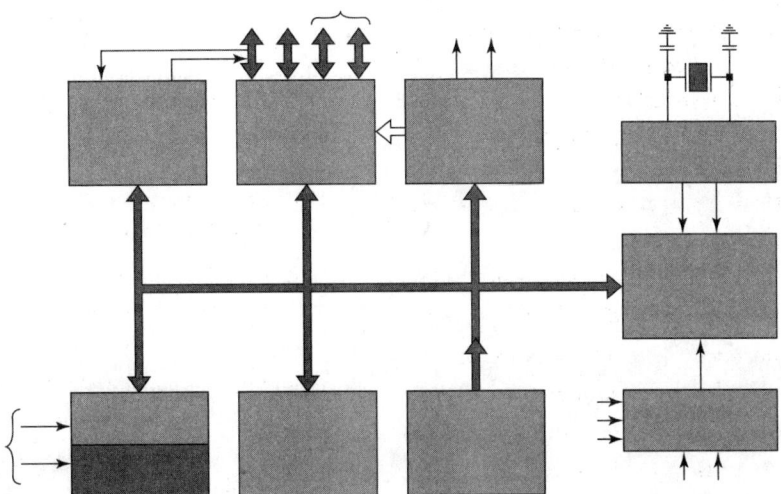

5. Is the following circuit sufficient for a power-on reset?

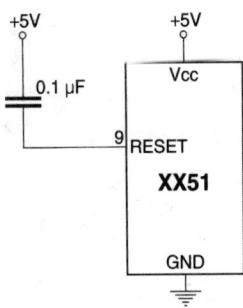

6. What is the maximum size of stack possible for any XX51 microcontroller? Would this size remain same for an XX52 microcontroller?

7. What is the width (number of bits) of the address bus for internal data memory of XX52?

8. What is the purpose of providing multiple numbers of register banks in XX51?

9. Why all ports output FFH after a system reset?

10. Why Princeton architecture was not used in MCS-51?

3 I/O PORTS AND SPECIAL FUNCTION REGISTERS

CHAPTER OBJECTIVES

In this chapter, the reader is introduced to some Special Function Registers (SFRs) and the overall architecture of the I/O ports of MCS-51. After completion of the chapter, the reader should be able to understand

- Names, locations (addresses), reset values and functions of all SFRs.
- Details of PSW, accumulator, B register, SP and I/O port registers.
- How to select any one of the four register banks of MCS-51.
- Bit addresses of different SFRs.
- Architecture and functioning of all four I/O ports of MCS-51.
- Difference between MCS-51 I/O ports and 8255 I/O ports.

3.1 | Introduction

In Chapter 2, it was mentioned that some of the internal memory locations (80H to FFH) of MCS-51 contain SFRs. In this chapter, we will examine details of a few of these SFRs. Later, we will discuss about the architecture of all four I/O ports of MCS-51.

3.2 | SFR Map

Fig. 3.1 shows the 128 bytes area reserved for SFRs in MCS-51. Note that the dark-shaded rectangles are imaginary memory locations and, therefore, physically not present. Out of these 128 locations, only 21 locations are used in MCS-51 for special functions. However, six more locations, from C8H to CDH (lightly shaded), are earmarked for special function registers related with timer 2, available only in XX52 and above. These SFRs would be discussed in Chapter 15 when we will discuss the timer 2 interrupt of XX52.

	FF	FE	FD	FC	FB	FA	F9	F8	F7	F6	F5	F4	F3	F2	F1	F0	
FF																B	F0
EF																A	E0
DF																PSW	D0
CF			TH2	TL2	RCAP2H	RCAP2L	T2MOD	T2CON									C0
BF								IP								P3	B0
AF								IE								P2	A0
9F							SBUF	SCON								P1	90
8F			TH1	TH0	TL1	TL0	TMOD	TCON	PCON				DPH	DPL	SP	P0	80
	8F	8E	8D	8C	8B	8A	89	88	87	86	85	84	83	82	81	80	

Figure 3.1 Map of SFRs of MCS-51 and XX52

3.3 | SFR Functions

Major functions of these 21 special function registers, their addresses and reset values are presented in Table 3.1. Bit-wise addressing options of relevant registers are also included in this table. Some of these SFRs would be discussed in this chapter and the rest in subsequent chapters at relevant places.

Table 3.1 MCS-51 SFR addresses, reset values and major functions

SFR name	Address (in Hex)	Reset value (in binary)	Bit addressable	Major function
ACC	E0	00000000	Yes	Accumulator
B	F0	00000000	Yes	For MUL and DIV operations and also for general purpose
DPH	83	00000000	No	Data pointer (high)
DPL	82	00000000	No	Data pointer (low)
IE	A8	0x000000	Yes	Interrupt enable
IP	B8	xx000000	Yes	Interrupt priority
P0	80	11111111	Yes	Port 0 latch/buffer
P1	90	11111111	Yes	Port 1 latch/buffer
P2	A0	11111111	Yes	Port 2 latch/buffer
P3	B0	11111111	Yes	Port 3 latch/buffer
PCON	87	0xxx0000	No	Power management
PSW	D0	00000000	Yes	Flags, register bank select
SBUF	99	xxxxxxxx	No	Serial I/O buffer
SCON	98	00000000	Yes	Serial communication control
SP	81	00000111	No	Stack pointer
TCON	88	00000000	Yes	Timer control
TH0	8C	00000000	No	Timer 0 counter (high)
TH1	8D	00000000	No	Timer 1 counter high)
TL0	8A	00000000	No	Timer 0 counter (low)
TL1	8B	00000000	No	Timer 1 counter (low)
TMOD	89	00000000	No	Timer mode select

If we closely observe the addresses of bit-addressable SFRs in Table 3.1, then we find a pattern in it. If the least-significant nibble of the SFR address is either 0 or 8, then only that SFR is bit-addressable. No SFR is bit-addressable if the least significant nibble of its address is neither 0 nor 8.

3.4 | Processor Status Word

Processor Status Word (PSW) is one of the important special function registers in MCS-51. Its importance lies on the fact that it is the only register through which the status of operations performed by the CPU is reflected. Functions of all bits of this register are presented in Fig. 3.2 and Table 3.2. As we can observe, two bits of this register, namely bit 3 (RS0) and bit 4 (RS1), help to select one of the four register banks #0, #1, #2 or #3.

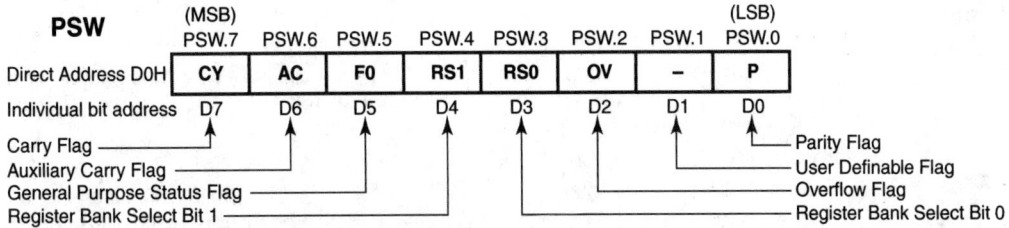

Figure 3.2 Details of PSW SFR of MCS-51

Table 3.2 Bit details of PSW SFR

Bit no.	Bit symbol	Designation	Direct bit address	Alternate address	Function
7	CY	Carry	D7	PSW.7	Set if data coming out from bit 7 of register A during ALU operations (arithmetic or Boolean) else cleared.
6	AC	Auxillary carry	D6	PSW.6	Set if data coming out from bit 3 to bit 4 of register A during addition or subtraction operations (mainly used for BCD operations).
5	F0	Flag 0	D5	PSW.5	User-defined flag for general-purpose operations.
4	RS1	Register bank select 1	D4	PSW.4	Register bank selects MSB. This bit along with RS0 selects the active register bank (one of four).
3	RS0	Register bank select 0	D3	PSW.3	

RS1	RS0	Select register bank
0	0	#0
0	1	#1
1	0	#2
1	1	#3

Bit no.	Bit symbol	Designation	Direct bit address	Alternate address	Function
2	OV	Overflow	D2	PSW.2	Overflow flag set by arithmetic operations if there is a carry from bit 6 but not from bit 7 or a carry from bit 7 but not from bit 6 of register A.
1	–	–	D1	PSW.1	User-definable flag for general-purpose operations.
0	P	Parity	D0	PSW.0	Set if register A has odd number of 1s, after an operation, else cleared.

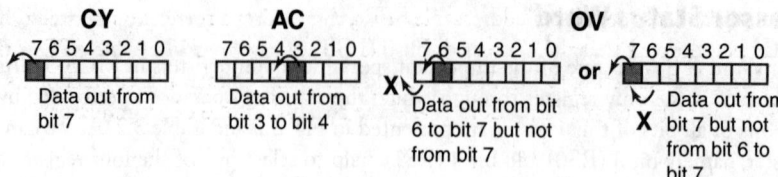

Figure 3.3 Conditions for setting of CY, AC and OV flags of PSW SFR

Fig. 3.3 explains the conditions of setting of carry (CY), auxiliary carry (AC) and overflow (OV) flags of the PSW SFR.

A simpler method to remember relative locations of the flags within) the PSW byte (Fig. 3.2) may be as follows;

Place register bank selecting bits (RS1 and RS0) at the middle of the byte. As carry is always from bit 7, therefore place CY in bit 7. As parity is a flag which is least used, therefore place it at the least significant bit. Leave second bit from right of both nibbles for flags (general purpose and user defined). Place auxiliary carry (AC) just after carry (CY). The remaining bit is left for OV flag.

3.4.1 | Comparison with 8085 Flags

If the reader is familiar with 8085 CPU, which offers five flags, namely CY, Z, S, parity (P) and AC, then it might be indicated here that CY, AC and P flags of both MCS-51 PSW and 8085 flag register (F) perform in identical manner. However, Z and S flags of 8085 are not present in MCS-51. On the other hand, OV flag of MCS-51 is not available in 8085.

The reason for not providing any sign flag (S flag) in MCS-51 is simple. As the accumulator of MCS-51 is bit-addressable, its bit 7 (MSB) may be tested directly whenever the sign of any integer in accumulator has to be checked. As this bit-addressing facility is not available in 8085, a separate sign flag was necessary there.

Although zero flag (Z flag) is not available in MCS-51 architecture, however, it executes instructions, like *Jump if Zero* or *Jump if not Zero*. In these cases, the concerned register is checked physically by the processor to ensure whether it is completely empty or not.

3.5 | Accumulator (Register A)

Accumulator or register A or ACC or simply 'A' is the most widely used register of the microcontroller. Results of most of the arithmetic and logical operations are available in this register. This special function register is both bit and byte addressable. It means that it may be addressed as a byte (direct address E0H) to access all 8 bits of this register at a time. Alternately, operations are also possible with a single bit of this SFR and each one of all eight bits has its individual bit address, as shown in Fig. 3.4. Register A cannot be addressed indirectly. As already mentioned, indirect addressing is achieved by placing the target address in any one of the two general-purpose registers, R0 or R1.

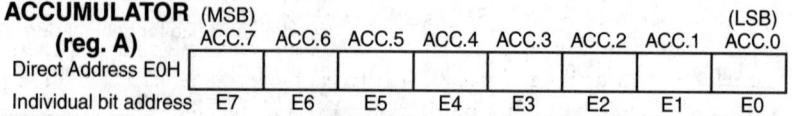

Figure 3.4 Details of the accumulator (SFR) of MCS-51

For accumulator and also for all bit-addressable SFRs, there exists a correlation between the byte and bit addresses. It may be observed that the address of bit 0 (LSB) of any bit-addressable SFR is the same as its byte address or vice versa. Referring to Fig. 3.4, we may observe that in case of accumulator, its direct address (for the whole byte) is E0H and the same E0H is also the address of bit 0 of accumulator.

3.6 | Register B

Fig. 3.5 shows the details of special function register B, which is indispensable for multiply (MUL) and divide (DIV) instructions. MCS-51 has single-byte instruction for unsigned 8-bit integer multiplication and division. One of the two operands of these instructions must be in register B. The other operand must be in register A.

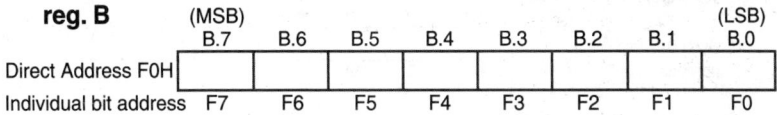

Figure 3.5 Details of register B (SFR) of MCS-51

Register B is also bit as well as byte addressable. Its direct address is F0H, which, as explained before, is also its bit-0's address. When not used for multiplication or division, this register may also be used for any other general-purpose operations, like any other general-purpose register. Register B cannot be addressed indirectly.

3.7 | Stack Pointer

Stack Pointer (SP) (direct address 81H) is not bit addressable (Fig. 3.6). It points to that address of the system stack where the last storage operation has taken place. To place a byte on the stack top, SP must first be incremented by one and then the storage command might be implemented.

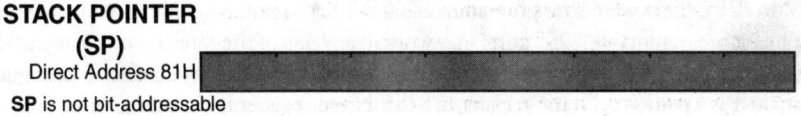

Figure 3.6 Details of SP (SFR) of MCS-51

Initially, after system reset, the SP is initialized as 07H so that the stack may grow from 08H onwards. However, the programmer may change this starting address of stack any time to locate the stack at any available memory location of the system.

Although it is possible for external expansion of data memory, however, in MCS-51, the system stack cannot be accommodated in the outside memory. As the maximum size of internal data memory is 256 bytes, therefore, the SP of MCS-51 is 8-bit and not a 16-bit register.

3.8 | Port Registers (P0, P1, P2 and P3)

Each one of the four I/O ports of MCS-51 has its own register, designated as the name of the concerned port (P0, P1, P2 and P3). The system communicates with these ports through these registers. Data should be written in these registers to send it out through ports. To get the incoming data from these ports, these port registers are to be read. All four port SFRs are bit addressable and their individual and bit-wise addresses are presented through Fig. 3.7.

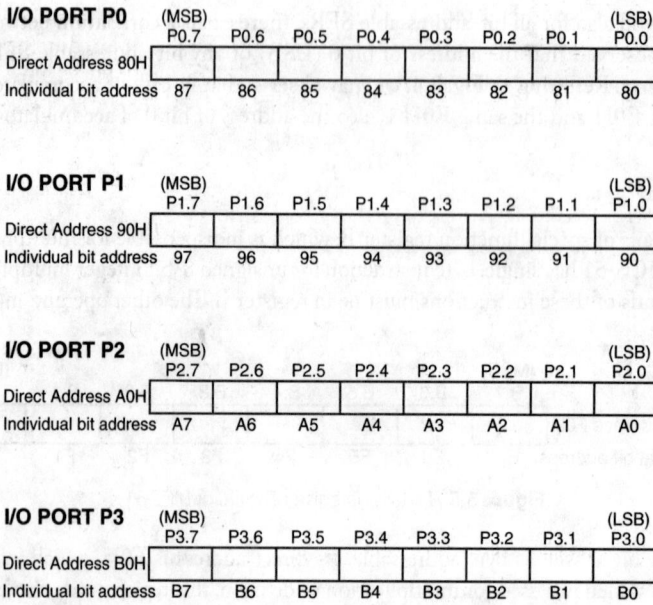

Figure 3.7 Details of port registers P0, P1, P2 and P3 (SFRs) of MCS-51

3.8.1 | Comparing 8255 PPI with MCS-51 Ports

If the reader has already studied 8085 or 8086 microprocessor, then 8255 I/O port would not be unfamiliar. However, there are some major differences between 8255 PPI and MCS-51 ports apart from the differences in their numbers and nomenclature.

In 8255, there are three ports named A, B and C whereas in MCS-51 there are four ports designated as P0, P1, P2 and P3. In 8255, ports were byte programmable (all 8 bits would be either input or output) whereas in MCS-51 they are bit programmable. 8255 ports may work in any one of the three modes, namely Mode 0, Mode 1 or Mode 2 (the last one only for group A). No such modes are available in MCS-51. To configure any port as input or output, there is a command status register in 8255. No such register is available or necessary in MCS-51 ports. To output any data, it is just to be written in that port register. Similarly, to read the data from a port, the concerned port register is to be read. However, in MCS-51, before reading any data through a port, first we must write 1s to that port. That means a port (or a port pin) must output logic high (1) to be capable of functioning as an input port. This is necessary because of the internal structure of the ports, which are depicted in Fig. 3.8.

3.8.2 | Architecture of MCS-51 Ports

Fig. 3.8 shows internal architecture of all four ports of MCS-51. There is a marginal difference between different ports because of the difference in their functions. As we already know, Port 0 is to output lower 8-bit address and send/receive 8-bit data apart from functioning as I/O port. Therefore, as and when required, it must be capable of handling multiplexed address-data bus. Port 2 is to output upper 8-bit address as and when necessary, apart from serving as I/O port. Port 3 is to handle alternate function signals. Only port 1 serves as purely I/O port without any other extra duties. The architecture of every port is related to its duties and, therefore, slightly different from the other. Out of these four ports, only port 0 may be taken as a truly bi-directional port. Remaining three ports are called quasi-bi-directional ports.

To understand the basic architecture and its operations, we take one bit of port 1 as a sample port bit, because all port bits are identical. The architecture of this port bit is shown in Fig. 3.8(b). To avoid initial complexity, we explain its operation through the following four steps.

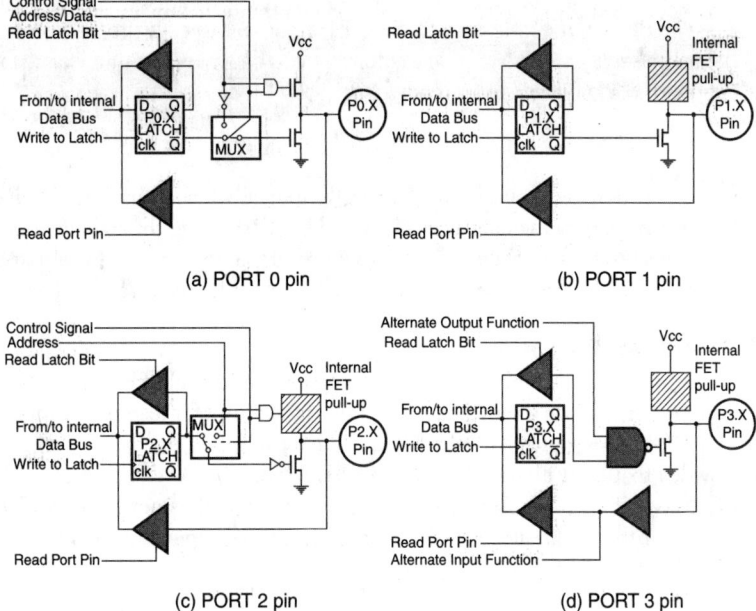

Figure 3.8 Internal architecture of MCS-51 ports

Step 1: Output latch

Fig. 3.9 shows the basic output architecture of a bit of port 1. It is a D-type flip–flop with two outputs, Q and $\overline{Q}$. The output data at D is latched at Q with the transition of the clock input. This latched output data would be available through Q and $\overline{Q}$ till a fresh clock transition with a different data.

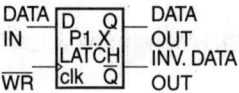

Figure 3.9 Output latch of one bit of port 1

Step 2: FET pull-up at output

The output from this latch is not directly available at the output pin. Fig. 3.10 shows the extra circuit, which is placed in between the D-flip–flop and the port pin. This extra circuit contains one Field-Effect Transistor (FET) (marked as F_L) and another circuit, which is marked as internal FET pull-up. Presently, we avoid the complexity of this 'internal FET pull-up' and may consider it to function just as a simple imaginary pull-up resistor. Note that the gate of the FET (F_L) is connected with the $\overline{Q}$ output of the D-flip–flop.

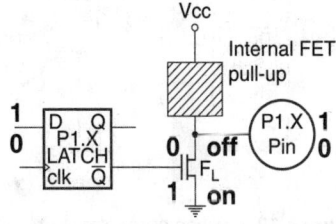

Figure 3.10 Output pin (showing 1 or 0 output) of port 1

 If the reader is not comfortable with a FET, assume it to be a simple NPN transistor. If its base is biased by high voltage sourcing, its collector-emitter continuity is established, else it remains discontinuous.

In this case, if the data written in the port latch is '1' the $\bar{Q}$ outputs '0' which turns the FET (F_L) off. Therefore, at output pin (represented as a circle) the data available would be '1' (sourcing through 'internal FET pull-up). On the other hand, for a latched data '0', the FET (F_L) would be turned on making the output pin at the same potential as ground reference (sinking through F_L). Therefore, the output in that case would be, as desired, a '0'.

Step 3: Read buffer for port pin

To implement the reading mechanism through the same port pin, an extra buffer is introduced as shown in Fig. 3.11, between the port pin and the internal data bus. There is a control for reading the input buffer, marked as 'READ PIN' in Fig. 3.11. When this 'read pin' command is activated, port-pin data is directly available at the internal data bus, provided that the FET (F_L) is off. If this FET is on, then irrespective of the input condition at the port pin, a logic '0' would be available at the internal data bus, which is not desirable. To switch off this FET, a '1' must be written at the D-flip–flop, before the port pin-reading operation. *This is very important.*

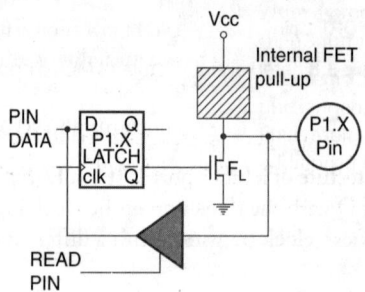

Figure 3.11 Reading from a port pin of port 1

Step 4: Read output latch

As many times, it is necessary to read the output data (for modification of present output or other purposes); another buffer is connected with the internal data bus of the system, which takes its input from the Q out-put of the D-flip–flop, as shown in Fig. 3.12. Note that 'READ LATCH' command is necessary to read the current value from this latch. Therefore, two types of read operations are possible in MCS-51 ports. One is reading the port pin and the other is reading the output latch.

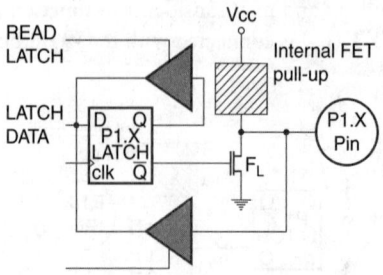

Figure 3.12 Reading output latch of port 1

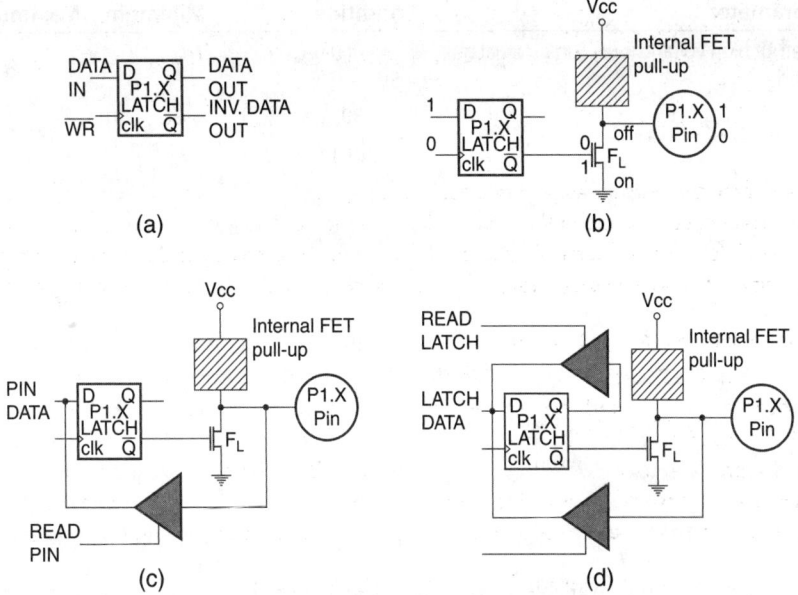

Figure 3.13 Port 1 (a) output latch, (b) output FET pull-up, (c) reading from port pin and (d) reading from port latch

In Fig. 3.13, all these four steps of explanation are presented in the form of (a), (b), (c) and (d) for easier comparison and understanding. Once the architecture of port 1 is understood, the remaining differences of the other three ports may easily be explained.

3.8.3 | DC Characteristics of MCS-51 Ports

Current- and voltage-related characteristics of any IC are generally designated as DC characteristics. AC characteristics indicate the time limits related with various timing diagrams. As far as 8051 is concerned,

Table 3.3 DC characteristics of AT89C51

Symbol	Parameter	Condition	Minimum	Maximum	Units
V_{IL}	Input low voltage	(Except $\overline{EA}$)	−0.5	0.2Vcc − 0.1	V
V_{IL1}	Input low voltage ($\overline{EA}$)		−0.5	0.2Vcc − 0.3	V
V_{IH}	Input high voltage	(Except XTAL1, RST)	0.2Vcc + 0.9	Vcc + 0.5	V
V_{IH1}	Input high voltage	(XTAL1, RST)	0.7Vcc	Vcc + 0.5	V
V_{OL}	Output low voltage (ports 1, 2, 3)*	I_{OL} = 1.6 mA		0.45	V
V_{OL1}	Output low voltage (port 0, ALE, $\overline{PSEN}$)*	I_{OL} = 3.2 mA		0.45	V
V_{OH}	Output high voltage (ports 1, 2, 3; ALE, $\overline{PSEN}$)	I_{OH} = -60 µA, Vcc = 5V+/−10%	2.4		V
		I_{OH} = −25 µA	0.75Vcc		V
		I_{OH} = −10 µA	0.9Vcc		V

(Continued)

Symbol	Parameter	Condition	Minimum	Maximum	Units
V_{OH1}	Output high voltage (port 0 in external bus mode)	$I_{OH} = -800\ \mu A$, Vcc = 5V+/−10%	2.4		V
		$I_{OH} = -300\ \mu A$	0.75Vcc		V
		$I_{OH} = -80\ \mu A$	0.9Vcc		V
I_{IL}	Logical 0 input current (ports 1, 2, 3)	$V_{IN} = 0.45$ volt		−50	μA
I_{CC}	power-supply current	Active mode 12 MHz		20	mA
		Idle mode 12 MHz		5	mA
	power-down mode	Vcc = 6V		100	μA
		Vcc = 3V		40	μA

Courtesy: Atmel Corporation.
*Under steady-state conditions, I_{OL} must be externally limited as follows:
Maximum I_{OL} per port pin 10 mA.
Maximum I_{OL} per 8-bit port: Port 0: 26 mA.
Ports 1, 2, 3: 15 mA.
Maximum total I_{OL} for all output pins: 71 mA.

as there are too many versions available in the market, it is not the correct place to accommodate the DC characteristics of all versions of 8051 in this section However, to get some idea, we discuss about the DC characteristics of a representative device, AT89C51. Vcc of this device is rated as 5V +/− 20%, which means it can be operated with a DC power source ranging between 4V and 6V. DC characteristics of this device are presented in Table 3.3.

If the reader is interested in fabricating a working system around 8051, then reading its original data sheets would save a lot of time for debugging the circuit.

3.9 | Power Management (PCON)

In Section 1.5, we were briefly introduced to power management features offered by microcontrollers. 8051 has a dedicated SFR, named PCON, for this purpose. As we can observe from Fig. 3.14, setting bit 0 of PCON puts the microcontroller in *idle mode* and setting bit 1 of PCON moves the processor to *power down mode*. The details of these power-saving modes are discussed in Chapter 24.

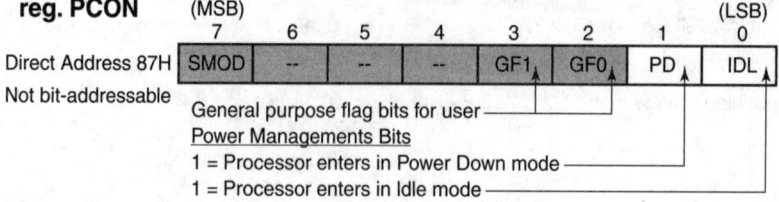

Figure 3.14 Power management bits of PCON SFR

3.10 | Solved Examples

Example 3.1

Purpose: To become familiar with the functioning of processor flags.

Problem

Explain how the flags of PSW would be affected if 01111111B or 7FH in accumulator is added with 01H.

Solution

Adding 01H with 7FH would produce 80H as a result and would be the final content of the accumulator. In binary form, 80H would be 10000000B. Following would be the changes in PSW flags:

1. CY flag (PSW.7) would be cleared as there is no overflow from bit-7 (MSB) of the accumulator.
2. AC flag (PSW.6) would be set as there is an overflow from bit-3 to bit-4 of accumulator during addition.
3. OV flag (PSW.2) would be set as there is a carry from bit-6 to bit-7 but no carry from bit-7 of accumulator.
4. P flag (PSW.0) would be set as accumulator containing odd number of 1s.

Example 3.2

Purpose: How to configure port bits of 8051.

Problem

How can we make bits 0, 1 and 5 of port 1 function as input pins and the remaining pins as output pins?

Solution

Any port pin would function as input pin when it is outputting 1. Therefore, in this case, we are to output 1 to bits 0, 1 and 5 of port 1 output latch. The output data may be as required for bits designated as output.

SUMMARY

SFRs are only directly addressable and located within MCS-51 between addresses 80H and FFH. Some of these SFRs are also bit addressable. In MCS-51, there are a total of 21 SFRs and six more for XX52 and upward versions.

PSW is a bit-addressable SFR containing four status flags, namely CY, AC, OV and P to reflect the status after various Arithmetic Logic Unit (ALU) operations. The register bank selection bits (RS0 and RS1) are also available within this SFR to select any one of the four general-purpose working register banks from 0 to 3. PSW is cleared after reset, selecting register bank #0 as default.

Accumulator is the most widely used SFR, which stores the results of all arithmetic and logical operations of the ALU. However, to implement multiply or divide instructions, another SFR, register B is also necessary.

These two registers are also bit addressable and both are cleared during reset.

SP is not bit addressable. This 8-bit register points to the last stored location of the internal data memory. During reset, SP is initialized as 07H so that system stack may grow from 08H onwards. System stack is always within internal data memory. MCS-51 does not allow the stack within externally interfaced data memory.

Four port registers (P0, P1, P2 and P3) are bit addressable and communicate with respective port pins. Every port contains an internal FET pull-up, a D-flip–flop to latch the output data, and two input buffers, one directly from the port pin and the other from the output latch. To read a port, it should be ensured that it is already outputting high.

POINTS TO REMEMBER

- Only for multiplication and division, B register, along with accumulator, would hold the result.

- To configure any port bit as input, ensure that 1 is already written in the corresponding bit of that port SFR.

REVIEW QUESTIONS

Evaluate Yourself

1. How many SFRs are there in MCS-51?

   (a) 20   (b) 21

   (c) 27   (d) None of these

2. What should be the values of RS1 and RS0 bits of PSW SFR to select register bank #2?

   (a) 1 and 0, respectively   (b) 1 and 1

   (c) 0 and 0   (d) None of these

3. How would you address bit 3 (bit 0 being LSB) of accumulator?

   (a) ACC.3   (b) E3H

   (c) Either of these   (d) None of these

4. What is the special function of register B?

   (a) Substitute accumulator

   (b) To perform 16-bit operations along with accumulator

   (c) Serve as a pointer register

   (d) None of these

5. What is the value of accumulator after system reset?

   (a) XXH   (b) FFH

   (c) 00H   (d) None of these

6. Which register bank is selected by default after system reset?

   (a) #0   (b) #1

   (c) #2   (d) None of these

7. What is the reset value of the SP?

   (a) 00H   (b) 70H

   (c) FFH   (d) None of these

8. If we write 0 in the output latch of a port bit before reading it, then at the time of reading that port pin, we would always read it as

   (a) 0   (b) 1

   (c) Undefined   (d) None of these

9. Which of the following ports of MCS-51 might be designated as a true bi-directional port?

   (a) Port 2   (b) Port 1

   (c) Port 0   (d) None of these

10. At output, all ports of MCS-51 have

    (a) Internal transistor pull-ups

    (b) Internal FET pull-ups

    (c) Internal resistance pull-ups

    (d) None of these

Search for Answers

1. Is there any correlation between the byte address and least significant bit address of the bit-addressable special function registers?

2. Apart from the accumulator, is there any other register involved in multiply and division instructions' execution?

3. What is the purpose of port registers?

4. What makes port 1 different from the remaining three ports of MCS-51?

5. What are the major differences between MCS-51 ports and 8255 ports?

6. How are MCS-51 ports configured to function as input or output?

7. Why is the SP of MCS-51 an 8-bit and not a 16-bit register?

8. Why is there no sign-flag in MCS-51?

9. How does MCS-51 ensure that the result of an operation is zero or not?

10. Name the flags which are identical in MCS-51 and 8085.

Think and Solve

1. In which microcontroller of the MCS-51 family is T2MOD SFR available?

2. What is the difference between CY and OV flags of PSW?

3. Why is the SP not bit-addressable?

4. Why FFH is the reset value of all ports of MCS-51?

5. If the SP is loaded with 23H, from which address would the system stack grow?

6. What is the purpose of 'MUX' (refer to Fig. 3.8) in ports 0 and 2?

7. Why was it necessary to provide two reading buffers in the ports of MCS-51?

8. Can the memory area unoccupied by SFRs be used as general-purpose storage area?

9. What would happen if 'internal FET pull-up' circuits of ports are replaced by a pull-up resistor?

10. Why may the system stack not be accommodated within external data memory of MCS-51?

4 ADDRESSING MODES AND DATA MOVE OPERATIONS

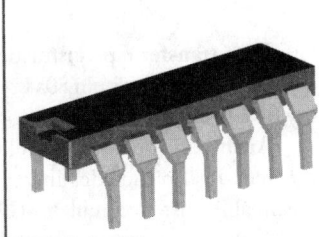

CHAPTER OBJECTIVES

In this chapter, the reader is introduced to simpler data-moving instructions of MCS-51 microcontrollers. Different addressing modes offered by MCS-51 are also discussed. After completion of this chapter, the reader should be able to understand:

- Five types of addressing modes offered by MCS-51.
- Mnemonics used for data move type instructions.
- All variations offered by the MOV instructions.
- Different syntax used in MOV instruction.
- Limitation of the MOV instruction.
- Difference between 8085 and MCS-51 data move instructions.

4.1 | Instructions and Instruction Groups of MCS-51

As we have made ourselves familiar with some essential parts of MCS-51 architecture, we may now concentrate on the instruction set of MCS-51. To start with, 49 instruction mnemonics of MCS-51 are divided into the following five groups as presented in Table 4.1.

Table 4.1 MCS-51 instruction groups and instructions

Data transfer	Arithmetic operations	Program branching	Logical operations	Boolean variable manipulation
MOV	ADD	LJMP	ANL	CLR
MOVC	ADDC	AJMP	ORL	SETB
MOVX	SUBB	SJMP	XRL	MOV
PUSH	INC	JZ	CLR	JC
POP	DEC	JNZ	CPL	JNC
XCH	MUL	CJNE	RL	JB
XCHD	DIV	DJNZ	RLC	JNB
	DA A	NOP	RR	JBC
		LCALL	RRC	ANL
		ACALL	SWAP	ORL
		RET		CPL
		RETI		
		JMP		

Data transfer type instructions deal with byte variables or constants, and their communication from one location to another within 8051 or from external program memory (like MOVC) or to/from external data memory (like MOVX). Only in case of XCHD instruction, it deals with digit variables that are four bits or a nibble.

Arithmetic operations include addition, subtraction, multiplication and division instructions using 8-bit integers. It also includes increment by one and decrement by one instruction and a special instruction, Decimal Adjust Accumulator (DA A), to deal with binary-coded decimal (BCD) arithmetic.

Next group contains **program-branching** instructions, which essentially control the flow of the program logic. All of these instructions, except No Operation (NOP), influence the PC in some or the other way.

Logical operations would be the next group containing byte-oriented logical operations like AND, OR, XOR, rotate, complement and clear. In this group, SWAP is the instruction, which interchanges two digits of accumulator.

Boolean variable manipulation is the group of instructions, which deal with bit variables. We already know that MCS-51 offers a bit-addressable area within its internal RAM, and some of the SFRs are also bit-addressable. In some cases, the CY flag is used as the result register for these Boolean operations.

In this chapter, we start with the instructions related to data transfer. Other types of instructions would be discussed in subsequent chapters. In each of these five groups we would find some instructions, which may not be essential for simple programming practices. All those instructions are placed together under the title 'Advanced Instructions' in Chapter 12.

Observe that in Table 4.1, MOV, CLR, CPL, ANL, ORL mnemonics appear twice. This is because in one case it is meant for byte variables and in the other case it is for bit variables.

4.2 | Addressing Modes

In general, data-transfer instructions are the most widely used instructions in any program. MCS-51 allows five types of addressing modes in these cases. What is an addressing mode? Addressing mode is the way to locate a target data. Another name of this target data is *operand*. Table 4.2 lists all five addressing modes used in MCS-51. The first four addressing modes are briefly discussed here. The last addressing mode, Indexed Addressing Mode, would be discussed along with the MOVC instruction in Chapter 12.

Table 4.2 MCS-51 addressing modes

Addressing mode	Explanation	Example	Refer to figure
Immediate	Data available in the instruction itself	MOV A, #data	4.1
Direct	Address of the data is available in instruction	MOV A, direct address	4.2
Register direct	Register name containing data is available in instruction	MOV A, R3	4.3
Register indirect	Instruction contains the concerned register's name which contains the address of the data	MOV A, @R1	4.4
Indexed	Address of the data is obtained by adding the contents of a base register and an offset register	MOVC A, @A+DPTR	12.2

In general, addressing modes are designated as per the method of targeting the *source data*. Rarely it is used for destination-data targeting purpose.

4.2.1 | Immediate Addressing Mode

Fig. 4.1 explains the details of execution of an immediate addressing mode instruction, MOV A, #data. As an example, the data was assumed as 5AH. Therefore, the instruction becomes MOV A, #5AH. Its opcode is 74H (refer to Appendix A). The instruction is of two bytes and executed in one cycle. No flags are affected after the execution. All these are documented within the small table at the top of Fig. 4.1. The left-hand side of Fig. 4.1 shows the program memory. It indicates that the address 01FDH is occupied by the opcode, 74H. The

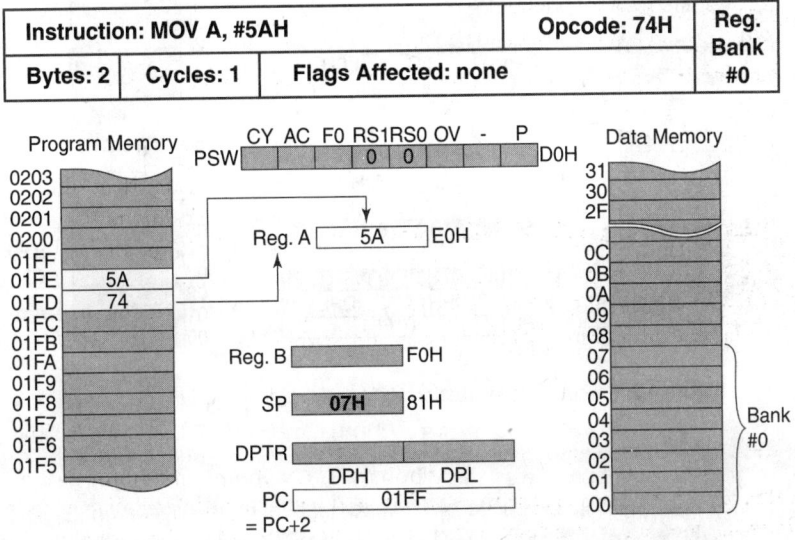

Figure 4.1 Execution of instruction MOV A, #data (immediate addressing mode)

next location, 01FEH, holds the immediate data, i.e. 5AH, which should be copied in the accumulator. There is no reason for these particular addresses, or data. These are included just for the sake of example. The arrow indicates the path of data, which is copied from program memory to the accumulator. Since it is a 2-byte instruction, the PC is incremented by two at the end of instruction execution, and it becomes 01FFH.

The reader may be reminded that the # symbol before the hexadecimal number indicates that it is a data. If the # symbol is not present, then the hexadecimal number would be taken as an address.

The reader may ask why PSW, DPTR, Reg. B, SP and data memory are included in Fig. 4.1. Indeed they do not take part in MOV A, #5AH instruction. However, as we shall use the same format of illustration for most of the instructions of MCS-51, therefore, a uniform general format of illustration is adopted for all instructions. Note that unused bytes or bits are shaded and only those bytes and bits which are directly related or altered with the execution of the instruction, are not shaded. In general we are assuming that register bank #0 is selected and the stack pointer is initialized by 07H.

4.2.2 | Direct Addressing Mode

In direct addressing mode, the address of the data is available in the instruction itself. An instruction, MOV A, 0F0H, is taken as the example case and presented in Fig. 4.2. Incidentally, this address, F0H, belongs to the SFR, register B. Just like the previous case, this is also a 2-byte instruction and the second byte contains the address. Opcode of this instruction is E5H, which forms the first byte of the instruction. Assuming before execution, register B contains the data, say 2CH; after execution, this would be copied to register A. Being a 2-byte instruction, the PC is incremented by 2 and no flags are affected.

Instruction: MOV A, 0F0H			Opcode: E5H	Reg. Bank #0
Bytes: 2	Cycles: 1	Flags Affected: none		

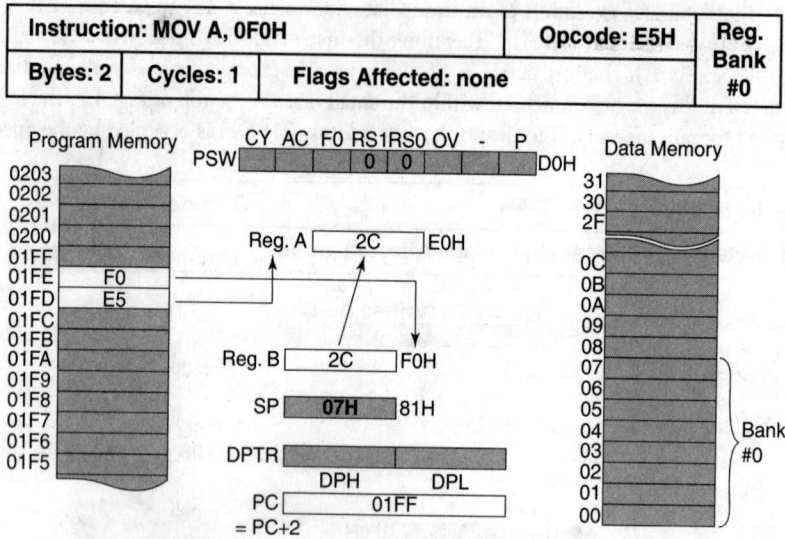

Figure 4.2 Execution of instruction MOV A, 0F0H (direct addressing mode)

What would happen if we write MOV A, #0F0H instead of MOV A, 0F0H? Well, in that case accumulator would be loaded by F0H and not by the present content of SFR B. Note that how the # symbol changes the whole thing. Omitting this # symbol is a common mistake in program development, which we frequently encounter.

4.2.3 | Register Direct Addressing Mode

In this addressing mode, instead of the address the name of the register containing the target data is indicated within the instruction. Using example instruction MOV A, R3, whose opcode is EBH, execution of the instruction is depicted in Fig. 4.3. Note that this is a 1-byte instruction. It is assumed that the present selected register bank is bank #0 and R3 of this bank contained F5H before execution of the instruction.

Being a 1-byte instruction, PC is incremented by 1 and F5 is copied from R3 to the accumulator as shown by the arrow. No flags are affected.

If register bank #1 is selected during the execution of MOV A, R3 instruction, then the data would be copied from the data memory location 0BH instead of 03H. We must be alert about the number of currently selected bank at the time of using any register oriented instruction.

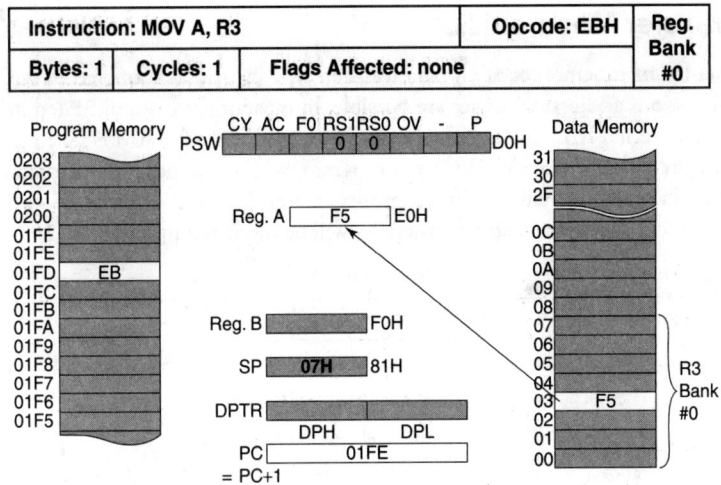

Figure 4.3 Execution of instruction MOV A, R3 (register direct addressing mode)

4.2.4 | Register Indirect Addressing Mode

For this addressing mode, the address of the data is available in a register. In MCS-51, two registers, namely R0 and R1 of the currently active register bank, are allowed to store such an address. For example, an instruction MOV A, @R1 is taken, whose opcode is E7H. It is assumed that before the execution of the instruction, register R1 of the active bank #0 was loaded with 30H and location 30H contained a data, say D2H.

Being a 1-byte one-cycle instruction, when executed, this instruction takes the content of register R1 (30H) as the address and the content of this address (D2H) is copied to the accumulator. No flags are affected and PC is incremented by one. Fig. 4.4 shows the details of execution of this instruction.

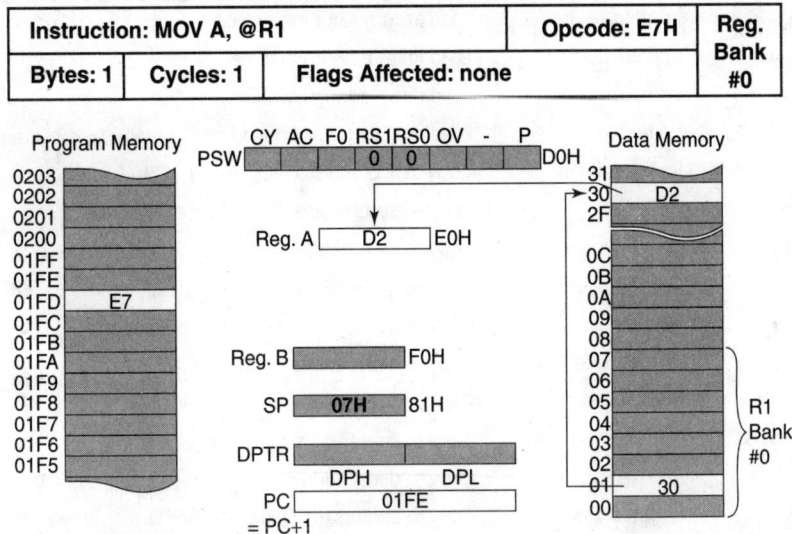

Figure 4.4 Execution of MOV A, @R1 (register indirect addressing mode)

4.3 | Data Transfer Mnemonics

Table 4.3 presents a list of mnemonics of all data transfer-type instructions offered in MCS-51. Note that for each mnemonic, various addressing modes are possible. In most of the cases indicated in Table 4.3, data is transferred within the internal RAM area of the microcontroller. External RAM is taken into account by the MOVX mnemonic. Only in one case (MOVC), the program memory comes into the picture.

Instructions PUSH and POP will be discussed in Chapter 8. In this chapter, we will discuss only the MOV instruction. Remaining data transfer instructions will be discussed in Chapter 12.

Table 4.3 Mnemonics related with data movements

Mnemonic	Brief description
MOV	General data loading and data transfer within internal RAM and SFRs
MOVC	Load accumulator by a byte from program memory
MOVX	Transfer a byte between external RAM and accumulator
PUSH	Store a byte over system stack
POP	Load a byte from system stack
XCH	Exchange bytes between two locations within internal memory
XCHD	Exchange digits (lower nibble) between two locations within internal memory

4.4 | Forms of MOV Instruction

Table 4.4 presents various forms of MOV instruction. These variations arise due to different addressing modes and operands. Note that 'n' of Rn varies from 0 to 7. That is, Rn might be any one of the eight registers, starting from R0 till R7. Using the instruction MOV A, Rn as an example case, all eight variations (for its eight possible operands) of this instruction are presented in Table 4.5.

Table 4.4 Addressing modes of MOV instructions

Instruction	Addressing mode	Function
MOV A, #data	Immediate	Load accumulator immediate
MOV Rn, #data	Immediate	Load Rn (R0–R7) immediate
MOV direct, #data	Immediate	Load direct address immediate
MOV @Ri, #data	Immediate	Load the location whose address is in Ri (R0 or R1) immediate
MOV A, Rn	Register direct	Copy to accumulator from Rn (R0–R7)
MOV A, direct	Direct	Copy to accumulator from direct address
MOV A, @Ri	Register indirect	Copy to accumulator from address in Ri (R0 or R1)
MOV Rn, A	Register direct	Copy to Rn (R0–R7) from accumulator
MOV Rn, direct	Direct	Copy to Rn (R0–R7) from direct address
MOV direct, A	Register direct	Copy to direct address from accumulator
MOV direct, Rn	Register direct	Copy to direct address from Rn (R0–R7)
MOV direct, direct	Direct	Copy to direct address from direct address
MOV direct, @Ri	Register indirect	Copy to direct address from address in Ri (R0 or R1)
MOV @Ri, A	Register direct	Copy to address in Ri (R0 or R1) from accumulator
MOV @Ri, direct	Direct	Copy to address in Ri (R0 or R1) from direct address
MOV DPTR, #data16	Immediate	Load DPTR (16 bit) immediate

Table 4.5 Variations of MOV A, Rn instruction

Instruction	Comments
MOV A, R0	Copy R0 to the accumulator
MOV A, R1	Copy R1 to the accumulator
MOV A, R2	Copy R2 to the accumulator
MOV A, R3	Copy R3 to the accumulator
MOV A, R4	Copy R4 to the accumulator
MOV A, R5	Copy R5 to the accumulator
MOV A, R6	Copy R6 to the accumulator
MOV A, R7	Copy R7 to the accumulator

The suffix 'i' in Ri varies from 0 to 1. So, Ri indicates the possibility of using either R0 or R1. This is explained through Table 4.6, which uses MOV @Ri, A as an example case.

Table 4.6 Variations of MOV @Ri, A instruction

Instruction	Comments
MOV @R0, A	Copy accumulator to the address pointed by R0
MOV @R1, A	Copy accumulator to the address pointed by R1

The term 'direct' indicates any valid direct address of internal RAM. In MCS-51, this address is 8 bit. Therefore, lower 128 bytes of internal RAM (00H to 7FH) and all SFRs may be addressed by this direct address. The term #data indicates 8-bit data and #data16 indicates 16-bit data. The @ sign in instructions of MCS-51 indicates indirect addressing. The location, indicated by this @ sign, would contain the address of the target data. First of the two operands denotes the destination of the data transfer. The second (or last) operand denotes either the data itself, in case of immediate addressing modes, or the source of the data to be transferred.

Note that during a data transfer operation, the data is just copied from the source location, leaving the source location unchanged. Another point we should note is that *no flags are affected in any data transfer operation*.

4.4.1 | 16-Bit Data Load

Instruction at the last row of Table 4.4 (MOV DPTR, #data16) is the only 16-bit immediate addressing instruction of MCS-51. It loads the 16-bit register (data pointer register, DPTR) by the second and third bytes of the instruction. The second byte of the instruction would be loaded in DPH and the third byte of the instruction would be loaded in DPL. For example, execution of the instruction

MOV   DPTR, #0F2C7H

would load DPH by F2H and DPL by C7H, and DPTR would have been F2C7H. Note that any assembly language instruction is divided in two fields as shown in Fig. 4.5. The leftmost part is called 'MNEMONIC' and the following part (if any) contains one or more operands. If more than one operand is used, they are generally separated by a comma. An instruction may or may not have any operands. But it must contain the mnemonic.

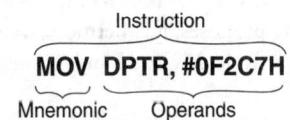

Figure 4.5 Fields of instructions in MCS-51

4.4.2 | 8-Bit Data Load

As we can observe, first four rows of Table 4.4 accommodate four immediate addressing mode type instructions. Examples of these instructions would be

MOV	A, #56H	; load accumulator by 56H
MOV	R7, #0FBH	; load R7 of current bank by FBH
MOV	0B0H, #74H	; load port P3 by 74H (output)
MOV	@R1, #08H	; load the address pointed by R1 by 08H

4.4.3 | Precautions in Hexadecimal Representations

In this context, two important points should be indicated related with the syntax of MCS-51. In MCS-51 instruction syntax, any data must be prefixed by a # sign, otherwise it would be taken as an address. If the third instruction of the above set of instruction is written as

MOV   0B0H, 74H

then it would load port P3 (direct address B0H) from the content of the location whose address is 74H. However, the # sign before 74H in above instruction indicates that it is a data and this data (74H) would be the output from port 3.

The second point we should always remember is that any data or address must start with a numeric character. We know that in hexadecimal notation, six alphabetical characters, namely A, B, C, D, E and F, are used. However, if we want to specify an address, say B0H, we must specify it as 0B0H. So, if B0H is a data, then it must be written as #0B0H and not as #B0H.

4.4.4 | Format of MOV Instruction

The format of the 'MOV' instruction is shown in Fig. 4.6. Fig. 4.6(a) shows data load type instructions, that is the instructions using immediate addressing mode. Other addressing modes use the syntax as shown in Fig. 4.6(b). Note that after the mnemonic 'MOV', the destination operand must be indicated followed by the source operand. There should be a comma (,) separating the destination from the source.

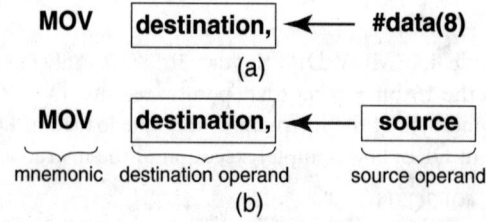

Figure 4.6 Format of (a) data load and (b) data copy instructions

4.4.5 | Comparison with 8085 Mnemonics

If the reader is familiar with 8085, then it may be mentioned that data load-type instructions of 8085 use different mnemonics, like MVI and LXI. The first one, MVI, is used for 8-bit data loading, while the other one, LXI, is used for 16-bit data loading purposes. Furthermore, there are separate mnemonics like LDA and STA to load and store accumulator directly. In MCS-51, only a single mnemonic, MOV, carries out all these operations.

4.4.6 | Operand Expressions

As indicated in Table 4.2, the source or destination operands [Fig. 4.6(b)] might be indicated in any one of the four ways for MOV instruction. They are

- register A or accumulator,
- any direct address, covering lower 128 bytes of internal RAM and SFR area,
- any one of the eight registers of the active register bank and
- indirectly by placing the address in R0 or R1 of active register bank.

Considering these four cases, all available options for data movements are presented schematically in Fig. 4.7. To start with, the circle at the top of Fig. 4.7 represents the data available within the instruction itself, which may be loaded in any one of the four destinations, using immediate addressing mode. Next, from any one of the four possible sources (marked by rectangles), data might be copied to another destination, with one exception. Note that both register direct and register indirect addressing modes would not be simultaneously available in any instruction, to specify source and destination. This is indicated by the absence of any bi-directional arrow (data movement) between two bottommost rectangular boxes. Another special point to be noted is that direct addressing is possible to specify both source and destination locations simultaneously. This data flow is indicated by a pair of semi-circular arcs at the top of the rectangular box indicating direct address.

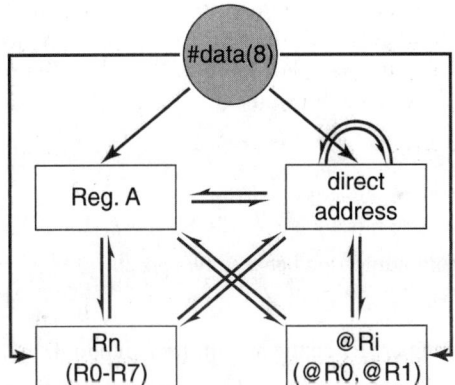

Figure 4.7 Schematic of MOV instructions of MCS-51

 Fig. 4.7 presents all data move type instructions in a compact way so that it is easier to remember. Just draw this figure once or twice without looking at the original and you will find that the data move instruction types are known to you in details.

4.4.7 | Restrictions in Addressing Modes

In Chapter 2 (refer to Fig. 2.6), we have briefly indicated the restriction of usage of addressing modes for SFRs and upper 128 bytes of internal RAM (available only in XX52 and upwards). We reaffirm here that SFRs are not indirectly addressable [Fig. 4.8(b)] and upper 128 bytes of internal RAM (available only in XX52 and higher) are not addressable by direct addressing [Fig. 4.8(a)]. For lower 128 bytes, both direct and indirect addressing modes are allowed. SFRs may only be directly addressed while upper 128 bytes of RAM may only be indirectly addressed.

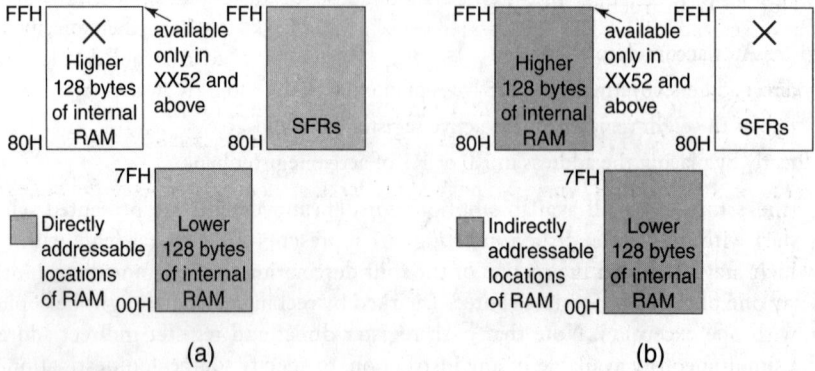

Figure 4.8 (a) Directly addressable and (b) indirectly addressable internal RAM area of MCS-51

4.4.8 | Reading from I/O Ports

In Chapter 3, it was mentioned that MCS-51 ports have two buffers, one for the input pin and another for the output latch. When the source of the MOV instruction is a port, then it reads from the input pins and not from the output latch. On the other hand, if the destination of a MOV instruction is a port, then it writes in the output latch. This is explained in Example 4.1 below. Note that the text after semicolons (;) following instructions are interpreted as non-executable comment statements by the assembler.

4.5 | Solved Examples

Example 4.1

Purpose: Alternatives of data communication between two ports.

Problem

How will the input data present in port P1 be the output through port 3?

Solution

We first configure P1 as input port. Then we may read this port and output the data through P3. The following three instructions are necessary for this purpose.

```
MOV    P1, #0FFH    ; configure port 1 as input port
MOV    A, P1        ; copy input pins of port 1 to accumulator
MOV    P3, A        ; copy accumulator to port 3 latch
```

The last two instructions may be replaced by the following single instruction.

```
MOV    P3, P1       ; copy from port 1 to port 3
```

Example 4.2

Purpose: Restriction in accessing data at some part of RAM.

Problem

A data is available in register R3 of bank #2, which is currently a selected register bank. How should this data be stored at internal RAM location 82H of a XX52 device?

Solution

Internal RAM location 82H is located at upper 128 bytes of internal RAM of XX52 devices. This area (upper 128 bytes) is only indirectly addressable. As the destination must be indirectly addressed, the register R3 · f bank #2 must also be directly addressed. Direct address of register R3 of bank #2 is 13H. Therefore, the follow ng two instructions are necessary which would copy the data from R3 of bank #2 to internal RAM location 82H.

```
MOV    R0, #82H     ; load pointer by destination address
MOV    @R0, 13H     ; copy R3 of bank #2 to RAM location 82H
```

Note that instead of R0, register R1 may also be used in identical manner producing the same result.

Example 4.3

Purpose: Alternate methods to move data bytes between two locations indicating constraints in some such cases.

Problem

Internal RAM location E0H of a XX51 microcontroller contains a data which should be copied to internal RAM location 18H. Indicate in how many ways it might be accomplished.

Solution

As the device is specified as XX51, the internal RAM location E0H cannot be at higher 128 bytes of internal RAM, which is non-existent in XX51 microcontroller. Therefore, internal RAM location E0H must be the address of accumulator (register A), located within the SFR area. Location 18H is also designated as register R0 of bank #3. The problem is to copy from register A to R0 of bank #3 (refer to Fig. 4.9). As nothing is indicated in the problem about the currently selected bank, therefore, any one of the following instructions might be used for this copy operation.

The following instruction is applicable irrespective of any selected bank, as it uses direct address of destination:

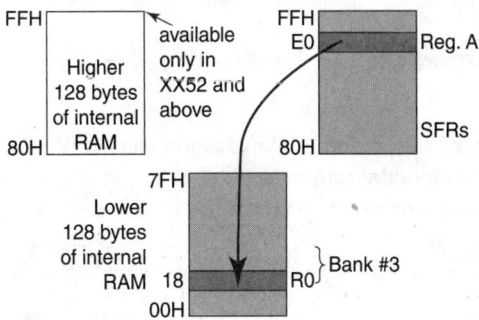

Figure 4.9 Copy from E0H to 18H

```
MOV    18H, A
```

The following instruction uses direct addresses for both destination and source, and again does not depend on the selected bank:

```
MOV    18H, 0E0H
```

Another method to solve the problem is to load a pointer register (R0 or R1) of the current register bank by 18H, the destination address. Therefore, any one of the following two instructions pairs may be used. The current bank may be any bank including bank #3. Note that R0 may always be replaced by R1, if it is used as a pointer.

MOV	R0, #18H	; load pointer
MOV	@R0, A	; copy data through the pointer
MOV	R0, #18H	; load pointer
MOV	@R0, 0E0H	; copy data through pointer

Assuming that register bank #3 is selected, any one of the following two instructions would perform the data transfer:

MOV	R0, A
MOV	R0, 0E0H

Example 4.4

Purpose: Restrictions of some addressing modes of 8051.

Problem

Assuming that register bank #0 is selected, how would a data in R7 of this bank be copied to R3 of the same bank (refer to Fig. 4.10)?

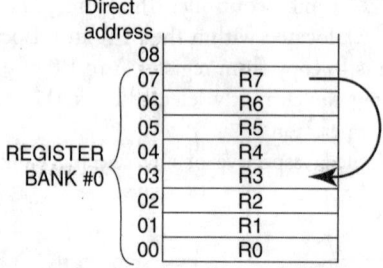

Figure 4.10 Copy from R7 to R3 of bank #0

Solution

At the first impulse, we might be tempted to coin instruction like MOV R3, R7 to serve the purpose. However, MCS-51 does not allow this addressing mode. Out of the source and destination, only one might be specified by the register name. The other one might be expressed only through its direct address. Note that the indirect addressing in one case is not allowed if the other one is expressed by the register name (register direct). Therefore, either one of the following two instructions would serve the purpose:

MOV	R3, 07H	; copy from R7 of bank #0 to R3 of bank #0
MOV	03H, R7	; copy to R3 of bank #0 from R7 of bank #0.

 As most of the readers have already gone through 8085 instruction set, therefore, MOV R3, R7 would be a common mistake for them as it is allowed in 8085. Of course, 8085 uses register names as A, B, C, D, E, H and L.

SUMMARY

MCS-51 offers five addressing modes to locate the target data. These five addressing modes are as follows.

- Immediate
- Direct
- Register direct
- Register indirect
- Indexed

For the first case, data is a part of the instruction itself, while for the second, address of the data is a part of the instruction. When the name of the register, where the target data is available, is indicated in the instruction itself, it is known as register direct addressing mode. Register indirect addressing mode indicates the name of the register where the address of the target data would be available.

MOV instruction of MCS-51 accomplishes both data loading and copying data from one location to another. No flags are affected by any MOV instruction. Format of this instruction is the mnemonic 'MOV' followed by destination and then source of the data. Source or destination might be the accumulator, any one of eight registers of the currently operational register bank, any direct address (comprising lower 128 bytes of RAM and SFRs) or any one of the pointer registers – R0 or R1 – of the current bank. In all above cases only a byte of data is transacted. However, the same MOV instruction is also capable of loading 16-bit data in the 16-bit register, DPTR.

In general, the source and the destination should have different addressing techniques, except for direct addressing. However, if the source or destination is any register other than the accumulator, then the destination or source cannot be expressed by indirect addressing. In other words, **there is no such instruction** as MOV R4, @R1 or MOV @R0, R6 in MCS-51.

Data must be represented by a # sign otherwise it is taken as an address in MCS-51. Moreover, a data should always start with a numerical value and never with any alphabets.

POINTS TO REMEMBER

- For any data copy instruction, destination register must be indicated first and then the source register.

- For all immediate addressing mode instructions, a # symbol is essential to indicate that the following information is not an address.

REVIEW QUESTIONS

Evaluate Yourself

1. What is the addressing mode of the instruction MOV R0, #03H?

    (a) Immediate   (b) Register direct

    (c) Register indirect   (d) None of these

2. What type of instruction is 'POP'?

    (a) Arithmetic   (b) Program branching

    (c) Logical operation   (d) None of these

3. What would be done by the following instruction, when executed?

    MOV   A, 0F0H

(a) Load A by F0H

(b) Copy from F0H in R2 of bank #1

(c) Copy register B to register A

(d) None of these

4. What would happen if the following instruction is executed?

    MOV   PSW, #18H

(a) R0 of bank #3 would be copied to PSW register

(b) Bank #3 would be selected

(c) Content of the location whose address is in R0 of bank #3 would be copied to PSW

(d) None of these

5. Assuming that bank #0 is selected and register R0 of this bank contains 80H, which of the following instructions would copy the data from port 0 to register B?

(a) MOV 0F0H, 80H

(b) MOV B, @R0

(c) MOV F0H, @R0

(d) None of these

6. Assuming that bank #2 is selected and register R0 of bank #2 contains E0H and register R1 of bank #2 contains 15H, which of the following instructions would not copy the data from location E0H to location 15H?

(a) MOV R5, A   (b) MOV @R1, 0E0H

(c) MOV 15H, @R0   (d) None of these

7. What would be the content of DPH after execution of MOV DPTR, #1234H instruction?

(a) 12H   (b) 34H

(c) 1234H   (d) None of these

8. Assume that port 0 output latch contains 00H and its port pins are receiving the data 77H from external source. Which of the following instructions would read port 0 pins and store it in the accumulator?

(a) MOV A, P0   (b) MOV 0E0H, P0

(c) MOV A, 80H   (d) None of these

9. What would be the content of the accumulator after execution of the instruction, MOV A, SP just after system reset?

(a) Undefined   (b) 07H

(c) 08H   (d) None of these

10. What would happen after execution of the following instruction?

MOV   AAH, #AAH

(a) Undefined

(b) Accumulator would have AAH

(c) Location AAH would have AAH

(d) None of these

Search for Answers

1. What is meant by 'Addressing Mode'? How many addressing modes are there in MCS-51? Give one example of each type.

2. What is the difference between MOVC and MOVX instructions?

3. Which part of the internal RAM is related with PUSH and POP instructions?

4. What is the difference between XCH and XCHD instructions?

5. Match the following instructions, in the left column, with the addressing modes on the right.

Instruction		Addressing mode
MOV	P2, #0FH	Register direct
MOV	P0, P2	Register indirect
MOV	A, R5	Immediate
MOV	A, @R0	Direct
MOV	@R0, 0F0H	Immediate
MOV	R0, #22H	Register direct
MOV	0FFH, @R1	Direct
MOV	B, R7	Register indirect

6. What is the difference between the immediate data loading instructions of 8085 and MCS-51?

7. (a) Does the MOV instruction read from the port pins or the output latch?

(b) Does the MOV instruction write in the port pins or the output latch?

8. What would happen if the # symbol is removed from a data in an MOV instruction?

9. What happens to the source location after its content is copied to a destination location using MOV instruction?

10. Which instruction would you use to select register bank #2?

Think and Solve

1. What is the difference between register direct and register indirect addressing modes?

2. Why is some area of internal RAM of MCS-51 only directly addressable and some other part only indirectly addressable?

3. Within the SFR area (from 80H to FFH) there are some locations where there is no SFR allotted. What would happen if data are stored in such an area by direct addressing?

4. Which register or counter of MCS-51 cannot be accessed with MOV instruction?

5. Out of accumulator, any register of the active register bank, SFRs and higher 128 bytes of RAM locations, which one can be addressed in maximum number of ways?

6. If port pin of P0 is receiving 44H and P1 is receiving 47H, then what would happen after the following three instructions are executed sequentially?

MOV	A, P0
MOV	R0, P1
MOV	@R0, A

7. What would happen after execution of the following four instructions?

MOV	A, #55H
MOV	11H, #77H
MOV	0D0H, #10H
MOV	@R1, A

8. What would be the content of locations 1FH and 1EH after execution of following six instructions?

MOV	DPTR, #1E1FH
MOV	A, 83H
MOV	B, 82H
MOV	PSW, #18H
MOV	R6, A
MOV	R7, B

9. Write a program for storing the current value in accumulator at location 23H and then storing the input data from port 0 in the accumulator?

10. Write a program to exchange the data between R0 and R1 of the current register bank?

5 ARITHMETIC OPERATIONS

CHAPTER OBJECTIVES

In this chapter, the reader is introduced to the instructions related with arithmetic operations of MCS-51. After completion of this chapter, the reader should be able to understand how to

- Use ADD, ADDC, SUBB, INC and DEC instructions.
- Perform unsigned integer addition and subtraction.
- Implement multiple byte addition and subtraction.

5.1 | Introduction

In Chapter 4, we learnt how to load and copy 8-bit data from various locations using different addressing modes of 8051. As a matter of fact, except the Program Counter (PC), data may be loaded anywhere and copied from any position to any location. However, all these would be meaningless unless we are able to do some processing with these data. In this chapter, we will see how simple arithmetic operations may be implemented in MCS-51.

MCS-51 offers a total of eight mnemonics for arithmetic operations. All of these are listed in Table 5.1. Being an 8-bit microcontroller, all of these operations are 8-bit operations. We observe from Table 5.1 that instructions are available for addition, subtraction, multiplication and division. These complete all basic arithmetic operations. The increment (by one) and decrement (by one) operations are generally related to the address manipulation and not data (although many times they are used for data processing). The last instruction, namely Decimal Adjust for Accumulator (DA A), is used for a special purpose, decimal number processing. In this chapter, we discuss the first five operations. The remaining three operations, i.e. MUL, DIV and DA A, will be discussed in Chapter 12.

Table 5.1 Mnemonics related with arithmetic operations

Mnemonic	Brief description
ADD	Add two operands
ADDC	Add two operands with carry
SUBB	Subtract one operand from another
INC	Increment operand by one
DEC	Decrement operand by one
MUL	Unsigned integer multiplication
DIV	Unsigned integer division
DA A	Decimal adjust for accumulator

5.2 | ADD Instruction

MCS-51 offers two mnemonics for arithmetic addition operations. They are ADD and ADDC. Both offer four addressing modes, namely immediate, direct, register direct, and register indirect. However, the difference between these two is the inclusion of carry during addition. In the first case, carry is not considered while for the second one, the condition of carry (CY) flag is taken into account at the time of addition. Therefore, for multiple bytes addition, the second one is applicable where carry plays an important role. Table 5.2 presents all addressing modes of ADD instruction. Note that in all cases the result would be available in the accumulator and all flags would be affected. If this instruction uses any port address, then the data would be taken from the port pins and not from the port latch.

Table 5.2 Addressing modes of ADD instruction

Instruction	Addressing mode	Function
ADD A, #data	Immediate	Add immediate with accumulator
ADD A, Rn	Register direct	Add content of specified register with accumulator
ADD A, direct	Direct	Add content of indicated address with accumulator
ADD A, @Ri	Register indirect	Add content of address indicated by the register with accumulator

Let us take an example case to explain an 8-bit addition without carry.

Example 5.1

Purpose: To visualize complete data flow along with data processing.

Problem

Assume that memory locations, 04H and 05H, contain two unsigned 8-bit numbers. Write a program to add these two numbers and store it in memory location 30H. Assume that register bank #0 is selected.

Solution

Locations 04H and 05H become R4 and R5 of the current bank, respectively. Therefore, following steps may be planned:

Step 1:   Copy one operand, say R4, in the accumulator.
Step 2:   Add the accumulator with the other operand, that is R5.
Step 3:   Copy the result in the accumulator to location 30H.

Following instructions are necessary to implement this objective.

```
MOV   A, R4        ; copy one operand in the accumulator
ADD   A, R5        ; add it with the second operand
MOV   30H, A       ; store result in desired location
```

To understand the functioning of the processor, let us take a closer look at these instructions' execution. Let us assume that using Appendix A, we have assembled this program and loaded it within the microcontroller from program memory location 01FCH onwards. We also assume that before execution of these instructions, registers R4 and R5 of bank #0 are loaded with D7H and 9EH, respectively.

Execution of the first instruction, MOV A, R4 is depicted in Fig. 5.1. This is a 1-byte instruction with the opcode as ECH. Execution of this instruction copies register R4 of bank #0 to the accumulator. Therefore, after execution, the accumulator contains D7H and no flags are affected. Note that PC is incremented by one and now pointing to the location 01FDH.

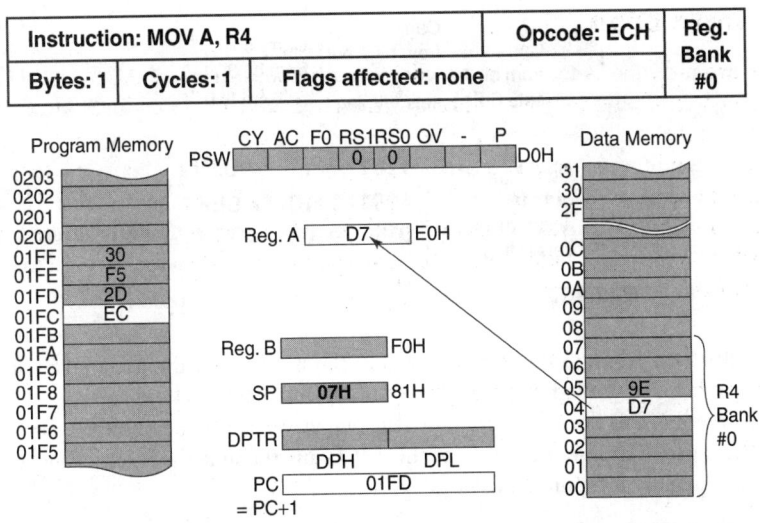

Figure 5.1 Processor condition after executing instruction MOV A, R4

The ADD A, R5 instruction is executed as the next step and shown in Fig. 5.2. Opcode of this instruction is 2DH, which takes the present content of R5 (i.e. 9EH) and adds with the accumulator content, producing the result of this addition, 75H. This result is placed in the accumulator itself. As it was already mentioned, this instruction (ADD) affects all four flags of the Program Status Word (PSW). How these flags are affected because of this addition, is shown in Fig. 5.3.

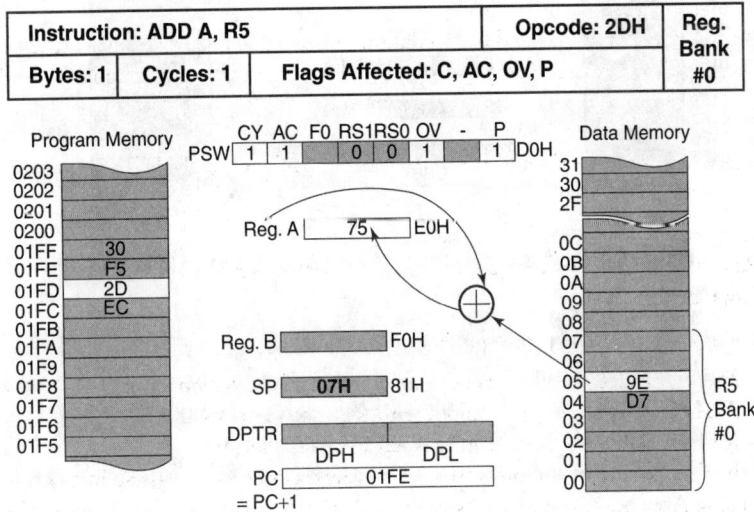

Figure 5.2 Processor condition after executing instruction ADD A, R5

Note that a carry out of bit 7 sets the CY flag and a carry out of bit 3 sets the auxiliary carry (AC) flag. Because there is a carry from bit 7 but not from bit 6, the overflow (OV) flag is set.

After the addition operation, the number of 1s in the accumulator is five, which is an odd number and, therefore, the parity is odd and the parity (P) flag is set.

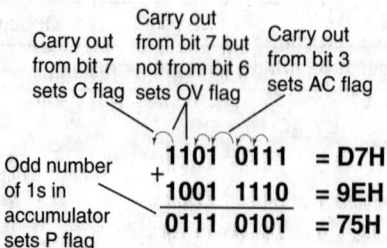

Figure 5.3 Condition of status flags after ADD instruction

The third instruction, MOV 30H, A is for storing the result from the accumulator in data memory location 30H. This is depicted through Fig. 5.4. Opcode of this instruction is F5H and it is a 2-byte instruction. Being a MOV instruction, this does not affect any flags. However, it may be observed that the condition of PSW remains unchanged even after execution of this MOV instruction. The result of the addition (75H) is now available in the memory location 30H.

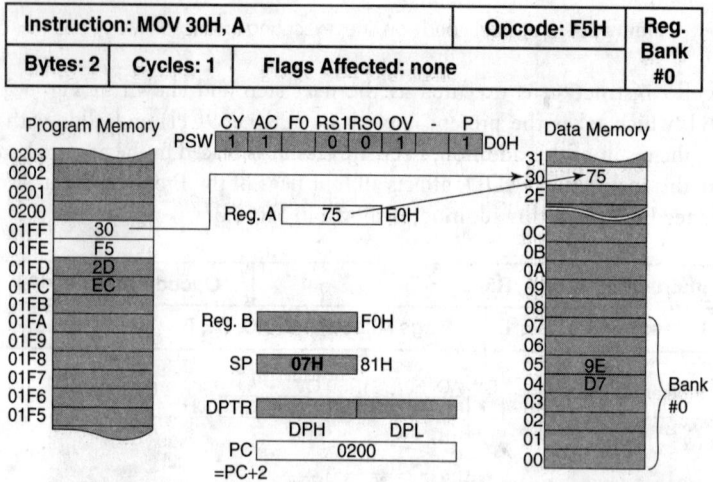

Figure 5.4 Processor condition after executing instruction MOV 30H, A

 Figures 5.1, 5.2 and 5.4 together present a small clip of the processor activities within a restricted limit. As a matter of fact, this type of 'get data from memory, add it with another data in memory and finally store back the result in memory' type program segments are frequently used to judge the efficiency of a processor.

5.3 | ADDC Instruction

As already indicated before, the ADDC instruction takes into account the condition of the CY flag at the time of performing the 8-bit addition. We have already observed how the instruction ADD A, R5 would work if the accumulator has D7H and R5 of bank #0 has 9EH. If we assume that the CY flag contained zero before execution of ADDC A, R5 instruction, then there would not be any difference in the result or final condition of any flag. However, if the CY flag was set before execution, then ADDC

A, R5 instruction would produce a result of 76H instead of 75H. Note that the carry condition is always added with bit 0. In this case also, all four flags would be set. When used for I/O ports, ADDC would get the data from the port pins and not from the port latch. ADDC instruction is useful for multiple-byte addition. All addressing modes of ADDC instruction are presented in Table 5.3. Let us now consider an example of it.

Table 5.3 Addressing modes of ADDC instruction

Instruction	Addressing mode	Function
ADDC A, #data	Immediate	Add immediate with accumulator with carry
ADDC A, Rn	Register direct	Add content of specified register with accumulator with carry
ADDC A, direct	Direct	Add content of indicated address with accumulator with carry
ADDC A, @Ri	Register indirect	Add content of address indicated by the register with accumulator with carry

Example 5.2

Purpose: Application of ADDC instruction and taking care of carry bit, generated by some previous addition operation.

Problem

Write a program to add two 16-bit numbers, say B7C2H and 549AH. The first 16-bit number is located in registers R2 and R3 of bank #0 (currently selected) with lower of the two bytes in R2 and the higher in R3. The second 16-bit number is located in memory location 09H and 0AH with its lower byte in 09H and higher byte in 0AH. Store the result in 2FH (lower byte) and 30H (higher byte).

Solution

Schematically, the problem might be depicted as in Fig. 5.5(a). The solution would need the following six steps.

Step 1: Copy R2 to the accumulator.
Step 2: Add the accumulator with the content of the address 09H.
Step 3: Store the result from the accumulator to the address 2FH [Fig. 5.5(b)].

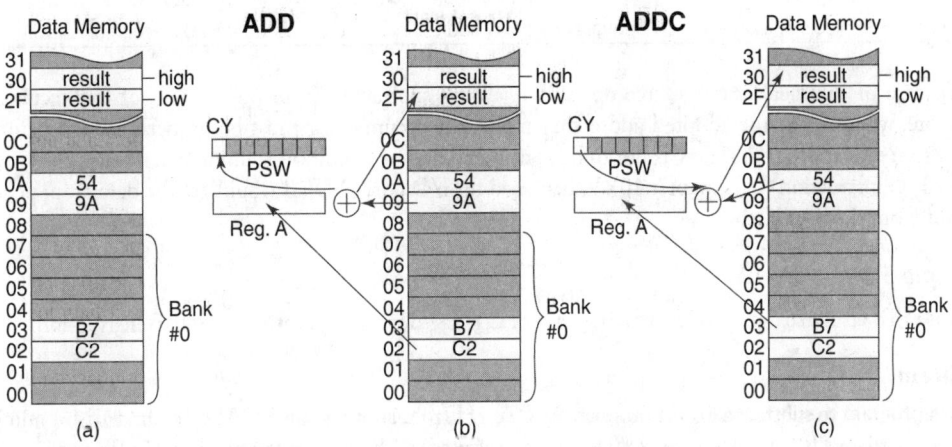

Figure 5.5 16-bit addition (a) problem definition, (b) lower byte result and (c) higher byte result

Step 4:   Copy R3 to the accumulator.
Step 5:   Add with carry the accumulator and the content of the address 0AH.
Step 6:   Store the result from the accumulator in the address 30H [Fig. 5.5(c)].

The program for it might have the following six instructions. It is assumed that register bank #0 is already selected and the numbers are already stored in their locations.

```
MOV    A, R2      ; get one LSB
ADD    A, 09H     ; add with the other, carry would be in CY
MOV    2FH, A     ; store lower byte of result, CY flag is not affected
MOV    A, R3      ; get the MSB
ADDC   A, 0AH     ; add with the other including CY flag condition
MOV    30H, A     ; save result, MSB
```

Here, the sequence of ADD and ADDC is important. For multiple bytes addition, the first addition should be performed with ADD and all subsequent additions with ADDC.

 If we like to replace the ADD instruction by ADDC instruction, then the carry flag must be cleared before that. The benefit of using ADDC in all steps of addition is that a subroutine may be developed for only one addition using pointers which may be used many times.

5.4 | SUBB Instruction

The SUBB instruction, unlike its counterpart, always considers the borrow (or CY flag condition) during its operation. Same four addressing modes are also available in this case as presented in Table 5.4. All flags are affected in identical manner.

Table 5.4 Addressing modes of SUBB instruction

Instruction	Addressing mode	Function
SUBB A, #data	Immediate	Subtract immediately from accumulator with borrow
SUBB A, Rn	Register direct	Subtract indicated register content from accumulator with borrow
SUBB A, direct	Direct	Subtract content of indicated address from accumulator with borrow
SUBB A, @Ri	Register indirect	Subtract content of address indicated by the register from accumulator with borrow

In case of subtraction, the source operand is always subtracted from the content of the accumulator. Therefore, whatever be the adopted addressing mode, the accumulator must be properly loaded before the subtraction operation. If the carry is not to be considered and its condition is unknown, it must be cleared by the CLR C instruction (see Chapter 10). When used for I/O ports, SUBB would get the data from port pins and not from the port latch.

Example 5.3

Purpose: To visualize data flow and taking care of CY flag during application of SUBB instruction.

Problem

Write a program to subtract a 16-bit number, say B7C2H from another, say, 549AH. The first 16-bit number is located in registers R2 and R3 of bank #0 (currently selected) with lower of the two bytes in R2 and the higher in R3. The second 16-bit number is located in memory location 09H and 0AH with its lower byte in 09H and higher byte in 0AH. Store the result in 2FH (lower byte) and 30H (higher byte).

Solution

Pictorially, the problem is presented through Fig. 5.6(a). Just like the previous example, here also two steps are necessary to perform two 8-bit subtractions, considering the borrow in the CY flag. However, before the first subtraction, the CY flag should be cleared. Although it is customary to use the instruction CLR C in this case, however, as the Boolean instructions are yet to be introduced, we may use the instruction MOV PSW, #00H to clear the CY flag. Note that this instruction also selects register bank #0 and clears all other flags of PSW, which is OK for our present purpose.

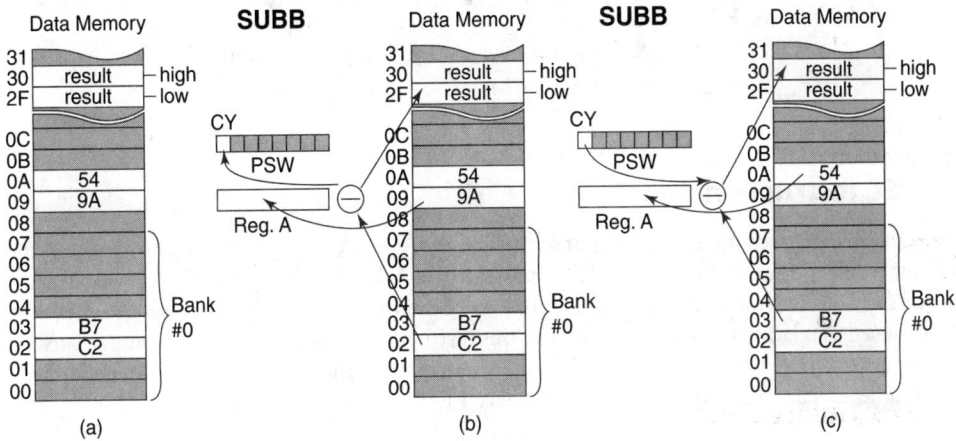

Figure 5.6 16-bit subtraction (a) problem definition, (b) lower byte result and (c) higher byte result

Following is the program for the 16-bit subtraction:

```
MOV   A, 09H
MOV   PSW, #00H    ; clear CY flag
SUBB  A, R2
MOV   2FH, A       ; store lower byte of the result
MOV   A, 0AH
SUBB  A, R3
MOV   30H, A       ; store higher byte of the result
```

5.5 | INC Instruction

INC instruction increments the content of the source operand by one. No flags are affected by this increment. Three addressing modes are available for this instruction and all of its variations are listed in Table 5.5.

Table 5.5 Addressing modes of INC instruction

Instruction	Addressing mode	Function
INC A	Register direct	Increment accumulator by one
INC Rn	Register direct	Increment content of indicated register by one
INC direct	Direct	Increment content of indicated address by one
INC @Ri	Register indirect	Increment content of address indicated by the register by one
INC DPTR	Register direct	Increment content of DPTR by one

INC DPTR is the only instruction that takes care of a 16-bit increment. If the original content of DPTR is FFFFH before execution of this instruction, then it would contain 0000H afterwards.

As a normal practice, this and the next instruction, that is DEC, are used for changing the address. That is the reason of leaving the flags undisturbed in these cases.

When the INC instruction is directly addressed to some I/O port (INC port address) then the data is read from the port's latch and not from its input pins.

DPTR is the only register available to the 8051 programmer which may be directly loaded by any 16-bit value and also might be incremented by 1 in a 16-bit format. These special features offered through DPTR may be helpful in some critical cases of software development.

5.6 | DEC Instruction

The DEC instruction decrements the content of the source operand by one, which is the opposite of the INC instruction. It offers three addressing modes as presented in Table 5.6. Note that there is no instruction to decrement the DPTR register in this case. Just like its counterpart, this instruction also does not affect any flag. When used in relation with any port address, this instruction reads from the port's output latch and not from its input pins. If an operand contains 00H before execution of this instruction, then an underflow would occur during its execution and the operand would have FFH, with no flags affected.

Table 5.6 Addressing modes of DEC instruction

Instruction	Addressing mode	Function
DEC A	Register direct	Decrement content of accumulator by one
DEC Rn	Register direct	Decrement content of indicated register by one
DEC direct	Direct	Decrement content of indicated address by one
DEC @Ri	Register indirect	Decrement content of address indicated by the register by one

If just after system reset, the very first instruction executed by 8051 is DEC A, then what would happen? No, the content of accumulator would not be undefined but it would have FFH, as a system reset clears the accumulator. Note that generally it takes two bytes to load accumulator by any data. These shortcuts, applicable only for special cases are sometimes adopted by expert programmers.

SUMMARY

Arithmetic instructions for addition (ADD and ADDC) and subtraction (SUBB) are introduced in this chapter. Four addressing modes are available for these instructions, namely immediate, direct, register direct and register indirect. All four flags, namely CY, AC, OV and P, are affected by these instructions. Out of these four flags, the CY flag plays the most important role and must be properly initialized before SUBB instruction. When a direct address is used, data is taken from input pins and not from output latch of any port.

INC and DEC instructions are used for address manipulation to get the next adjacent address. No flags are affected by these two instructions. Three addressing modes, namely direct, register direct and register indirect modes, are available. If any I/O port is incremented or decremented by these instructions, then the data is taken from the output latch of that port and not from the input pins.

The INC instruction may also be used to increment the 16-bit register, DPTR, by one. Three more instructions, namely MUL, DIV and DA A, are indicated and would be discussed in details in Chapter 12.

POINTS TO REMEMBER

- There is no Z flag in 8051. The accumulator must be physically checked by the processor for any zero result.

- There is no DEC (decrement by one) instruction applicable for DPTR. Only INC is available for this register.

REVIEW QUESTIONS

Evaluate Yourself

1. Which one of the following addressing modes is not available for ADD, ADDC and SUBB instructions?

   (a) Indexed   (b) Direct

   (c) Register indirect (d) None of these

2. What would be the result if 0110 0010B is added with 0001 0001B using ADDC instruction and the CY flag was cleared before the execution of the instruction?

   (a) 0111 0100B   (b) 0111 0011B

   (c) 1000 0110B   (d) None of these

3. Which of the following instructions would clear the CY flag?

   (a) CLR PSW   (b) CLR ALL

   (c) MOV PSW, #00H   (d) None of these

4. What would be the content of register R7 after execution of INC R7 instruction, if initially it contained 2FH?

   (a) 3FH   (b) 2EH

   (c) 20H   (d) None of these

5. What would be the content of DPTR after execution of INC DPTR instruction, if initially it contained F0FFH?

   (a) 00FFH   (b) F000H

   (c) F100H   (d) None of these

6. If 3DH in accumulator is added with F1H in register B using ADDC instruction and the CY flag was set before the execution of the instruction then which of the following flags would not be set?

   (a) OV and AC flags   (b) CY and AC flags

   (c) P and OV flags   (d) None of these

7. Which of the following flags would be set when R2, originally having 00H, is decremented by one using the instruction DEC R2?

   (a) AC flag   (b) CY flag

   (c) OV flag   (d) None of these

8. Assuming the CY flag is set (1) and accumulator contains 00H, what would be the content of accumulator after executing SUBB A, #00H followed by ADD A, #00H?

   (a) 00H   (b) 01H

   (c) FFH   (d) None of these

9. Assuming accumulator contains 01H, CY flag is 0 and data memory location 30H contains FFH, which of the following instructions, when executed, would set the CY flag?

   (a) INC 30H   (b) ADD A, 30H

   (c) DEC A   (d) None of these

10. What would be the content of the accumulator when the following program is executed?

    MOV   PSW, #00H

    MOV   R4, A

    SUBB   A, R4

    (a) Undefined   (b) 00H

    (c) 01H   (d) None of these

Search for Answers

1. What is the difference between ADD and ADDC instructions?

2. What is the purpose of the CY flag?

3. How many addressing modes are available for SUBB instruction?

4. What precaution should we take before using SUBB instruction?

5. How many addressing modes are available for DEC instruction?

6. Why are flags not affected for INC and DEC instructions?

7. What are the conditions for setting the OV flag?

8. What are the conditions for clearing the AC flag?

9. What are the conditions for setting the P flag?

10. How are 16-bit or 32-bit additions implemented using MCS-51 instructions?

Think and Solve

1. What is the purpose of the AC flag?

2. Which of the following two instructions would be more appropriate to increment the accumulator by one? INC A or ADD A, #01H? Justify your answer.

3. Why is there no provision of 'DEC DPTR' instruction in MCS-51?

4. Is it possible to subtract the content of a register from the accumulator without using SUBB instruction?

5. If you need to decrement DPTR by one, how would you do it?

6. Assuming bank #0 is selected, how many addressing modes of ADD instruction might be applicable if it is to add two numbers stored in locations E0H and 00H?

7. After executing the following program, the accumulator contained 00H. What was the initial content of register R2?

MOV   A, R2
ADDC   A, R2
INC A

8. All other conditions remaining the same, if the second instruction of above program is changed to SUBB A, R2 then what would be the final content of the accumulator?

9. Write a program to add content of R2 with R3 and then subtract R4 and R5 from it.

10. If the register contents of the above problem remain unchanged, would there be any difference in the final result if R4 is subtracted from R2, followed by addition of R3 and finally, subtraction of R5?

6 PROGRAM BRANCHING

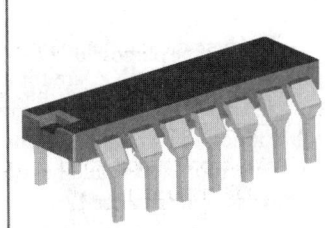

CHAPTER OBJECTIVES

Controlling the flow of the program is one of the important aspects of program development. This is achieved by conditional and unconditional branching instructions, which are discussed in this chapter. After completion of the chapter, the reader should be able to understand

- Limitation of different types of unconditional jump instructions.
- Different types of conditional branching instructions.
- How to control the program flow using instructions like JZ, CJNE and DJNZ.

6.1 | Program Branching Instructions

Program branching plays a very important role in software development. In Chapter 1, we have introduced a microcontroller as a device with the duty of continuously fetching and executing instructions (refer to fig. 1.2). This may be designated as a sequential execution of instructions. However, many times the processor is to branch from one sequence to another, known as conditional branching. There may be some cases of unconditional jumps also. MCS-51 offers several mnemonics for these operations as listed in Table 6.1.

Table 6.1 Mnemonics related with program branching operations

Mnemonic	Brief description
LJMP	Unconditional long jump (within 64K)
AJMP	Unconditional absolute jump (within 2K)
SJMP	Unconditional short jump (relative to PC)
JZ	Jump if accumulator is zero (relative to PC)
JNZ	Jump if accumulator is not zero (relative to PC)
CJNE	Compare and jump if not equal (relative to PC)
DJNZ	Decrement and jump if not zero (relative to PC)
NOP	No operation
LCALL	Unconditional long call (within 64K)
ACALL	Unconditional absolute call (within 2K)
RET	Return from call
RETI	Return from interrupt
JMP	Jump to the address indicated by DPTR and A

The first three mnemonics of this table are for unconditional jumps. Next four mnemonics, namely JZ, JNZ, CJNE and DJNZ, are for conditional branching. NOP is an instruction for no operation. We will discuss all these eight mnemonics in this chapter. Mnemonics LCALL, ACALL and RET are associated with subroutines and would be discussed in Chapter 8 (Subroutines and Stack). Instruction RETI is associated with interrupts and would be introduced in Chapter 12. JMP is a special instruction of loading PC and will be discussed in Chapter 12.

6.2 | Unconditional Jumps

MCS-51 provides three different instructions for unconditional jump. They are LJMP, AJMP and SJMP. The difference between these three is the range limits attached with them (see Table 6.2). While it is the maximum for LJMP (anywhere within 64K of program memory), the limit shrinks to only 256 bytes for SJMP instruction. AJMP covers an intermediate range of 2K limit within the current page, *starting from first byte of next instruction after AJMP*. Fig. 6.1 highlights these range limits.

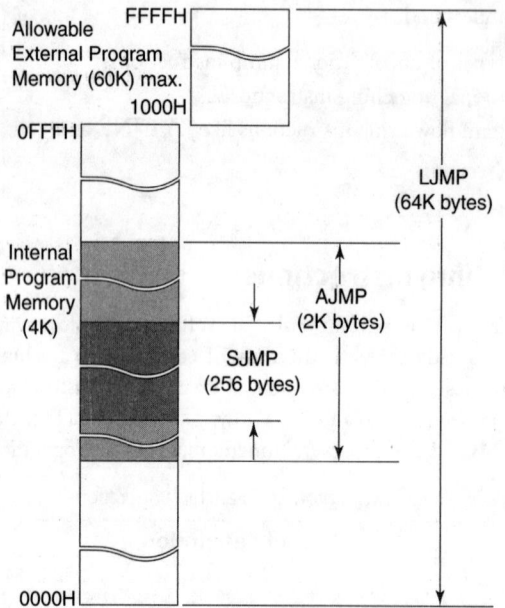

Figure 6.1 Range of jump instructions in program memory

This choice is offered as the length of these instructions is different and so also their capabilities. We now discuss all these three instructions individually.

6.2.1 | LJMP Instruction

LJMP is a 3-byte unconditional jump instruction and capable of transferring the program control anywhere within 64K of program memory. The second and the third bytes of this instruction contain the address where the control would be transferred. The second byte of the instruction is loaded in PCH and the third byte in PCL for transferring the control. So, the second byte contains the higher order and the third byte contains the lower order jump address. Two thin curved arrow lines in Fig. 6.2 indicate this. No flags are affected and it takes two cycles to complete the transfer.

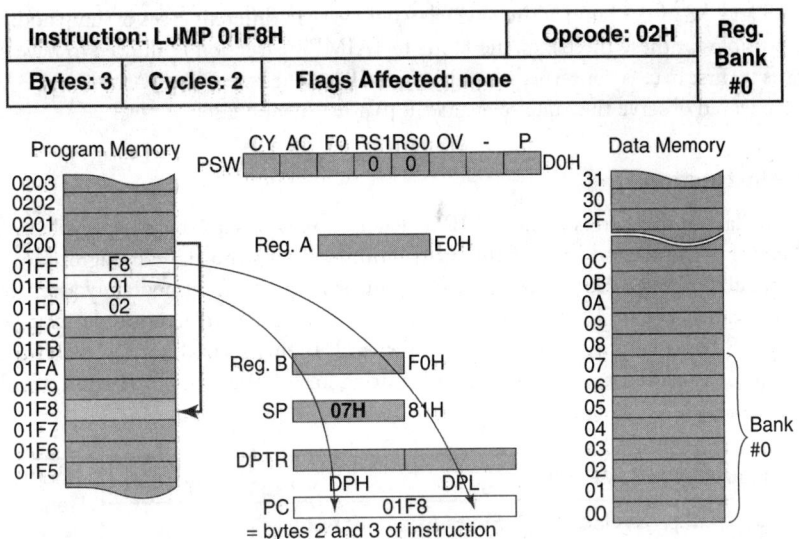

Figure 6.2 Execution of LJMP instruction

Fig. 6.2 also indicates the transfer of control to location 01F8H by a thick arrow line. One very important point to be observed in this case is that this arrow starts from location 0200H. This is because of the fact that the third byte of LJMP instruction is located at 01FFH, and after fetching the third and last byte of the ongoing instruction, the PC is automatically incremented to the first byte of the next instruction. However, decoding and execution of the instruction lead to filling up the PC with proper values (01F8H in this example case) so that the next executable instruction is fetched from the location as indicated now by the PC. This point is elaborated here as it plays a crucial role in the execution of next two jump instructions, namely AJMP and SJMP.

6.2.2 | AJMP Instruction

As compared with LJMP, which is a 3-byte instruction, AJMP is a 2-byte two-cycle instruction capable of performing unconditional jumps. However, unlike LJMP, which can change the PC to any address within 64K of program memory, AJMP's domain is limited within the current 2K page limit as indicated by the PC. This mechanism is explained through Fig. 6.3. At its top we find the 2-byte opcode containing 11 bits of the jump address A10–A0). During the execution of the AJMP instruction, these 11 address bits are loaded in the PC as indicated by the arrows. Note that highest five bits of the PC remain unchanged. We should remember that before copy, PC was incremented to indicate the next byte's address after the 2-byte AJMP instruction.

Another interesting fact related with the opcode of this instruction is that unlike the opcode of other instructions, which is fixed, the opcode changes with the destination address of the jump. This may easily be understood by observing the structure of bits of the first byte of opcode in Fig. 6.3. Its highest three bits

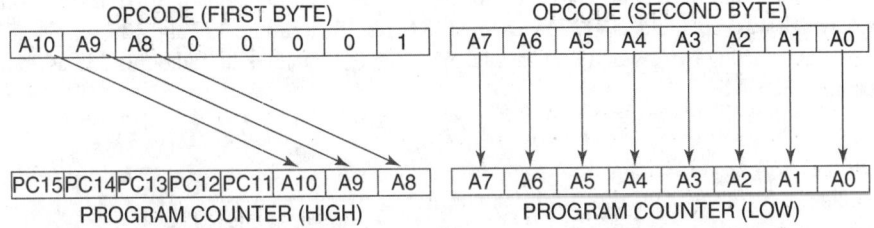

Figure 6.3 Mechanism of AJMP instruction

contain A10, A9 and A8 of the jump address which would vary in different cases of jump addresses. As eight variations are possible for these three bits, therefore, the AJMP instruction is allowed to have eight different opcodes as far as its first byte is concerned. As a matter of fact, when we will discuss the ACALL instruction in Chapter 8, we would observe the same speciality in that instruction also.

6.2.3 | SJMP Instruction

The third and the last in this series is the SJMP instruction which is a 2-byte two-cycle instruction having the opcode 80H. The second byte of this instruction contains the signed magnitude of the relative displacement calculated from the starting byte of the next instruction immediately after it. Because this second byte, also called the offset byte, is in two's complement signed magnitude form, therefore, the jump might be 128 bytes backward or 127 bytes forward. If this second byte is FEH, then the SJMP instruction would be executed again and again generating an infinite loop. Execution of this instruction is depicted in Fig. 6.4.

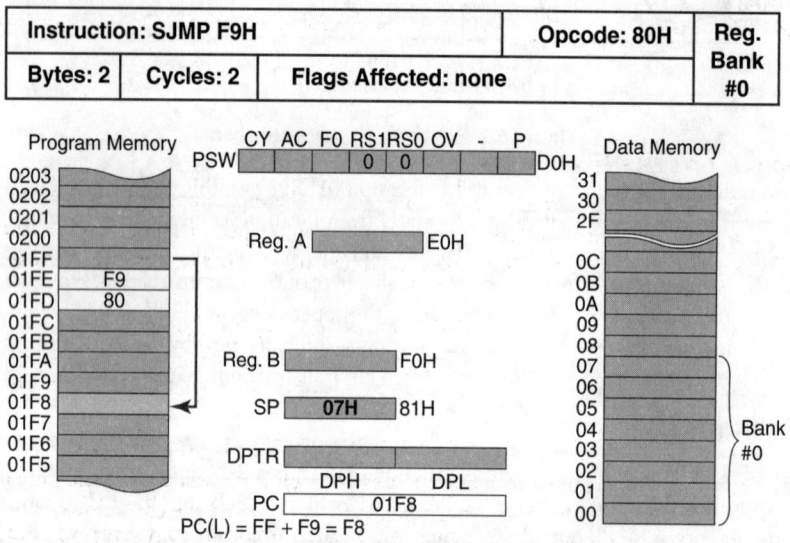

Figure 6.4 Execution of SJMP instruction

Note that in this case, the offset byte (second byte of instruction) is F9H (−7 in decimal). This is added with FFH available in lower byte of PC producing F8H. Instead of this, for a forward jump of, say, 2 bytes, the instruction should have been SJMP 02H, which would have changed the PC value to 01FF + 2 = 0201H. Therefore, SJMP 02H would have fetched the next executable byte from 0201H. A comparison of three types of unconditional jumps is presented in Table 6.2.

One minor point may need some clarification here. In Table 6.2, the range of SJMP instructions is given as 256 bytes, which was explained as +127 bytes to −128 bytes. If we add up 127 and 128, we get 255 and not 256. However, the starting byte, which is always taken as the 0th byte would make this 255 as 256, which is the correct range of SJMP.

Table 6.2 Comparison between LJMP, AJMP and SJMP instructions

Parameters	LJMP	AJMP	SJMP
Bytes	3	2	2
Cycles	2	2	2
Range	64K	2K	256 bytes (+127 or −128 bytes)
Width of jump address	16 bit	11 bit	8 bit
Mode of changing the PC	Direct and complete loading	Direct but partial loading	Adding LSB
Both forward and backward jumps	Possible	Possible	Possible
Hand assembly	Easier	Difficult	Difficult
Address of the first byte of the next instruction	Unimportant	Important	Important
To be used when there is	No space restriction	Space restriction	Space restriction

6.2.4 | Application of AJMP and SJMP

As both AJMP and SJMP are 2-byte two-cycle instructions, therefore apparently, it may be concluded that SJMP is an inferior instruction as it offers a shorter range with respect to AJMP. Indeed the application of SJMP does not save any memory area or increase the execution speed of the software. Then what is the reason of offering this instruction?

If we observe the performance of AJMP instruction carefully, then we will see that in border areas of 2K bytes, this instruction places some restriction as it cannot cross the current 2K page limit. For the sake of examples, let us consider two cases of application of AJMP instruction presented in Figs 6.5(a) and (b).

In the first case [Fig. 6.5(a)], the instruction AJMP 0803H is placed at location 07FDH. However, as the first byte of the next instruction is located at 07FFH, which is the last byte of the present 2K boundary, the usage of this AJMP 0803H instruction would generate an error and the control would never perform the unconditional jump to 0803H. Of course, had the same instruction (AJMP 0803H) been placed at 07FEH, the jump would have been performed properly, as the first byte of the next instruction falls within the next 2K page limit. An identical problem would be faced by the instruction AJMP 07FBH, if it is located within upper 2K page limit of 8051 [Fig. 6.5(b)].

However, this 2K boundary problem is not faced by the SJMP instruction, as shown in Figs 6.5(c) and (d). Note that SJMP is also a 2-byte instruction and located in the same place as the AJMP instruction. If the single byte offset is properly calculated, then in both cases, control would be unconditionally transferred to the proper target address.

Therefore, the standard programming practice is to select SJMP instruction as the first choice. If the target address is beyond its range (+127 or −128 bytes) then it may be replaced by AJMP instruction, which is the second choice. If AJMP is facing the 2K page boundary problem then, as the final choice, instruction LJMP comes into rescue, but at the cost of 3-byte memory space.

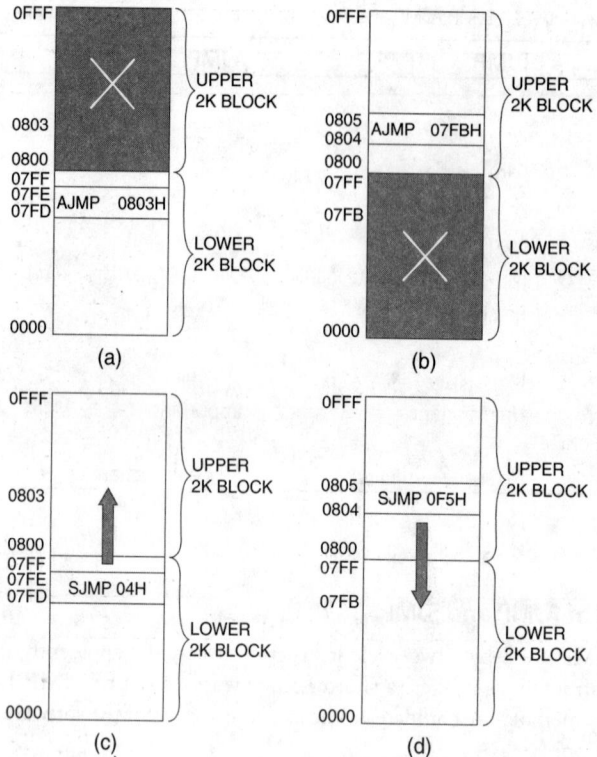

Figure 6.5 AJMP instructions cannot cross 2K boundaries (a) and (b) while SJMP has no such boundaries (c) and (d)

 If the reader is a beginner in programming then it is safer to start with only LJMP instruction in all cases. At a later stage, after using a few conditional jump instructions, the relative addressing technique would be easier to understand and then SJMP instruction may be picked up. However, for AJMP, be sure not to cross the present 2K limit. In all cases try to use labels as jump locations. A good assembler would do the rest for you.

6.3 | Conditional Branching

In general, conditional branching instructions are more widely used in the software and play more critical roles in software development. Only four of such conditional branching instructions, offered by MCS-51, are discussed in this chapter. The remaining will be discussed in Chapter 10. It is interesting to note that all conditional branching instructions of MCS-51 use the offset for PC-related branching, which is more often referred as relative addressing.

6.3.1 | JZ Instruction

JZ is a conditional branch instruction, which operates only with the accumulator. Just like the SJMP instruction, the second byte of this instruction also contains the displacement in two's complement form, making the instruction capable of jumping forward by 127 bytes or backward by 128 bytes. Note that there is no Z flag in the PSW and the instruction checks the accumulator physically before taking the decision. If the accumulator is not zero then the next instruction is executed, otherwise the relative jumping takes place. Figs 6.6(a) and (b) explain both conditions. No flags are affected in either condition.

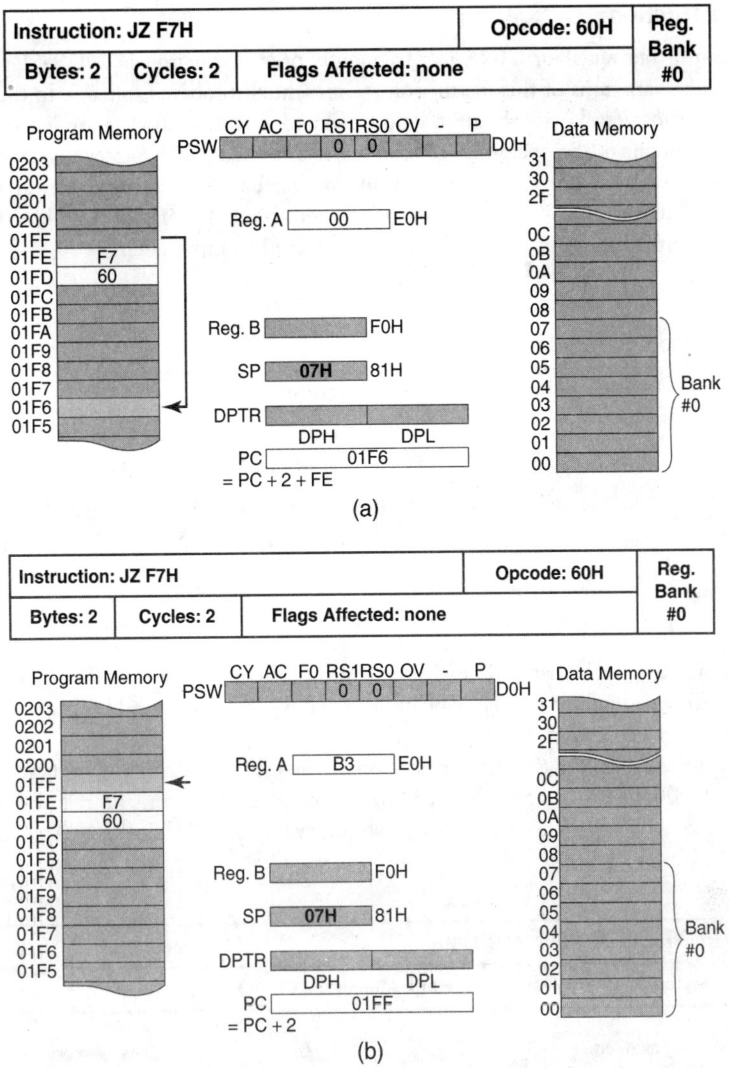

Instruction: JZ F7H			Opcode: 60H	Reg. Bank #0
Bytes: 2	Cycles: 2	Flags Affected: none		

Program Memory

CY AC F0 RS1 RS0 OV - P

PSW [0 0] D0H

Data Memory

```
0203
0202
0201
0200
01FF
01FE   F7
01FD   60
01FC
01FB
01FA
01F9
01F8
01F7
01F6
01F5
```

Reg. A [00] E0H

Reg. B [____] F0H

SP [**07H**] 81H

DPTR [_____]
        DPH   DPL

PC [01F6]
= PC + 2 + FE

```
31
30
2F

0C
0B
0A
09
08
07
06
05
04
03
02
01
00
```

Bank #0

(a)

Instruction: JZ F7H			Opcode: 60H	Reg. Bank #0
Bytes: 2	Cycles: 2	Flags Affected: none		

Program Memory

CY AC F0 RS1 RS0 OV - P

PSW [0 0] D0H

Data Memory

```
0203
0202
0201
0200
01FF
01FE   F7
01FD   60
01FC
01FB
01FA
01F9
01F8
01F7
01F6
01F5
```

Reg. A [B3] E0H

Reg. B [____] F0H

SP [**07H**] 81H

DPTR [_____]
        DPH   DPL

PC [01FF]
= PC + 2

```
31
30
2F

0C
0B
0A
09
08
07
06
05
04
03
02
01
00
```

Bank #0

(b)

Figure 6.6 Execution of JZ when accumulator is (a) zero and (b) non-zero

6.3.2 | JNZ Instruction

JNZ is the counterpart of the JZ instruction and performs a relative jump if the accumulator is not zero. Otherwise, the execution proceeds sequentially. Opcode of this instruction is 70H and this is also a 2-byte two-cycle instruction. The second byte of the instruction contains the signed offset value to be added with the PC.

In 8051, JZ or JNZ instruction may be used immediately after loading the accumulator by an immediate data. However, this is not possible in case of 8085 as these instructions are dependent upon the condition of Z flag of 8085, which is not affected by any data load instruction.

6.3.3 | CJNE Instruction

'Compare and jump if not equal' or CJNE is a frequently used instruction in MCS-51 programming. All addressing modes and variations of this instruction are presented through Table 6.3. In this case, the source operand is *virtually subtracted* from the destination operand and the action is implemented. During this, virtual subtraction contents of the operands are assumed as unsigned positive integer values. As it is a virtual subtraction, neither the source nor the destination operand is changed. However, the carry flag reflects the status of this virtual subtraction. No other flags are affected. Note that if a port is directly addressed, this instruction reads from the input pins of the port and not from the output latch.

Table 6.3 Addressing modes of CJNE instruction

Instruction	Addressing mode	Function
CJNE A, direct, rel	Direct	Compare A with direct address and jump if not equal
CJNE A, #data, rel	Immediate	Compare A with immediate data and jump if not equal
CJNE Rn, #data, rel	Immediate	Compare register with immediate data and jump if not equal
CJNE @Ri, #data, rel	Immediate	Compare content of address indicated by the register with immediate data and jump if not equal

CJNE is a 3-byte two-cycle instruction with the third byte containing the offset byte to be added with the PC. Assuming the accumulator having 7BH and internal RAM location 06H contains 82H, execution of CJNE A, 06H, 0FBH is depicted in Fig. 6.7.

As we may observe from Fig. 6.7, after execution of the CJNE instruction, both accumulator as well as memory location 06H (R6 of bank #0) remain unchanged. However, the result of the subtraction (7BH-82H) generated a carry, which was stored in the carry flag of PSW as 1 (set). The unequal condition

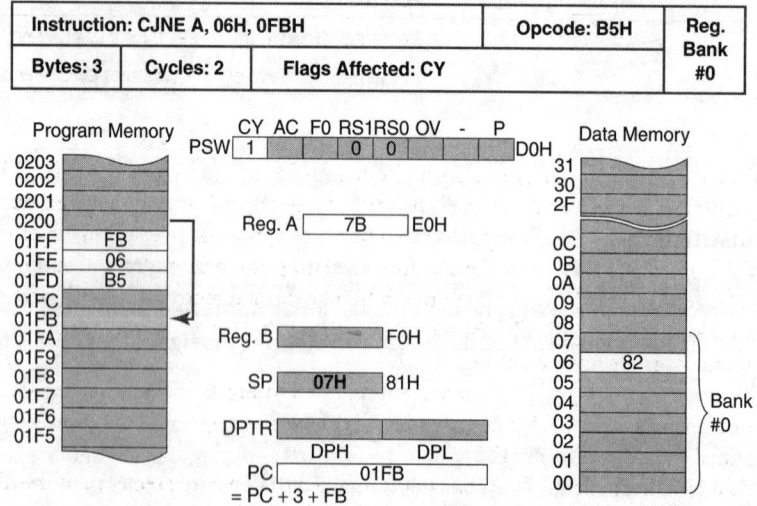

Figure 6.7 Execution of CJNE A, direct, rel instruction

forced the branching to the address obtained by adding the third byte of the instruction (FBH) with the incremented PC value (0200H) producing 01FBH from where the next executable instruction would be fetched. Had the content of the accumulator and location 06H been equal, the carry flag would have been cleared and the next instruction from 0200H would have been executed.

The purpose of reflecting the result of virtual subtraction in the carry flag is to perform further branching, using JC or JNC instructions, which we will discuss in Chapter 10.

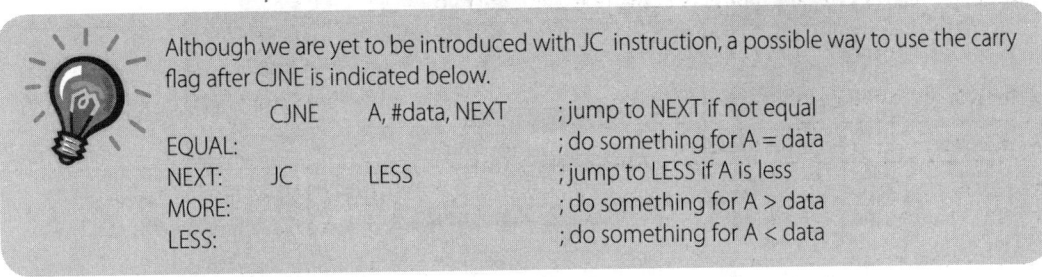

Although we are yet to be introduced with JC instruction, a possible way to use the carry flag after CJNE is indicated below.

```
        CJNE    A, #data, NEXT    ; jump to NEXT if not equal
EQUAL:                            ; do something for A = data
NEXT:   JC      LESS              ; jump to LESS if A is less
MORE:                             ; do something for A > data
LESS:                             ; do something for A < data
```

6.3.4 | DJNZ Instruction

This is another conditional branching instruction but having a very special purpose. Many a times in iterative procedures, some specific number of times the iteration is to be performed. In general, we use a counter, which is decremented for each iteration and when the counter is zero, we assume that the process is complete. The instruction DJNZ is tailor-made for this purpose, which combines the *decrement by one* instruction and *jump if not zero* instruction to a single instruction. Addressing modes of this instruction are presented in Table 6.4. Note that by using its direct address, the accumulator might also be used as a counter.

Table 6.4 Addressing modes of DJNZ instruction

Instruction	Addressing mode	Function
DJNZ Rn, rel	Register direct	Decrement register by one and jump if not zero
DJNZ direct, rel	Direct	Decrement content of direct address by one and jump if not zero

DJNZ always takes two cycles to perform. When direct address is used, it takes 3 bytes otherwise it is a 2-byte instruction. Moreover, if a port is directly addressed, data would be taken from its output latch and not from the input pins. As already described previously, the last byte of this instruction contains the offset byte, which is to be added with the PC for calculating the next address of execution if the result of decrement is not zero. It may be noted here that whether it is a directly addressed location or any register, the decremented location is physically checked for its zero value before taking the decision. No flag is used in any case and no flags are affected either.

Pictorially, execution of DJNZ instruction is presented in Fig. 6.8. Note that register R7 is assumed to contain 3CH before execution of this instruction. As a decrement by one is not producing zero, the relative jump is performed to the location 01F6H.

What would have happened if R7 of bank #0 contained 01H before execution of this DJNZ instruction? Definitely after execution, R7 would have contained 00H and the branching would not have taken place. Therefore, the next instruction at program memory address 01FFH would have been executed.

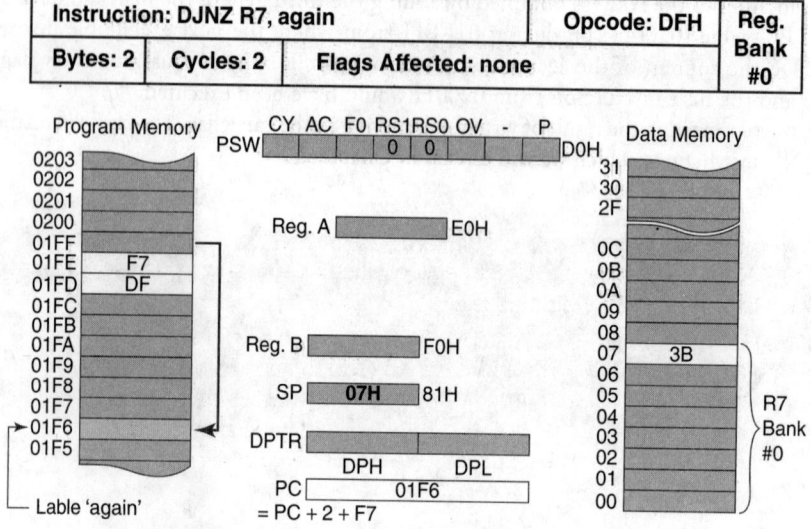

Figure 6.8 Execution of DJNZ R7, rel instruction

 Be careful to limit your iterative instructions within 128 bytes' backward range of DJNZ, remembering that the instruction itself consumes two to three bytes depending upon the addressing mode and the range is calculated from the first byte of next instruction after DJNZ. For a larger loop-body, an intermediate SJMP instruction may be introduced.

6.3.5 | Labels in Program Listing

So far in our examples, we have used the complete target address or offset byte for jump instructions like CJNE A, 06H, 0FBH. Although it is a valid format, however, calculating offset bytes or including direct addresses may not make a program readable or easier to debug.

For the ease of software development, it is customary to use labels to mark different locations of the program memory. Labels are alphanumeric characters, starting with alphabets. They are terminated by a colon (:). For example, execution of the following instructions

```
Label1:   MOV   R7, #02H
          DJNZ  R7, Label1
Label2:   MOV   A, #0FFH
```

would keep on looping the control within first two instructions and the accumulator would never be loaded by FFH. However, if the first instruction is changed to

```
Label1:   MOV   R7, #01H
```

then the control would execute the DJNZ instruction only once and then would load the accumulator by FFH.

When the program listing is assembled using an assembler, these labels are assigned appropriate addresses and related offset bytes (or displacement bytes) are automatically calculated by the assembler itself.

6.4 | NOP Instruction

NOP is an instruction that is generally used for fine-tuning the delay loops. This 1-byte one-cycle instruction does not do anything except killing time of the processor. We will use this instruction as an idle (or filler) instruction in Chapter 12.

6.5 | Solved Examples

Example 6.1

Purpose: To be familiar with indirect addressing and controlling any program loop using DJNZ instruction.

Problem

Write a program to clear lower 128 bytes of internal RAM.

Solution

In this case, we may select bank #0 and use register R0 as a pointer. The program may be written as follows:

```
START:  MOV   A, #00H      ; clear accumulator
        MOV   PSW, A       ; select bank #0
        MOV   R0, #7FH     ; load pointer with highest address
CLEAR:  MOV   @R0, A       ; clear 1 byte pointed by R0
        DJNZ  R0, CLEAR    ; point next lower address
```

Incidentally, this program would clear the lowest addressed location, that is R0 by DJNZ instruction and not by MOV @R0, A instruction. However that serves our purpose. Note that how DJNZ instruction controls the loop counter, R0, which is also used as a pointer.

Example 6.2

Purpose: To understand the limitation of indirect addressing and DJNZ instruction.

Problem

Write a program to fill lower 128 bytes of internal RAM by FFH.

Solution

With respect to the previous example, here the first instruction should load the accumulator by FFH for filling. R0 of bank #0 may again be selected as the pointer. However, the last address, i.e. 00H or R0, of bank #0 must be filled up by FFH by an extra instruction. The complete program may be developed as follows:

```
START:   MOV   A, #0FFH      ; load accumulator by fill data
         MOV   PSW, #00H     ; select bank #0
         MOV   R0, #7FH      ; point at the highest address
FILLUP:  MOV   @R0, A        ; fill one location by FFH
         DJNZ  R0, FILLUP    ; point to next location
         MOV   R0, A         ; fill the last location by FFH
```

Note that unlike the previous example, here an extra instruction (MOV R0, A) is necessary at the end to fill R0 (internal RAM address 00H) by FFH.

Example 6.3

Purpose: How to utilize CJNE instruction.

Problem

Sixteen consecutive locations, starting from 50H to 5FH, of internal RAM contain 16 unsigned positive random numbers. Write a program to check if the number 5AH is located in any one of these bytes. If yes then load accumulator by 5AH otherwise clear accumulator.

Solution

Like previous examples, we can use R0 as the pointer and initialize it by the address of the highest location, 5FH. We will need a counter for counting 16 comparisons and use R7 of the current bank for that purpose. If a match is found, then A is loaded immediately with 5AH and the execution terminates at an infinite loop. Otherwise A is cleared and the program is terminated in the same way.

```
START:      MOV  R0, #5FH          ; point to highest address
            MOV  R7, #10H          ; decimal 16 for number of iterations
COMPIT:     CJNE @R0, #5AH, NEXT   ; compare with one location
            MOV  A, #5AH           ; match found
            SJMP FOUND
NEXT:       DEC  R0                ; point to next location
            DJNZ R7, COMPIT        ; keep matching
NOMATCH:    MOV  A, #00H           ; no match, clear accumulator
FOUND:      SJMP FOUND             ; stop at infinite loop
```

Note that after execution, this program enters within an infinite loop.

SUMMARY

Controlling the program flow plays an important part in software development. In this chapter, unconditional jump instructions and conditional branching instructions are discussed which are generally used for program-flow control.

Unconditional jump instructions, namely LJMP, AJMP and SJMP, perform identical operations but in different manner. The difference lies in the method of assigning the target address for jumping. In LJMP, a complete 16-bit jump address is provided with the instruction while for AJMP a truncated 11-bit address is provided. In case of SJMP, only the offset byte is indicated in two's complement format. Ranges of these unconditional jumps are 64K, 2K and 256 bytes, respectively.

Both JZ and JNZ instructions check the accumulator physically and perform the conditional branching as indicated through the relative displacement byte. Instruction CJNE compares the content of two operands and jumps if they are not equal. A virtual subtraction is performed for this comparison, which is reflected through the carry flag but the operands remain unaltered. DJNZ instruction decrements by one the content of the indicated location, register or directly addressed memory location and branches if the location does not contain zero after this decrement. NOP instruction does nothing except killing time.

POINTS TO REMEMBER

- AJMP instruction cannot cross its present 2K boundary, and there is no such boundary for relative jump instructions.

- Relative jump's offset byte need not be calculated if an assembler is used. Simple labels are sufficient in these cases.

REVIEW QUESTIONS

Evaluate Yourself

1. What is the limit of LJMP instruction?

    (a) 256 bytes   (b) 2K bytes

    (c) 64K bytes   (d) None of these

2. How many bits of the jumping address are available in AJMP instruction?

    (a) 11 bits   (b) 10 bits

    (c) 9 bits   (d) None of these

3. The second byte of the SJMP instruction indicates the displacement in

    (a) signed magnitude form

    (b) two's complement form

    (c) either one of the above

    (d) none of these

4. JZ instruction is applicable

    (a) only for registers

    (b) only for SFRs

    (c) only for register A

    (d) none of these

5. How many addressing modes are available for JNZ instruction?

    (a) 1   (b) 2

    (c) 3   (d) None of these

6. How many flags of PSW are affected by CJNE instruction?

    (a) None   (b) Only carry flag

    (c) All four flags   (d) None of these

7. What is the common point between JZ, JNZ, CJNE and DJNZ instructions?

    (a) All are 3-byte instructions

    (b) All use displacement bytes for branching

    (c) All depend on the Z-flag's status

    (d) None of these

8. If register R7 contains 00H before execution of the instruction DJNZ R7, 02H, from which address would the next instruction after it be executed? Assume that the DJNZ instruction is located at 005FH.

    (a) 005FH   (b) 0061H

    (c) 0063H   (d) None of these

9. Which of the following instructions affects any flag of PSW?

    (a) SJMP   (b) DJNZ

    (c) NOP   (d) None of these

10. What would the processor do after executing the instruction SJMP 0FEH?

    (a) Execute from 0FEH

    (b) Execute the same instruction again

    (c) Execute from 0002H

    (d) None of these

Search for Answers

1. How many opcodes are possible for AJMP instruction? What is its reason?

2. What are the similarities between three unconditional jump instructions of MCS-51? What are their differences?

3. How the decision to jump or not to jump is taken by the processor in case of JZ instruction?

4. Does the CJNE instruction change any register or flags of the processor?

5. How many bytes of forward jump are possible for any DJNZ instruction?

6. What is the purpose of NOP instruction?

7. What are the addressing modes offered in DJNZ instruction?

8. Does the DJNZ instruction read from the output latch or from the input pins of a port, when directly addressed?

9. Which part of internal RAM cannot be accessed by the DJNZ instruction?

10. When the CJNE instruction is associated with a directly addressed port, from which part of the port would it access its data?

Think and Solve

1. Is it possible to implement the DJNZ instruction using register A? Justify your answer.

2. DJNZ instruction offers the usage of an 8-bit counter. How would it be possible to use a 16-bit counter for *decrement and jump if not zero* procedure?

3. Codes 01H and 55H are first and second bytes of a 2-byte AJMP instruction. Assuming that the first byte is located at address 07FFH, what would be the jumping address for this AJMP instruction?

4. Why are the lengths of backward and forward jumps for SJMP, JZ, DJNZ, etc., unequal?

5. Both SJMP and AJMP are of 2 bytes and take two cycles to be executed. Why two different mnemonics are provided for the same purpose?

6. Out of CJNE, DJNZ and SJMP instruction, which one may be used to terminate a program?

7. Assuming that register bank #0 is currently selected and active, what would happen if the following program segment is executed?

THERE:	MOV	R3, 01H
HERE:	MOV	A, 00H
AGAIN:	DJNZ	R3, THERE
	CJNE	A, 03H, HERE
	SJMP	AGAIN

8. How is the CJNE instruction used to compare with a variable and not with a constant?

9. Instead of using DJNZ 0E0H, 0FH, which decrements the directly addressed accumulator; if the following two instructions are used then what would be the difference between these two approaches?

| DEC | A |
| JNZ | 0FH |

10. What should be the content of the counter to perform maximum number of iterations through the DJNZ instruction?

7

PROGRAMMING EXAMPLES-I

CHAPTER OBJECTIVES

In the last three chapters, the reader was introduced to a few MCS-51 instructions like data moving and loading, arithmetic operations and program branching. In this chapter, we will discuss applications of these instructions through a few example cases of programming. After completion of the chapter, the reader should be able to understand how these instructions may be properly utilized to

- Manipulate data in different ways using pointers.
- Use counters for iterations.
- Plan any program flow and necessary branching.
- Perform some basic arithmetic operations.

7.1 | Introduction

So far, we have made ourselves familiar with some, though not all, instructions of MCS-51 in the previous three chapters. Using these instructions, we will now develop a few programs in this chapter to solve some simple problems. Although these programming examples are very elementary, the reader will gain some insight into software development and will also be familiar with the applications of the knowledge gained so far.

In every example, after presenting the problem, basic solution techniques are discussed followed by the algorithm or flowchart. Complete program listing is placed after that, with adequate comment statements for easier understanding of the programming logic.

7.2 | Copy Block

Example 7.1

Purpose: How to handle pointers during indirect addressing mode.

Problem

Copy a block of 20 bytes of data available from address 60H to 73H to the location starting from 40H.

Solution

Schematically, the problem is explained through Fig. 7.1(a). The data available between 60H and 73H of the internal RAM area, shown at left, is to be copied in the same sequence to RAM location 40H onwards.

As the addresses are sequential in this case, it is preferable to use pointers, which can be manipulated easily. Therefore, R0 and R1 may be initialized by the starting addresses of source and destination area. Another register, say R7, may be used as a counter to keep a track of number of copy operation and indicate when to terminate. As the MOV instruction does not allow both source and destination to be indirectly addressed, therefore, the accumulator may be used for in-between storage of the data. This is schematically explained through Fig. 7.1(b). Following are the algorithm and program listing.

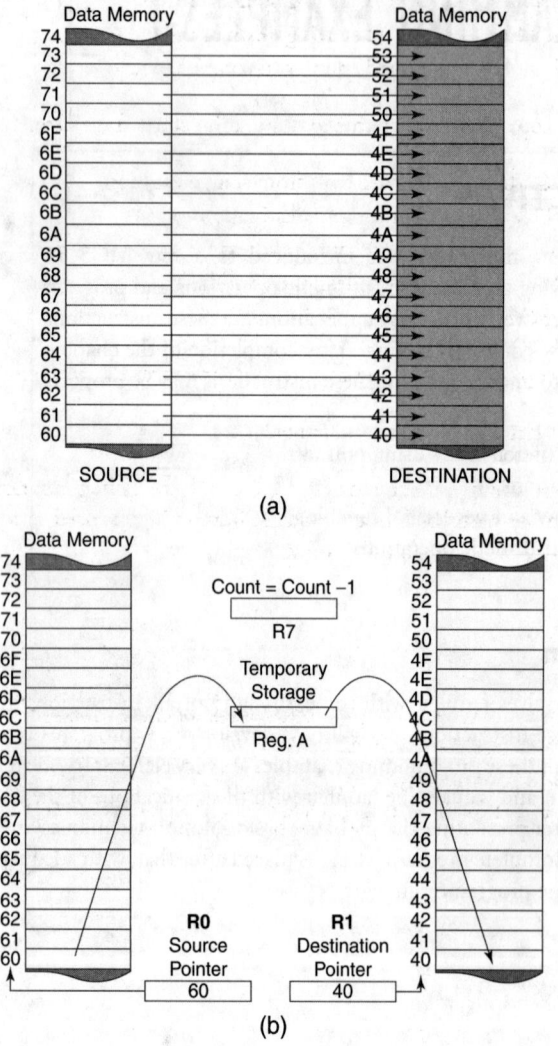

Figure 7.1 Schematics of (a) problem and (b) solution technique

Algorithm

Step 1: Initialize R0 as source and R1 as destination pointers and load R7 by 20d to serve as the counter.

Step 2: Copy a byte from source to destination, using R0 and R1 through accumulator (as a temporary storage). Update pointers after copying.

Step 3: Decrement counter by one. Continue at Step 2 if the counter is not zero.

Step 4: Terminate the process.

Program

```
; Program to copy 20 bytes starting from 60H to location 40H onwards.
; First three instructions initialize both source and destination pointers and the counter.

START:  MOV   R0, #60H      ; load starting source address
        MOV   R1, #40H      ; load starting destination address
        MOV   R7, #14H      ; load counter to copy 20 bytes

; Iterative procedure to copy data from source to destination starts from here.

COPYIT: MOV   A, @R0        ; get 1 byte from source location
        MOV   @R1, A        ; store it in destination
        INC   R0            ; next source address
        INC   R1            ; next destination address
        DJNZ  R7, COPYIT    ; keep copying till over

; R7 = 0 indicates data copy over. Following instruction terminates the program.

OVER:   SJMP  OVER         ; terminate here
```

Generally, this type of routine is terminated by the RET (return to the calling program) instruction. However, as we are yet to become familiar with RET instruction, an infinite loop is used to terminate the program. A system reset is to be used to come out from this infinite loop.

A common mistake by beginners is to write the third instruction of this type of program as:
        MOV   R7, #20H
Generally they overlook that H stands for hexadecimal numbers and 20H indicates 32 in decimal. A safer practice is to write it as:
        MOV   R7, #20d
so that the assembler takes it as a decimal number and converts it correctly.

C-version

```c
#include <regx51.h>
void main(void)
        {
                register unsigned *src, *dest;
                src = 0x60;
                dest = 0x40;
                        do
                                {
                                        *dest = *src;
                                        dest++;
                                        src++;
                                }while (src!= 0x74);
        }
```

7.3 | Shift Block

Example 7.2

Purpose: To understand the restrictions related with data shifting.

Problem

Shift a block of 8 bytes of data, presently located from 50H to 57H, 1 byte up, so that the data is available from 51H to 58H.

Solution

The difference between copy block and shift block is that there is no overlapping in former, which exists in the latter. Moreover, shift-up and shift-down are to be treated in a slightly different manner. Otherwise, the usage of pointers and the counter is identical (refer to Fig. 7.2) in both cases. Here also R0 and R1 are used as pointers and R7 is used as the counter. Following are the algorithm and program listing.

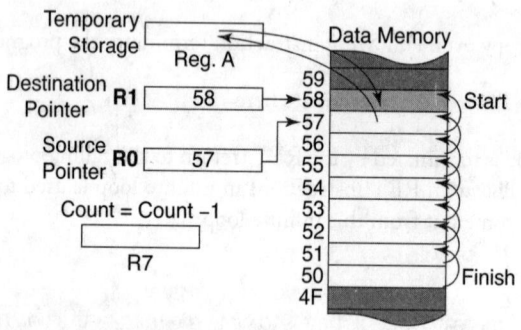

Figure 7.2 Schematic of solution for shift-block operation

Algorithm

Step 1: Initialize R0 as source and R1 as destination pointers for the *highest* addresses and load R7 by 08 to serve as the counter.

Step 2: Copy a byte via accumulator from source to destination, using R0 and R1, through accumulator. Decrement both pointers by one after copying.

Step 3: Decrement counter by one. Continue at Step 2 if the counter is not zero.

Step 4: Terminate the process.

Program

```
; Program to shift 8 bytes by 1 byte up.
; Following three instructions initialize both pointers and the counter.

START:  MOV  R0, #57H        ; point last location of source
        MOV  R1, #58H        ; point last location of destination
        MOV  R7, #08H        ; initialize counter for eight shift operations

; Shifting operation of 8 bytes starts from here.

SHIFT:  MOV  A, @R0          ; get a byte from source
        MOV  @R1, A          ; store it in its destination
```

```
        DEC    R0             ; point to next source
        DEC    R1             ; point to next destination
        DJNZ   R7, SHIFT      ; loop on eight times
```

; R7 indicates that the operation is over.
; Next instruction is used to terminate the program.

```
OVER:   SJMP   OVER           ; terminate here.
```

Please note that had it been a case of shifting down, pointers would have been loaded for the other end (50H and 4FH) and would have been incremented after every iteration.

C-version

```c
#include <regx51.h>

void main(void)
{
        unsigned char *src, *dst;
        signed char ptr; // note we take signed data type here

        src = 0x50; // point first location of source
        dst = 0x51; // point first location of destination

        for (ptr = 7; ptr >= 0; ptr --)
        {
                dst[ptr] = src[ptr];
        }
        while(1);
}
```

7.4 | Count No. of Nulls

Example 7.3

Purpose: Application of CJNE instruction in its indirect addressing mode.

Problem

Twenty bytes of data are stored in location from 7FH to 6CH of internal RAM. Count the number of those bytes, which contain 00H, and store this number of null bytes in RAM location 6BH.

Solution

It is assumed that the null bytes might be present anywhere and be of any number. All 20 locations might contain null [Fig. 7.3(a)] or, in opposite case, not even a single location might be having it [Fig. 7.3(b)]. Any general case might be in between these two as shown in Fig. 7.3(c). Therefore, the solution technique assumes that the number of null bytes is zero and checks every byte. Using register indirect addressing mode of CJNE instruction, the procedure checks every byte and stores the final outcome of this checking from the accumulator to location 6BH [Fig. 7.3(d)]. Note that whenever any null byte is detected, accumulator content is incremented by one. Following are the algorithm and listing of the program.

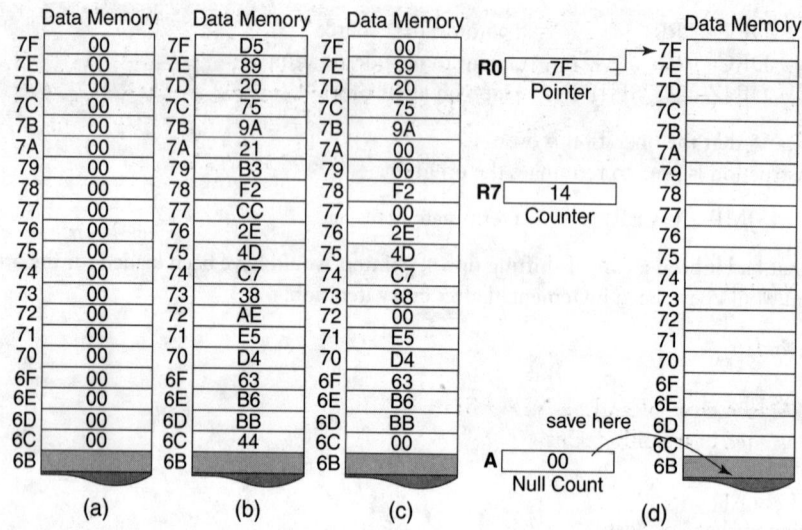

Figure 7.3 Possibilities of location of null bytes (a–c) and schematic solution technique (d)

Algorithm

Step 1: Initialize R0 as source pointer for the highest address and load R7 by 14H to serve as the counter for 20 numbers. Clear accumulator.

Step 2: Compare a byte with zero through pointer. If it is zero then increment accumulator by 1.

Step 3: Decrement both pointer and counter by one. Continue at Step 2 if the counter is not zero.

Step 4: Save accumulator content through the pointer. Terminate the process.

Program

```
; Program to count number of null bytes within 20 bytes starting from 6CH.
; Following three instructions initialize the pointer and the counters.

START:   MOV   R0,#7FH          ; point to highest address
         MOV   R7,#14H          ; counter for 20d bytes
         MOV   A,#00H           ; to count number of null bytes

; Scanning of 20 bytes starts from here.

MAIN:    CJNE  @R0,#00H,NOTNUL   ; check for null byte
         INC   A                 ; one more null byte found
NOTNUL:  DEC   R0                ; point to next location to be tested
         DJNZ  R7,MAIN           ; loop on 20 times

; R7 indicates that the scanning is over.
; Following two instructions store the number of null bytes and terminate the program.

         MOV   @R0,A             ; save number of null bytes in 6BH
OVER:    SJMP  OVER              ; terminate here
```

 An alternate method would be to use JZ or JNZ instruction after assigning another register as the counter. However, as we are interested in application of CJNE instruction, therefore we follow the present method.

C-version

```
#include <regx51.h>

void main(void)
{
        unsigned char *pRamLoc, *pLoc6BH;
        unsigned char nullByteCnt=0;

        pLoc6BH = 0x6B;

        for (pRamLoc=0x6C, nullByteCnt=0; pRamLoc<=0x7F; pRamLoc++)
        {
                if (0==*pRamLoc)
                {
                        nullByteCnt += 1;
                }
        }
        *pLoc6BH = nullByteCnt;

        while(1);
}
```

7.5 | Find Checksum

Example 7.4

Purpose: Multiple additions through indirect addressing mode.

Problem

Sixteen consecutive bytes starting from 50H have unsigned integers. Develop a program to add all these 16 integers and store the 8-bit sum in memory location 60H.

Solution

Checksum is generally used to ensure the correctness of a dataset after some communication. It is generally taken as 8 bits, neglecting the carry. Therefore, ADD instruction is adequate here to find the sum. Indirect addressing using R0 as the pointer helps to minimize number of instructions of the program. R7 was used as the counter for 16 locations and the checksum is initially stored in the accumulator. Following are the algorithm and program listing.

Algorithm

Step 1: Initialize R0 as source pointer and load R7 by 16d to serve as the counter. Also clear the accumulator to calculate the sum.

Step 2: Add 1 byte with accumulator as pointed by R0. Then increment pointer by one.

Step 3: Decrement counter by one. Continue at Step 2 if the counter is not zero.

Step 4: Store accumulator as pointed by R0. Terminate the process.

Program

```
; Program to calculate checksum for 16 bytes from location 50H.
; Following three instructions complete the initialization process.

START:   MOV    R0, #50H      ; point to first number
         MOV    R7, #10H      ; counter for 16 additions
         MOV    A, #00H       ; clear checksum

; Iterative procedure, for addition of 16 numbers, starts from here.

ADDIT:   ADD    A, @R0        ; add one number with the accumulator
         INC    R0            ; point to next number
         DJNZ   R7, ADDIT     ; cover all 16 numbers

; Addition is complete. Store the result or checksum.

         MOV    @R0, A        ; save the checksum
OVER:    SJMP   OVER          ; terminate here
```

C-version

```c
#include <regx51.h>

void main(void)
{
        register unsigned char *src, *dest;
        unsigned short int sum = 0, i;
        src = 0x50, dest = 0x60;
                for (i = 16; i > 0; I --)
                {
                        sum += *src;
                        src++;
                }
                *dest = sum;
}
```

7.6 | Sum of Natural Numbers

Example 7.5

Purpose: How to accomplish multiple duties within a single loop.

Problem

Write a program to generate and store natural numbers starting from 1 to 'N' terms and also find the sum of these numbers. Assume that the value of 'N' is stored in location 30H. Store generated natural numbers from 40H. Leave the sum in the accumulator.

Solution

We select bank #0 and use R6 for generating the natural numbers. These generated numbers are stored at target locations from R6 using R0 as the pointer. Bank #0 is selected through PSW as direct addressing for source to be used for saving the generated number through indirect addressing. R7 counts down from N, which was loaded from location 30H. It is assumed that N is not zero. The accumulator is used to calculate the sum after each cycle of iteration. Fig. 7.4 explains the scheme of register allotment. Following are the adopted algorithm and the program listing.

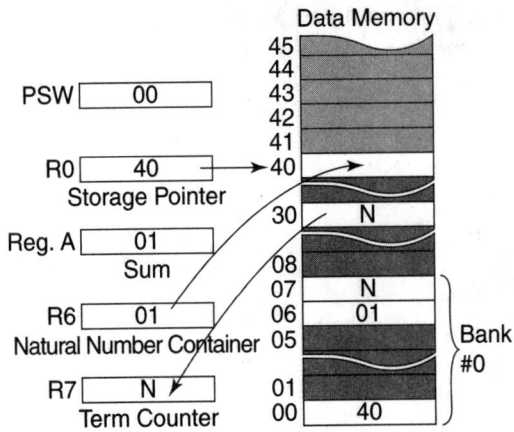

Figure 7.4 Schematic for natural number generation

Algorithm

Step 1: Select bank #0. Initialize R0 as destination pointer for storage of generated natural numbers and load R7 from 30H by N to serve as the counter. Also clear R6 and the accumulator to generate the numbers and calculate the sum.

Step 2: Increment R6 by one to generate the next natural number. Save it through R0 and add it with accumulator. Then increment R0 to point the next storage location.

Step 3: Decrement counter R7 by one. Continue at Step 2 if the counter is not zero.

Step 4: Terminate the process.

Program

; Program to generate and add N natural numbers. Value of N is in location 30H.
; Following five instructions complete the initialization procedure.

```
START:  MOV   PSW, #00H    ; select bank #0
        MOV   A, #00H      ; initialize the sum
        MOV   R6, #00H     ; to start generation of natural numbers
        MOV   R0, #40H     ; point to start of storage location
        MOV   R7, 30H      ; counter for N terms of the series
```

; Main loop for natural number generation, storing and addition starts from here.

```
MAIN:   INC   R6          ; generate next natural number
        MOV   @R0, 06H     ; save generated number in R6 to its space
        ADD   A, R6        ; find the sum and store it in register A
        INC   R0          ; point to next storage location
        DJNZ  R7, MAIN     ; continue up to N terms
```

; R7 indicates that the process is over. Following instruction terminates the program.

    OVER: SJMP   OVER   ; terminate here

> If we are interested in 16-bit sum, then instead of accumulator, two registers to be initialized as zero and ADD followed by ADDC instructions to be used.

C-version

```c
#include <reg51.h>

void main(void)
{
        unsigned char *pLocN;
        unsigned char sum, ntrlNum;

        // initialize

        pLocN = 0x30;
        sum = 0x00;

        if (*pLocN > 0)
        {
                for (ntrlNum=1; ntrlNum<=(*pLocN); ntrlNum++)
                {
                        sum += ntrlNum;
                }
        }

        // store result in the accumulator

        ACC = sum;

        while(1);
}
```

7.7 | Sum of a Series

Example 7.6

Purpose: To illustrate conditional application of ADD and SUBB instruction and taking care of carry flag before using SUBB instruction.

Problem

Write a program to find the sum of the series $1 - 2 + 3 - 4 + \ldots$ up to N terms. Assume the non-zero value of N is available in location 30H. Store the sum in the accumulator.

Solution

Every series follows a particular sequence. The present series is of natural numbers starting from 1. Sign of all odd terms are positive and all even terms are negative. Therefore, we may use the same technique for natural number generation as we have followed in the previous example. However, the addition is to be performed only for odd terms and subtraction must be carried out for even terms. Moreover, as N might be either even or odd which is unknown to the program, the decrement of iteration counter must be carried out in two steps.

Another point must be noted that before execution of every SUBB instruction, the carry flag of PSW should be cleared. As a matter of fact, CLR C is the standard instruction for it. However, as we are yet to discuss about this instruction, therefore, we should find an alternative for it. In this case we can do it by loading PSW with 00H. Therefore, it is preferable to use bank #0 so that this PSW clearing for carry does not alter the selected bank. Following are the algorithm and the program listing.

Algorithm

Step 1: Select bank #0. Load R7 by N from location 30H to serve as the counter. Also clear R6 and the accumulator to generate the numbers and calculate the sum.

Step 2: Increment R6 by one to generate the next (odd) natural number. Add it with accumulator.

Step 3: Decrement counter R7 by one. Continue at Step 4 if the counter is not zero. Otherwise continue at step 6.

Step 4: Increment R6 by one to generate the next (even) term. Clear carry flag by loading 00 in PSW. Then subtract the generated number from the accumulator.

Step 5: Decrement counter R7 by one. Continue at Step 2 if it is not zero.

Step 6: Terminate the program.

Program

; Program to calculate the sum of the given series up to N terms. N is in location 30H.
; First four instructions perform the initialization.

```
START:  MOV   PSW, #00H      ; select bank #0
        MOV   A, #00H        ; clear sum
        MOV   R6, #00H       ; to generate terms of the series
        MOV   R7, 30H        ; number of terms in R7 as counter
```

; Start generating numbers and keep on adding for odd terms.

```
MAIN:   INC   R6             ; get next odd number
        ADD   A, R6          ; odd numbers to be added
        DJNZ  R7, NEXT       ; N terms not over yet
        SJMP  OVER           ; N terms over
```

; Now generate an even term and subtract from the sum.

```
NEXT:   INC   R6             ; get next even number
        MOV   PSW, #00H      ; to clear carry flag
        SUBB  A, R6          ; even numbers to be subtracted
        DJNZ  R7, MAIN       ; continue if necessary
```

; R7 indicates that all terms were generated. Following instruction terminates the program.

```
OVER:   SJMP  OVER           ; terminate here, sum in register A
```

 In the present problem, the sign is toggling, that is alternately changing between positive and negative states. A better way to tackle this type of toggling problems is to use a one-bit flag, which may be complemented after every iteration. This we shall discuss after being familiar with Boolean variable manipulation instructions.

C-version

```c
#include <regx51.h>

#define ODD          1
#define EVEN         0

void main(void)
{
        unsigned char *pLocN;
        unsigned char ntrlNum, evenOddState;
        signed char sum;
        // initialize

        pLocN = 0x30;
        sum = 0x00;
        evenOddState = ODD;

        if (*pLocN > 0)
        {
                for (ntrlNum=1; ntrlNum<=(*pLocN); ntrlNum++)
                {
                        if (ODD==evenOddState)
                        {
                                sum += ntrlNum;
                        }
                        evenOddState ^= 1;
                }
        }

        // store result in the accumulator

        ACC = sum;

        while(1);
}
```

7.8 | Fibonacci Series

Example 7.7

Purpose: How to use multiple data pointers in synchronization with the program flow.

Problem

Generate Fibonacci series up to N terms and save from location 50H onwards. Assume N being available in location 4FH. Also assume the value of N to be greater than 3.

Solution

Fibonacci series generates any term by adding its two previous terms. The first and second terms of this series are always 0 and 1, respectively, and the series starts as follows:

Fibonacci series   0, 1, 1, 2, 3, 5, 8, 13d, ...

To count the number of terms to be generated, we may use R7 as the counter, which should be loaded from location 4FH. R1 may be used as the pointer for storing a freshly generated term while R0 may be used to point previous two terms one after another as shown in Fig. 7.5.

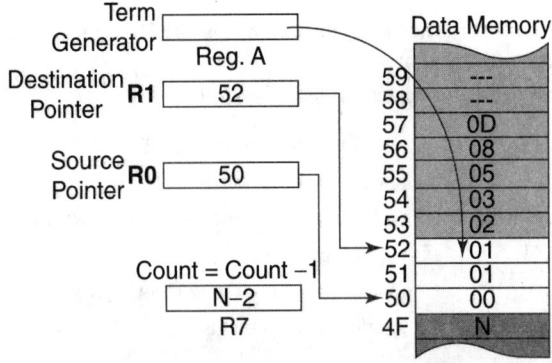

Figure 7.5 Schematic for Fibonacci series term generation

The program loop should take over from the generation of the third term of the series while first two terms may simply be loaded in the storage area during initialization. Following are the algorithm and the program listing.

Algorithm

Step 1: Load R7 by number of terms to be generated.

Step 2: Load R0 by second term's location and store 01 there.

Step 3: Decrement R0 by 1 to point the location of the first term and store 00 there as the first term.

Step 4: Initialize R1 by the location of the third term. Decrement R7 by two as two terms are generated and stored.

Step 5: Load the term pointed by R0 in the accumulator. Increment R0 by one.

Step 6: Add to the accumulator the term pointed by R0. Store this new term from accumulator to its storage location pointed by R1. Then increment R1 by one to point the next storage location.

Step 7: Decrement R7 by one and continue at Step 5 if it is not zero.

Step 8: Terminate the program.

Program

; Program to generate and store N terms of Fibonacci series.
; First two instructions initialize the counter (R7) and storage pointer (R1).

```
START:   MOV    R7, 4FH        ; R7 has total terms count N
         MOV    R0, #51H       ; storage location of second term
```

; Store first two terms of the series. This is done by following three instructions.

```
         MOV    @R0, #01H      ; save second term
         DEC    R0             ; location for first term
         MOV    @R0, #00H      ; save first term
```

; Initialize the storage pointer (R1) and adjust the counter (R7) for two terms, already stored.

```
         MOV    R1, #52H       ; R1 pointing location for third term
         DEC    R7             ; reduce counter value by 2 as
         DEC    R7             ; two terms already generated
```

; Now, R0 pointing to the first of the two generated terms and R1 pointing to the storage area of the
; third term, which is yet to be generated.
; Iterations for generating and storing terms start from here.

```
MAIN:    MOV    A, @R0         ; get first term in register A
         INC    R0             ; point to second term
         ADD    A, @R0         ; add first two terms to get the next one
         MOV    @R1, A         ; store new term
         INC    R1             ; location for next new term, if any
         DJNZ   R7, MAIN       ; generate all N terms
```

; R7 indicates all N terms that are generated and stored. Terminate the program.

```
OVER:    SJMP   OVER           ; terminate here
```

C-version

```c
#include <regx51.h>

void main(void)
{
        register unsigned char *n, *dest;
        unsigned char count;
        n = 0x4F;
        dest = 0x50;
        *dest = 1;
        dest++;
        *dest = 1;
        dest++;
            for (count = (*n); count > 2; count --)
            {
                    *(dest) = (*(dest – 1) ) + (*(dest – 2));
                    dest++;
            }
}
```

7.9 | Generate a Series

Example 7.8

Purpose: Application of nested loops.

Problem

Generate and store the following series up to N terms. Value of N is available in location 30H. The series is presented using decimal number system.

$$1, 2, 3, 11, 12, 13, 21, 22, 23, 31, \ldots \text{ up to } N \text{ terms.}$$

Solution

A careful study reveals that the series contains only natural numbers with some regular gaps. After first three terms, there is a gap of seven numbers and the pattern repeats. Therefore, two consecutive loops may be planned whose first one to be repeated three times and the second one for seven times. Moreover, there must be a counter for keeping track of N terms of the series. Following are the flowchart (Fig. 7.6) and program listing.

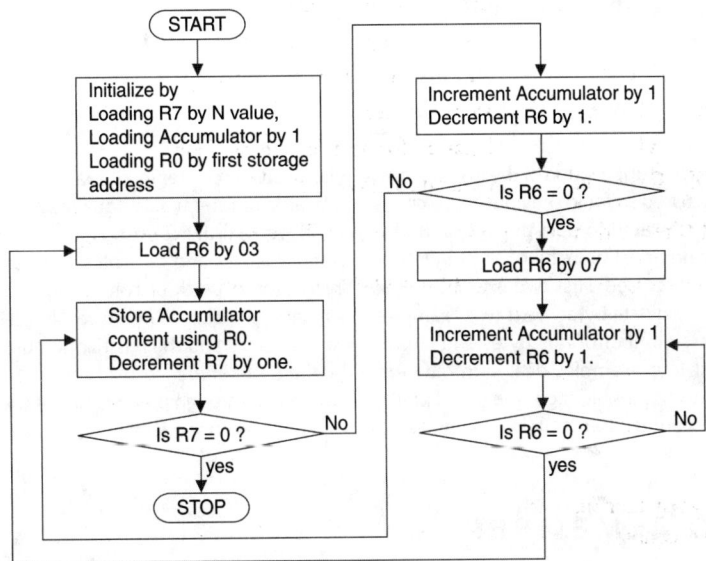

Figure 7.6 Flowchart for Example 7.8 (series generation)

Program

```
; Program to generate a series starts.
; First three instructions complete initialization process.
START:   MOV   R7, 30H      ; counter for N terms
         MOV   A, #01        ; first term in register A
         MOV   R0, #40H      ; storage area pointer
; Iteration to generate and store the terms of the series starts from here.
```

```
MAIN:   MOV    R6,#03H      ; term counter
MAIN1:  MOV    @R0, A       ; save one term
        DJNZ   R7, NEXT     ; continue if more terms are necessary
OVER:   SJMP   OVER         ; all terms are generated, terminate here
NEXT:   INC    A            ; next number
        DJNZ   R6, MAIN1    ; continue for maximum three times
```

; Following part generates, but not stores, the unwanted numbers.

```
        MOV    R6,#07H      ; get number of gaps
NEXT1:  INC    A            ; find next number
        DJNZ   R6, NEXT1    ; count seven gaps
        SJMP   MAIN         ; continue
```

 Another way to solve this problem is to add 7 with every third generated number, However, in that case only one loop would be sufficient.

SUMMARY

This chapter deals with the application of the instructions, which have been discussed so far in previous three chapters. Using instructions related to data moving, arithmetic operations and conditional branching, eight different examples have been discussed. First two examples deal with copying and shifting data bytes. Next two examples show how to calculate number of null bytes and checksum of an array. Last four examples deal with different types of series. These examples highlight the use of pointers (R0 and R1) and counters. To maintain uniformity, R7

was used as counter in all examples. However, any other available register would equally serve that purpose.

In general, DJNZ instruction is used to process the counter values and terminate the iteration process. CJNE instruction is used for comparing two numbers and taking some decision as per its result. CJNE instruction offers various forms and the most applicable one to be selected by the programmer. Instead of direct addressing, indirect addressing through pointers makes the algorithm simpler, especially when sequential addresses are to be dealt with.

POINTS TO REMEMBER

- There is no overlapping for data-copy operation while overlapping exists for data-shift operations.

- Depending upon shift-up or shift-down requirement, the start address of shifting must be initialized by the programmer.

REVIEW QUESTIONS

Evaluate Yourself

1. What would be the value of R0 and R1 at termination of the copy-block program of Example 7.1?

   (a) 73H and 53H   (b) 74H and 54H

   (c) 00H and 00H   (d) None of these

2. In Example 7.3, what would have been the content of location 6BH at termination of the program, if all bytes from 6CH to 7FH contained 00H?

   (a) 00H   (b) 20H

   (c) 14H   (d) None of these

3. In Example 7.5, what might be the maximum value of N if the program remains unchanged?

   (a) FFH (b) 7FH

   (c) 16H (d) None of these

4. In Example 7.5, what would happen if the first instruction of the program (MOV PSW, #00H) is removed?

   (a) No change

   (b) Incorrect result

   (c) Depends upon current bank

   (d) None of these

5. The program in Example 7.6 would work properly only if the value of N is

   (a) odd (b) even

   (c) any value (d) none of these

6. If all other parameters remain unchanged, what might be the maximum value of N in Example 7.7?

   (a) 10d (b) 14d

   (c) 255d (d) None of these

7. An alternate method to calculate the sum of N natural numbers is

   (a) $0.5*N*(N+1)$ (b) $0.5*N^2$

   (c) $N*(N-1)$ (d) none of these

8. If the value of N is 8 then the sum of the series in Example 7.6 would be

   (a) positive (b) zero

   (c) negative (d) none of these

9. In Example 7.8, what would be the value of R6 at the termination of the program?

   (a) Undefined (b) 03H

   (c) 07H (d) None of these

10. In Example 7.8, what would be the value of R7 at the termination of the program?

    (a) Undefined (b) 03H

    (c) 07H (d) None of these

Search for Answers

1. What is the difference between copy block and shift block?

2. What is the purpose of calculating checksum of an array?

3. What was the purpose of selecting bank #0 in Example 7.6?

4. What should be the initialization value of R7 in Example 7.2, if 16 bytes are to be shifted 2 bytes down?

5. Apart from CLR C instruction, what is the alternative way to clear the carry flag?

6. How is the Nth term of Fibonacci series generated?

7. What is the difference between an algorithm and a flowchart?

8. Find an alternative method to calculate the sum of the series in Example 7.6 if N is always an even number.

9. What would happen if the ADD instruction, in Example 7.4, is replaced by ADDC instruction?

10. What is the maximum size of an array that might be pointed by R0 or R1?

Think and Solve

1. Draw the flowchart of Fibonacci series generation problem.

2. Instead of using two pointers, R0 and R1, can the copy-block operation (Example 7.1) be completed using a single pointer?

3. Apart from using the SJMP instruction to jump at the same place, is there any other way to terminate any general-purpose program?

4. Why are there two pointers (R0 and R1) in MCS-51?

5. What is the limitation of DJNZ instruction?

6. Eight random numbers are stored in locations from 40H to 47H. Write a program to reverse their order in the same place from 40H to 47H. Do not use any other memory area except current register bank and accumulator.

7. Two arrays, each of 16 bytes, are stored from 40H to 4FH and 50H to 5FH. Write a program to compare these two arrays and store 00H in location 30H if they are identical (same number and in same sequence). Otherwise, store a non-zero value in location 30H.

8. Sixteen random numbers are stored in location 40H to 4FH. Write a program to take out alternate numbers, starting from the first number, and create a second array of eight numbers from 50H to 57H.

9. Extend the previous program so that the first array is compacted within 8 bytes from 40H to 47H after shifting eight of its entries.

10. Sixteen random numbers are stored in an array, starting from location 40H. Write a program to count the number of non-zero elements in this array and store it in location 30H.

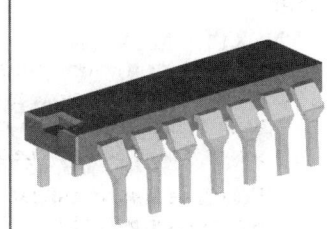

8 SUBROUTINES AND STACK

CHAPTER OBJECTIVES

Usage of subroutines makes the program size optimum and offers a scope of its modular development. During the use of subroutines, stack is used for related data storage. In this chapter, the reader is introduced to these two aspects and related instructions details. After completion of the chapter, the reader should be able to understand

- How subroutines and stack work.
- When and how to incorporate subroutines.
- Functioning of instructions like LCALL, ACALL, RET, PUSH and POP.
- Importance and usage of stack and stack pointer (SP).

8.1 | Need of Subroutines

Imagine a situation where an 8051 microcontroller is to read thermal values of three different locations of a furnace using three ports, say P0, P1 and P2; check whether any one of those has exceeded the preset limit (say AAH) or not, and, if any, then output through the fourth port, P3, a warning signal, say 55H. Fig. 8.1 is a schematic representation of this furnace-monitoring system. Following is a sample program for it.

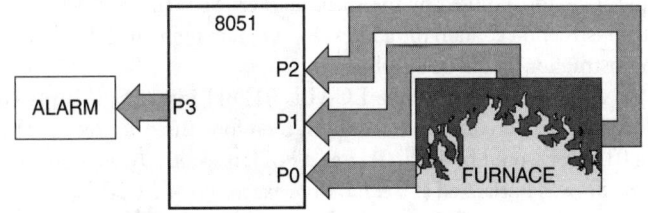

Figure 8.1 Schematic of a furnace monitoring system

; Sample program to explain the need of subroutines.
; Read temperature from three ports, P0, P1 and P2.
; If the temperature exceeds AAH then output 55H through P3 to raise an alarm.

```
START:  MOV   A, P0        ; read thermal value through port 0
        CLR   C            ; clear carry flag
        SUBB  A, #0AAH      ; check for threshold
        JC    NEXT1         ; within limits, check next port
        MOV   P3, #55H      ; limit crossed, raise alarm
```

```
NEXT1:  MOV    A, P1          ; read thermal value through port P1
        CLR    C              ; clear carry flag
        SUBB   A, #0AAH       ; check for threshold
        JC     NEXT2          ; within limits, check next port
        MOV    P3, #55H       ; limit crossed, raise alarm
NEXT2:  MOV    A, P2          ; read thermal value through port P2
        CLR    C              ; clear carry flag
        SUBB   A, #0AAH       ; check for threshold
        JC     NEXT3          ; within limits, check next port
        MOV    P3, #55H       ; limit crossed, raise alarm
NEXT3:  SJMP   START          ; loop on
```

This program would function properly and keep on checking all three ports in the same sequence and raising an alarm, when the preset threshold limit of AAH exceeds. However, we can readily observe that the same four instructions (in bold and italics) were repeated thrice for threshold limit checking of port inputs. This provides an ideal opportunity to use a subroutine to optimize the program length without altering its efficiency. A subroutine may be defined as *a program segment terminated by a 'return' instruction, so that it may be accessed from anywhere of the main program body*. Therefore, we have to be familiar with the mechanism of 'call' and 'return' instructions, which are discussed now. After that, in Section 8.5, we will resume further discussions and modifications of the same example program.

Strictly speaking, JC NEXT1, JC NEXT2 and JC NEXT3 are three different instructions. However, all of them fall under the same group of identical condition branching.

8.2 | LCALL Instruction

LCALL is a long call instruction used in programs to jump to a subroutine, placed anywhere within 64K of program memory, after storing the return address of the next executable instruction (immediately following the LCALL instruction) on the stack top, pointed by the stack pointer (SP). This is a 3-byte instruction and the second and third bytes of the instruction contain the address for program branching. Its operation does not affect any flags. Execution of this instruction would be as follows.

Let us assume that the first byte of the 3-byte LCALL 01F6H instruction is placed in the program memory address 01FDH. As the opcode of LCALL is 12H, therefore, three consecutive bytes starting from program memory location 01FDH would be 12H, 01H and F6H, respectively, as shown in Fig. 8.2.

We also assume that the SP was initialized at 07H. Before execution of LCALL, 3 bytes of opcode would be fetched through the program counter (PC) and the PC would then point to the first byte of the next instruction that is 0200H. After decoding the LCALL instruction, the processor would implement following steps:

Step 1: SP would be incremented by one and the lower byte of the present value in PC, that is 00H, would be stored on the stack at address 08H.

Step 2: SP would be again incremented by one and the higher byte of the PC, that is 02H, would be stored over the stack at address 09H.

Step 3: PC would be loaded by the second and third bytes of the instruction, that is 01H and F6H, so that the next fetch and execution starts from 01F6H.

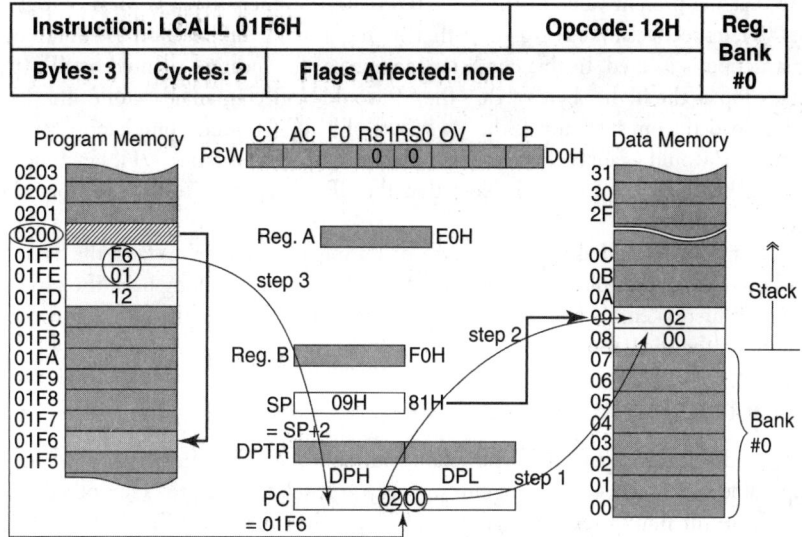

Figure 8.2 Schematic of execution of LCALL 01F6H instruction

It is expected that the return address, stored on the stack top, would be used by a RET instruction at the termination of the subroutine, which would force the control to resume the operation from the location 0200H.

8.3 | RET Instruction

RET is a 1-byte two-cycle instruction, which does not affect any flag, and is, generally, used as the terminator of any subroutine. Schematic execution of this instruction is presented through Fig. 8.3.

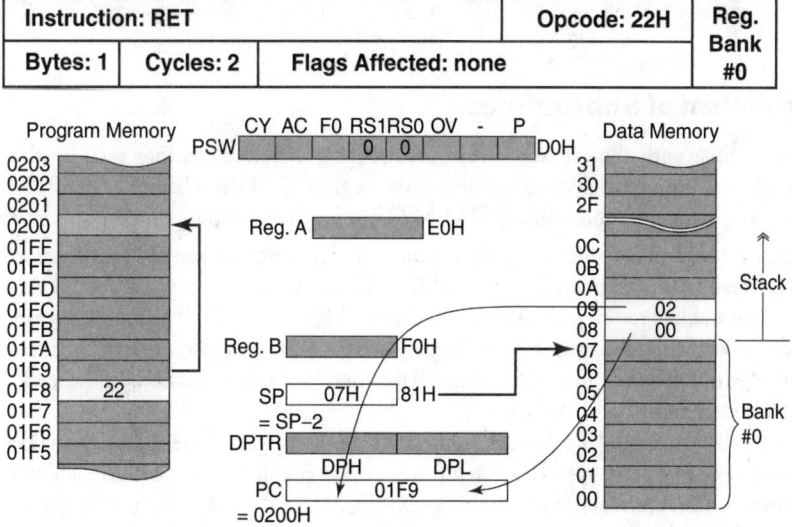

Figure 8.3 Schematic of execution of RET instruction

Let us assume that the present value of the SP is 09H and the internal RAM location 09H and 08H contains 02H and 00H, respectively. We also assume that at program memory location 01F8H, 1-byte opcode (22H) of RET instruction is located. In this condition, content of the location pointed by SP (02H) would be loaded from stack top as the higher byte of PC. The SP would be decremented by one and the next byte, that is 00H, would be loaded from stack top as the lower byte of PC. This would completely erase the original PC content (01F9H). SP would again be decremented by one so that it becomes 07H. Fetching of the next instruction would be, therefore, from 0200H. Note that the SP comes back to the location where it was storing the return address before.

In this context, it may be reminded that at the time of storing a byte over the stack, the SP is first incremented by one and then the data is placed on the stack top. At the time of taking back the saved data from the stack top, the data is first taken out and then the SP is decremented by one. The SP always points towards the last saved location on the stack top.

8.4 | ACALL Instruction

ACALL instruction is identical with LCALL instruction with only one difference. This is a 2-byte instruction and, therefore, applicable within 2K of the program memory area related with the current value of the PC. The second byte of the instruction stores least significant 8 bits of the call address. The first byte of instruction accommodates A8–A10 of the call address. Most significant 5 bits of PC remain unchanged during the call address generation (Fig. 8.4). This address modification is identical with that of the AJMP instruction, which we have discussed in Section 6.2.2.

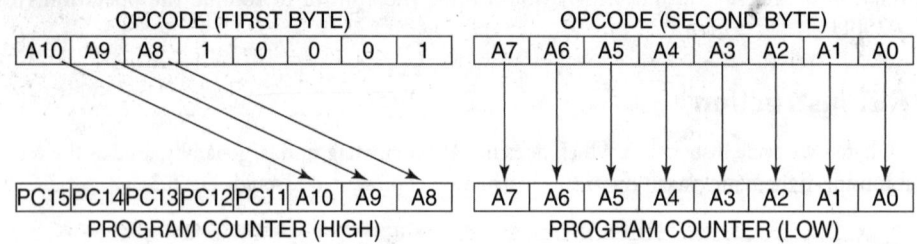

Figure 8.4 Mechanism of ACALL instruction

8.5 | Mechanism of Subroutines

As we are now familiar with the call and return instructions, therefore, we may pick up the thread of the example program, which we have been discussing at the beginning of this chapter. Using LCALL and RET instructions and introducing a subroutine (CHKLMT), we may rewrite the program as is given in Fig. 8.5.

In this case, the program is reduced to 12 instructions and 26 bytes instead of 16 instructions and 32 bytes, previously. Three more bytes might be reduced if LCALL instructions are replaced by ACALL instructions. Observe the first call and first return paths. When compared with the second call and second return paths, we find that after the return instruction the second return is going back to a different place than the first return. How is this possible? It is because of the storage of return address by the concerned call instruction, on the stack top. Let us take a closer look at this mechanism.

Whenever a call instruction is executed, address of the first byte of the next instruction after the call instruction is stored on the stack top. This stack top is always pointed by the SP. The address stored on the stack top is known as the return address.

Execution of LCALL instruction may take the control anywhere within the program memory area. However, whenever the return instruction is encountered, the return address is taken from the stack top and

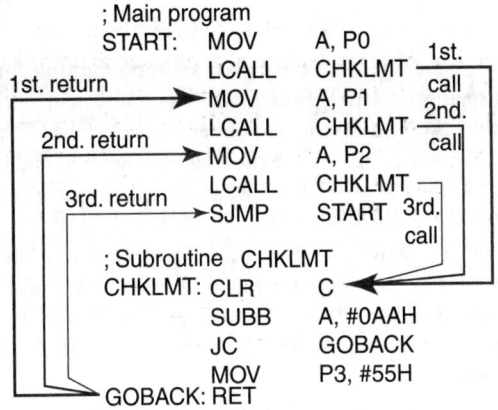

Figure 8.5 Program flow with a subroutine

loaded on the PC. Therefore, the control automatically comes back to the same location, which it had left to execute the LCALL instruction. The only point is that any subroutine, thus called, must be terminated by a return instruction. We should remember that the range of RET instruction is 64K like the LCALL instruction.

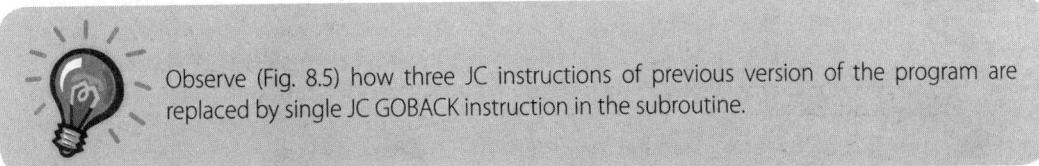

Observe (Fig. 8.5) how three JC instructions of previous version of the program are replaced by single JC GOBACK instruction in the subroutine.

8.6 | Nesting of Subroutines

Multiple subroutines may also be nested in their usual configuration. It means one subroutine may call another subroutine and that subroutine may call a third one (refer to Fig. 8.6), and so on. The process may grow in this way, which is limited only by the size of the stack, which we would discuss now.

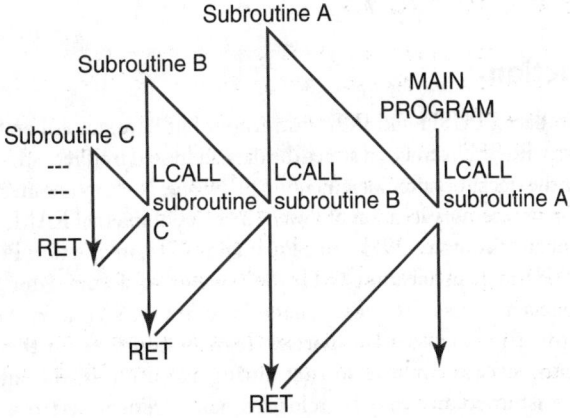

Figure 8.6 Nesting of subroutines

8.7 | Stack

We can understand from above discussions that stack plays an important role for call and return instructions. In general, an area of system memory (RAM) is earmarked as the system stack, which is not used for any other purpose. Generally, stack is accessed through a pointer known as the SP. We should also be careful for nested subroutine calls so that the stack does not overflow causing error in functioning of the system. The SP always indicates the stack top and the last saved data byte at that place. Stack always works with the principle of last-in-first out or LIFO.

SP is initialized by 07H after system reset and the stack grows from 08H onwards. However, it may also be initialized at any other location by the programmer. As the stack is addressed by indirect addressing, therefore, for XX52 and similar microcontrollers, stack may grow within upper 128 bytes of internal RAM also. However, external data memory cannot accommodate any stack in MCS-51, even if it is interfaced with the system.

Apart from storing the return addresses, stack may also be used for storing important data available in registers. Registers or other directly addressable memory locations may be saved on the stack by PUSH and then collected from the stack by POP instructions, which are discussed now.

The term *stack top* is a bit confusing for any beginner as *top* is a relative term. This top and bottom are dependent on how the addresses of system RAM are depicted. In 8051, these addresses increase upwards, that is, lowest RAM address are generally shown towards the bottom of the figure. On the other hand, for 8085 based systems, RAM addresses are presented in such way in sketches so that they increase downwards (higher address at bottom).

Now, for both cases, the data is placed on the stack top, but in a different manner. For 8051, the SP is incremented and then the data is placed on the stack top. Therefore SP value increases for every storage. For 8085, data is placed on the stack top and then SP is decremented by two. Therefore SP value decreases for every storage. However, the way RAM addresses are shown in sketches, we find that in both cases data goes to the top of the stack. The best way to realize is to draw two figures of system stack and compare the data storing operation (PUSH).

8.8 | PUSH Instruction

In Chapter 4, we have introduced PUSH and POP instructions but have not elaborated their details. These are 2-byte two-cycle instructions. PUSH instruction stores the data addressed by direct addressing through the second byte of the instruction, over the stack top, after incrementing SP by one. No flags are affected. The directly addressed location may be any SFR or any memory location of lower 128 bytes of internal RAM.

Assuming that accumulator contains 29H and SP contains 07H, instruction PUSH A would increment SP by one (making it 08H) and, in location 08H, the content of accumulator (29H) would be copied (Fig. 8.7). Note that the direct address of the accumulator becomes the second byte of the instruction.

In general, it is customary to store all important register contents on the stack top using PUSH instruction at the beginning of a subroutine so that during execution of the subroutine these flags and register contents are not disturbed and may be reloaded before returning from the subroutine by POP instruction.

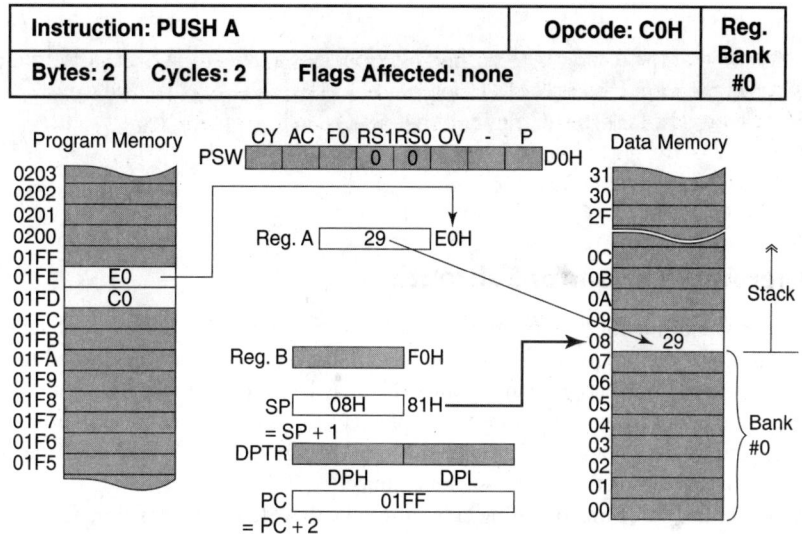

Figure 8.7 Execution of PUSH A instruction

8.9 | POP Instruction

As already mentioned, POP instruction restores a byte of data from the stack top, pointed by SP, to its location as directly addressed by the second byte of the instruction. After that SP is decremented by one. Assuming that SP has 08H and location 08H of internal RAM contains 29H, execution of POP B instruction would load register B by 29H and SP would be decremented by one to make it 07H (Fig. 8.8). No flags are affected unless POP PSW instruction is executed. POP instruction is generally used to restore important data before returning from a subroutine.

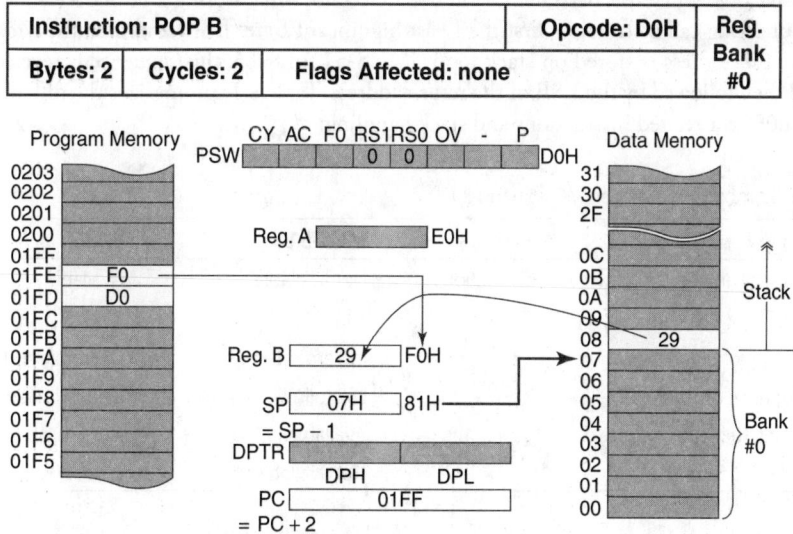

Figure 8.8 Execution of POP B instruction

 Note that DPTR cannot be pushed or popped as a sixteen-bit register. We are to do it in two parts by using PUSH DPH, followed by PUSH DPL. Similarly we are to use POP DPL followed by POP DPH, if they have been saved in the same order.

8.10 | Other Applications of Subroutine

Replacing multiple sets of identical instructions sequence is not the only application of subroutines. They are also adopted to make the program structure modular. A large program may be divided to several small-sized subprograms and for each subprogram a subroutine may be developed. Finally, a main program might be written which simply calls these subroutines to implement the task. This modular approach improves the efficiency of the software design and its development, as several groups may work simultaneously, one with each subroutine, for faster completion of the software project.

However, one limitation of incorporating subroutines is, they demand extra time for executing related CALL and RETURN instructions. Therefore, to assure a faster program response, usage of subroutines should be minimized or eliminated.

8.11 | Comparison with 8085

Conditioned call and conditioned return instructions, offered in 8085, are not available in MCS-51 instruction set. All call and return instructions of MCS-51 are unconditional. Moreover, in 8085, the SP is a 16-bit register, while in MCS-51 it is an 8-bit register. Therefore, theoretically, the height of MCS-51 stack is limited to a maximum of 255 bytes with respect to the maximum size of 64 KB RAM, possible in 8085-based system. Finally, no such 2-byte ACALL-type instruction is available in 8085.

MCS-51 and 8085 stack operation shows a few variations. SP is always incremented for storing data over stack in MCS-51, while it is decremented in 8085. Second, while storing a 16-bit return address during the execution of a long call instruction, first the Least Significant Byte (LSB) and then the Most Significant Byte (MSB) of the address is stored on stack top in MCS-51. In 8085, this sequence is reversed by storing MSB first and then followed by the LSB of the return address. Table 8.1 summarizes the differences between MCS-51 and 8085 for related instructions and stack handling.

Table 8.1 Comparison between MCS-51 and 8085

	MCS-51	8085
Conditional call and conditional return instructions	Not available	Available
SP register	8 bits	16 bits
During storage of data, stack address is	Incremented	Decremented
Saving sequence of 16-bit address	LSB followed by MSB	MSB followed by LSB
2-byte call instruction	Available	Not available
1-byte call instruction	Not available	Available

8.12 | Solved Examples

Example 8.1

Purpose: To understand how parameters are passed in both ways through subroutines.

Problem

Memory locations 31H onwards, several pairs of random integers are stored. The number of pairs is available in location 30H. Develop a program to find the sum of the numbers obtained by multiplying these pairs. Multiplications are to be performed without using the MUL instruction of MCS-51. Assume that the sum would never exceed 8 bits.

Solution

Fig. 8.9 explains the problem graphically. Here a0, a1, a2, etc. are random integers and sum of the product of the pairs is to be calculated.

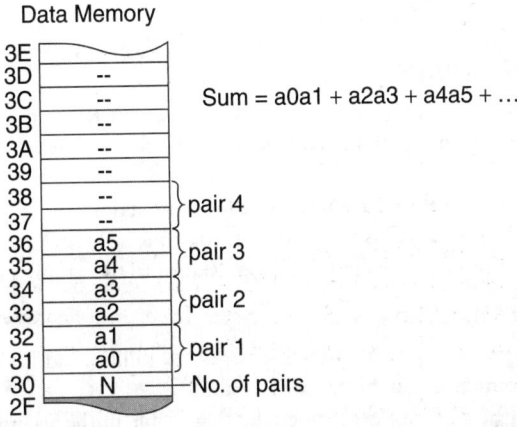

Figure 8.9 Problem definition for Example 8.1

This may be done even without the usage of MUL instruction of MCS-51. We may write a subroutine to multiply any two given integers using the method of repeated additions. However, as the numbers are not sequential, therefore, subroutine must be used in this case. Following are the suggested algorithms and program listings of the related main program and the subroutine.

Algorithm for the main program

Step 1: Load R5 from 30H by count value N. Load R0 by 31H to point the first integer of first pair. Clear R4 to initialize the sum.

Step 2: Call subroutine PRODCT to multiply two integers as pointed by R0. At return, product is in the accumulator and R0 becomes R0 + 2. Add the accumulator with R4 and store back the result from the accumulator in R4.

Step 3: Decrement R5 by one. If R5 is not zero then continue at Step 2.

Step 4: R5 indicates that the series calculation is over. Terminate the program.

; Main program to calculate the sum of product of pairs up to N pairs.
; Registers (any bank) used:

```
; R0 = pointer to stored numbers
; R4 = final sum (within 8 bits) of products of the pairs
; R5 = counter for N pairs
; R6 = temporary storage
; R7 = temporary counter
; A = computed results

MAIN:    MOV    R5, 30H      ; load counter by N
         MOV    R0, #31H     ; point to first integer of first pair
         MOV    R4, #00H     ; clear sum
ADDUP:   ACALL  PRODCT       ; multiply one pair, product in A, R0 = R0 + 2
         ADD    A, R4        ; add with previous sum
         MOV    R4, A        ; store new sum in R4
         DJNZ   R5, ADDUP    ; continue if any more pair pending
OVER:    SJMP   OVER         ; terminate program
```

Algorithm for the subroutine PRODCT

Step 1: Store first of the pair as pointed by R0 in R7. Increment R0 by one and store the next number as pointed by R0 in R6. Again increment R0 by one for next iteration. Clear the accumulator to calculate the product.

Step 2: Add accumulator with R6. Leave result in the accumulator.

Step 3: Decrement R7 by one. If R7 is not zero then continue at Step 2.

Step 4: R7 indicates that product calculation is over. Return to the calling program

```
; Name:        PRODCT
; Function:    Calculates product of two unsigned integers by multiple additions
; Input:       R0 must point at the first integer of a pair
; Output:      Product of the pair (not exceeding 8 bits) available in the accumulator
; Calls:       None
; Uses:        Flags, A, R0, R6, R7

PRODCT:  MOV    A, @R0       ; get first number
         MOV    R7, A        ; save it in R7
         INC    R0           ; point to next number
         MOV    A, @R0       ; get second number
         MOV    R6, A        ; save it in R6
         INC    R0           ; point to next number for next iteration
         MOV    A, #00H      ; clear product
PLOOP:   ADD    A, R6        ; keep on adding as long as
         DJNZ   R7, PLOOP    ; the counter is not zero
         RET                 ; product in the accumulator, R0 = R0 + 2
```

Discussions

Let us take a closer look at the mechanism of this program. To start with, the subroutine PRODCT needs only R0 with correct value as its input. R0 should point to the first of the random integer pair. It loads this value as a counter in R7 and gets the next integer through R0 (after incrementing R0) and saves in R6. The accumulator is cleared to store the product. The iterative procedure for multiplication by repeated additions starts next. Note that we do not consider carry for addition, as the result would not exceed 8 bits. Every time R6 is added with the accumulator, R7 is decremented by one. When R7 becomes zero, the process terminates and the subroutine returns the control to the calling program with the product of two integers in the accumulator and R0 pointing to the first of the next integer pair.

The main program loads the number of integer pairs (N) in R5 and loads R0 by the address of the first of the integer pair. Register R4 is cleared to store the final sum. Main program then calls subroutine PRODCT and, upon its return, adds the content of R4 with the return value available in the accumulator and stores it back at R4 itself. Here also carry condition is neglected, as the result would be within 8 bits. Then R5 is decremented by one and till it becomes zero, the calling of the subroutine, PRODCT and execution of related instructions are continued. Finally, the result is available in R4 as well as in the accumulator at the termination of the main program.

Example 8.2

Purpose: To understand the limitation of ACALL instruction.

Problem

Discuss the criticality of placing an ACALL instruction at the end of any 2K-program memory segment. Given that a 2K segment starts with its least significant 11 bits as all zeros and ends with its least significant 11 bits as all ones.

Solution

Fig. 8.10(a) explains 2K-program memory segments of MCS-51, related to ACALL instruction. Within 4K-program memory, there are two such 2K segments from 0000H to 07FFH and from 0800H to 0FFFH. If any ACALL instruction is placed at the boundary, that is, at address 07FEH or at 07FFH, then its jurisdiction would be the next 2K segment and not the current one.

To explain this, let us assume that we have placed a subroutine at location 07F0H. To access this routine, if we place an ACALL instruction at 07FE H [Fig. 8.10(b)] then during its execution, the PC would first be incremented to 0800H and then 11 bits of PC would be loaded with the bits, available in the instruction bytes. This would lead to generation of the branching address 0FF0H, which is not the desired location for branching.

However, if the same ACALL instruction with unchanged opcode is placed at location 07FDH, then it would branch properly to the target location 07F0H, as the current page boundary was not violated even after incrementing the PC to the starting address of the next instruction [Fig. 8.10(c)].

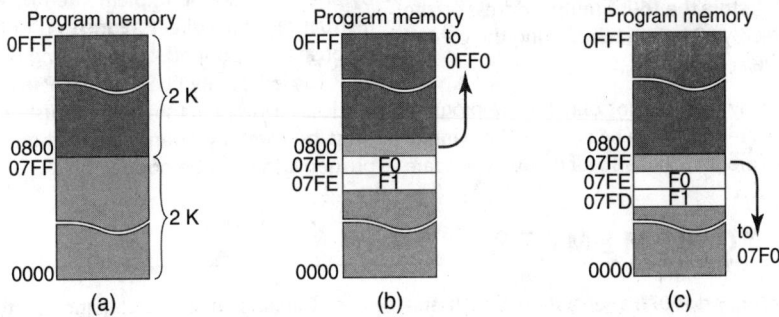

Figure 8.10 (a) 2K blocks for ACALL, (b) malfunction of ACALL at 2K boundary and (c) correct functioning of ACALL, away from boundary

8.13 | Parameter Passing in Subroutines

One important point to be noted through Example 8.1 is the way different parameters were passed from the main program to the subroutine and vice versa. The subroutine needs a correct value at R0, which was initially taken care of by the main program. After the first iteration, the subroutine itself updates this register so that, at every iteration, it starts with a correct value in R0.

Similarly, the product from the subroutine is passed to the main program through the accumulator. After receiving it, the main program processes it and releases the accumulator to the subroutine for further usage.

This parameter passing must be carefully planned in case of subroutines, and the registers to be used by it should not be disturbed by the calling program. Similarly, the subroutine should also not disturb the registers and flags used by the main program. As already indicated, one safe method is to push necessary registers over stack at the start of a subroutine and pop those from stack before leaving the subroutine. In this case, care must be taken to plan the size of the stack so that there is no overflow of it.

To avoid any mistake in parameter passing, a good practice is to write names of relevant registers with associated variable names at the start of a subroutine development. When the calling program is being developed, indicate those registers which are associated with parameter passing.

SUMMARY

A subroutine is a program segment terminated by a return instruction, which may be accessed from different location(s) of the program memory. The return instruction collects the return address from the stack top and helps the control to go back to the calling location. LCALL is a 3-byte instruction capable of calling a subroutine placed anywhere within the 64K-program memory area. ACALL is a 2-byte version of this instruction capable of communicating with only those subroutines that are placed within the same 2K zone with respect to the location of the first byte of the next instruction immediately after the ACALL instruction. Both LCALL and ACALL store the 16-bit return address on the stack top pointed by SP before transferring the control to the subroutine.

Subroutines may be used not only for the program segments, which are allowed to be accessed from multiple locations of a program, but also for modular program development. However, usage of subroutines demands some extra execution time and limited or no subroutine runs a program faster.

Stack is a preplanned area by the programmer within the internal data memory where the return addresses and other important information are stored. Stack always works with the principle of LIFO. SP always points to the location of the last saved data. SP must be incremented by one before pushing a new byte on the stack top. Similarly, after popping a byte from the stack top, SP must be decremented by one. This may be implemented by PUSH and POP instructions with direct addressing of target bytes to be pushed or popped. Incrementing or decrementing SP is carried out by PUSH and POP instructions, automatically. However, the order of pushing and popping must be carefully arranged and the data, which was pushed last, must be popped first.

POINTS TO REMEMBER

- As SP is initialized as 07H after system reset, therefore, it must be re-initialized if register bank #1 is to be used in the program.

- Conditional call and conditional return instructions are not available in 8051 and to be substituted by multiple instructions, properly sequenced.

REVIEW QUESTIONS

Evaluate Yourself

1. From which address would the next executable instruction be fetched by the LCALL 0123H instruction, if the first byte of it is located at 0023H?

   (a) 0123H
   (b) 2301H
   (c) 0026H
   (d) None of these

2. If the first byte of an ACALL instruction is located at 0123H and before its execution SP contained 23H, then what would be the content of the internal RAM location 24H immediately after execution of the ACALL instruction?

   (a) 01H
   (b) 23H
   (c) 25H
   (d) None of these

3. How many flags are affected by the execution of the following instruction?

   POP 0D0H

   (a) No flags are affected
   (b) Only C flag is affected
   (c) All flags are affected
   (d) None of these

4. What would happen if the following subroutine is executed?

   ```
   SUBRT0:   PUSH   A
             PUSH   B
             POP    A
             POP    B
             RET
   ```

   (a) Nothing happens
   (b) Data present in registers A and B are interchanged
   (c) Returns to the address by combining registers A and B
   (d) None of these

5. In a case of nested subroutines, main program calls subroutine 1, which calls subroutine 2. If subroutine 2 was developed as follows, what would happen after the execution of subroutine 2?

   ```
   SUBRT2:   DEC   81H
             DEC   81H
             RET
   ```

   (a) Return to main program
   (b) Return to subroutine 1
   (c) Stack underflow
   (d) None of these

6. What is the main drawback of using subroutines?

   (a) Needs extra space
   (b) Needs extra time
   (c) Hampers program flow
   (d) None of these

7. What is the maximum number of nesting of subroutines that is possible in 8051?

   (a) 127
   (b) 64
   (c) 63
   (d) None of these

8. The following main program and subroutine were developed to calculate the sum of natural numbers up to *N* terms. The integer value of *N* was stored in register R7 before evoking the main program. At the end of the execution of the program, which of the following registers would contain the sum?

   ```
   MAIN:     MOV    A,#00H
             MOV    R4,#00H
   LOOP:     LCALL  NXTNUM
             ADD    A,R4
             MOV    R4,A
             DJNZ   R7,LOOP
   OVER:     SJMP   OVER
   ; Subroutine NXTNUM
   NXTNUM:   INC    A
             RET
   ```

   (a) Accumulator
   (b) Register R4
   (c) Register R7
   (d) None of these

9. What would be the content of the accumulator after execution of the following main program?

   ```
   MAIN:     MOV    R7,#22H
             MOV    A,#00H
             LCALL  LOAD33
             ADD    A,1FH
   OVER:     SJMP   OVER
   LOAD33:   PUSH   PSW
             MOV    PSW,#18H
   ```

```
MOV    R7, #33H
POP    PSW
RET
```

(a) 1FH   (b) 33H

(c) 22H   (d) None of these

10. What would happen when the following program is executed?

```
MAIN:    MOV    A, #01H
         MOV    B, #0FFH
         PUSH   B
         PUSH   A
         RET
```

(a) It would jump at the program memory location 01FFH

(b) It would execute a call at location FF01H

(c) Undefined

(d) None of these

Search for Answers

1. How many flags are affected by the LCALL instruction?

2. What types of addressing modes are allowed for PUSH and POP instructions?

3. What might be the reason for introduction of a subroutine in a program?

4. What is the difference between a program and subroutine?

5. What are the differences between the LCALL and ACALL instructions?

6. What happens when the ACALL instruction is placed at the location 07FEH?

7. What are the differences between call or return instructions of 8085 and MCS-51?

8. What is the purpose of the stack?

9. What precautions are to be taken for nested subroutines?

10. How parameters may be passed from any main program to its subroutine in 8051?

Think and Solve

1. What would happen to the stack if a different register bank is selected?

2. If the 2-byte ACALL instruction contains only 11 bits of call address, then how is this instruction capable of storing 16-bit return address on the stack top?

3. What might happen if the single-byte RET instruction is placed at the end of a 2K-program memory segment?

4. Is it possible for any one subroutine to have multiple numbers of return instructions? If yes, then what might be its requirement?

5. What would be the value of bit 7 (MSB) of the accumulator after execution of the following two instructions?

```
PUSH    PSW
POP     ACC
```

6. Is it possible to use the bit-addressable area of internal data memory for the purpose of stack?

7. How can a byte of data, located at the upper 128 bytes of data memory of a XX52 microcontroller, be saved on stack top by PUSH instruction?

8. Is it possible for a subroutine to check and change its own return address? If yes, then how?

9. Is it possible to substitute the LCALL and RET instructions by LJMP, PUSH and POP instructions? Assume that INC and DEC may also be used to service the SP.

10. Can a subroutine call itself? Justify your answer.

9 LOGICAL OPERATIONS

CHAPTER OBJECTIVES

Apart from arithmetical operations, which we have discussed in Chapter 5, logical operations play an important role in data processing in many cases. In this chapter, the reader is introduced to the related instructions for logical operations offered in MCS-51. After going through the chapter, the reader should be able to understand

- Addressing modes and syntax of instructions used in logical operations.
- Applications of these instructions in different problem-solving software routines.

9.1 | Introduction

Before we start our discussions on logical operations possible with MCS-51, the reader may be reminded that MCS-51 is capable of performing logical operation with both byte and bit variables. Operations possible with the bit variables would be discussed in the next chapter ('Boolean Variable Manipulation'). In this chapter, we focus our attention on logical operations with byte variables only.

Table 9.1 represents all the mnemonics related to byte-oriented logical instructions available in MCS-51. Out of a total of 10 mnemonics, provisions are there for important logical operations, such as AND, OR, XOR, and for complementing the operands. Instructions are also available for rotating the accumulator content circular

Table 9.1 Mnemonics related with logical operations

Mnemonics	Brief description
ANL	Logical AND operation
ORL	Logical OR operation
XRL	Logical XOR operation
CLR	Clear accumulator
CPL	Complement accumulator
RL	Rotate accumulator left
RLC	Rotate accumulator left through carry
RR	Rotate accumulator right
RRC	Rotate accumulator right through carry
SWAP	Swap nibbles within the accumulator

or through carry bit. Finally, there is a special instruction, SWAP, to interchange the place of the nibbles of the accumulator. By the term 'nibble', a group of four bits is indicated; that is, a byte is composed of two nibbles.

Various addressing modes, like those offered for arithmetical operations, are also available for the first three logical operations. In arithmetical operations, we have observed that the result of any operation is generally available in the accumulator itself. However, in the case of logical operations, there is an extra provision for making the result of operation available at a directly addressed location, though in restricted addressing modes. Details of the first nine mnemonics of Table 9.1 are discussed in the subsequent sections. Mnemonic SWAP is discussed in Chapter 12.

Nibble plays in important part as it is capable of dealing with BCD as well as hexadecimal digits. Earlier processors, developed specially for calculators, were mostly nibble-oriented.

9.2 | ANL Instruction

Four addressing modes in six variations are available for logical ANDing of byte variables. These addressing modes are: immediate, direct, register direct and register indirect. All six variations of AND operations are presented in Table 9.2.

Table 9.2 Addressing modes of ANL instruction

Instruction	Addressing mode	Function
ANL A, Rn	Register direct	Logically AND accumulator with register
ANL A, direct	Direct	Logically AND accumulator with direct address
ANL A, @Ri	Register indirect	Logically AND accumulator with address indicated by the register
ANL A, #data	Immediate	Logically AND accumulator with immediate data
ANL direct, A	Register direct	Logically AND direct address with accumulator. Result in direct address
ANL direct, #data	Immediate	Logically AND direct address with immediate data. Result in direct address

In the first four instruction types, it is in the accumulator that the final result would be available. However, in the last two instruction types, it is the directly addressed memory location which would store the final result of the AND operation.

The execution of instruction ANL A, B is illustrated as an example case in Fig. 9.1. Assuming original contents of the registers A and B to be 57H and A9H, respectively, after the AND operation, the accumulator would have 01H as the result. No flags would be affected. Note that this is a 2-byte instruction indicated by the opcode 55H. Direct address of the SFR B (F0H) becomes the second byte of the instruction.

On the other hand, if we take as an example case the execution of the instruction ANL B, A as illustrated in Fig. 9.2, then we find that the result of the ANDing operation is available in register B. In this case also, we have assumed that before execution, registers A and B contained 57H and A9H, respectively.

Allowing the result of logical operations to be available in target memory locations rather than in accumulator helps in making the program work faster. Otherwise, in general, the result needs to be copied from the accumulator to the target location, which needs an extra instruction.

If any I/O port is directly addressed in any ANL operation, data from the output latch of the port, not from its input pins, would be used for the AND operation. That means, ANL A, P2 instruction would get the data from the output latch of port 2, which would be logically ANDed with the accumulator. Instructions like ANL P2, A would use the output latch of port 2 as the source of one operand as well as the destination of the result.

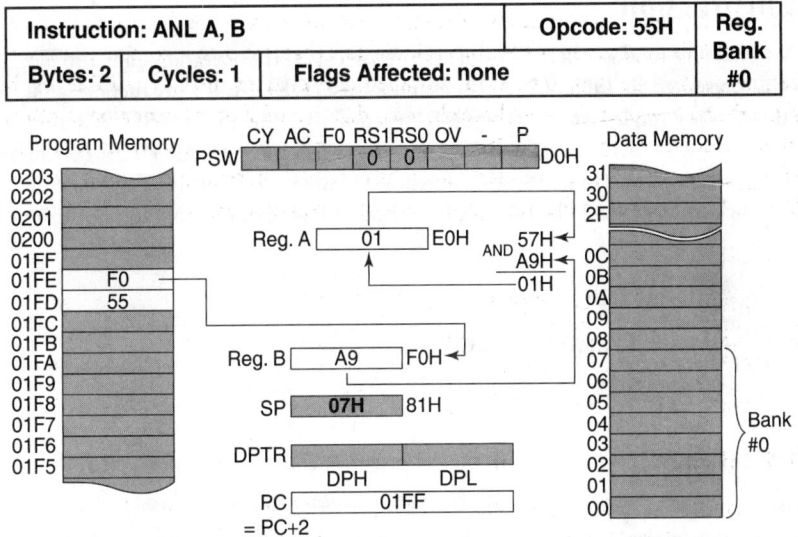

Figure 9.1 Execution of ANL A, direct instruction

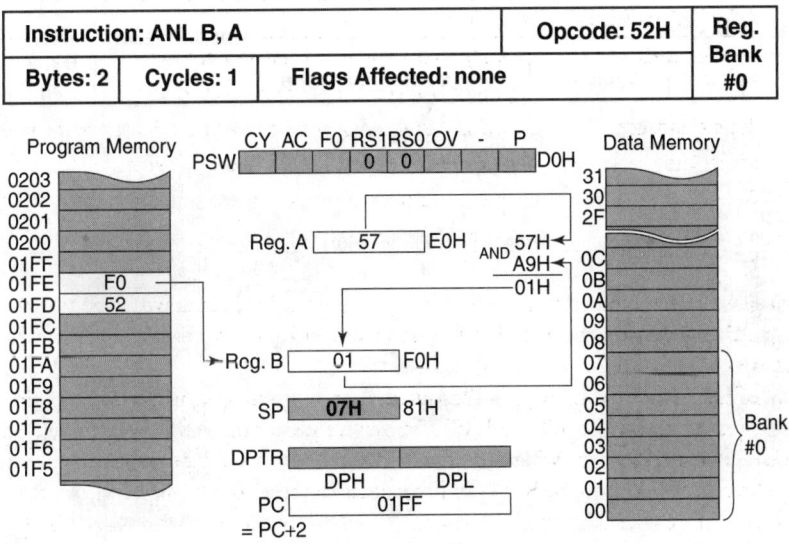

Figure 9.2 Execution of ANL direct, A instruction

 In 8085, results of all operations, arithmetic or logical, are available only in its accumulator. Therefore, those readers who have already gone through 8085, would find the storage of result in a directly addressed memory location as a new concept. However, this type of instructions are extremely helpful for developing a compact program.

9.3 | ORL Instruction

Variations and addressing modes, offered in the case of logical AND operations, are also available for logical OR operations as presented in Table 9.3. As an example case, ORL A, R3 instruction is illustrated through Fig. 9.3. Note that this is a single-byte single-cycle instruction with the opcode 4BH. Assuming that before the ORing operation, the accumulator contained 57H and the register R3 of bank #0 contained A9H, a result of FFH would be available in the accumulator after execution of the instruction. No flags would be affected. If any port is directly addressed, data would be taken from the output latch of the port and not from its input pins.

Table 9.3 Addressing modes of ORL instruction

Instruction	Addressing mode	Function
ORL A, Rn	Register direct	Logically OR accumulator with register
ORL A, direct	Direct	Logically OR accumulator with direct address
ORL A, @Ri	Register indirect	Logically OR accumulator with address indicated by the register
ORL A, #data	Immediate	Logically OR accumulator with immediate data
ORL direct, A	Register direct	Logically OR direct address with accumulator. Result in direct address
ORL direct, #data	Immediate	Logically OR direct address with immediate data. Result in direct address

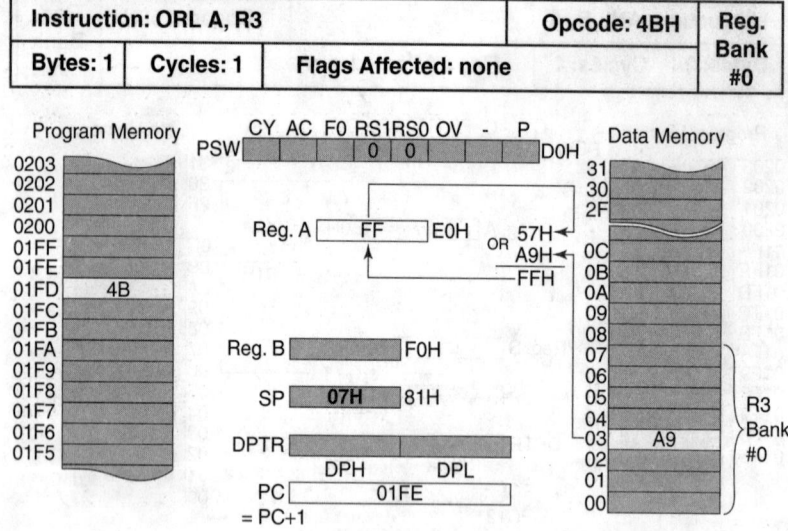

Figure 9.3 Execution of ORL A, Rn instruction

Note that in 8051, logical operations do not affect any status flag. This is another new feature for 8085-oriented readers.

9.4 | XRL Instruction

Just as one of the many usage of AND operation is to clear some target bits of a register or a memory location and that of OR operation is to set some target bits of a register or a memory location, similarly, in the case of XOR operation, one of its usages is to toggle some bits of a register or a memory location.

Just like the previous two cases, the XOR operation of MCS-51 also allows six variations and four addressing modes as explained in Table 9.4. As an example case, XRL A, @R1 instruction is illustrated in Fig. 9.4. This is a 1-byte, one-cycle instruction with the opcode 67H. Assuming that the accumulator contained 57H before the operation, register R1 contained 30H, and the memory location 30H contained A9H, execution of the instruction would leave FEH in the accumulator. No flags would be affected. If any port is directly addressed, then data would be taken from its output latch and not from its input pins.

Table 9.4 Addressing modes of XRL instruction

Instruction	Addressing mode	Function
XRL A, Rn	Register direct	Logically XOR accumulator with register
XRL A, direct	Direct	Logically XOR accumulator with direct address
XRL A, @Ri	Register indirect	Logically XOR accumulator with address indicated by the register
XRL A, #data	Immediate	Logically XOR accumulator with immediate data
XRL direct, A	Register direct	Logically XOR direct address with accumulator. Result in direct address
XRL direct, #data	Immediate	Logically XOR direct address with immediate data. Result in direct address

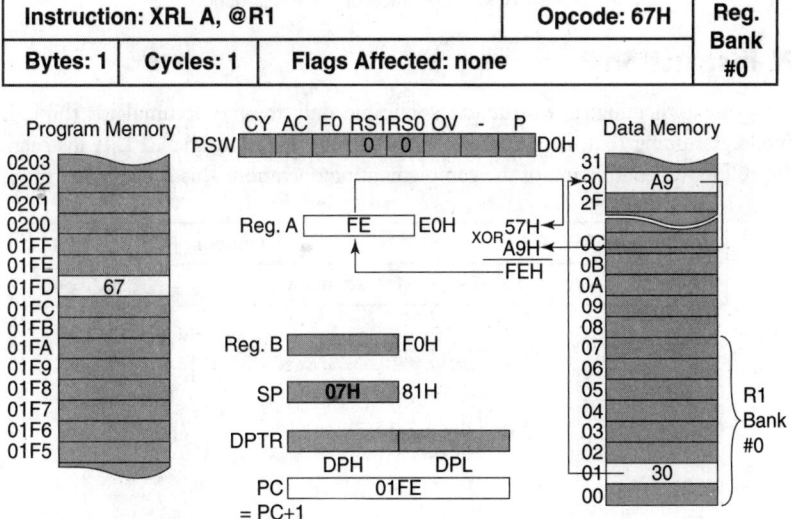

Figure 9.4 Execution of XRL A, @Ri instruction

 To clear a bit, AND it with 0, to set a bit, OR it with 1. Finally, to toggle a bit, XOR it with 1. These are three widely used purpose of AND, OR and XOR operators.

9.5 | CLR Instruction

This 1-byte, one-cycle instruction clears the accumulator. No flags are affected. This instruction is equivalent to MOV A, #00H instruction, which is a 2-byte instruction. Therefore, using CLR A, the programmer may save 1 byte of coding. In general, this instruction is useful during any initialization process. Fig. 9.5 illustrates the execution of this instruction.

Instruction: CLR A			Opcode: E4H	Reg. Bank #0
Bytes: 1	Cycles: 1	Flags Affected: none		

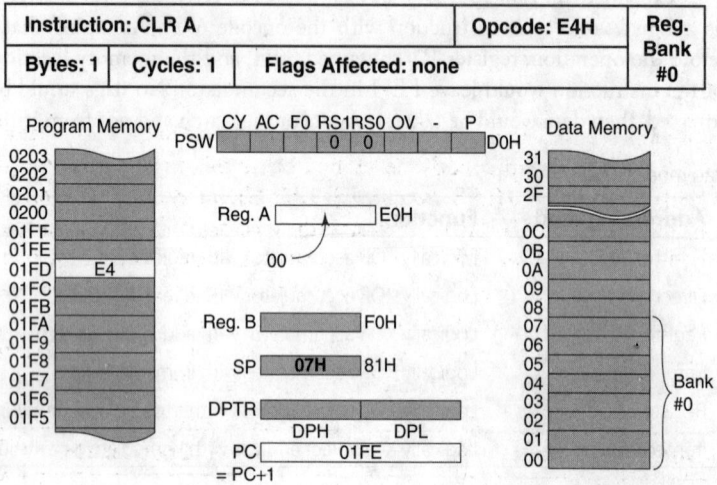

Figure 9.5 Execution of CLR A instruction

9.6 | CPL Instruction

Complementing is another logical instruction applicable only for the accumulator through register direct-addressing mode. Assuming that the accumulator contains 4DH before the CPL A instruction's execution, it would contain B2H after completion of the complementing operation. This is illustrated through Fig. 9.6. This

Instruction: CPL A			Opcode: F4H	Reg. Bank #0
Bytes: 1	Cycles: 1	Flags Affected: none		

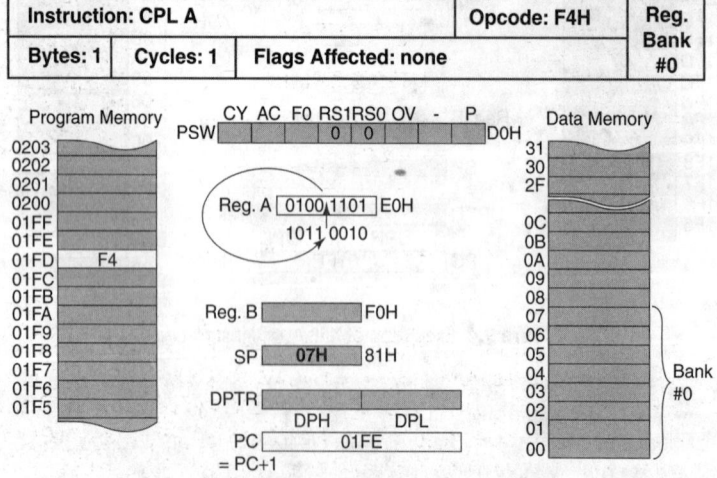

Figure 9.6 Execution of CPL A instruction

1-byte, one-cycle instruction does not affect any flags. If this instruction is followed by the INC A instruction, then it generates the two's complement of any number stored in the accumulator.

An alternate way of generating the complement of any number is to XOR it with FFH. For example, if 55H is XORed with FFH, it would produce its complement, i.e. AAH. This method of complementing a memory location takes the same number of bytes of opcode as compared with the traditional method of getting the byte in the accumulator, complementing the accumulator and then finally storing back the complemented data in the original memory location if direct addressing is not used. Otherwise, XORing with FFH would take lesser bytes of coding. See Example 9.2 for related discussions.

9.7 | RL Instruction

The RL and the next three instructions rotate bits of a data byte stored in the accumulator. In RL A instruction, the accumulator data bits are shifted towards the left by 1 bit. In this process, what was previously bit 7 (MSB) becomes bit 0 (LSB). Similarly, bit 0 (LSB) becomes bit 1, bit 1 becomes bit 2 and so on (Fig. 9.7). If initially the accumulator contained 4DH or 0100 1101B, then after the execution of the RL A instruction it would contain 9AH or 1001 1010B. As bit 7 becomes bit 0, this is designated as a type of circular rotation. No flags are affected. This is a 1-byte, one-cycle instruction with only one type of addressing mode. No other register, SFR or memory location may be used to hold the operand for the purpose of rotation like this.

Although in arithmetic, rotation has a very limited usage, in Boolean algebra this rotation plays an important role. Moreover, in control operations, this type of instruction is very helpful.

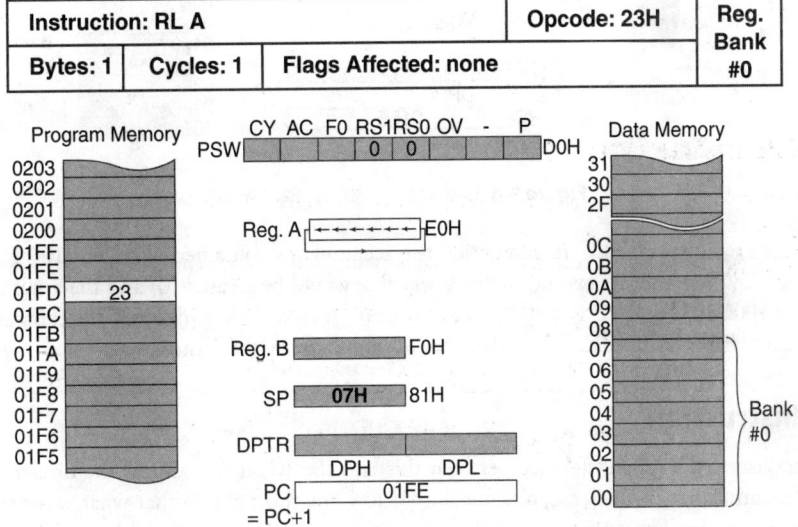

Figure 9.7 Execution of RL A instruction

In 8051, no flags are affected by circular rotation instructions (RL or RR). However, in 8085, even circular rotation of accumulator changes its carry flag. Attention is drawn to this difference for those readers who are already familiar with 8085. Moreover, for 8085, mnemonic for rotate circular would be RLC or RRC. For 8051, RLC or RRC indicates rotate through carry, *not rotate circular*.

9.8 | RLC Instruction

Instruction RLC is very similar to the previous instruction, i.e. RL, with only one difference. As in the previous instruction, in this case too the content of the accumulator is rotated towards the left by 1 bit. But here bit 7 (MSB) from the accumulator is shifted to the location of the Carry flag within the PSW register and the previous content of the Carry flag is shifted to bit 0 (LSB) of the accumulator. That is the reason for its nomenclature 'Rotate Left through Carry' (RLC). Fig. 9.8 illustrates the operation of the 1-byte, one-cycle instruction, which does not affect any other flags except the Carry flag.

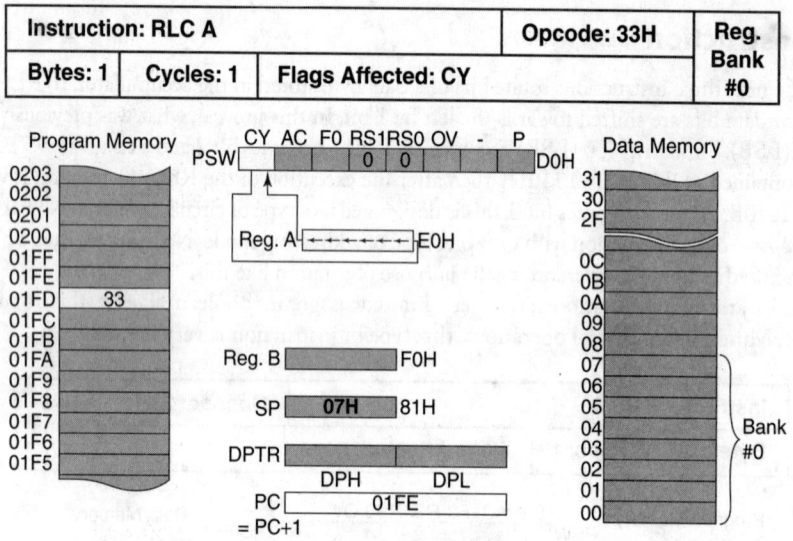

Figure 9.8 Execution of RLC A instruction

If before the execution of RLC A instruction the accumulator contained 4DH or 0100 1101B and the Carry flag is set (1), then after its execution the Carry flag would be cleared (0) and the accumulator would contain 9BH or 1001 1011B. RLC is useful in many operations, like 16-bit addition. The carry generated and stored in the Carry flag may be placed at LSB of the accumulator by this instruction.

9.9 | RR Instruction

As its name indicates, the only difference between this and the RL instruction is the direction of shifting of bits of the accumulator. In this case, bits are shifted towards the right so that what was previously bit 7 becomes bit 6, and so on. The previous bit 0 (LSB) comes out in this process and is placed as bit 7 (MSB). No flags are affected by this 1-byte, one-cycle instruction. Fig. 9.9 illustrates the execution of this instruction.

If before the execution of RR A instruction the accumulator contained 4DH or 0100 1101B, then after its execution the accumulator would contain A6H or 1010 0110B.

9.10 | RRC Instruction

Fig. 9.10 illustrates the execution of the last of these four 'Rotate accumulator' operations, which is termed Rotate Right through Carry (RRC). Only the Carry flag is affected by this operation; it is occupied by the previous bit 0 (LSB) content of the accumulator. The previous content of the Carry flag is shifted to bit 7 (MSB) of the accumulator. All other bits of the accumulator are shifted 1 bit right.

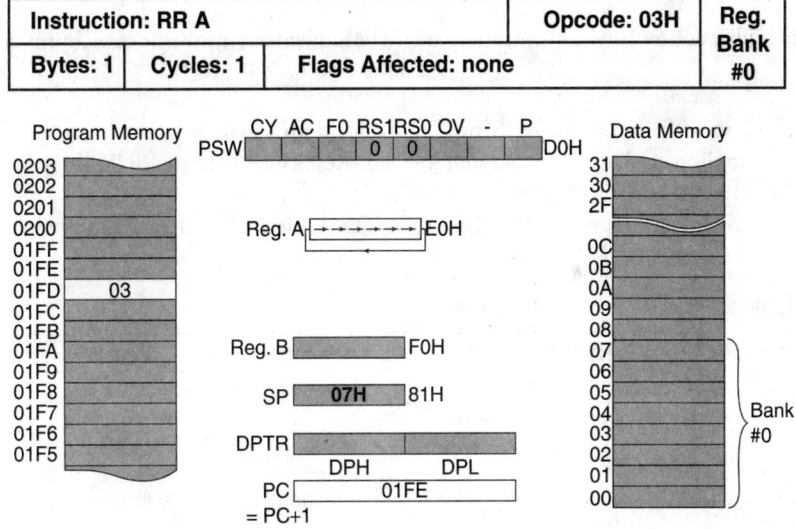

Figure 9.9 Execution of RR A instruction

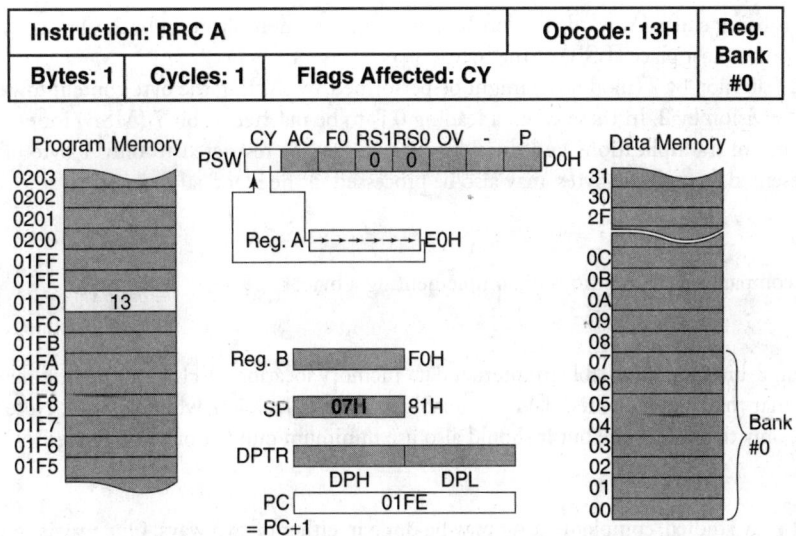

Figure 9.10 Execution of RRC A instruction

If before the execution of RRC A instruction the accumulator contained 4DH or 0100 1101B and the Carry flag is cleared (0), then after its execution the Carry flag would be set (1) and the accumulator would contain 26H or 0010 0110B. Incidentally, multiplication and division operations by 2 are easier to implement through these rotate instructions. This point is discussed in Example 9.1 of the following section of the solved examples.

9.11 | Solved Examples

Example 9.1

Purpose: To investigate about extra outcomes of rotate instructions.

Problem

Discuss how multiplication and division operations might be implemented, in some cases, by rotate instructions.

Solution

Many times we are to multiply an integer N by 2, 4, 8 and so on. Assuming that N is equal to 9, different values of these multiplications are presented in Table 9.5. Results of these multiplications are presented in their decimal, hexadecimal and binary representations.

If we observe the binary representation (fourth column), then we find that the original binary value

Table 9.5 Successive multiplication by 2

Number	Decimal	Hex	Binary	Remarks
N	9	09H	00001001B	
2N	18d	12H	00010010B	Shift left one bit
4N	36d	24H	00100100B	Shift left one bit
8N	72d	48H	01001000B	Shift left one bit
16N	144d	90H	10010000B	Shift left one bit

(00001001B) is being shifted towards the left by 1 bit for every multiplication by 2. On each shift, a zero is inserted to the rightmost place (LSB) of the byte.

Similarly, a division by 2 (modulo 2) might be performed, by shifting the byte content towards the right, 1 bit for every division by 2. In these cases, a leading 0 is to be inserted at bit 7 (MSB) for every right shift. These techniques of multiplications and divisions by two are not restricted to only 1 byte of data. Larger integers, represented by multiple bytes, may also be processed in the identical manner.

Example 9.2

Purpose: To compare several methods of complementing a byte.

Problem

Assume that an 8-bit data is available in internal data memory location 7FH. Find the best method to complement it so that the complemented data is available in its own location, which is 7FH. Note that the best method is not only the fastest one but it should also use minimum number of coding bytes.

Solution

As we have already studied, complementing may be done in either of two ways. One way is to copy the data in the accumulator, then complement it by CPL A instruction and, finally, store back the data to its original place. Alternatively, the data itself can be XORed with FFH using XOR direct, #0FFH instruction. We write instructions necessary for all possible methods and then evaluate and compare.

Method A

Load the data in the accumulator, complement it and then store it back to its original place. These instructions take 5 bytes of coding and need three cycles to be executed.

```
START:  MOV  A, 7FH      ; get data in the accumulator
        CPL  A           ; complement it
        MOV  7FH, A      ; store back complemented data
```

Method B

XOR FFH with the data in its own address. The result would be available in the original address itself. This instruction needs 3 bytes and two cycles to be executed.

START: XRL   7FH,#0FFH   ; complement the data and store result at 7F

Method C

Load the accumulator by FFH, and using indirect addressing, XOR the data with the accumulator. Then save it back to its original location by the same indirect addressing. These instructions need 6 bytes and four cycles to be executed.

```
START:   MOV   R0,#7FH    ; load pointer
         MOV   A,#0FFH
         XRL   A,@R0
         MOV   @R0,A       ; store result
```

Method D

Load the accumulator by FFH and then XOR the data with the accumulator so that the result is available at the original address of the data. It needs two instructions. These two instructions need 4 bytes and two cycles to be executed.

```
START:   MOV   A,#0FFH
         XRL   7FH,A       ; result in location 7FH
```

We now present the summary of these four methods through Table 9.6.

A comparison of these methods indicates that Method B is optimum, taking 3 bytes and two cycles.

Table 9.6 Summary of four methods

	No. of bytes	No. of cycles	Remarks
Method-A	5	3	
Method-B	3	2	Optimum
Method-C	6	4	
Method-D	4	2	

Example 9.3

Purpose: To compare two methods of subtraction.

Problem

Write a program to subtract an unsigned 8-bit integer in R2 from another unsigned 8-bit integer in R3. The result should be available in the accumulator. Assume that register bank #0 is active.

Solution

This problem may be solved by two methods.

Method A

We may use the SUBB instruction directly. To clear the Carry flag we may load PSW by 00H. The program would have the following three instructions:

```
START:   MOV   A,R3
         MOV   PSW,#00H    ; clear Carry flag
         SUBB  A,R2
```

Method B

Alternatively, we may calculate two's complement of the integer to be subtracted, after loading it in the accumulator; it may then be added with the other integer. The program would need the following four instructions:

```
START:    MOV    A, R2
          CPL    A
          INC    A
          ADD    A, R3
```

SUMMARY

MCS-51 offers 10 mnemonics for logical operations out of which nine are discussed in this chapter. Instructions are available for byte operators for AND, OR and XOR operations on using four addressing modes. In these cases, apart from the accumulator, the directly addressed locations are also capable of holding the results.

Four instructions are available to rotate the accumulator content by 1 bit to the right or to the left. These operations may either avoid or incorporate the Carry flag and as a practice used for checking the status of multiple bits of a byte, especially in control operations. Apart from other purposes, these instructions may also be used to multiply or divide any number by 2. Finally, instructions CPL A and CLR A are also available for complementing and clearing the accumulator.

POINTS TO REMEMBER

- Any directly addressable memory location, including SFRs, can hold the result of any logical operation such as AND, OR, XOR.

- Execution of RL A and RR A rotates the accumulator content circular, and the Carry flag remains unaffected during the execution of these two instructions.

REVIEW QUESTIONS

Evaluate Yourself

1. The AND instruction is generally used to

   (a) Set a few bits

   (b) Clear a few bits

   (c) Complement a few bits

   (d) None of these

2. The OR instruction is generally used to

   (a) Set a few bits

   (b) Clear a few bits

   (c) Complement a few bits

   (d) None of these

3. The XOR instruction is generally used to

   (a) Set a few bits

   (b) Clear a few bits

   (c) Complement a few bits

   (d) None of these

4. Which of the following instructions, when executed, would leave the accumulator unchanged?

   (a) ANL A, #0FFH   (b) ORL A, #00H

   (c) Either of these   (d) None of these

5. Execution of which one of the following instructions affects the Carry flag?

   (a) RR A   (b) CPL A

   (c) CLR A   (d) None of these

6. Assuming that Carry flag is set, which of the following instructions, when executed, would double the accumulator content if its original content is less than 127d?

   (a) RL A   (b) RLC A

   (c) RR A   (d) None of these

7. Assuming accumulator contains 80H and the Carry flag is cleared, which of the following instructions, when executed, would set the Carry flag and clear the accumulator?

   (a) RRC A   (b) RLC A

   (c) RL A   (d) None of these

8. Which of the following instructions, when executed four times, would interchange (swap) the nibbles of the accumulator?

   (a) RR A   (b) RL A

   (c) Either of these   (d) None of these

9. Instead of using the instruction ANL A, #00H, which of the following instructions may be used to get the same effect?

   (a) CPL A   (b) CLR A

   (c) Either of these   (d) None of these

10. Instead of using the instruction XRL A, #0FFH, which of the following instructions may be used to get the same effect?

    (a) CPL A   (b) CLR A

    (c) Either of these   (d) None of these

Search for Answers

1. How many mnemonics for logical operations are offered by MCS-51? What are they?

2. What are the addressing modes offered for AND operations by MCS-51?

3. What is the difference between XRL A, direct and XRL direct, A instructions?

4. From where would the data be used (output latch or input pins) if ORL A, P0 instruction is executed?

5. How many ORL instructions of MCS-51 offer immediate addressing mode?

6. What is the difference between RL A and RLC A instructions?

7. How many flags are affected by RRC A instruction?

8. What is the easiest way to divide a number in the accumulator by 4?

9. Which bits of the accumulator would contain 0 after executing the following instruction?

   ANL A, #88H

10. How many bits of the accumulator would be toggled after execution of the instruction XRL A, #0FFH?

Think and Solve

1. What would be the content of bit 3 of the accumulator after execution of RL A instruction if the accumulator originally contained 29H?

2. Is it possible to implement NOR and NAND operations through MCS-51 instruction set? Justify your answer.

3. Which instruction you may directly use to toggle the previous status of an output port?

4. How many flags would be affected by the execution of XRL PSW, A instruction?

5. What would be the accumulator content after executing the following instructions?

   START: CLR   A

                CPL   A

6. What would be the accumulator content after execution of the XRL A, 0E0H instruction?

7. Which of the following two instructions may be used to find the input of bit 0 (LSB) of a port after reading the port and placing its value in the accumulator? Justify your answer.

   RLC A   RRC A

8. What would be the content of the accumulator after execution of the following program?

   START: CPL   A

                RR   A

                CPL   A

                RL   A

9. What is the purpose of the following sub-routine?

   START: MOV   R7, #04H

   LOOP: RL   A

                DJNZ   R7, LOOP

                RET

10. How can the XRL operation be implemented using any combination of ANL, ORL and CPL instructions?

10 BOOLEAN VARIABLE MANIPULATION

CHAPTER OBJECTIVES

In the previous chapter, it was mentioned that apart from manipulating byte variables, MCS-51 can also manipulate bit variables. This chapter discusses the instructions dealing with Boolean variable manipulation. After completion of the chapter, the reader should be able to understand

- Instructions related with Boolean data-load and data-move operations.
- Program-branching instructions related to Boolean variables.
- Logical operations possible with Boolean variables.
- Application areas of some of these instructions.

10.1 | Introduction

All the instructions we have discussed so far are related to byte variables. However, MCS-51 also offers a bunch of instructions, which are capable of manipulating bit variables. These instructions are extremely useful in control operations where informations are generally received bit-wise and commands are also sent in the similar way. For example, it needs only 1 bit to store the information received from a light-dependent resistor (LDR) indicating whether it is light or dark ambient. Similarly, a pump may be switched on or off through a single bit command.

Mnemonics related to Boolean variable operations offered by MCS-51 are presented in Table 10.1. The first 3 mnemonics are related to data load and movement. The next five deal with program-branching

Table 10.1 Mnemonics related to Boolean variable manipulation

Mnemonic	Brief description
CLR	Clear bit
SETB	Set bit
MOV	Move bit
JC	Jump if carry
JNC	Jump if no carry
JB	Jump if bit is set
JNB	Jump if bit is not set
JBC	Jump if bit is set and clear bit
ANL	Logically AND bit with CY
ORL	Logically OR bit with CY
CPL	Complement bit

operations. The last three offer the scope of logical operations. In subsequent sections of this chapter, we will discuss all these instructions in detail, except JBC, which would be discussed in Chapter 12.

10.1.1 | Source and Destination of Boolean Operations

In case of byte variables, it is the accumulator which stores the result of most of the operations. For bit variables, it is the carry (CY) flag of PSW, which serves the same purpose. However, bit variables may be stored or manipulated only from a limited area of internal RAM. Either it should be available within the bit-addressable area of the internal RAM or within the bit-addressable SFRs. Just like the addresses of byte locations of the internal RAM, these bits also have their unique addresses as already indicated before (Fig. 2.9). In general, it may be noted that addresses of bits within the bit-addressable memory location vary from 00H to 7FH, and addresses of bits located within the bit-addressable SFRs vary from 80H to FFH, with some discontinuity.

Fig. 10.1 shows all bit-addressable SFRs with their bit addresses. Note that no addressable bits are present within bit addresses between C0H and C7H, D8H and DFH, E8H and EFH and F8H and FFH. As already mentioned, addressing these non-existent bits would produce unpredictable results.

B | F7 | F6 | F5 | F4 | F3 | F2 | F1 | F0 | F0

ACC | E7 | E6 | E5 | E4 | E3 | E2 | E1 | E0 | E0

PSW | D7 | D6 | D5 | D4 | D3 | D2 | D1 | D0 | D0

T2CON | CF | CE | CD | CC | CB | CA | C9 | C8 | C8

IP | BF | BE | BD | BC | BB | BA | B9 | B8 | B8   P3 | B7 | B6 | B5 | B4 | B3 | B2 | B1 | B0 | B0

IE | AF | AE | AD | AC | AB | AA | A9 | A8 | A8   P2 | A7 | A6 | A5 | A4 | A3 | A2 | A1 | A0 | A0

SCON | 9F | 9E | 9D | 9C | 9B | 9A | 99 | 98 | 98   P1 | 97 | 96 | 95 | 94 | 93 | 92 | 91 | 90 | 90

TCON | 8F | 8E | 8D | 8C | 8B | 8A | 89 | 88 | 88   P0 | 87 | 86 | 85 | 84 | 83 | 82 | 81 | 80 | 80

Figure 10.1 Bit-addressable SFRs with bit addresses (in Hex)

 Remember that for Boolean operations, carry flag (CY) of 8051 serves the purpose of accumulator.

10.2 | Boolean Data Loading and Movement

MCS-51 offers three mnemonics for bit-wise data load and movement: CLR, SETB and MOV. These are discussed as follows.

10.2.1 | CLR Instruction

Table 10.2 lists the variations of CLR instruction. Possible operand locations in this case would be either the CY flag or any addressable bit. This instruction clears the addressed location (a bit) by loading it with 0.

Table 10.2 Variations of CLR instruction

Instruction	Function
CLR C	Clear carry
CLR bit	Clear addressed bit

CLR C is a 1-byte, 1-cycle instruction to clear the CY flag. No other flags are affected. Irrespective of the previous condition of the CY flag, after execution of this instruction, it would contain 0 as illustrated in Fig. 10.2.

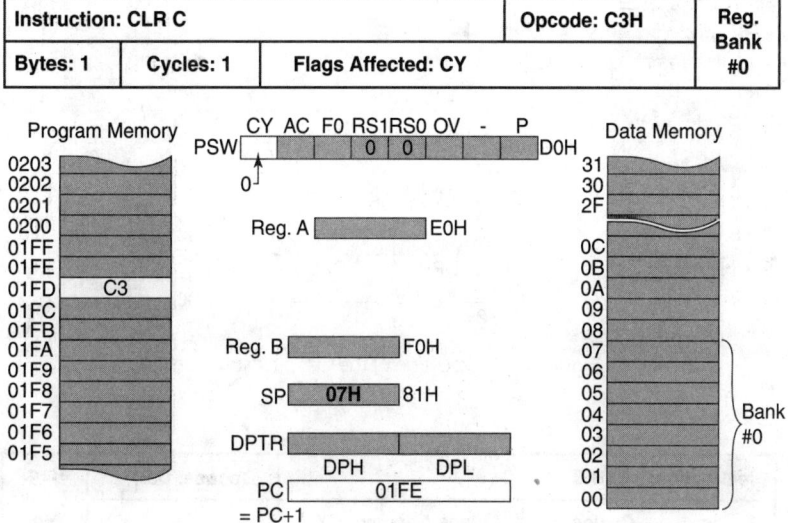

Figure 10.2 Execution of CLR C instruction

CLR bit is a 2-byte, 1-cycle instruction and loads the bit with 0, which is addressed by the second byte of the instruction. Fig. 10.3 illustrates the execution of this instruction. For example, the least significant bit of bit-addressable byte 2FH is shown, the bit address of which would be 78H. If CLR 78H instruction is executed (opcode C2H) then this bit would contain 0. No other bits or any flags would be affected. It may be noted that any port bit (output latch) may also be cleared by this instruction, as all port-SFRs are bit-addressable. CLR 0D7H is a 2-byte counterpart of CLR C instruction as the address of CY flag is D7H.

10.2.2 | SETB Instruction

Instead of loading a bit by 0, this instruction loads it by 1. Either the CY flag or any addressable bit may be the destination, as indicated in Table 10.3.

Execution of the instruction SETB C is illustrated in Fig. 10.4. Execution of this 1-byte instruction sets the CY flag as 1. No other flags are affected.

Table 10.3 Variations of SETB instruction

Instruction	Function
SETB C	Set carry as 1
SETB bit	Set addressed bit as 1

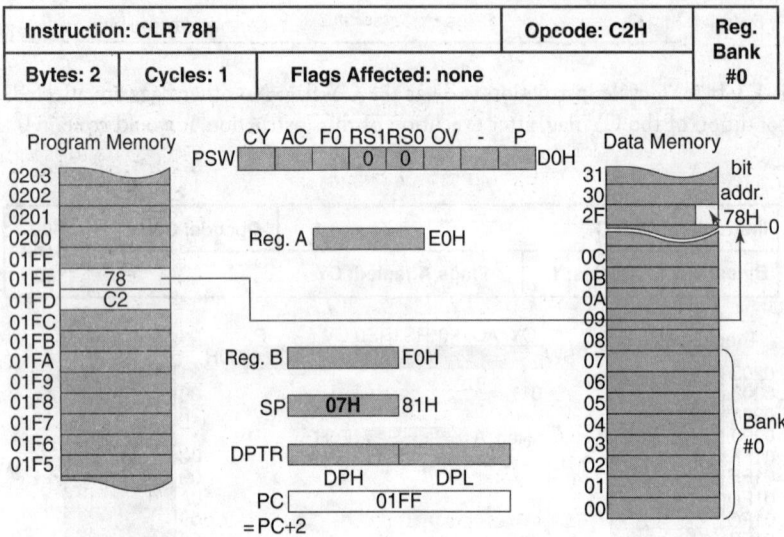

Figure 10.3 Execution of the CLR bit instruction

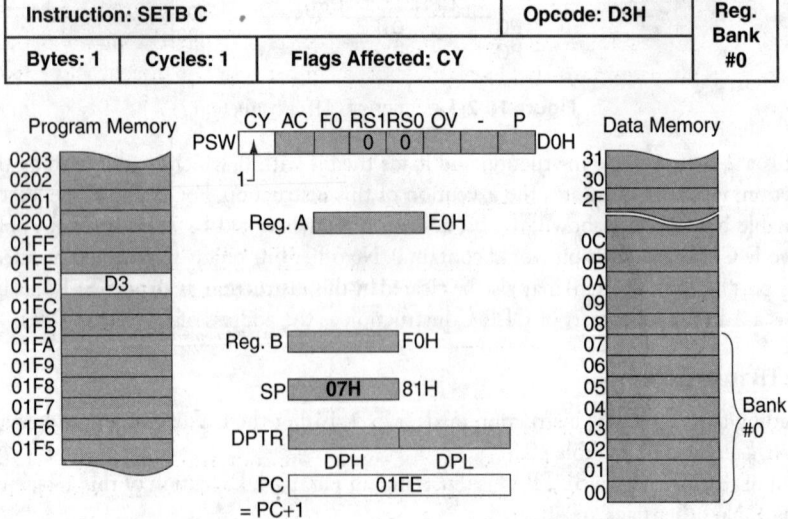

Figure 10.4 Execution of SETB C instruction

SETB bit is a 2-byte instruction to load any directly addressable bit by 1. Execution of this instruction, for example, is shown in Fig. 10.5, with a bit address of 78H within bit-addressable area. After execution of the SETB 78H instruction, this bit would contain 1. No other bits or any flags would be affected. Note that the bit address becomes the second byte of the instruction.

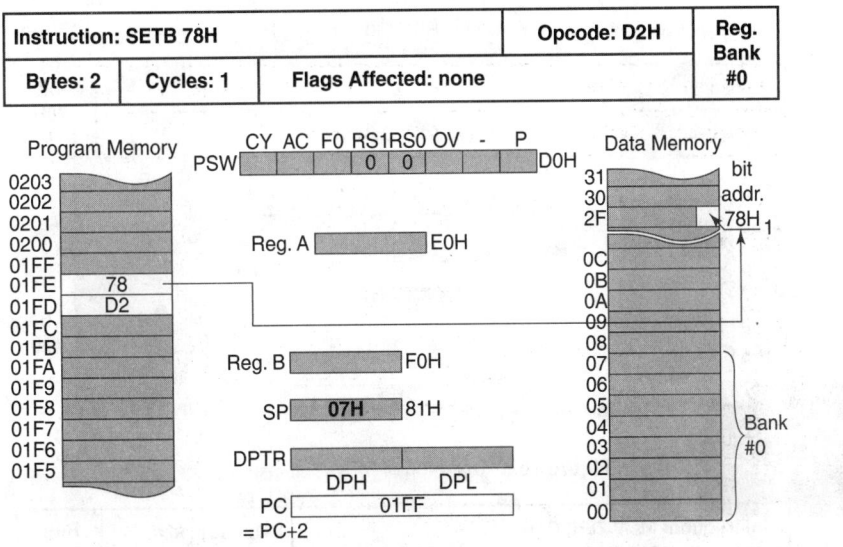

Figure 10.5 Execution of SETB bit instruction

10.2.3 | MOV Instruction

Table 10.4 lists the instructions available for bit-variable movements. Note that all transactions are either to or from the CY flag, making it either a destination or a source. On the other hand, any directly addressed bit from SFR or bit-addressable area may serve the purpose of sending or receiving the data. It may also be noted that there is no instruction which allows data movement between two directly addressed bits as source and destination.

Table 10.4 Variations of MOV instruction

Instruction	Function
MOV C, bit	Copy addressed bit to carry
MOV bit, C	Copy carry to addressed bit

Execution of MOV C, bit instruction is illustrated through Fig. 10.6. For example, a bit within the bit-addressable area of internal RAM is taken as the source. The address of this bit is 78H. The instruction is of 2 bytes and takes 1 cycle to be executed. The second byte of the instruction stores the bit address. No flags, except CY flag, are affected. If the addressed bit is an I/O port bit, then data would be taken from the input pin, and not from the output latch.

Fig. 10.7 illustrates execution of the MOV bit, C instruction. Note that unlike its counterpart, this instruction takes 2 cycles to be executed. This is a 2-byte instruction and does not affect any flag.

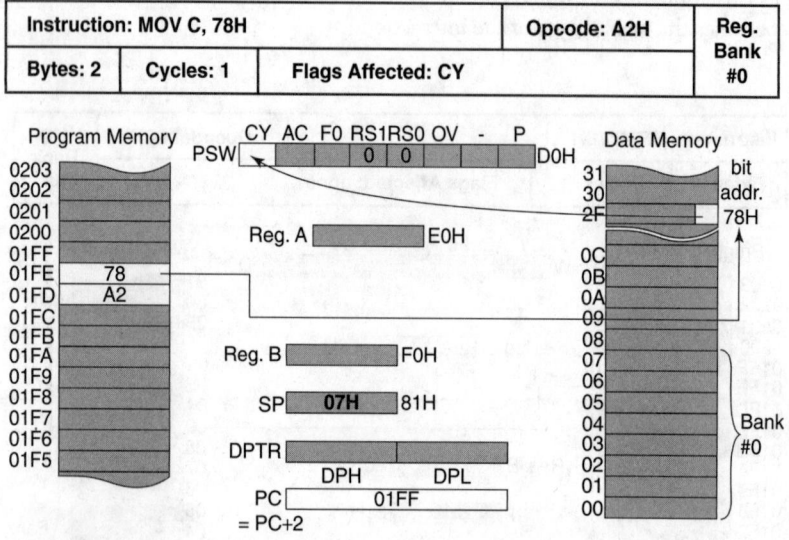

Instruction: MOV C, 78H			Opcode: A2H	Reg. Bank #0
Bytes: 2	Cycles: 1	Flags Affected: CY		

Figure 10.6 Execution of MOV C bit instruction

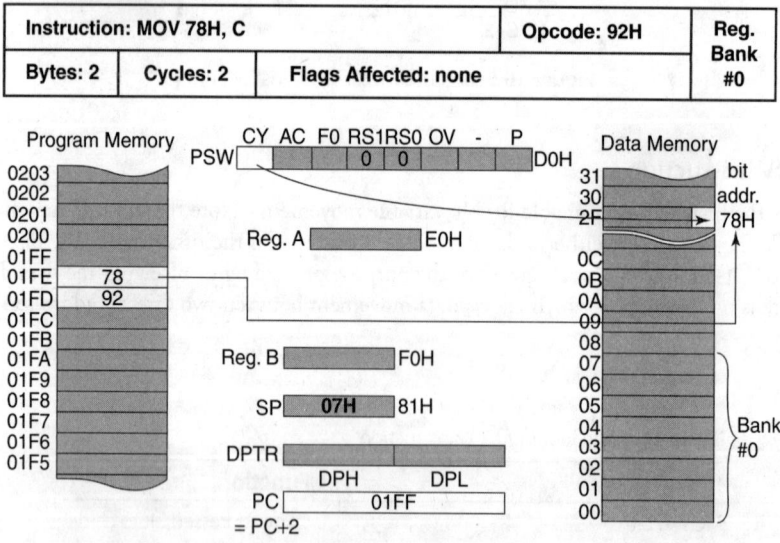

Instruction: MOV 78H, C			Opcode: 92H	Reg. Bank #0
Bytes: 2	Cycles: 2	Flags Affected: none		

Figure 10.7 Execution of MOV bit, C instruction

It is a common mistake for beginners of 8051 to coin up instructions like:

MOV 18H, 0E0H

to copy accumulator's least significant bit to bit-addressable area's bit address 18H. Although for byte variables, 8051 allows both source as well as destination to be directly addressed, for bit variables such a type of addressing is *not* allowed.

10.3 | Bit-Oriented Program Branching

In Chapter 6, we have discussed all conditional program-branching instructions of MCS-51, which are byte related. Few more similar instructions, such as JC, JNC, JB, JNB and JBC, are offered by MCS-51, which are bit oriented. By the term 'bit' we mean either the CY flag or any *addressable* bit. All of these instructions use the relative jumping mode, similar to the SJMP instruction, as discussed in Chapter 6. Out of these five instructions, JBC would be discussed in Chapter 12.

10.3.1 | JC Instruction

JC or 'Jump if Carry is set' is a 2-byte, 2-cycle instruction, which performs a relative branching operation with respect to the current value of the program counter (PC), as per the condition of the CY flag. The offset byte, used to

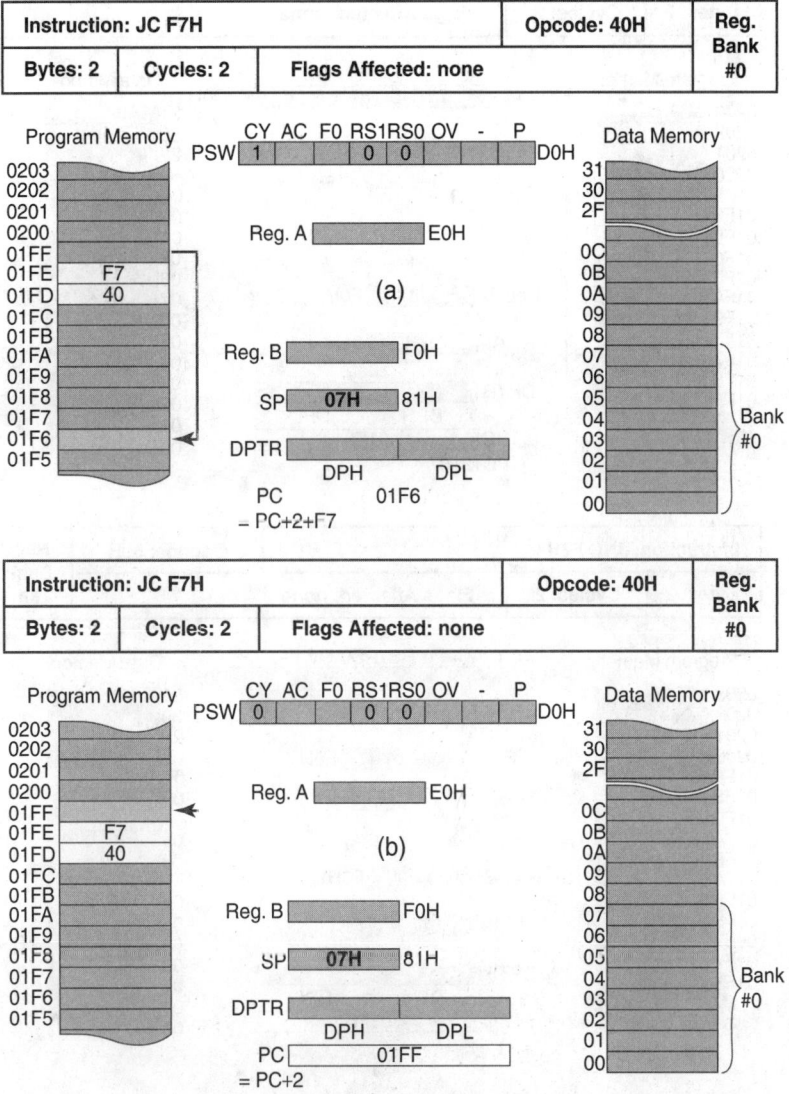

Figure 10.8 Execution of JC instruction (a) if CY = 1 and (b) if CY = 0

calculate the branching address, is the second byte of the instruction. If CY is set [Fig. 10.8(a)], then this branching takes place. Otherwise, the next instruction after it is executed [Fig. 10.8(b)]. Note that the condition of the CY remains unchanged after execution of the instruction. The jump may be forward or backward depending on the value in the second byte of the instruction. If it is a forward one then it is within 127 bytes. Otherwise, it is within 128 bytes, starting from the next address after the JC instruction.

10.3.2 | JNC Instruction

JNC or 'Jump if No Carry' performs almost in identical way as its counterpart JC. The only difference is the condition for branching which is opposite of JC. Figs 10.9(a) and (b) illustrate both of these conditions of JNC instruction.

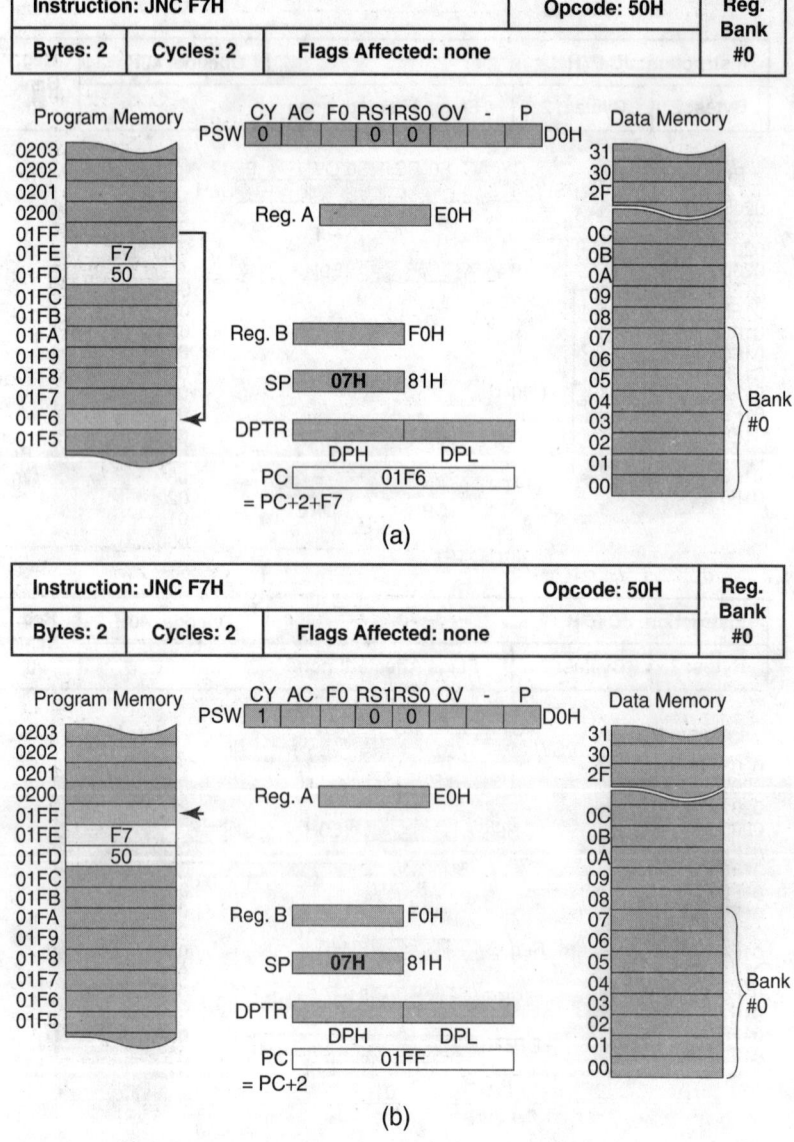

Figure 10.9 Execution of JNC instruction (a) if CY = 0 and (b) if CY = 1

10.3.3 | JB Instruction

JB or 'Jump if Bit is set' checks the bit addressed by the second byte of the instruction and if the condition is true then the jump address is calculated by adding the third byte of the instruction after incrementing PC to the first byte of the next instruction [Figs 10.10(a) and (b)]. The addressed bit may either belong to the bit-addressable area of RAM or within any bit-addressable SFR. The condition of the addressed bit remains unchanged after execution of the instruction.

As PSW is a bit-addressable SFR, this instruction is also applicable for all bits of SFR. That means, all four flags of SFR, namely CY, AC, OV and P, may be used for conditional branching through this instruction. However, application of JC or JNC would be better against JB for CY flag as they are 3-byte instructions.

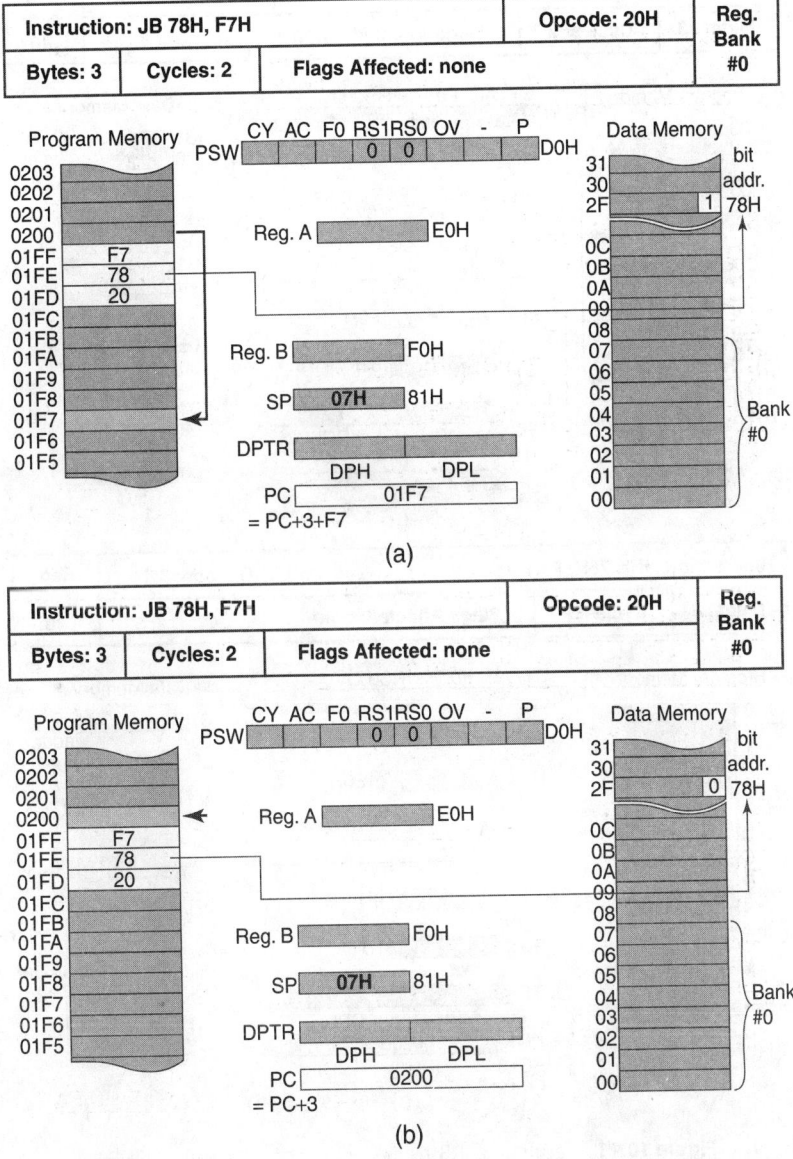

Figure 10.10 Execution of JB instruction (a) if bit = 1 and (b) if bit = 0

If the bit of any I/O port is addressed by this instruction, then the data is taken from the input pin of the port, and not from its output latch.

10.3.4 | JNB Instruction

JNB or 'Jump if Not Bit set' is identical with JB with the difference in performing the jump as per opposite bit conditions. In this case, a branching would take place if the addressed bit is zero. Otherwise, it is similar to the JB instruction in all respects. If any I/O port bit is addressed, then the data would be taken from its input pin, and not from its output latch. Figs 10.11(a) and (b) illustrate the execution of JNB instruction for either bit conditions.

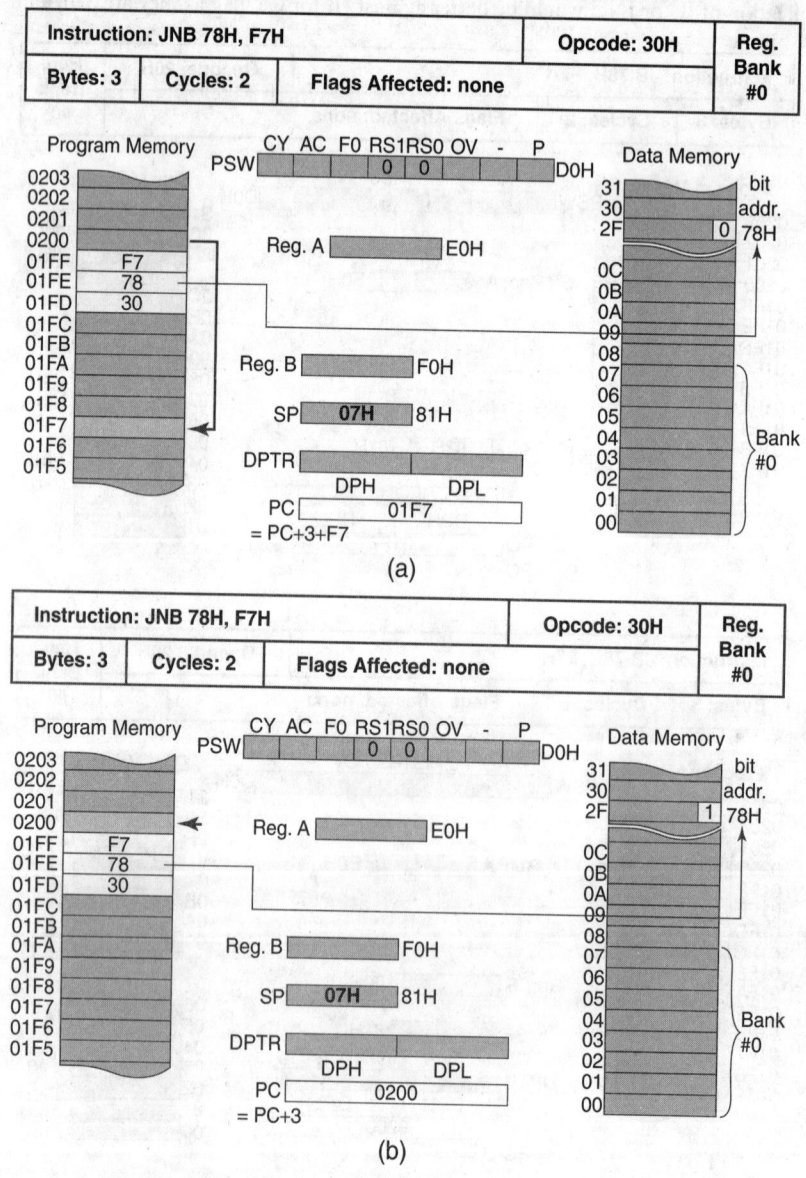

Figure 10.11 Execution of JNB instruction (a) if bit = 0 and (b) if bit = 1

10.4 | Bit-Oriented Logical Operations

Three logical operations are allowed with bit variables: AND, OR and NOT. When two operands are used, one of them must be within the CY flag, which would also accommodate the result of the operation. The other may be in any addressable bit as already explained.

10.4.1 | ANL Instruction

Two variations of logical AND operation are presented in Table 10.5. The first one performs a standard AND operation with CY flag and the addressed bit. This is illustrated through Fig. 10.12. This is a 2-byte, 2-cycle instruction and does not affect any other flags except the CY flag, which stores the result of the AND operation.

Table 10.5 Variations of ANL instruction

Instruction	Function
ANL C, bit	Logically AND carry with addressed bit
ANL C, /bit	Logically AND carry with complement of addressed bit

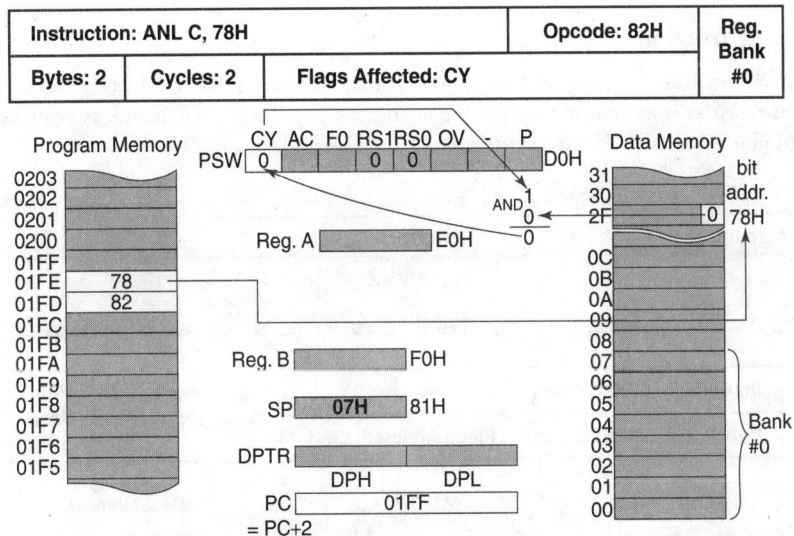

Figure 10.12 Execution of ANL C, bit instruction

The second byte of the instruction stores the address of the bit. As an example case, the least significant bit of memory location 2FH is taken, the bit address of which is 78H. Assuming that this bit contains 0 and CY flag 1, a logical AND operation would produce 0, which would be available in the CY flag at the end of the execution of the instruction. When any I/O port bit is addressed, data would be collected from its input pin, and not from the output latch.

The other version of ANL also works in identical way except that it complements the addressed bit before performing the logical AND operation. Original condition of the addressed bit remains unchanged and the result of the operation is available in the CY flag. Fig. 10.13 illustrates this operation. As we may observe from this diagram, original content of the addressed bit, that is 0, is complemented to 1 and then the AND operation with CY was performed. Assuming CY to be 1, the result 1 is stored in the CY flag itself. No other flags are affected by this 2-byte, 2-cycle instruction.

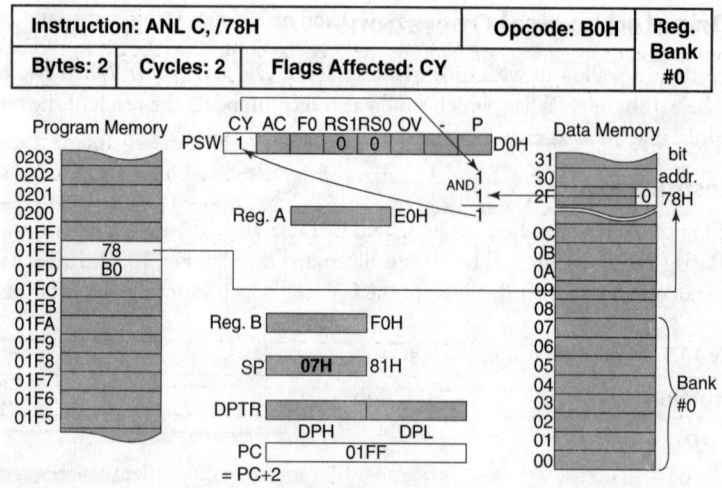

Figure 10.13 Execution of ANL C, /bit instruction

10.4.2 | ORL Instruction

Just like the AND instructions, logical OR operation may also be performed in two ways, either with the directly addressed bit as it is, or inverting the bit, as indicated in Table 10.6. Result is again available in CY flag. Fig. 10.14 illustrates ORL C, bit instruction with the bit address of 78H.

Table 10.6 Variations of ORL instruction

Instruction	Function
ORL C, bit	Logically OR carry with addressed bit
ORL C, /bit	Logically OR carry with complement of addressed bit

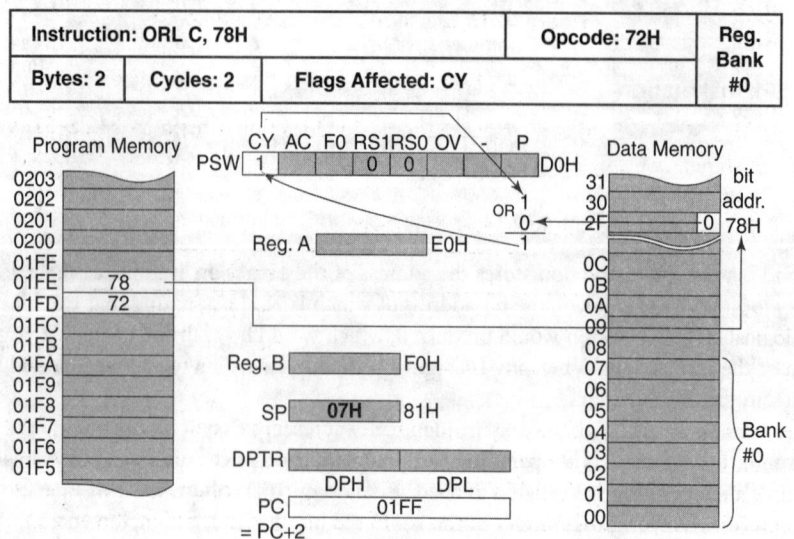

Figure 10.14 Execution of ORL C, bit instruction

Assuming that the original content of this bit is 0 and of CY flag 1, this instruction would store 1 in the CY flag after completion of its execution. Neither the original content of the bit would be changed nor would any other flags be affected. If any port pin is addressed then the data would be collected from its input pin, and not from its output latch.

Fig. 10.15 illustrates the other version of ORL instruction, which complements the addressed bit before performing the OR operation. If the original content of the addressed bit is 0 and CY flag contained 1 then the instruction would leave 1 in CY flag after its execution. No other flags would be affected by this 2-byte, 2-cycle instruction. In case of a directly addressed bit of any I/O port, data would be collected from its input pin, and not from its output latch.

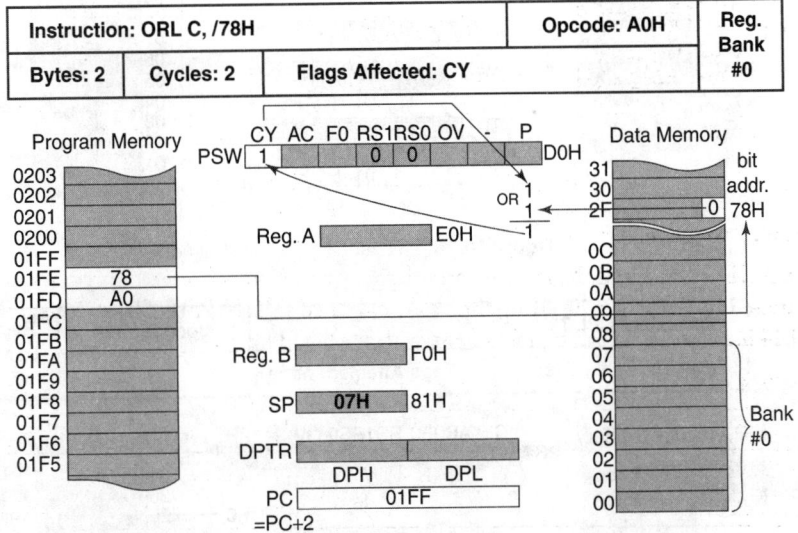

Figure 10.15 Execution of ORL C, /bit instruction

10.4.3 | CPL Instruction

Complementing a bit is the third bit-oriented logical instruction offered by MCS-51. Variations of this instruction are shown in Table 10.7.

Table 10.7 Variations of CPL instruction

Instruction	Function
CPL C	Complement carry
CPL bit	Complement addressed bit

The first of these two, CPL C instruction, complements original content of the CY flag without changing the status of any other flag. This 1-byte instruction takes 1 cycle to get executed as illustrated in Fig. 10.16.

Fig. 10.17 shows the execution of CPL bit instruction using an assumed bit address, 78H. If the original content of this bit is 0, after completion of execution of the 2-cycled instruction, it would contain 1. The second byte of this 2-byte instruction contains address of the bit to be complemented. No other flags are affected. If any I/O port pin is addressed then the data would be collected from its output latch, and not from its input pin.

Instruction: CPL C			Opcode: B3H	Reg. Bank #0
Bytes: 1	Cycles: 1	Flags Affected: CY		

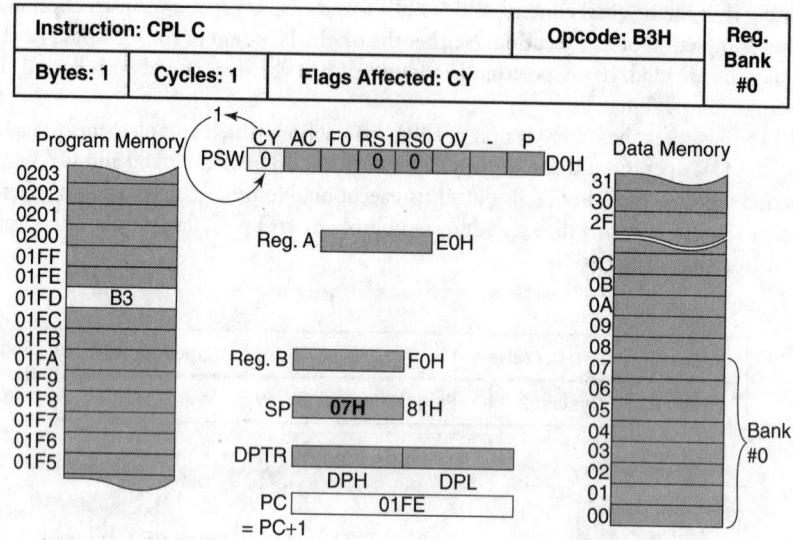

Figure 10.16 Execution of CPL C instruction

Instruction: CPL 78H			Opcode: B2H	Reg. Bank #0
Bytes: 2	Cycles: 1	Flags Affected: none		

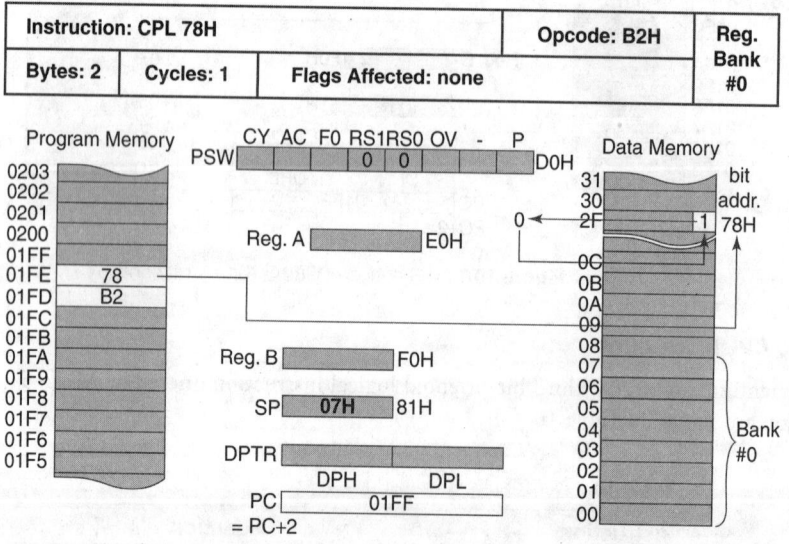

Figure 10.17 Execution of CPL bit instruction

In many a cases during intermediate stages of operations with Boolean variables, it is necessary to store some temporary results in form of one bit only. For beginners it is recommended that two bits available in PSW (bit 1 and bit 5) may be used for this purpose. These two bits may be addressed as PSW.1 and PSW.5, respectively. This scheme relieves the programmer from the burden of allowing extra space within bit-addressable area of RAM.

10.5 | Solved Examples

Example 10.1

Purpose: To keep a sub-routine ready for XORing 2 bits as no single instruction is available. It also shows the usage of ANL C,/bit and ORL C,/bit instructions.

Problem

Develop a sub-routine to XOR 2 directly addressable bits and store the result in CY flag. Assume the bit addresses to be 78H and 79H, and condition of these 2 bits should remain unchanged.

Solution

As direct XOR with bit variables is not available in the instruction set of MCS-51, we may use ANL C,/bit instruction twice by interchanging bits and then logically OR the results. A sample sub-routine is presented below, which uses bit 1 of PSW as temporary storage.

; Program to find XOR of two bit variables located in 78H and 79H, respectively.
; The result is available in CY flag. Destroys bit 1 of PSW.

```
XORBIT: MOV   C, 78H        ; get 1 bit
        ANL   C, /79H       ; AND with complement of other bit
        MOV   0D1H, C       ; temp save within PSW.1
        MOV   C, 79H        ; get the other bit
        ANL   C, /78H       ; AND with complement of the bit
        ORL   C, PSW.1      ; OR with previous result in PSW.1
        RET                 ; result in CY flag
```

Example 10.2

Purpose: To indicate how hardware gates may be substituted through software.

Problem

Develop a subroutine to turn off a heater by outputting 0 through P3.4 if and only if the inputs from bit 0, 2 and 5 of port 1 are 0, 0 and 1, respectively.

Solution

Schematically, the problem is presented through Fig. 10.18. To implement this logic, following sub-routine may be used.

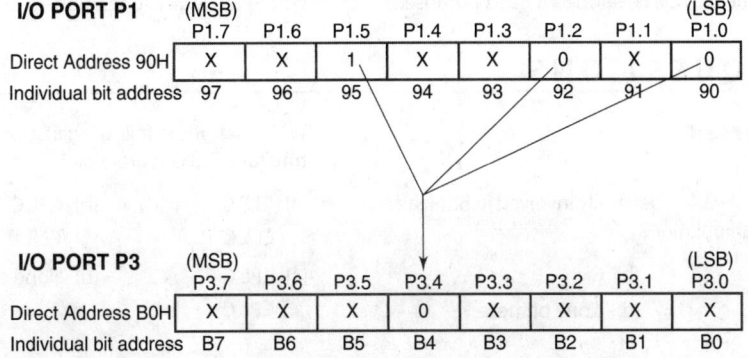

Figure 10.18 Schematic representation of the problem of Example 10.2

; Program to clear P3.4 if P1.0 and P1.2 are both 0 and P1.5 is 1.

```
START:  SETB  C            ; load CY by 1
        ANL   C,/P1.0      ; CY is 0 if P1.0 is 1
        ANL   C,/P1.2      ; CY is 0 if P1.2 is 1
        ANL   C,P1.5       ; CY is 0 if P1.5 is 0
        CPL   C            ; complement the result
        MOV   P3.4,C       ; output to P3.4
        RET
```

Just like the expressions of Boolean expressions may be written in different ways, similarly these programs may also be developed in different manners. The reader may write another version which might be equally effective as the present one.

SUMMARY

MCS-51 instructions that are capable of manipulating Boolean variables are discussed in this chapter. Three instructions, namely CLR, SETB and MOV, are available for data load and movements. All bit movements are allowed either from or to the CY flag, making it either the source or destination. The other side may be any directly addressable bit from bit-addressable internal RAM location or bit-addressable SFRs.

Conditional-branching instructions, such as JC, JNC, JB and JNB, are available to perform relative jumps of forward 127 bytes or backward 128 bytes from the present PC position, which is always the next byte after the branching instruction. There is another such instruction, JBC, which would be discussed in Chapter 12.

Three logical instructions (ANL, ORL and CPL) may be used for logical operations between CY flag and any addressed bit. Versions of ANL and ORL also allow to complement the directly addressed bit before AND or OR with the CY flag.

POINTS TO REMEMBER

- Instructions for direct XOR operation with bit variables are not available in 8051 instruction set and a sub-routine may be developed for this purpose.

- Versions of AND and OR instructions also allow the bit variable to be or not to be complemented before the logical operation.

REVIEW QUESTIONS

Evaluate Yourself

1. Which flag of PSW is generally involved in Boolean variable manipulation?

    (a) AC
    (b) OV
    (c) P
    (d) None of these

2. Which set of the following instructions is a substitute for SETB C instruction?

    (a) CLR C
        CPL C
    (b) CPL C
        CLR C
    (c) CPL C
        CPL C
    (d) None of these

3. Which of the following instruction may substitute CLR C instruction?

   (a) CLR 0D0H   (b) CLR 0D7H

   (c) CLR 0D8H   (d) None of these

4. Which byte of the JNB instruction contains the bit address?

   (a) First   (b) Second

   (c) Third   (d) None of these

5. If a JB instruction is placed at the address 0200H and the addressed bit is not set, from which address the next instruction would be executed?

   (a) 01FFH   (b) 0201H

   (c) 0203H   (d) None of these

6. Which one of the following can perform a NAND operation?

   (a) CPL C followed by ANL C, bit

   (b) ANL C, /bit

   (c) ANL C, bit followed by CPL C

   (d) None of these

7. With respect to MOV C, bit instruction, MOV bit, C instruction takes

   (a) one cycle less   (b) same cycles

   (c) one cycle more   (d) none of these

8. In which of the following conditions ANL C, bit and ANL C, /bit instructions would produce identical results for the same values of C and addressed bit?

   (a) If CY = 0   (b) If CY = 1

   (c) If bit = 0   (d) None of these

9. In which of the following conditions ORL C, bit and ORL C, /bit instructions would produce identical results for the same values of C and addressed bit?

   (a) If CY = 0   (b) If CY = 1

   (c) If bit = 0   (d) None of these

10. Which of the following instructions may replace JNC instruction?

   (a) CPL C   (b) CLR C

       JC addr   JC addr

   (c) SETB C   (d) None of these

       JC addr

Search for Answers

1. Name an application area of Boolean variable manipulation instructions. Justify your answer.

2. How many flags are affected by SETB C instruction?

3. During the execution of MOV C, P3.0, would the bit value be taken from the input pin or from the output latch?

4. What would be the condition of the CY flag after execution of the JNC instruction?

5. How many instructions discussed in this chapter are capable of reading from input pins of a port, and not from its output latch?

6. Is there any bit-move instruction available in MCS-51 specifying bit addresses of both the source and destination?

7. How many bytes are necessary to accommodate CPL bit instruction?

8. How many bits of a XX51 device may be complemented by the CPL bit instruction?

9. Is it possible to implement the ORL bit instruction through the use of ANL bit and CPL bit instructions? Justify your answer.

10. Which instruction may be used to ensure that both CY are AC flags are 0?

Think and Solve

1. How can we perform arithmetic addition between 2 bits?

2. How useful would be the instruction JC 00H? Justify your answer.

3. What would happen if the second byte of the JC instruction is specified as FEH?

4. Is it possible to implement a NOR operation using MCS-51 instruction set for Boolean variables? If yes, then how would it be possible?

5. Define XNOR and write a program to perform this operation when both inputs are available in 2 bit addresses, 00H and 01H.

6. Develop a sub-routine to turn on a heater by outputting 1 through P1.0 if and only if the input from bit 0,3 and 7 of port 3 are 1, 1 and 0, respectively.

7. Develop a program to implement the following logic with inputs from port 1.

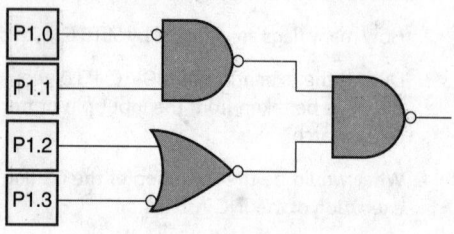

8. Why do some instructions read data from input pins and others from output latch in case of a directly addressed I/O port bit?

9. Develop an instruction, which performs a relative jump if parity bit of PSW is set, otherwise proceeds sequentially.

10. Develop a program to complement all bits of the bit-addressable area.

   [HINT: Instead of using CPL bit instruction 128 times, find a shorter method.]

11 PROGRAMMING EXAMPLES-II

CHAPTER OBJECTIVES

In the last three chapters, we have discussed about MCS-51 instructions related to subroutines, logical operations and Boolean variable operations. In this chapter, we will study a few example programs dealing with their applications. After completion of the chapter, the reader should be able to understand

- Application areas of subroutines.
- Application techniques of logical and Boolean instructions.
- Some important solution techniques such as packing and unpacking of BCD numbers and bubble sorting.

11.1 | Introduction

As we have discussed about most of the instructions of MCS-51, it would be appropriate now to solve few more example problems highlighting applications of these instruction set. Problems in this chapter are little more complex than those presented earlier in Chapter 7. After presenting the problem, basic solution techniques are discussed followed by the algorithm or flowchart. Complete program listing is placed thereafter with adequate comment statements for easier understanding of the programming logic.

11.2 | Count 1s in a Byte

Example 11.1

Purpose: Application of rotate instruction.

Problem

Find out number of 1s in a byte, available in internal data memory location 30H. Store the result in the accumulator itself.

Solution

We can use either RRC A or RLC A instruction to count number of 1s (or even 0s) in any byte. The LSB (or MSB, as the case might be) would be shifted to the carry (CY) flag helping the program to decide whether the bit is 0 or 1 (Fig. 11.1). The byte must be loaded in the accumulator for this purpose. We are to use one register as a counter for rotating the byte eight times. A second register is necessary for storing the result.

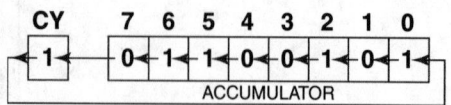

Figure 11.1 Schematic of set bit counting using RLC A instruction

Algorithm

Step 1: Initialize by loading the accumulator with the target byte, R7 by bit counter, that is 8, and clear R6 to store the result.

Step 2: Rotate the accumulator left through carry. Bit 7 (MSB) is shifted from the accumulator to the CY flag.

Step 3: If CY is 0 then go to Step 5.

Step 4: Increment R6 by 1.

Step 5: Decrement R7 by 1. If R7 is not 0 then go to Step 2, otherwise return.

```
; Subroutine to count number of 1s in a byte.
; First three instructions complete the initialization procedure.

COUNT1:    MOV   A, 30H        ; get target byte in the accumulator
           MOV   R7, #08H      ; counter for 8 bits
           MOV   R6, #00H      ; for temporary result storage

; Iteration to count number of 1s in the accumulator starts from here.

SHIFT1:    RLC   A             ; shift MSB in to the CY flag
           JNC   NEXT1         ; this bit is not 1
           INC   R6            ; the bit is 1, increment count
NEXT1:     DJNZ  R7, SHIFT1    ; continue for 8 bits

; Iteration complete. Result in R6. Copy it to the accumulator.

           MOV   A, R6         ; result in the accumulator
           RET
```

How to develop the logical steps for solving any similar problem is a general dilemma faced by most students at their initial stage of practice sessions. I have come across many good students unable to solve this type of problems at their early stage of software development.

Software development is a skill which needs constant practice. There is no general rule to develop program logic for all types of problems. At the initial stage, going through various types of solved examples may give you a good start. Then try to develop same programs with minor alternations and start comparing. Next, take a few unsolved problems and try to solve by yourself. Finally, never be afraid to make mistakes. In most of the cases we learn from our mistakes.

C-version

```c
#include <regx51.h>

void main(void)
{
        register unsigned char *n, i = 0, count = 0, *acc;
        unsigned char number, mask;
        n = 0x30;
        number = *n;
        mask = 0x01;
        acc = 0xE0;            // address of accumulator is E0H
            for (i = 8; i > 0; i --)
            {
                    if (number & mask)
                    count++;
                    number = number >>1;
            }
        *acc = count;
}
```

11.3 | Unpack a BCD Number

Example 11.2

Purpose: How to unpack a BCD number and an application of ANL and rotate instructions.

Problem

A packed BCD number is in location 30H. After unpacking, store it in R6 (lower digit) and R7 (higher digit).

Solution

BCD numbers vary from zero to nine, and they always occupy 4 bits. Therefore, two BCD numbers may be accommodated within a byte. The upper and lower nibbles (4 bits) of the byte may be separated by using AND operation. There should always be a leading zero before any unpacked BCD number (Fig. 11.2). Note that 9 of 95 (BCD MS digit) become 09 after unpacking.

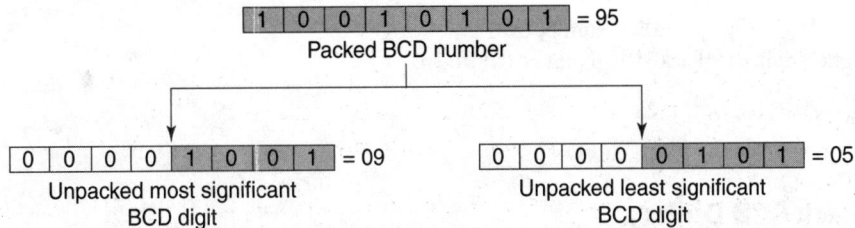

Figure 11.2 Packed and unpacked BCD numbers

Algorithm

Step 1: Load the accumulator by packed BCD number to be unpacked.

Step 2: Take out LS BCD digit by ANDing the accumulator with 0FH. Store it in R6.

Step 3: Reload the accumulator by packed BCD number.

Step 4:   Take out MS BCD digit by ANDing the accumulator with F0H.

Step 5:   Rotate the accumulator four times to shift the digit to lower nibble.

Step 6:   Store it in R7 and return.

; Subroutine to unpack a BCD number.

```
UNPACK: MOV    A, 30H        ; get packed BCD number in the accumulator
        ANL    A, #0FH       ; take out lower nibble
        MOV    R6, A         ; store lower nibble in R6
        MOV    A, 30H        ; get unpacked BCD number again
        ANL    A, #0F0H      ; take out higher nibble
        RR     A             ; shift it to the lower nibble
        RR     A
        RR     A
        RR     A
        MOV    R7, A         ; store it in R7
        RET
```

 Later we shall find that the SWAP instruction may replace four RR A instructions.

C-version

```c
#include <regx51.h>

void main(void)
{
        unsigned char *pPackedBcdNum;
        unsigned char lowerDigit, higherDigit;

        // initialize

        pPackedBcdNum = 0x30;

        lowerDigit = *pPackedBcdNum & 0x0F;
        higherDigit = (*pPackedBcdNum & 0xF0)>>4;

        while(1);
}
```

11.4 | Pack BCD Digits

Example 11.3

Purpose: How to pack two BCD digits and application of ORL and rotate instructions.

Problem

Two unpacked BCD digits are available in location 30H (MS digit) and 31H (LS digit). Write a program to pack these two and store in R7.

Solution

The method to be adopted is just opposite of the method adopted in the previous example.

Algorithm

Step 1: Load pointer by address of MS BCD digit.
Step 2: Load MS BCD digit in the accumulator.
Step 3: Increment pointer by 1 to target LS BCD digit.
Step 4: Rotate the accumulator left four times to shift MS BCD digit to upper nibble.
Step 5: OR accumulator with LS BCD digit as pointed by the pointer.
Step 6: Store the accumulator in R7 and return.

```
; Subroutine to pack two BCD digits.

PACKIT: MOV    R0,#30H      ; load pointer by first address
        MOV    A,@R0        ; get MS digit in the accumulator
        INC    R0           ; point to next location
        RL     A            ; shift digit to upper nibble
        RL     A
        RL     A
        RL     A
        ORL    A,@R0        ; merge with lower digit
        MOV    R7,A         ; save packed BCD number in its location
        RET
```

C-version

```
unsigned char msDigit _at_ 0x30;
unsigned char lsDigit _at_ 0x31;

#include <regx51.h>

void main(void)
{
        unsigned char packedBcdNum;

        packedBcdNum = (msDigit<<4) | (lsDigit);

        while(1);
}
```

11.5 | Pack Array of Unpacked BCD Digits

Example 11.4

Purpose: How to handle multiple arrays using only one pointer.

Problem

Using only one pointer, R0, pack two arrays of BCD digits to create a third array. The higher digits are available from 30H to 3FH. Lower digits are available from 40H to 4FH. Packed BCD numbers are to be stored from 50H to 5FH.

Solution

As this problem deals with the three arrays, three different pointers are generally expected. However, even with a single pointer it may be solved easily as some symmetry exists in the data structure of this problem.

Fig. 11.3 illustrates the problem with a few sample inputs and outputs. As an example case, we may indicate that 08 to be taken from address 30H and 02 may be copied from address 40H. After packing these two, the packed number 82 is to be saved in the location 50H. It may be observed that last 4 bits of all these three addresses are identical (30H, 40H, 50H). Only MS 4 bits are changing as 3, 4 and 5. Utilizing this special case, we may develop the program using only one pointer, R0. The algorithm and complete program would be as follows.

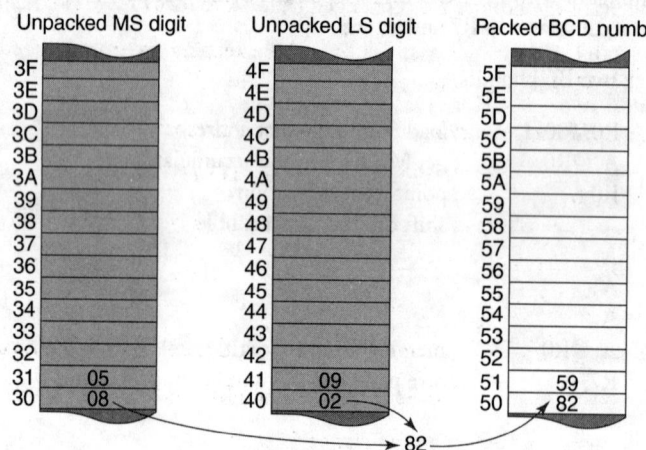

Figure 11.3 Problem definition of Example 11.4

Algorithm

Step 1: Select bank #0, load pointer by 30H and counter by 16d (10H).
Step 2: Load higher digit in the accumulator through the pointer. Shift it to the higher nibble.
Step 3: Change MS digit of pointer to 4.
Step 4: Logically OR accumulator with lower digit through the pointer.
Step 5: Change MS digit of the pointer to 5.
Step 6: Save packed BCD number from the accumulator to its storage location as pointed by the pointer.
Step 7: Restore MS digit of the pointer to 3.
Step 8: Increment the pointer by 1.
Step 9: Decrement counter by 1. If counter is not 0 then go to Step 2. Otherwise return.

; Program to pack two arrays of unpacked BCD digits using only one pointer (R0).
; Source address varies from 30H to 3FH for MS digits and 40H to 4FH for LS digits.
; Destination address for packed BCD digits varies from 50H to 5FH.
; First three instructions complete the initialization process.

```
PACKAR: MOV   PSW, #00H    ; select register bank #0
        MOV   R0, #30H     ; load the pointer with the first address of the first array
        MOV   R7, #10H     ; load counter for sixteen sets of operations
```

; Iterative procedure for packing BCD digits starts.
; To start packing of next digit, the loop should start from here.

```
LOOP:   MOV   A, @R0        ; get one higher digit
        RL    A             ; shift to higher nibble
        RL    A
        RL    A
        RL    A
```

; Now process the pointer to target the next array of same LS address nibble.
; This is done by the following two logical operations with R0 of bank #0.
; Results are available in the directly addressed location (R0 of bank #0).
; Note that the content of the pointer R0 is changed from 3x to 4x by these two instructions.

```
        ANL   00H, #0FH     ; clear the MS nibble keeping lower unchanged
        ORL   00H, #40H     ; insert 4 in the MS nibble
```

; Now pointer is targeting the second array. Use the pointed number.

```
        ORL   A, @R0        ; the accumulator has packed number
```

; Now process the pointer to target the third array, for storage (make 4x as 5x).

```
        ANL   00H, #0FH     ; clear MS nibble
        ORL   00H, #50H     ; insert 5 in it
```

; Pointer is now showing the saving location. Save the packed number from the accumulator.

```
        MOV   @R0, A        ; store the packed number
```

; Process the pointer to target the first array (change 5x to 3x).

```
        ANL   00H, #0FH     ; clear the MS nibble
        ORL   00H, #30H     ; restore 3 in it
```

; Increment pointer to target next location of the first array. Then continue packing.

```
        INC   R0            ; update pointer to target next higher digit
        DJNZ  R7, LOOP      ; continue for all digits
        RET                 ; over
```

For any beginner, it would be a good practice to visualize the input and output stages of any problem by drawing a sketch similar to Fig. 11.3. Against such a sketch, try to visualize the data flow. It would definitely make the software development easier.

C-version

```c
#include <regx51.h>

void main(void)
{
        unsigned char *pMsDigit, *pLsDigit, *pPackedBcdNum;
        unsigned char i;
```

```
// initialize

pMsDigit = 0x30;
pLsDigit = 0x40;
pPackedBcdNum = 0x50;
i = 0;

for(i=0; i<=0x0F; i++)
{
        pPackedBcdNum[i] = (pMsDigit[i]<<4 | (pLsDigit[i]);
}

while(1);
}
```

11.6 | Find Largest and Smallest Integers of an Array

Example 11.5

Purpose: How to implement multiple operations in a single pass.

Problem

Find the largest and smallest from *N* unsigned integers. Assume the value of *N* to be available in the internal data memory location 30H. The array starts from location 31H. Store the maximum integer in R4 and minimum in R3.

Solution

Both maximum and minimum values of any array may be identified by a single pass. To start with, register R4 is loaded with the worst maximum integer that is 00H. Similarly, R3 is initialized as FFH to represent the worst minimum integer. Register bank # is selected as 0 only to use direct addressing for registers. CJNE instructions are used to get the status of the comparison reflected in the CY flag. If the present integer is larger than the integer stored in R4 then R4 is replaced by the current integer. In identical manner, the minimum number is selected in R3. R7 works as a counter. The flowchart is shown in Fig. 11.4. The program listing is given below.

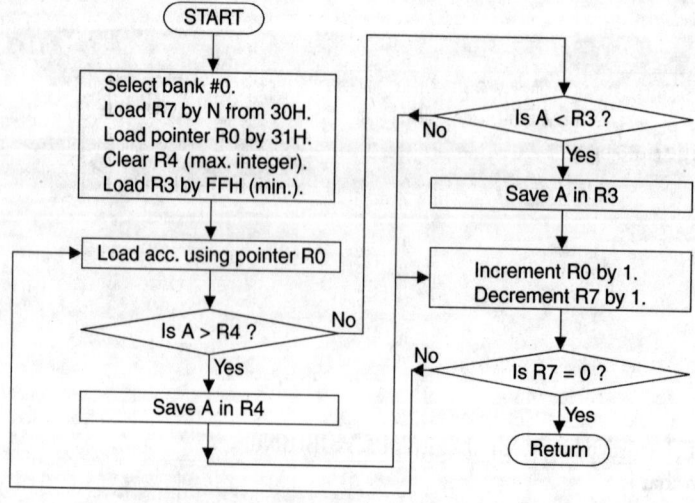

Figure 11.4 Flowchart of Example 11.5

```
; Program to find the maximum and minimum numbers within an array.
; First five instructions complete the initialization procedure.

MAXMIN:    MOV     PSW, #00H       ; select register bank #0
           MOV     R7, 30H         ; load counter by N value
           MOV     R0, #31H        ; load pointer by starting address of the array
           MOV     R4, #00H        ; worst maximum number for comparison
           MOV     R3, #0FFH       ; worst minimum number for comparison

; The pass to find largest and smallest unsigned integers starts from here.
; To check the next integer of the array, the procedure must start from here.
; Following five instructions check and store the maximum integer, so far.

CHKMAX:    MOV     A, @R0          ; get one term of the array
           CJNE    A, 04H, NEXT1   ; compare with R4, CY = 1, if R4 is greater
NEXT1:     JC      CHKMIN          ; R4 is greater than the current term
           MOV     R4, A           ; update R4 by the greater term
           SJMP    NXTNUM          ; get next term

; The term is still in the accumulator, unchanged.
; Following three instructions check and store the minimum integer, so far.

CHKMIN:    CJNE    A, 03H, NEXT2   ; compare with R3, CY = 1, if R3 is greater
NEXT2:     JNC     NXTNUM          ; R3 is less than the current term
           MOV     R3, A           ; update R3 by the smaller term
; Checking of one term is over. Point to the next term and loop on, if necessary.

NXTNUM:    INC     R0              ; point to the next term, if any
           DJNZ    R7, LOOP        ; continue till the last term
           RET
```

 If you are a beginner and solve this type of problem with two passes instead of one pass, there is nothing wrong in it. Rather it is better at the initial stage to go for less complicated program logic. Later your skill would definitely improve with practice.

C-version

```c
#include <regx51.h>

void main(void)
{
        unsigned char *pLocN, *pArray, i;
        unsigned char maxNum, minNum;

        // initialize

        pLocN = 0x30;
        pArray - 0x31;
```

```
        maxNum = *pArray;
        minNum = *pArray

        if(*pLocN > 0)
        {
                for(i=0; i<*pLocN; i++)
                {
                        if(pArray[i]>maxNum
                        {
                                maxNum = pArray[i];
                        }
                        else if (pArray[i]<minNum)
                        {
                                minNum = pArray[i];
                        }
                }
        // store results if needed
        }
        while(1);
}
```

11.7 | Bubble Sorting

Example 11.6

Purpose: Example of nested loops.

Problem

A few unsigned integers are stored from the location 31H onwards of the internal data memory area. Arrange these integers in ascending order. The number of terms of the array is available in the location 30H. Store the arranged integers from 31H itself.

Solution

The term ascending order means larger number should be in higher address and smaller number in lower address. In MCS-51, generally higher addresses are shown at the top with respect to the lower addresses. However, for the sake of easier understanding, in the following three illustrations, higher addresses are shown at lower levels.

One of the many methods to rearrange a random array of integers in ascending or descending order is known as 'bubble sorting' technique. If there are N integers then there should be $N–1$ passes, and in every pass there should be one or more comparison and interchanging of places if not found in order. Using a set of four integers, this method is illustrated through the following three diagrams.

Fig. 11.5(a) shows the initial condition of an array, which is to be arranged in the ascending order. To start with, the top two integers, namely F6H and 94H, are checked for correct order. As they are not, their positions are interchanged, which is shown in Fig. 11.5(b). Now the next two integers, namely F6H and 48H, are checked for the correct order. As they are not, their positions are interchanged (known as swapping) and their condition was achieved as shown in Fig. 11.5(c). The third and the final comparison takes place between the last two numbers, i.e., F6H and 07H. As they are not in the correct order, they are also swapped, and the final position after the first pass is achieved as shown in Fig. 11.5(d).

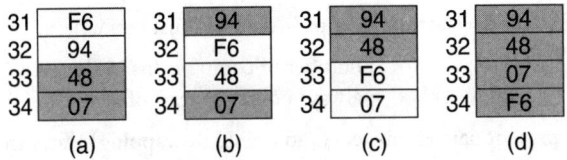

Figure 11.5 Pass 1 steps (a) first, (b) second, (c) third comparisons and (d) final condition

The second pass starts with the condition shown in Fig. 11.6(a). After comparing first two numbers and swapping (they were not in correct order), the position at which it arrives is shown in Fig. 11.6(b). Similarly, the second checking takes place, and we arrive at the condition shown in Fig. 11.6(c). At this stage, the third comparison indicates that the numbers are in the correct order. Therefore, they were left as they were, and we arrive at the final position of the second pass shown in Fig. 11.6(d).

	(a)		(b)		(c)		(d)
31	94	31	48	31	48	31	48
32	48	32	94	32	07	32	07
33	07	33	07	33	94	33	94
34	F6	34	F6	34	F6	34	F6

Figure 11.6 Pass 2 steps (a) first, (b) second, (c) third comparisons and (d) final condition

Fig. 11.7(a) presents the starting condition of pass #3. After checking the order, first two integers were interchanged within their places, and the condition of the array becomes as shown in Fig. 11.7(b). However, the remaining two comparisons find the numbers that are in order resulting in no change of their respective positions. Fig. 11.7(d) shows the final condition of the array at the end of pass 3. Here, we find that all four integers are properly ordered in their respective places to indicate an ascending order list.

	(a)		(b)		(c)		(d)
31	48	31	07	31	07	31	07
32	07	32	48	32	48	32	48
33	94	33	94	33	94	33	94
34	F6	34	F6	34	F6	34	F6

Figure 11.7 Pass 3 steps (a) first, (b) second, (c) third comparisons and (d) final condition

If it is observed then we may find that more the iteration progresses, lesser would be the number of comparisons required. For example, in Fig. 11.6(c) we find that the last comparison is not necessary, and in Figs 11.7(b) and (c), the last two comparisons may be avoided. As a matter of fact, the number of comparisons required is inverse of the pass number, which decreases continuously.

Although the description of the procedure takes larger space, when this programming logic is implemented, it does not make the program too long. Flowchart of the technique is shown in Fig. 11.8 and the program listing would be as follows.

```
; Program for bubble sorting for ascending order.
; First three instructions complete the initialization process.

BUBBLE:    MOV    PSW, #00H    ; select register bank #0
           MOV    R7, 30H      ; load counter by number of terms, N
           DEC    R7           ; number of passes required would be N–1
```

; Iteration for sorting starts. To start the next pass, control must come here.

| PASS: | MOV | 06H, R7 | ; number of comparisons in this pass |
| | MOV | R0, #31H | ; always point the start of array |

; Comparison for the present pair of integers (and eventual swapping) starts from here.

COMP2:	MOV	A, @R0	; get first term
	MOV	R4, A	; temp save
	INC	R0	; point to the second term
	MOV	A, @R0	; second term in the accumulator
	CLR	C	; necessary before subtraction
	SUBB	A, R4	; subtract R4 from A, if A < R4 then CY = 1
	JNC	NEXT1	; present order is correct

; Compared integers are not in order. Interchange their places.

	MOV	A, @R0	; get the second term in A
	DEC	R0	; point to the first term's location
	MOV	@R0, A	; store the second term there
	INC	R0	; point the second term's location
	MOV	@R0, 04H	; store the first term there

; Following instruction checks for remaining numbers of the comparison necessary.

| NEXT1: | DJNZ | R6, COMP2 | ; continue, if necessary |

; Following instruction checks whether any more pass is pending.

| | DJNZ | R7, PASS | ; continue, if necessary |
| | RET | | ; sorting complete. |

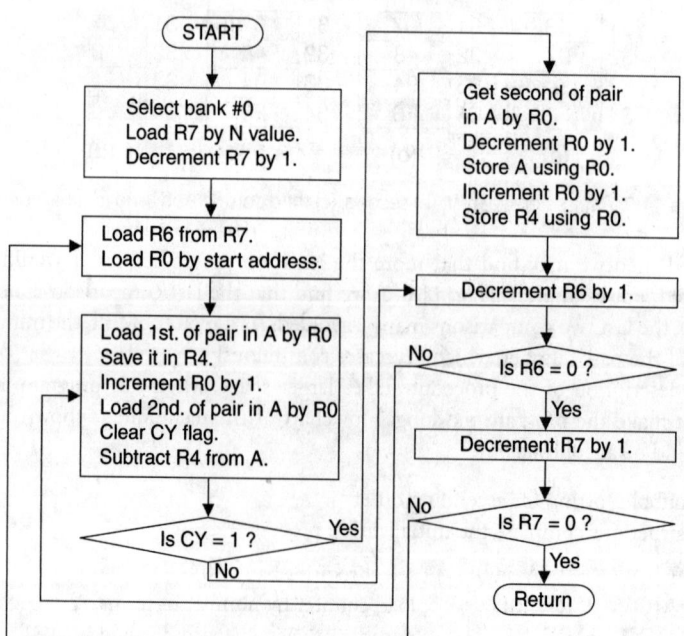

Figure 11.8 Flowchart using bubble sorting for ascending order

For any beginner, this may be apparently a complex problem, to be avoided initially. However, if the problem is broken to smaller modules, it would become much easier to comprehend. For example, after practicing in long hand the bubble sorting technique with a few random integers, start with the swapping technique. Once this is mastered, think of how to manipulate the pointers during any one pass. This approach would help you to solve simpler modules one at a time which would lead to the final solution.

11.8 | Find the Sum of Factorials

Example 11.7

Purpose: Example of nested subroutines.

Problem

A few random unsigned integers are stored from the internal data memory location 31H onwards. Number of terms (N) is available in location 30H. Assuming that none of these numbers is greater than 5, find the factorials of these integers and then find their sum. Assume that the sum would not exceed 8-bit value.

Solution

Factorial of any integer, say X, is calculated by multiplying all integers from X to 1. Although MCS-51 offers instruction for multiplication, that is MUL AB, we are yet to discuss about this instruction, which would be done in Chapter 12. However, multiplication may also be done by repeated addition, as we have discussed in Example 8.1. So, without using MUL instruction, a subroutine may be developed to find the product of the two integers by repeated additions.

Another subroutine may be planned for calculating the factorial of any integer. The main program is to call this subroutine and calculate the sum of factorial of the random integers of the array. The flowchart of the main program is shown in Fig. 11.9.

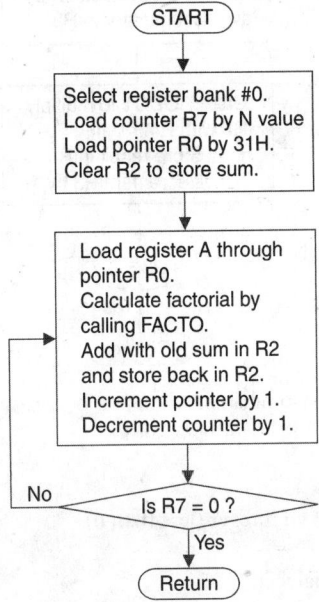

Figure 11.9 Flowchart for the main program (sum of factorials)

; Program to calculate the sum of factorials of the random integers of an array.
; At return, the sum would be in R2.
; Registers (bank #0) used:
; R0 = pointer to stored integers
; R2 = sum of factorials of the integers
; R7 = integer counter
; A = computed results

```
START:   MOV     PSW, #00H       ; select register bank #0
         MOV     R7, 30H         ; load counter
         MOV     R0, #31H        ; load pointer
         MOV     R2, #00H        ; clear sum
LOOP:    MOV     A, @R0          ; get one term
         ACALL   FACTO           ; calculate its factorial, value in A
         ADD     A, R2           ; calculate sum
         MOV     R2, A           ; save it back
         INC     R0              ; point next term, if any
         DJNZ    R7, LOOP        ; continue till end
         RET                     ; sum in R2
```

Flowchart of the subroutine FACTO to calculate factorial is given in Fig. 11.10. Note that both R4 and R3 were used for this purpose. R3 is decremented in steps of 1 to get the next lower integer, which was multiplied with the product, in R4. At return, result would be in the accumulator.

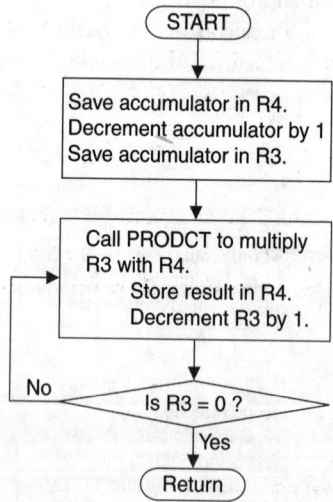

Figure 11.10 Flowchart of subroutine for calculating factorial

; Name: FACTO
; Function: Calculates factorial of an integer (less than 6)
; Input: A contains the integer
; Output: A contains the factorial value
; Calls: PRODCT

```
; Uses:      A, R3, R4

FACTO:   MOV      R4, A
         DEC      A          ; next lower number
         MOV      R3, A      ; next lower number cum counter
MULTI:   ACALL    PRODCT
         MOV      R4, A
         DJNZ     R3, MULTI
         RET
```

Fig. 11.11 shows the flowchart of the subroutine to multiply two integers by repeated additions. The subroutine is placed after that.

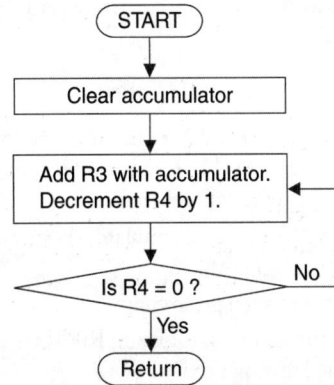

Figure 11.11 Flowchart of the subroutine for multiplying two integers

```
; Name:      PRODCT
; Function:  Calculates the product of the two unsigned integers by multiple additions.
; Input:     R3 and R4 having two integers
; Output:    Product available in the accumulator. R3 remains unchanged. R4 cleared.
; Calls:     None
; Uses:      A, R3, R4

PRODCT:  CLR      A
ADDIT:   ADD      A, R3
         DJNZ     R4, ADDIT
         RET
```

For a beginner, this is another example of breaking a problem to a few simpler modules and sub-modules. Also observe the parameter passing technique from one subroutine to another.

11.9 | Sort Out Numbers Divisible by 4

Example 11.8

Purpose: To substitute division operation by the logical operation in special cases.

Problem

Create a new array by removing only those integers that are perfectly divisible by 4 from an array, starting from 31H. Location 30H contains number of terms of this array. The new array is to be created from the location 60H. At return, the accumulator should indicate number of terms found. Original locations with digits divisible by 4 should be replaced by null.

Solution

For an unsigned integer to be divisible by 4, bits 0 and 1 must be zero. The algorithm to solve this problem is given below.

Algorithm

Step 1: Initialize by loading counter (R7) by N value from the location 30H. Load pointer R0 by the starting address of source array (31H). Load pointer R1 by the starting address of the destination array (60H). Clear the term counter (R6).

Step 2: Get the term pointed by R0 in the accumulator. Logically AND it with 03H to clear MS 6 bits.

Step 3: If the accumulator is not zero then go to Step 5.

Step 4: Load the term again in the accumulator using R0. Then store it using R1. Increment both R1 and R6 by one. Clear the location pointed by R0.

Step 5: Increment R0 and decrement R7 by one. If R7 is not zero then go to Step 2.

Step 6: Load the accumulator from R6 and return.

; Program to find terms divisible by four.

```
START:    MOV   R7, 30H      ; store N value as counter
          MOV   R0, #31H     ; point at the start of array
          MOV   R1, #60H     ; destination pointer
          MOV   R6, #00H     ; counter of terms divisible by four
LOOP:     MOV   A, @R0       ; get one term
          ANL   A, #03H      ; clear MS 6 bits
          JNZ   NXTERM       ; not divisible by four
          MOV   A, @R0       ; divisible, get the term
          MOV   @R1, A       ; save in new array
          INC   R6           ; one more found
          INC   R1           ; point to the next storage
          MOV   @R0, #00H    ; replace term by null
NXTERM:   INC   R0           ; point to the next term
          DJNZ  R7, LOOP     ; loop on till end
          MOV   A, R6        ; load the accumulator by the number found
          RET
```

SUMMARY

Eight example programs dealing with the applications of logical instructions, Boolean instructions and subroutine calls were discussed in this chapter. Examples are taken from the counting number of set bits in any byte, packing and unpacking BCD numbers, finding maximum and minimum terms of any random integer array, bubble sorting method, calculation of factorials and finding out numbers divisible by 4.

POINTS TO REMEMBER

- After unpacking the MS BCD digit, it must be shifted to the lower nibble of the byte.

- An easier way to multiply or divide an unsigned integer by 2, 4, 8 etc. is to rotate the byte left or right, respectively.

REVIEW QUESTIONS

Evaluate Yourself

1. In Example 11.1, if the location 30H contained 57H then at return the accumulator would contain

   (a) 05H   (b) 03H

   (c) 57H   (d) none of these

2. What would happen if the RRC A replaces the fourth instruction of Example 11.1?

   (a) Count no. of 0s   (b) No change

   (c) Count both 1s and 0s   (d) None of these

3. If a packed BCD number, say 28, is unpacked, which of the following would be its unpacked most significant digit?

   (a) 02   (b) 0002

   (c) Either of these   (d) None of these

4. In Example 11.5, what would be the value of the third byte of the first CJNE instruction?

   (a) 04H   (b) 02H

   (c) 00H   (d) None of these

5. In Example 11.7, if the stack pointer was initialized at 07H, what would be its value during execution of the subroutine PRODCT?

   (a) 0CH   (b) 0BH

   (c) 009H   (d) None of these

6. What would be the hexadecimal value of the factorial of 4?

   (a) 05H   (b) 0AH

   (c) 18H   (d) None of these

7. What would be the characteristics of any integer perfectly divisible by 8?

   (a) LS 3 bits 0   (b) LS bit 0

   (c) LS 4 bits 1000   (d) None of these

8. To change 33H to 53H, we should AND 33H with

   (a) FFH and then OR with 53H

   (b) 0FH and then OR with 50H

   (c) F0H and then OR with 50H

   (d) none of these

9. If the bubble sorting technique is applied to an array with four integers then in the worst case, the total number of comparisons would be

   (a) 6   (b) 9

   (c) 10   (d) none of these

10. In Example 11.7, if four integers of the array were one, two, three and four then what would be the value of the accumulator at the end of the main program?

    (a) 18H   (b) 33H

    (c) 21H   (d) None of these

Search for Answers

1. Apart from the accumulator, does any other register offers the option of rotating it circular or through the CY flag?

2. Make a list of all instructions capable of counting the cleared bits in a byte.

3. What is the difference between the packed and the unpacked BCD numbers?

4. In Example 11.5, what is the reason to store 00H in the place where the maximum number would be placed?

5. What is meant by the term 'pass' in the case of sorting?

6. What is meant by the term 'swapping'? In which example it was implemented?

7. If an array of randomly placed 50 integers are to be sorted for descending order, what should be the maximum number of passes required in the bubble sort method?

8. Is it possible to develop a routine which would jump to a relative address if the accumulator is having even parity? Justify your answer.

9. How is it possible to extend the program in Example 11.8 so that after completion of deletion the deleted array may be compacted?

10. Find a method to generate prime numbers up to 50d.

Think and Solve

1. Is it possible to regain the original accumulator content by repeated execution of the RLC A instruction? If yes, then how many times it should be executed?

2. What modifications are necessary in the program given in Example 11.1 so that it is able to count number of 0s in a byte?

3. Is it possible to solve Example 11.4 using only one pointer if the addresses of the three arrays are changed to 30H to 3FH, 42H to 51H and 55H to 64H, respectively? Justify your answer.

4. Is it possible to solve Example 11.7 without using subroutine calls? Assume that MUL instruction may be used for the multiplication of any two integers.

5. In Example 11.4, the pointer was changed for other array by first AND followed by OR operations. What would happen if these operations were interchanged in sequence?

6. In Example 11.5, is it possible to calculate the average value of all numbers without using any extra pass?

7. In Example 11.7, if the array was in the ascending order in the beginning then also $N-1$ passes were executed by the program. Is there any method to reduce the number of redundant passes in such cases?

8. Find out another method of sorting, and compare its efficiency with the bubble sorting method.

9. Develop a subroutine to check whether any given unsigned integer is perfectly divisible by 7 or not. Using this subroutine, write a program to delete all integers divisible by 7 from a given array starting from 31H. Location 30H stores the number of terms of the array. At the end of the program, R7 should contain the number of deleted terms.

10. An array from 30H to 3FH contains unpacked MS digits, and the corresponding unpacked LS digits are stored from 40H to 4FH. Write a program to pack the BCD numbers and store them from 50H to 5FH. Then call a subroutine to arrange these packed BCD numbers in descending order.

12 ADVANCED INSTRUCTIONS

CHAPTER OBJECTIVES

In the last few chapters, some mnemonics were introduced but not discussed in details. In this chapter, we discuss those undiscussed mnemonics. After completion of the chapter, the reader should be able to understand how to

- Implement table look-up method.
- Communicate with the external data memory.
- Interchange bytes by a single instruction.
- Perform multiplications and divisions using MUL or DIV instructions.
- Get proper entry point among multiple subroutines.
- Perform BCD arithmetic.

12.1 | Introduction

Table 12.1 lists a few of mnemonics, which were introduced in the previous chapters. However, their discussions were left pending for this chapter. Here, we will discuss those instructions and also identify their application areas. It may be pointed out here that by the term 'advanced' we mean simple programming may be carried out without using these instructions.

Table 12.1 Mnemonics to be discussed in this chapter

Mnemonic	Brief description	Introduced in chapter
MOVC	Load the accumulator by a byte from the program memory	4
MOVX	Transfer a byte between external RAM and the accumulator	4
XCH	Exchange bytes between two locations within the internal memory	4
XCHD	Exchange digits (lower nibble) between two locations within the internal memory	4
MUL	Unsign integer multiplication	5
DIV	Unsign integer division	5
DA A	Adjust decimal for the accumulator	5
JMP	Jump to the address indicated by the DPTR and A	6
RETI	Return from the interrupt	6
SWAP	Swap nibbles within the accumulator	9
JBC	Jump if bit is set and clear the bit	10

12.2 | MOVC Instruction

Variations of the MOVC instruction are indicated in Table 12.2. This instruction loads a target byte from the program memory to the accumulator. The target address is obtained by adding the content of the accumulator (offset byte) with a 16-bit base register, which may be the program counter (PC) or the DPTR. This addressing mode is known as the indexed addressing mode, and it was mentioned in Chapter 4. Both variations of the MOVC instruction are described one after another.

Execution of MOVC A, @A+DPTR instruction is illustrated in Fig. 12.1. In this case, DPTR is the base register containing a 16-bit address, which is added with the current content of the accumulator (unsigned integer) to generate the target address. The content of DPTR remains unchanged. The accumulator is loaded with the content of the targeted byte from the program memory. No flags are affected. The addition is a 16-bit addition, which means, if a carry is produced from bit 7, it would be propagated to bit 8.

For example, if DPTR contains 01FFH and the accumulator contains 03H, then a target address of 0202H would be generated and the content of this address of the program memory would be copied in the accumulator. The original content of the accumulator (03H) would be destroyed. However, DPTR would still contain its original value, that is 01FFH.

Table 12.2 Variations of MOVC instruction

Instruction	Function
MOVC A, @A+DPTR	Load the accumulator by the code byte from the program memory addressed by A and DPTR
MOVC A, @A+PC	Load the accumulator by the code byte from the program memory addressed by A and PC

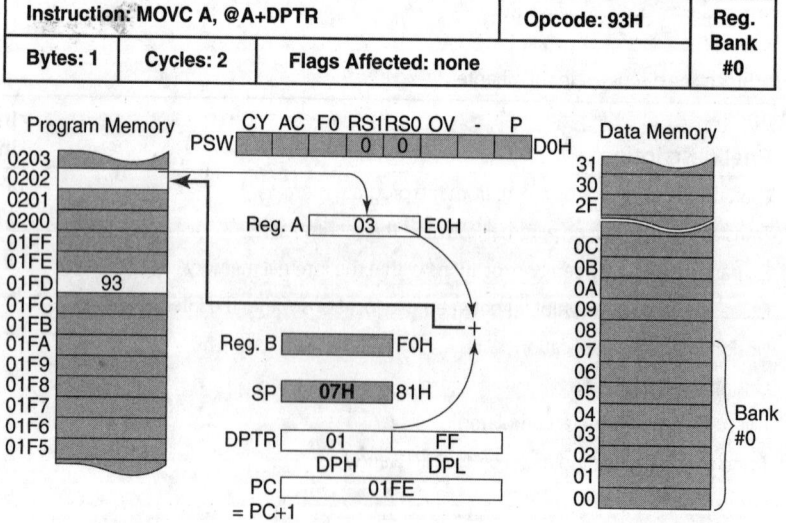

Figure 12.1 Execution of MOVC A, @A+DPTR instruction

Instruction MOVC A, @A+PC works in an identical way to get a byte from the program memory. However, in this case, the accumulator content is added with the current value of the PC. Note that at the time of this addition, the PC would be pointing to the first byte of the next instruction located immediately after MOVC instruction. This addition does not physically change the value of the PC. The value (16-bit sum) is used to copy the target byte from the program memory to the accumulator. An example case of this instruction is illustrated in Fig. 12.2. Assuming that the 1-byte MOVC instruction exists in the program

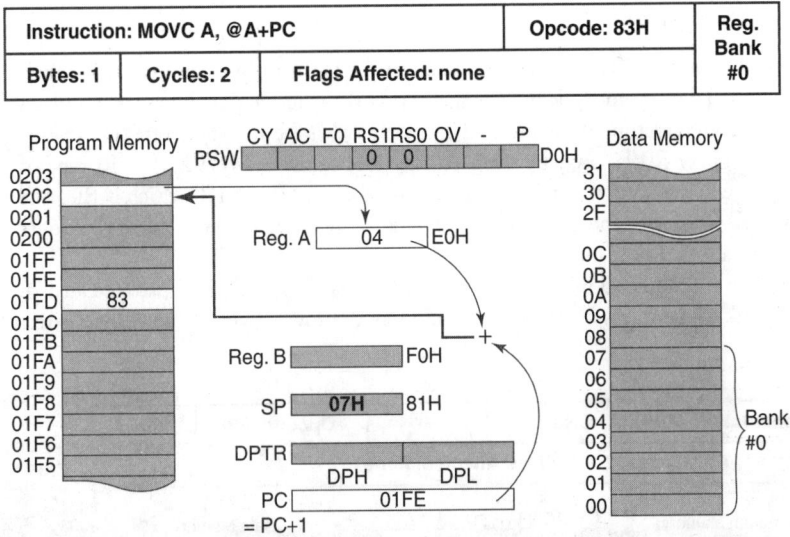

Figure 12.2 Execution of MOVC A, @A+PC instruction

memory at location 01FDH and the accumulator contains 04H, data from the program memory location 0202H would be copied to the accumulator. This 0202H is obtained by incrementing the PC to the start of the next instruction, that is 01FEH, and then adding 04H to it.

MOVC instruction is useful for getting entries from look-up tables. For an appropriate use of MOVC instruction, refer to Example 12.1 of this chapter.

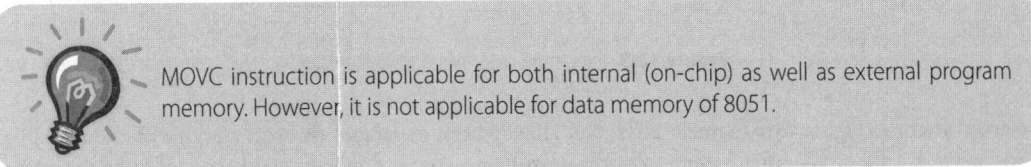

MOVC instruction is applicable for both internal (on-chip) as well as external program memory. However, it is not applicable for data memory of 8051.

12.3 | MOVX Instruction

MOVX instruction communicates between the accumulator and the external data memory only. There is no scope of communicating with any other type of memory (program memory or internal data memory) through this instruction. Variations of this instruction are presented in Table 12.3.

Table 12.3 Variations of MOVX instruction

Instruction	Function
MOVX A, @Ri	Load the accumulator by a byte from the external data memory using indirect 8-bit addressing
MOVX A, @DPTR	Load the accumulator by a byte from the external data memory using indirect 16-bit addressing
MOVX @Ri, A	Store the accumulator in the external data memory using indirect 8-bit addressing
MOVX @DPTR, A	Store the accumulator in the external data memory using indirect 16-bit addressing

As we may observe from Table 12.3, the accumulator is the only location that is capable of transacting data with the external data memory. It may be noted that the external data memory is addressed by the 16-bit address. Addressing architecture of the internal data memory is 8 bit. To generate this 16-bit external address, either DPTR or register R0/R1 may be used. At the time of using MOVX A, @Ri type of instruction, the lower 8 bits of the external data memory addresses are stored in R0 or R1, which is the output through port P0 in a multiplexed form with the incoming 8-bit data. To generate upper 8 bits of the external data memory address, help from another port may be asked.

As an example case, execution of the instruction MOVX A, @R0 is illustrated in Fig. 12.3. Presently, register R0 of bank #0 contains F9H, which is used as the lower byte of the 2-byte external address. It is

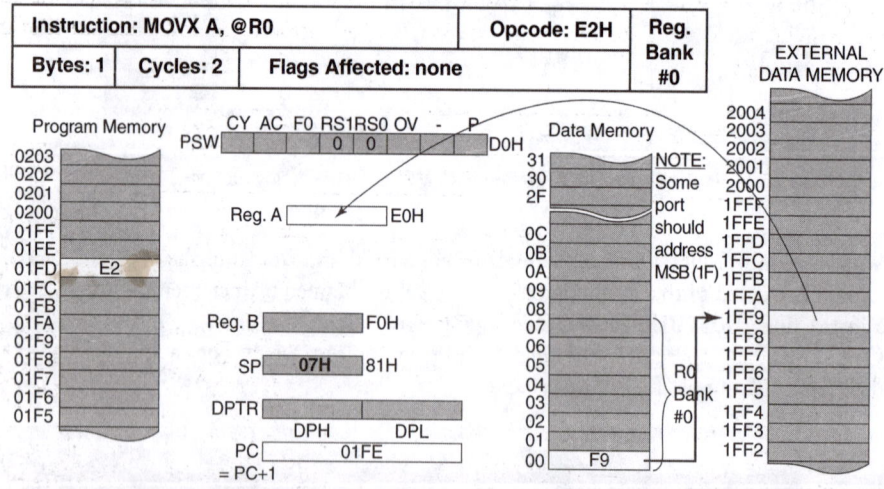

Figure 12.3 Execution of MOVX A, @R0 instruction

assumed that the higher byte, namely 1FH, had already been generated through any one of the remaining three ports, P1, P2 or P3. After execution of this instruction, the accumulator would be loaded by the present content of the external data memory location 1FF9H. No flags would be affected, and the content of register R0 would remain unchanged.

An identical effect may be achieved by MOVX A, @DPTR instruction as illustrated in Fig. 12.4. In this case, the complete 16-bit address 1FF9H must be stored within DPTR before the execution of the instruction. This is emitted through ports P0 (lower byte) and P2 (higher byte) and the target data from the external data memory that is copied to the accumulator. No flags are affected, and the DPTR remains unchanged. There is no need, in this case, to entrust any port to output any of the address bytes.

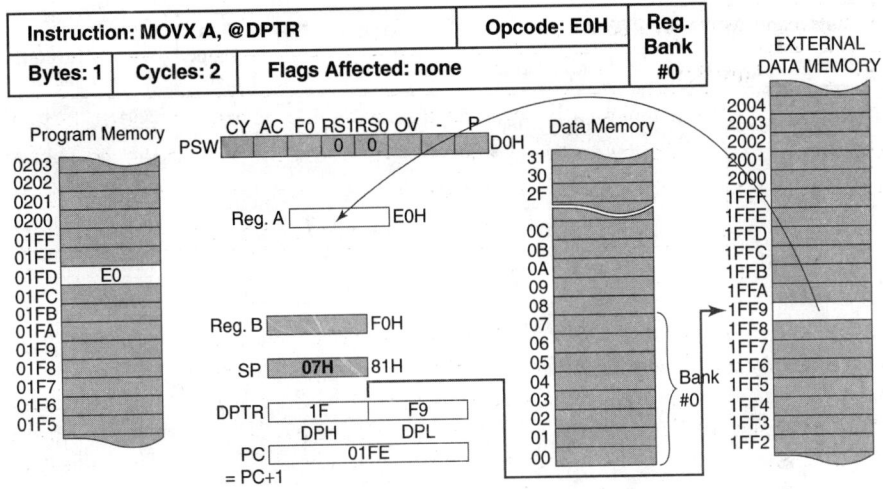

Figure 12.4 Execution of MOVX A, @DPTR instruction

To store data from the accumulator to the external data memory, MOVX, @Ri, A instruction may be used. As an example case, an illustration of the execution of the MOVX @R1, A instruction is presented in Fig. 12.5. In this case, R1 has the lower byte of the 16-bit address (F9H), and the higher byte is assumed to be the output through any other port except port P0. After the execution of the instruction, the accumulator content is copied to the external data memory location. No flags are affected, and the content of the register R1 remains unchanged.

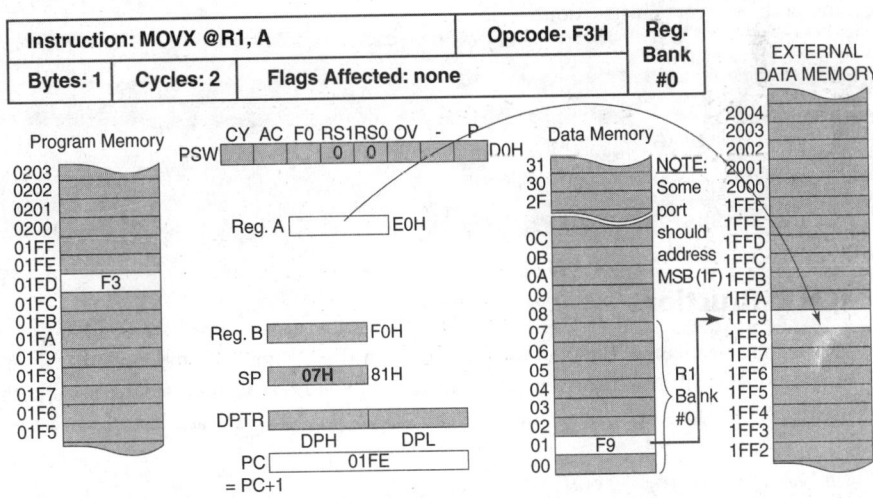

Figure 12.5 Execution of MOVX @R1, A instruction

DPTR may also be used for storing the accumulator content in the external data memory as illustrated in Fig. 12.6. The complete 16-bit address must be available in the DPTR before the execution of this instruction. Just like the previous case of MOVX A, @DPTR, here also, ports P0 and P2 would be used to hold the 16-bit address for the external data memory till the data byte transaction is complete. Port P0 would also transmit the 8-bit data from the accumulator to the external location by multiplexing. No flags would be affected, and the original content of the DPTR would remain unchanged.

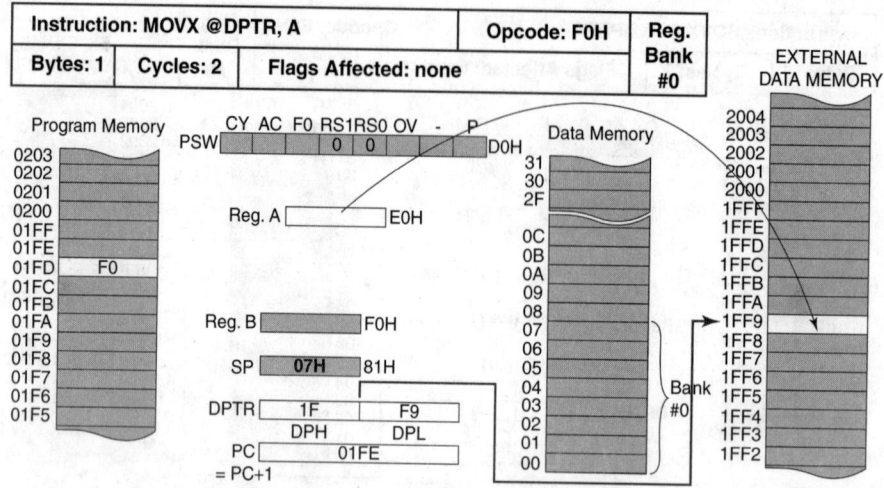

Figure 12.6 Execution of MOVX @DPTR, A instruction

To conclude the discussions about communication between the external data memory and the accumulator, it may be pointed out that the operation becomes more straightforward when the DPTR is used. However, DPTR may only be incremented and cannot be decremented by 16-bit. On the other hand, usage of R0 or R1 would demand a prior use of another port other than P0 to hold the higher byte of the 16-bit address. However, in this case, the registers offer both incrementing and decrementing features, which might be appropriate for some data transactions within a limit of 256 bytes. Otherwise, both types of instructions need two cycles, and both are 1-byte instructions.

Note that data flow in MOVX instruction may be of two types: either loading or storing. However, MOVC allows only loading option.

12.4 | XCH Instruction

The XCH instruction interchanges the byte variables between the accumulator and any other location. Three addressing modes available for this instruction are presented in Table 12.4. As indicated in this table, the possible addressing modes are: register direct, direct and register indirect. The details of each instruction are as follows.

Table 12.4 Addressing modes of XCH instruction

Instruction	Addressing mode	Function
XCH A, Rn	Register direct	Interchange the accumulator and the register contents
XCH A, direct	Direct	Interchange the accumulator and the direct address contents
XCH A, @Ri	Register indirect	Interchange the accumulator and the indirectly addressed contents

As an example the execution of XCH A, R4 instruction is presented in Fig. 12.7. Assuming that bank #0 is currently selected, register R4 of bank #0 and the accumulator would interchange their contents when the instruction is executed. No flags would be affected, and the execution would take only one cycle to complete this 1-byte transaction.

Any directly addressed location's content may also be interchanged with the accumulator using XCH A, direct instruction. As an example, direct address of 30H is used in the instruction XCH A, 30H, the execution of which is illustrated in Fig. 12.8. This is a 2-byte instruction, and the second byte of the opcode contains the direct address. The interchanging operation takes one cycle, and no flags are affected. If any port is addressed, then the data would be taken from its input pins and not from its output latches.

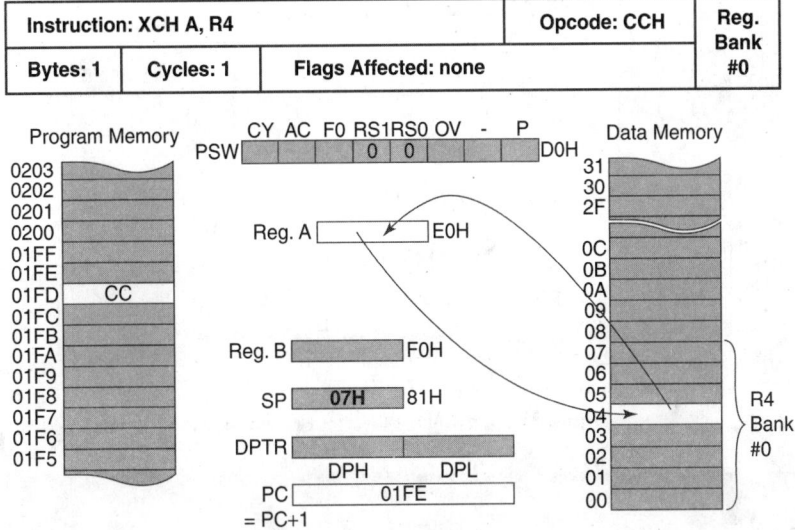

Figure 12.7 Execution of XCH A, R4 instruction

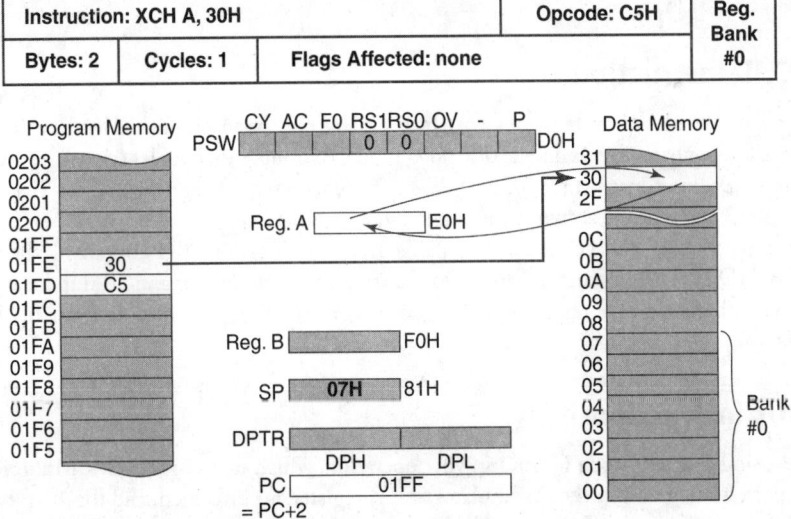

Figure 12.8 Execution of XCH A, 30H instruction

An example of indirect addressing of XCH instruction is presented in Fig. 12.9. The XCH A, @R1 instruction is shown here. Assuming that register bank #0 is selected and register R1 of this bank contains 30H, the illustration shows the interchange of data between the data memory location 30H and the accumulator, which takes only one cycle without affecting any flag.

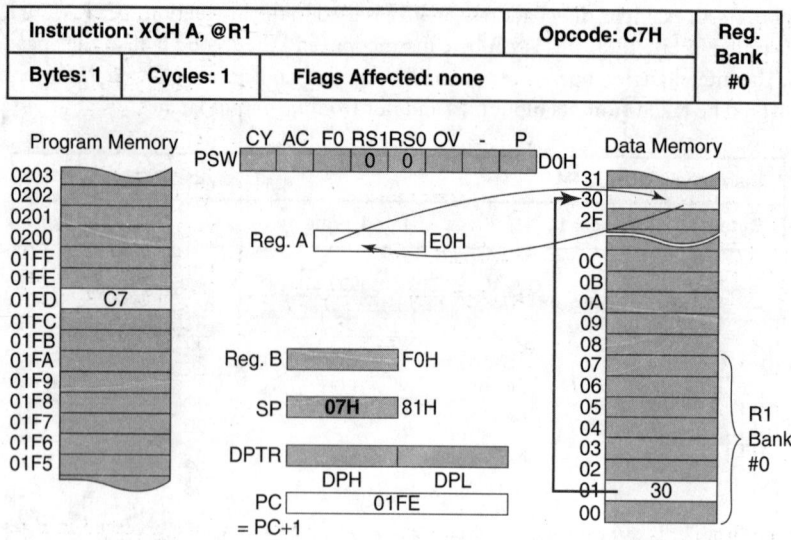

Figure 12.9 Execution of XCH A, @R1 instruction

 Remember the bubble sorting technique of the previous chapter? Try to modify the program by using this XCH instruction.

12.5 | XCHD Instruction

Interchanging of only the lower nibble instead of a byte variable may be accomplished by the instruction XCHD A, @Ri, which allows only register indirect addressing mode. Assuming that bank #0 is selected and register R0 contains 30H, execution of instruction XCHD A, @R0 is illustrated through Fig. 12.10. The exchange takes place between bits 3 and 0 of the accumulator and the indirectly addressed location 30H. No flags are affected.

As an example case, if the accumulator contains 3BH and location 30H contains 25H, then after the execution of XCHD A, @R0 instruction would leave 35H in the accumulator and 2BH in location 30H. Note that the respective higher nibble remains unchanged. This instruction is mostly used for the BCD number manipulations.

12.6 | MUL Instruction

MCS-51 offers single-byte opcode for multiplying operation, which takes four cycles. Multiplication of two unsigned integers, located within the accumulator and B register, is performed, and the 16-bit result is made available in the accumulator and B register. Note that no other registers may be used for this operation. The

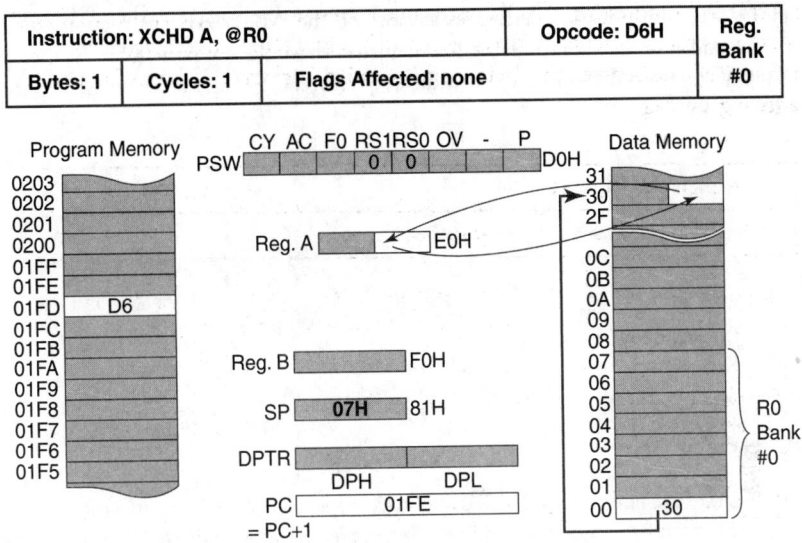

Figure 12.10 Execution of XCHD A, @R0 instruction

lower byte of the product is available in the accumulator, and its higher byte is placed in B register. The CY flag is always cleared by this instruction. If the product is greater than FFH (255d), then the OV flag is set; otherwise it is cleared. No other flag would be changed.

As an example case, the execution of MUL AB instruction is illustrated in Fig. 12.11. It is assumed that the original content of the accumulator, before execution of the instruction, was 04H. Register B

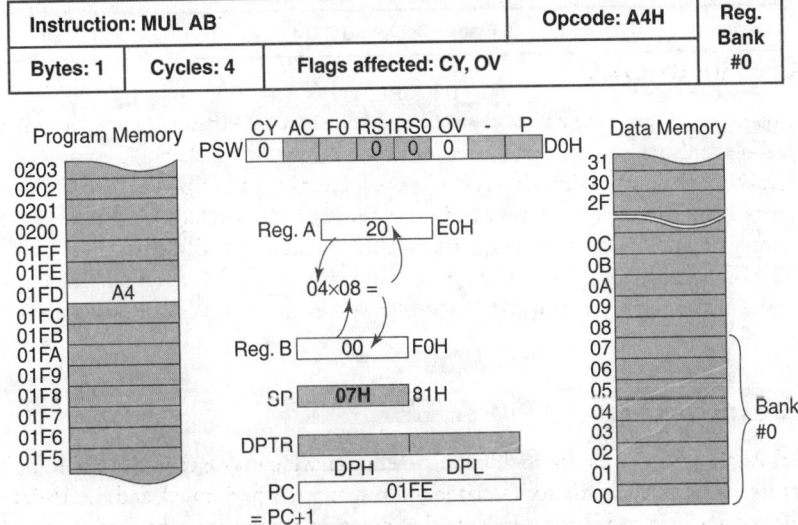

Figure 12.11 Execution of MUL AB instruction

contained 08H before multiplication. After execution of the MUL AB instruction, the product 20H (32d) would be available in the accumulator and register B would contain 00H. As the product is not more than FFH, OV flag would be cleared. As indicated before, CY flag would always be cleared. No other flags would be affected.

 It is a good practice to check the OV flag of PSW before using the result of any MUL instruction. If OV is set, do not forget to use the upper byte of the result available in register B.

12.7 | DIV Instruction

Allowing the DIV AB instruction, MCS-51 offers the scope of dividing an unsigned integer by another unsigned integer, both of an 8-bit value. The integer in the accumulator is divided by the integer in register B. The quotient is available, after execution, in the accumulator, and the remainder may be found in register B. Both CY and OV flags are cleared. No other flags are affected. If it becomes a case of division by zero (register B containing zero before execution), then the OV flag would be set thus indicating the division by zero error. In such a case, the accumulator and register B contents would be undefined.

As an example case, Fig. 12.12 shows the execution of DIV AB instruction assuming that the accumulator contained 21H (33d) and register B contained 10H (16d) before the execution of DIV AB instruction. As illustrated, after the execution of the instruction, the accumulator would contain 02H and register B would be having 01H, indicating the quotient and the remainder, respectively. Both CY and OV flags would be cleared without affecting any other flags.

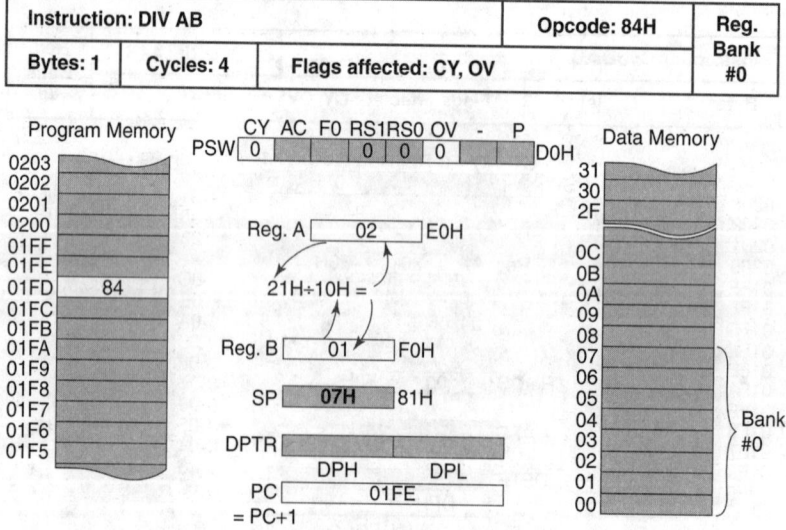

Figure 12.12 Execution of DIV AB instruction

12.8 | DA A Instruction

To perform BCD arithmetic, MCS-51 offers DA A instruction. Execution of this instruction needs some prerequisites. To start with, DA A may only be used immediately after executing either ADD or ADDC instruction of any addressing mode. DA A would be of no use even after a SUBB instruction. Loading any number in the accumulator and then executing DA A instruction would not change the accumulator contents to its BCD equivalent (refer to Fig. 12.13).

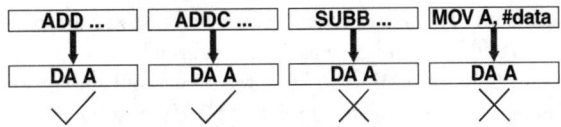

Figure 12.13 Scope of DA A instruction

Secondly, if only BCD operands are used prior to ADD or ADDC instruction, then and then only would DA A generate correct results. This needs some clarification. If we load the accumulator by 3B, then obviously it is a hexadecimal number. However, if we load the accumulator by 32, then it may be interpreted either as a BCD or as a hexadecimal number. As already indicated, DA A instruction is applicable only with BCD arithmetic addition, and 32 would be interpreted as a BCD number in this case.

DA A is a 1-byte, one-cycle instruction with the opcode D4H (Fig. 12.14). If after any addition of BCD numbers any BCD digit within the accumulator is changed to a non-BCD value, then the execution of this instruction would change it back to its correct BCD value. If the digits are within their BCD limits, the accumulator would remain unchanged. In essence, the DA A instruction adds any one of 00, 06, 60 or 66 with the accumulator, depending upon the content of the accumulator and PSW register. These conditions are illustrated in Fig. 12.15.

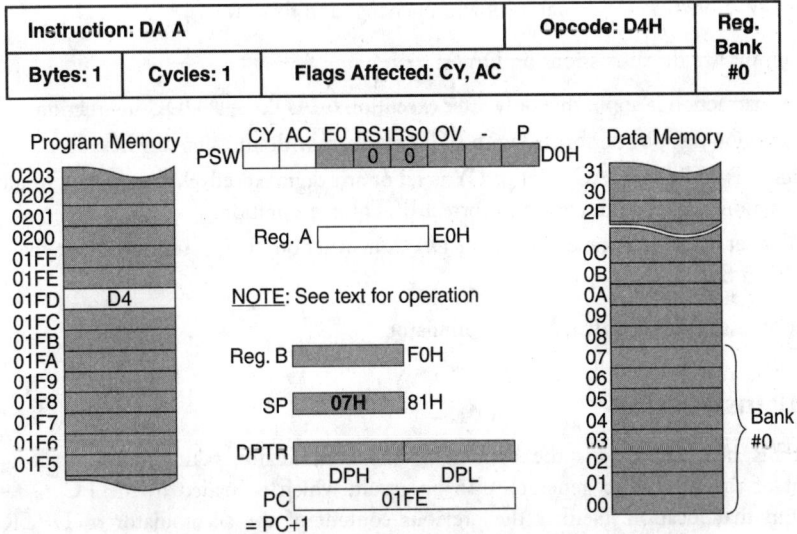

Figure 12.14 Execution of DA A instruction

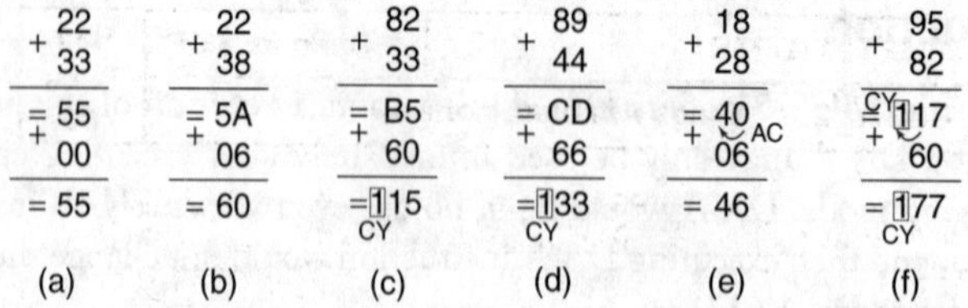

$$
\begin{array}{cccccc}
\begin{array}{r}22\\+\ 33\\\hline\end{array} &
\begin{array}{r}22\\+\ 38\\\hline\end{array} &
\begin{array}{r}82\\+\ 33\\\hline\end{array} &
\begin{array}{r}89\\+\ 44\\\hline\end{array} &
\begin{array}{r}18\\+\ 28\\\hline\end{array} &
\begin{array}{r}95\\+\ 82\\\hline\end{array}
\end{array}
$$

| (a) | (b) | (c) | (d) | (e) | (f) |

$$
\begin{array}{r}=55\\+\ 00\\\hline\end{array}\quad
\begin{array}{r}=5A\\+\ 06\\\hline\end{array}\quad
\begin{array}{r}=B5\\+\ 60\\\hline\end{array}\quad
\begin{array}{r}=CD\\+\ 66\\\hline\end{array}\quad
\begin{array}{r}=40\ _{AC}\\+\ 06\\\hline\end{array}\quad
\begin{array}{r}^{CY}117\\+\ 60\\\hline\end{array}
$$

$$= 55 \qquad = 60 \qquad \underset{CY}{=\!115} \qquad \underset{CY}{=\!133} \qquad = 46 \qquad \underset{CY}{=\!177}$$

(a)   (b)   (c)   (d)   (e)   (f)

Figure 12.15 Different conditions of executions of DA A instruction

Fig. 12.15 presents six different conditions where DA A instruction may be executed. Note that in all cases two two-digit BCD numbers are added, and immediately after this addition, DA A instruction is executed. In the first case [Fig. 12.15(a)], the result of the initial addition was 55 without any carry or auxiliary carry. Therefore, DA A adds 00 with the sum to produce the unaltered result of 55 in the BCD.

The second one is the case of adding BCD numbers 22 with 38, which results in a sum of 5A in the accumulator. The execution of DA A instruction in this case would add 06 with the sum, as the LS digit of the sum is more than 9 [Fig. 12.15(b)]. This produces a result of 60, which is the correct BCD representation of the sum of 22 and 38, both being BCD.

The third case [Fig. 12.15(c)] shows that MSB (B) of the sum B5 invited the DA A instruction to add 60 with the sum to generate the result of 115. Note that in this case, CY is set, indicating that the result of the BCD addition is more than 99 in the BCD.

In the fourth case [Fig. 12.15(d)], the sum becomes CD, and DA A must add 66 with this to generate the correct result of 133, again indicating a carry. These first four cases show that if any digit exceeds 9, then 6 must be added with that digit during DA A execution.

However, in the fifth case, addition of 18 and 28 (both BCD numbers) produce 40 [Fig. 12.15(e)], and neither 4 nor 0 is more than 9. Then why is DA A adding 06 with it? Because there is a carry from bit 3 to bit 4 during this addition, as indicated by the AC flag. That is the reason for adding 06 to get the correct result of 46.

In the last case, addition of 95 and 82 generates 117 with the carry set. Therefore, although none of the digits is more than 9, 60 is then added with the accumulator by the DA A instruction to indicate the correct BCD result of 177. Note that CY remains set and unchanged in this case.

Therefore, to summarize the discussions on DA A:

  (i)   DA A instruction is applicable only after execution of ADD or ADDC instruction.
  (ii) This addition must be performed with BCD numbers without using any hexadecimal number.
  (iii) During this addition, if AC is set or CY is set or any digit exceeds 9, then DA A would take appropriate action to change the result to correct BCD representation.
  (iv) If CY is set even after the execution of DA A instruction, then it indicates that the result is more than 99 in BCD.

DA A instruction is applicable only for the accumulator.

12.9 | JMP Instruction

JMP @A+DPTR instruction adds the content of the accumulator, taking it as an unsigned integer, to the content of the DPTR to generate a 16-bit result, which is loaded in the PC to fetch the next instruction from that location. Neither the previous content of the accumulator or DPTR is changed, nor is any flag affected. This 1-byte instruction takes two cycles to be executed, which is illustrated in Fig. 12.16. Assuming that the DPTR contains 0200H and the accumulator contains 01H, control would

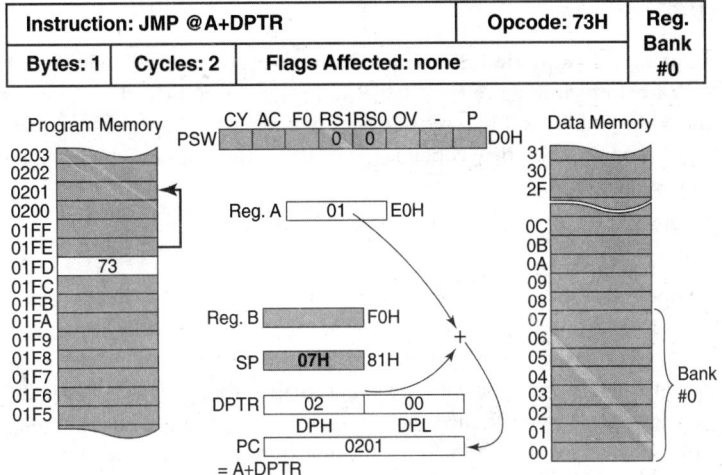

Figure 12.16 Execution of JMP @A+DPTR instruction

branch unconditionally to the location 0201H after the execution of this instruction. This instruction is widely used to perform branching operations such as 'case' statements in high level languages (refer to solved Example 12.2).

12.10 | SWAP Instruction

The execution of SWAP A instruction is illustrated in Fig. 12.17. This 1-byte, one-cycle instruction interchanges the content of the upper and the lower nibbles of the accumulator. No flags are affected. Considering as an example, if the accumulator originally contained 2FH, the accumulator would contain F2H after execution of the SWAP A instruction without affecting any flags. This instruction is equivalent to four successive RR A or RL A instructions.

SWAP is applicable only for the accumulator and not for any other register or memory location.

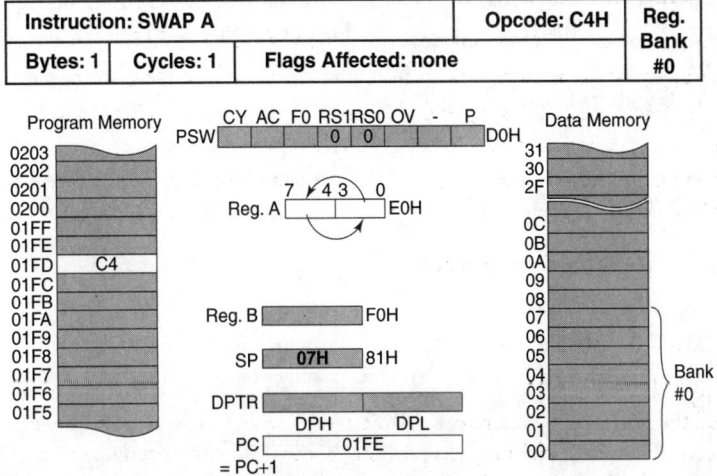

Figure 12.17 Execution of SWAP A instruction

12.11 | JBC Instruction

The 3-byte, two-cycle instruction JBC (opcode 10H) is identical with the JB instruction in all respects, except that it clears the addressed bit before jumping to the relative address if the addressed bit is set. Otherwise, it proceeds sequentially if the addressed bit is 0 (cleared). It does not affect any flag of PSW. The second byte contains the address of the bit, and the third byte contains the relative offset for jumping, having the range of 127 bytes forward or 128 bytes backward.

12.12 | RETI Instruction

The 1-byte, two-cycle (opcode 32H) instruction RETI (return from interrupt service routine) is identical with the instruction RET (return from subroutine) in all aspects except one. RETI enables the disabled interrupt. Whenever any interrupt is acknowledged and the control branches to its service routine, that particular interrupt is automatically disabled for the time being by the processor itself so that multiple false interrupt branching do not take place. However, the concerned interrupt would not function properly against fresh interrupt requests unless it is enabled again before returning from that interrupt's service routine. This is done by RETI, which should be placed at the end of any interrupt service routine. RETI does not affect any flag of the PSW.

12.13 | Solved Examples

Example 12.1

Purpose: Application of MOVC instruction in table look-up method.

Problem

Register R4 would be loaded by an unsigned integer, which may vary from 0 to 5. Find the square of the integer and store it in R5.

Solution

Using MOVC instruction and a table of squares, the problem may be solved in the following way:

```
START:    MOV      A, R4          ; get integer in the accumulator
          ACALL    SQUARE         ; use subroutine SQUARE for squaring it
          MOV      R5, A          ; store result
          RET
```
; Subroutine SQUARE with its table SQTAB.
; This table must be located immediately after the subroutine.

```
SQUARE:   INC      A              ; to get around the next 1 byte RET instruction
          MOVC     A, @A+PC       ; get square in the accumulator using SQTAB
          RET
SQTAB:
          DB       00H            ; square of 0
          DB       01H            ; square of 1
          DB       04H            ; square of 2
          DB       09H            ; square of 3
          DB       10H            ; square of 4
          DB       19H            ; square of 5
```

The INC A instruction in subroutine SQUARE is not required if the range of the input started from 1 and not from 0.

Example 12.2

Purpose: Application of the JMP instruction.

Problem

Write a program to perform addition, subtraction, multiplication or division as per accumulator content of 1, 2, 3 or 4, respectively. Assume that two related operands are available in R2 and R3.

Solution

This program may be developed with DJNZ acc, rel instruction also. However, using JMP @A+DPTR makes the program compact and more efficient. This is more so when the range of the offset is larger.

The first instruction doubles the accumulator content. This is necessary to get the proper entry point using AJMP instruction. Note that AJMP instruction takes 2 bytes. The second instruction loads DPTR with the address of the program memory location indicated by the label 'BRANCH'. The third instruction gets the next executable instruction's address, also known as 'branching address', by adding DPTR to the accumulator content while taking both as unsigned integers. Because JMP is a 1-byte instruction, an NOP instruction is placed to maintain a gap of 2 bytes.

Depending upon the original content of the accumulator, one of the four AJMP instructions would be performed to get the entry point to the relevant subroutine.

```
START:   RL     A              ; multiply the accumulator content by two
         MOV    DPTR,#BRANCH
BRANCH:  JMP    @A+DPTR        ; branch to get proper entry point
         NOP
         AJMP   ADDIT          ; add two integers and return
         AJMP   SUBIT          ; subtract and return
         AJMP   MULIT          ; multiply and return
         AJMP   DIVIT          ; divide and return
; Subroutine to perform addition.
ADDIT:   MOV    A, R2
         ADD    A, R3
         RET
; Subroutine to subtract one register content from the other.
SUBIT:   MOV    A, R2
         CLR    C
         SUBB   A, R3
         RET
; Subroutine to perform multiplication of two unsigned integers.
MULIT:   MOV    A, R2
         MOV    B, R3
         MUL    AB
         RET
; Subroutine to perform unsigned integer division.
DIVIT:   MOV    A, R2
         MOV    B, R3
         DIV    AB
         RET
```

Example 12.3

Purpose: Application of XCH instruction.

Problem

Write a program to interchange data between data memory locations 30H and 31H.

Solution

This may be performed in various ways, and one of them may be as follows. The original content of the accumulator remains unchanged.

```
START;   XCH     A, 30H        ; accumulator gets data from 30H
         XCH     A, 31H        ; data at 30H goes to 31H, 31H data in A
         XCH     A, 30H        ; 31H data placed in 30H, original data in A
         RET                   ; register A unchanged, 30H and 31H exchanged
```

 We shall discuss a few more examples of application of some of these instructions in the next chapter.

SUMMARY

Instructions such as MOVC, MOVX, XCH, MUL, DIV and DA A were discussed in this chapter. These are very powerful instructions to perform BCD arithmetic, unsigned integer multiplication and division, to communicate with external data memory, to get a constant from the program memory.

POINTS TO REMEMBER

- DA A is applicable after ADD or ADDC instructions only if its operands are already in BCD.

- After execution of MUL AB instruction, the lower byte of the 16-bit product would be available in the accumulator and the higher byte in the B register.

REVIEW QUESTIONS

Evaluate Yourself

1. Which of the following instructions, when executed four times, would perform the same duty as of the SWAP A instruction?
   (a) RLC A   (b) RR A
   (c) RRC A   (d) None of these

2. Communication with the external data memory may be achieved only through
   (a) The accumulator   (b) B register
   (c) PSW   (d) None of these

3. How many addressing modes are offered by XCHD instruction?
   (a) One   (b) Two
   (c) Three   (d) None of these

4. After execution of MUL AB instruction, the higher byte of the product would be available in
   (a) The accumulator   (b) B register
   (c) PSW   (d) None of these

5. After execution of DIV AB instruction, the quotient is available in

   (a) The accumulator   (b) B register

   (c) PSW   (d) None of these

6. DA A instruction is applicable only for the following number type?

   (a) Binary   (b) Hexadecimal

   (c) BCD   (d) None of these

7. DA A instruction may be used immediately after the execution of

   (a) SUBB   (b) MUL

   (c) DIV   (d) None of these

8. The instruction JMP @A+DPTR is capable of transferring the control within

   (a) 256 bytes   (b) 2K bytes

   (c) 64K bytes   (d) None of these

9. The base address of MOVC A, @A+PC instruction is

   (a) PC   (b) PC+1

   (c) PC+1+A   (d) None of these

10. The base address of MOVC A, @A+DPTR instruction is

    (a) DPTR   (b) DPTR+1

    (c) DPTR+1+A   (d) None of these

Search for Answers

1. Which addressing mode is used by MOVX instruction?

2. How is the 16-bit address generated during execution of MOVX A, @Ri instruction?

3. What is the difference between XCH and XCHD instructions?

4. Is it possible to exchange a byte between R2 and R4 of the current register bank, using XCH instruction?

5. What would happen if you tried to divide a number by zero, using DIV AB instruction?

6. How many flags are affected by DA A instruction?

7. What are the restrictions in using DA A instruction?

8. How does DA A instruction indicate that the result is more than 99 in BCD?

9. Is it possible to perform a backward jump using JMP @A+DPTR instruction?

10. What changes would be necessary in Example 12.1 to get the cube of the integer input?

Think and Solve

1. How can Example 12.3 be modified so that data in 30H is shifted to 31H, data in 31H is shifted to 32H and data in 32H is shifted to 30H?

2. What is the limitation of the size of the table that may be looked up using MOVC A, @A+PC instruction?

3. During the usage of MOVX A, @DPTR instruction, would the higher byte of the 16-bit external data memory address be latched in P2 so that the subsequent MOVX A, @Ri instruction is sufficient to get the correct data in the accumulator?

4. What would be the branching address if the accumulator contains 00H during the execution of JMP @A+DPTR instruction?

5. How can an infinite loop be generated using JMP @A+DPTR instruction?

6. In Example 12.2, what changes would be necessary if all AJMP instructions are replaced by LJMP instructions?

7. Is it possible to multiply a 16-bit unsigned integer by an 8-bit unsigned integer using MUL instruction? Justify your answer.

8. Assume there is a display of a digital clock in BCD format with a range of 12 h. Write a program that can be called at every minute to change its hour and minute counters, which are also BCD counters.

9. Write a program to copy 16 bytes of data from the external data memory location 2000H onwards to internal data memory location 30H. Assume that port 1 is interfaced with the higher 8-bit address of the external data memory.

10. Sixteen-bit values of the square of integers are stored in a table that starts from program memory location 0200H onwards. The lower byte of the 16-bit number is at the lower address, and the higher byte is at the higher address. Assuming that the accumulator contains the integer whose square is required, write a routine to get the result using table look-up procedure (MOVC instruction).

13 PROGRAMMING EXAMPLES-III

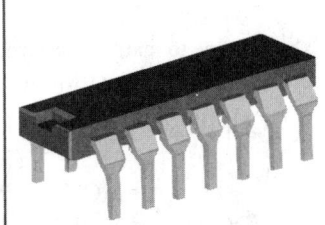

CHAPTER OBJECTIVES

As we have completed discussions of all MCS-51 instructions, it may not be inappropriate to study a few more example programs highlighting certain software development techniques. After completion of the chapter, the reader should be able to understand

- How to communicate with external data memory using MOVX instruction.
- How to use table look-up method and MOVC instruction.
- How to convert a hexadecimal number to its BCD form using DA A instruction.
- How to use XCH, XCHD and SWAP instructions.
- How to use flag bits and bit-addressable area of internal data memory.

13.1 | Compare with External Array

Example 13.1

Purpose: How to access an array located within external data memory.

Problem

An array of 10 bytes is available in the external data memory location from 2000H onwards. Compare it with the array of 10 bytes available in the internal data memory location from 30H onwards. If both are identical in all respects, then return with carry (CY) flag cleared; otherwise, set CY flag at return.

Solution

In general, two arrays are taken as identical if both have equal number of terms and if individual array elements match in their position. For example, an array of 1, 2, 3 would match with another array of 1, 2, 3 but would not with a third array of 3, 1, 2.

   As one array is located at the external data memory, it may be copied to the accumulator with MOVX instruction through DPTR. The internal array may be pointed through a pointer, R0. Under these conditions, comparison may be carried out using indirect addressing mode. However, as a CJNE instruction does not allow comparison between the accumulator and any indirect address, SUBB instruction is more appropriate to check whether both terms are matching or not. The CY flag may be adjusted accordingly before returning from the routine. It may be pointed out that whenever the first mismatch is found, the routine need not compare the remaining terms and should quit with its CY flag as set. The following is the program listing.

; Subroutine to match an external array with an internal one.
; If both are matching then, at return, CY flag is cleared otherwise it is set.

```
COMPEX:   MOV    DPTR, #2000H   ; initialize external array pointer
          MOV    R0, #30H       ; initialize internal array pointer
          MOV    R7, #0AH       ; initialize counter for 10 bytes
LOOP:     MOVX   A, @DPTR       ; get one external term
          CLR    C              ; compare with corresponding
          SUBB   A, @R0         ; internal term
          JZ     NEXT1          ; equal, then jump at NEXT1
          SETB   C              ; unequal, no match, set CY
          RET                   ; and go back
NEXT1:    INC    DPTR           ; point next external term
          INC    R0             ; point next internal term
          DJNZ   R7, LOOP       ; any more to compare?
          CLR    C              ; no, all term matched, clear CY
          RET                   ; and go back
```

13.2 | Find Sum of a Series

Example 13.2

Purpose: To highlight the difference in subroutines for table look-up method with data starting from 0 and 1.

Problem

Write a subroutine to find the sum of the following series up to N terms. N is stored in location 30H. At return, the sum should be available in the accumulator. Assume that the value of N would not be more than 5.

$$(\text{Term})_n = n^3 - (n-1)^2$$
$$\text{Sum} = (1^3 - 0^2) + (2^3 - 1^2) + (3^3 - 2^2) + \dots \text{ up to } N \text{ terms}$$

Solution

As the value of N is available, it may be used as a counter, and individual terms may be calculated during iteration. Table look-up method to find the square and the cubes of natural numbers would simplify the program-development process. The following is the program listing.

; Program to calculate the sum of the given series.
; Value of N is available in location 30H.
; Result (sum) would be available in the accumulator at return.
; R5 holds temporary sum, R6 the term, and R7 is used as the counter.
; This routine uses two more subroutines, GETSQ and GTCUBE.

```
SERIES:   MOV    R7, 30H        ; load counter by N value
          CLR    A              ; to initialize
          MOV    R6, A          ; first term − 1
          MOV    R5, A          ; clear sum
LOOP:     MOV    A, R6          ; get term − 1
          ACALL  GETSQ          ; square it
          MOV    R4, A          ; temp. save for deduction
          INC    R6             ; make it term
```

```
        MOV        A, R6        ; load term in A
        ACALL      GTCUBE       ; cube it
        CLR        C
        SUBB       A, R4        ; subtract saved number
        ADD        A, R5        ; add with the previous sum
        MOV        R5, A        ; save back
        DJNZ       R7, LOOP     ; continue for all terms
        MOV        A, R5        ; copy the sum in the accumulator
        RET                     ; done
```

; Subroutine to supply square of integer input in A using table look up.
; Table used is SQTAB. INC A is necessary as table starts from 0.

```
GETSQ:  INC        A            ; to take care of RET
        MOVC       A, @A+PC     ; get square of integer in accumulator
        RET
SQTAB:
        DB         00H          ; square of 0
        DB         01H          ; square of 1
        DB         04H          ; square of 2
        DB         09H          ; square of 3
        DB         10H          ; square of 4
```

; Subroutine to supply cube of integer input in A using table look up.
; Table used is CUBETB. INC A is not necessary as table starts from 1.

```
GTCUBE: MOVC       A, @A+PC     ; get cube of integer in accumulator
        RET
CUBETB:
        DB         01H          ; cube of 1
        DB         08H          ; cube of 2
        DB         1BH          ; cube of 3
        DB         40H          ; cube of 4
        DB         7DH          ; cube of 5
```

If the look-up table starts from 0 then INC A must be used before MOVC instruction. However, if the look-up table starts from 1, INC A instruction to be avoided before MOVC instruction.

C-version

```c
#include <regx51.h>

// function proto

unsigned char SeriesFunc(unsigned char n);
```

```
void main(void)
{
        unsigned char *pLocN, i;
        unsigned char sum;

        // initialize

        pLocN = 0x30;
        sum = 0;

        if(*pLocN > 0)
        {
                for(i=1; i <= *pLocN; i++)
                {
                        sum = sum + SeriesFunc(i);
                }

                // store results

                ACC = sum;
        }
        while(1);
}
```

// Function Definition

```
unsigned char SeriesFunc(unsigned char n)
{
        unsigned char term;

        term = n*n*n – (n-1)*(n-1);
        return term;
}
```

13.3 | Reverse an Array

Example 13.3

Purpose: Application of XCH instruction through indirect addressing.

Problem

An array of random integers is placed from internal data memory location 31H onwards. The number of terms (N) of the array is available in the location 30H. Develop a program to place the entire array in reverse order in the same memory area.

Solution

Schematically, the problem is illustrated in Fig. 13.1. Assuming only 15 terms in the array (N = 15d), it is shown that the term in location 31H is to be shifted to 3FH, 32H to 3EH and so on. Following this order, the last change would be from location 3FH to 31H. Note that as the number of terms is odd in this case, the term in location 38H would remain unchanged. For an array with even terms, this would not happen. A simpler or uncomplicated solution technique would be to create another array in reverse order and copy that array to the present location. However, using XCH instruction, the interchanging operation may be performed easily without the need of any additional storage area.

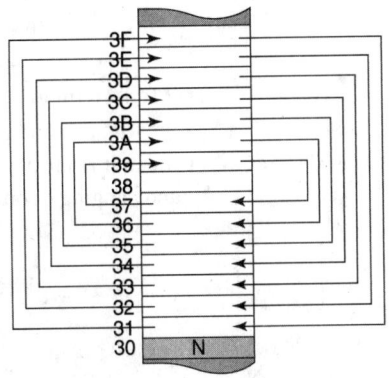

Figure 13.1 Schematic for problem definition

We may use R0 and R1 as two pointers and develop a subroutine in the line of Example 12.3. The number of place interchanges would be half of N, which is the number of terms, if N is even. If it is odd, then the number of iterations (place interchanges) required would be $(N-1)/2$. This may easily be achieved by performing the division by rotating the accumulator right, one bit, after clearing CY flag. To find the last address of the array, N value may be added with the base address 30H. The following is the program listing, including the subroutine CHANGE.

; Program to reverse the order of an array of N terms.
; Value of N is available in location 30H. The array itself starts from 31H.
; First four instructions divide N by 2 to get the number of iterations to store in R7.

```
REVRSE:     MOV     A, 30H          ; get N value
            CLR     C
            RRC     A               ; divide by 2
            MOV     R7, A           ; store in counter
```

; Following four instructions load R0 and R1 by start and end addresses of the array. End address is calculated
; from base address and the number of terms.

```
            MOV     R0, #31H        ; start address pointer
            MOV     A, 30H          ; get N value
            ADD     A, #30H         ; add with base address
            MOV     R1, A           ; store end address in pointer
```

; Iteration for reversing the order of array starts from here.

```
XCHNG:      ACALL   CHANGE          ; exchange places of two terms
            INC     R0              ; next term near start of array
            DEC     R1              ; next term near end of array
```

```
            DJNZ        R7, XCHNG    ; continue for all terms
            RET                      ; over
; Name:      CHANGE
; Function:  Interchange contents of two locations as pointed by R0 and R1
; Input:     R0, R1 point the pair to be interchanged
; Output:    Pointed contents interchanged, the accumulator remains unchanged
; Calls:     None
; Uses:      A, R0, R1
    CHANGE:  XCH        A, @R0       ; get front term in A
             XCH        A, @R1       ; store it at rear and get rear term in A
             XCH        A, @R0       ; store it at front
             RET                     ; interchanging of one pair complete
```

 Observe how all important details of the subroutine are documented at its beginning.

13.4 | HEX to BCD Conversion

Example 13.4

Purpose: Application of DIV and SWAP instructions.

Problem

Some random hexadecimal numbers are stored from location 31H onwards. Number of terms (N) is available in location 30H. Convert all numbers to their corresponding BCD forms and store in their original locations. Assume no stored number is more than 63H.

Solution

A simple way to convert any hexadecimal number to its BCD equivalent is to use the table look-up procedure. However, this needs a 256-byte table, which consumes substantial memory space. Another method is to divide the hexadecimal number by powers of 10. As the problem states that the hexadecimal number would not be more than 63H or 99d, the given hexadecimal number may be divided by 10d or 0AH. The quotient becomes the MS, and the remainder becomes the LS BCD digits. The following is the program for HEX to BCD conversion within the BCD limit of 99.

```
; Program to convert an array of hexadecimal digits to their BCD equivalent.
; Assumed all entries of input array are less than 64H (100d).
; Converted BCD numbers replace original hexadecimal numbers by destroying them.
    HXBCD:    MOV    R7, 30H      ; load N in counter
              MOV    R0, #31H     ; set pointer at start of array
    LOOP:     MOV    A, @R0       ; get one hexadecimal number
              ACALL  HEXBCD       ; convert to BCD, result in the accumulator
              MOV    @R0, A       ; save in the original location
```

```
            INC     R0          ; point next location
            DJNZ    R7, LOOP    ; continue for all entries
            RET                 ; over
; Name:      HEXBCD
; Function:  Calculates BCD equivalent of any HEX number less than 64H (100d)
; Input:     accumulator contains the HEX number
; Output:    accumulator contains converted BCD number
; Calls:     none
; Uses:      flags, A, B
    HEXBCD: MOV     B, #0AH     ; load B by 10d
            DIV     AB          ; quotient in the accumulator and reminder in B
            SWAP    A           ; place MS BCD digit in its place
            ORL     A, B        ; merge with LS BCD digit
            RET
```

13.5 | Update Clock Display

Example 13.5

Purpose: Application of DA A instruction for a real-time clock (RTC).

Problem

Develop a subroutine to update the display of a clock that can be called at every minute. The clock should display hours and minutes in BCD format. After displaying 23.59, the display should show 00.00. Assume that the hour count is stored at location 31H and the minute count in location 30H, both in packed BCD format.

Solution

A schematic representation of the problem is presented in Fig. 13.2. On the left-hand side, the display is indicating hour and minute of the day in 24-h format. On the right-hand side, the concerned memory locations act as counters for hour and minute in packed BCD format.

31	2	3	Hour Count
30	5	9	Minute Count

DISPLAY   MEMORY CONDITION

Figure 13.2 Schematic of clock display counter

A routine to increment the minute counter should process both memory locations. A program for that, using DA A instruction, may be developed as follows.

```
; Program to update the BCD minute (and hour, if necessary) counter of a clock.

    UPCLK:  MOV     A, 30H      ; old BCD minute count in A
            ADD     A, #01H     ; increment it by one
            DA      A           ; adjust for BCD increment
```

MOV	30H, A	; update minute count
CJNE	A, #60H, GOBAK	; yet to complete an hour
MOV	30H, #00H	; 1 h over, set minute as 00
MOV	A, 31H	; get hour count in A
ADD	A, #01H	; increment hour by one
DA	A	; adjust for BCD increment
MOV	31H, A	; store back hour count
CJNE	A, #24H, GOBAK	; no need to reset hour
MOV	31H, #00H	; reset hour count also
GOBAK: RET		; over

13.6 | Display Shift for Right Entry

Example 13.6

Purpose: To highlight the coordination of XCH, XCHD and SWAP instructions.

Problem

A four-digit BCD display to be shifted left by one digit in order to accommodate a freshly entered BCD digit available in the accumulator. Develop a subroutine to accomplish this task, assuming that locations 31H and 30H contain the higher and lower order numbers, respectively, in packed BCD format.

Solution

The schematic representation of the problem is illustrated in Fig. 13.3. As we may observe from this illustration, this problem is oriented on a 4-bit left shift of a 16-bit number system. However, implementation of RLC A instruction would not result in a compact and efficient routine.

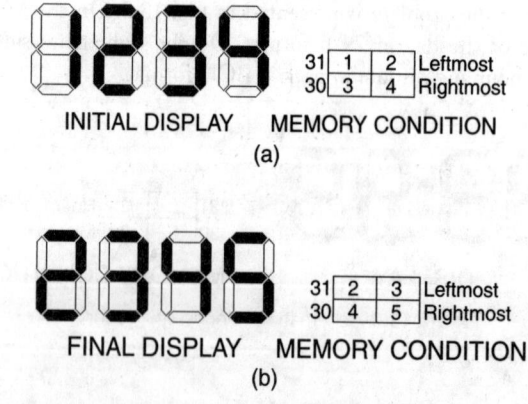

| 31 | 1 | 2 | Leftmost |
| 30 | 3 | 4 | Rightmost |

INITIAL DISPLAY   MEMORY CONDITION
(a)

| 31 | 2 | 3 | Leftmost |
| 30 | 4 | 5 | Rightmost |

FINAL DISPLAY   MEMORY CONDITION
(b)

Figure 13.3 Schematic of right entry operation (a) before and (b) after entry

An alternate method is to use XCH, XCHD and SWAP instructions in proper sequence to achieve the target. Assuming that R0 is used to point the lower byte (30H) and R1 is used to point the higher byte (31H), Fig. 13.4 presents the program listing along with pictorial comments against the execution of every instruction.

RENTRY: MOV R0, #30H   31 | 1 | 2 | 30 | 3 | 4 |
      MOV R1, #31H   R1 | 31 | reg. A | 0 | 5 | R0 | 30 |

XCH A, @R1   31 | 0 | 5 | 30 | 3 | 4 |
              reg. A | 1 | 2 |

SWAP A   31 | 0 | 5 | 30 | 3 | 4 |
              reg. A | 2 | 1 |

XCH A, @R1   31 | 2 | 1 | 30 | 3 | 4 |
              reg. A | 0 | 5 |

XCH A, @R0   31 | 2 | 1 | 30 | 0 | 5 |
              reg. A | 3 | 4 |

SWAP A   31 | 2 | 1 | 30 | 0 | 5 |
              reg. A | 4 | 3 |

XCHD A, @R1   31 | 2 | 3 | 30 | 0 | 5 |
              reg. A | 4 | 1 |

XCHD A, @R0   31 | 2 | 3 | 30 | 0 | 1 |
              reg. A | 4 | 5 |

XCH A, @R0   31 | 2 | 3 | 30 | 4 | 5 |
RET   reg. A | 0 | 1 |

Figure 13.4 Instructions and execution of the right entry software

In this subroutine, the first two instructions load two pointers R0 and R1 by the addresses of the two target memory locations. The next five instructions perform 4-bit left shift of both memory locations using SWAP instruction. The next three instructions insert new BCD digits in both memory locations as necessary for a left shift with the new entry as the rightmost digit. The last instruction terminates the subroutine.

Note that it is the SWAP A instruction which performs the left shift. As swapping is possible only if the number is in the accumulator, XCH A, @Ri instructions are used at proper places. Finally, XCHD instruction inserts a new digit by exchanging it with another one. It may also be noted that the indirect addressing mode, used in this subroutine, is responsible for maintaining the code length at its minimum level. In Fig. 13.4, communication or action takes place only between or within unshaded locations. All locations, which are shaded, remain passive during the execution of the concerned instruction.

 The reader may develop the same solution using RLC A instruction. In that case accumulator to 30H and 30H to 31H shifting to be performed four times. Of course the accumulator nibbles must be swapped before that.

13.7 | Count Number of Words

Example 13.7

Purpose: To scan an array of string and to identify location of words.

Problem

A portion of a written text is stored in the internal data memory location from 40H to 7FH so that it occupies 64 bytes. The text is in the form of American Standard Code for Information Interchange (ASCII) and contains several words. ASCII character 'space' of code 20H separates any two words in the text. The text may or may not start with a space and may or may not end with a space. Multiple spaces are also possible in between words and at the start and the end. Develop a program to count the number of words within the text, and store this number in the accumulator.

Solution

To start with, let us assume two extremes of the condition spectrum. In one worst case, the entire memory (64 bytes) may not have a single letter. On the other hand, it might be possible that there is no 'space' (ASCII code 20H) at all and all locations contain some alphanumeric characters. Therefore, the algorithm must be robust enough to tackle any such situation and produce correct results. It cannot be assumed that there would be only one 'space' between any two words or the array would always start with a word and not with a 'space'.

One simple method to tackle this situation is to use a flag. A flag is generally a bit which can have only two states, namely 0 and 1 or cleared and set. In the present case, it may be assigned to indicate that the scanning is passing through a 'space' by its cleared state. In other words, if the flag bit is 0, then the last checked location contained a 'space'. On the other hand, if the flag bit is showing 1, then the last scanned location was not a 'space' but a character. Content of current location under study and content of flag bit indicating last studied condition can generate any one of the four combinations, as shown in Table 13.1. Within bit-addressable area, bit location 00H is designated as flag bit.

Table 13.1 Indications from flag bit and current location content combinations

Condition	Flag bit content	Current location content	Remarks
1	0	Space	Scanning through space
2	0	Character	Start of a word
3	1	Space	End of a word
4	1	Character	Scanning through a word

As the table indicates, the start and end of any word would be having their characteristic features indicated by the combination of flag bit and current location content. This may be carefully used to count the number of words in an array. A flowchart representing the program logic is shown in Fig. 13.5. One more point may be noted here. Instead of using a bit, we may also assign a byte to indicate the same flag. The byte may contain 00H to indicate a 'space' and FFH to indicate the presence of a character. However that would be a waste of memory space.

```
; Program to calculate number of words in an array.
; The array contains ASCII characters (character 20H represents space).
; Within bit-addressable area, bit location 00H is designated as flag bit.

        WCOUNT:  MOV   R0, #40H            ; point to start the array of ASCII codes
                 MOV   R7, #04H            ; counter for 64d bytes
                 MOV   R6, #00H            ; counter for number of words found
                 CLR   00H                 ; clear flag bit
        LOOP:    CJNE  @R0, #20H, CHAR1    ; not space but one character
```

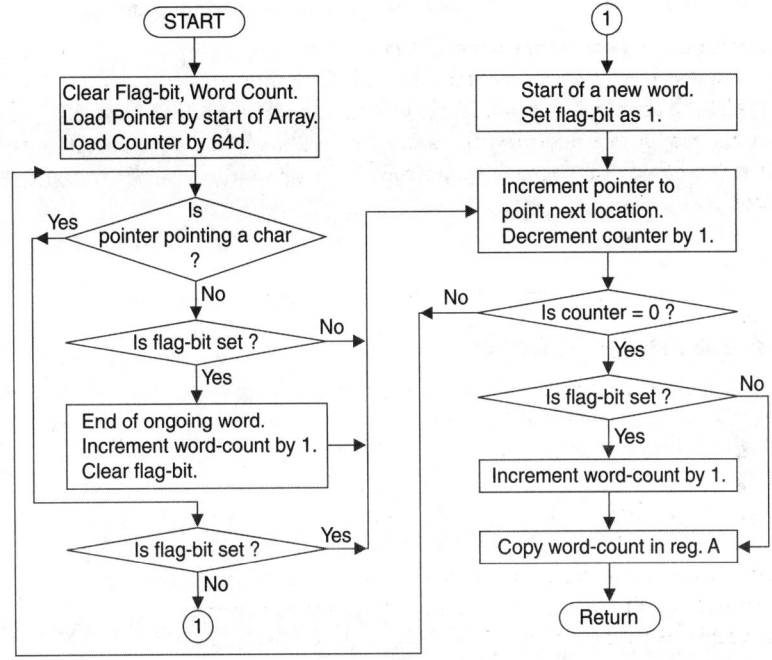

Figure 13.5 Flowchart of word counting subroutine

; Present ASCII code is a space. If flag bit is set, then it is the end of the ongoing word. In that case, word
; count (R6) is to be incremented by 1, and flag bit is to be cleared. Otherwise, if flag bit is already cleared,
; then no action is required. Check next byte, if any.

```
SPACE:  JNB     00H, NEXT1      ; check next if flag bit is cleared
        INC     R6              ; flag bit set, one more word found
        CLR     00H             ; found a space, clear flag bit
        SJMP    NEXT1           ; check next location, if any
```

; Present ASCII code is a character.
; If flag bit is cleared, then this is the start of a new word.
; Set flag bit in that case and continue.

```
CHAR1:  JB      00H, NEXT1      ; continuation of the current word
        SETB    00H             ; start of new word, set flag bit
```

; Check for any more locations pending.

```
NEXT1:  INC     R0              ; point to the next location
        DJNZ    R7, LOOP        ; continue for all locations
```

; If flag bit is set, then increment word count (R6) by 1, else quit.
; This is necessary when array terminates with a character other than space.

```
        JNB     00H, OVER       ; flag bit is cleared
        INC     R6              ; add up the last word
```

OVER:	MOV	A, R6	; the accumulator collects word count
	RET		; done

> The problem becomes much simpler if it is assumed that there would be no space at the beginning and there would be only one space between any two words. The reader may rewrite the program with these assumptions.

13.8 | Generate Prime Numbers

Example 13.8

Purpose: Highlighting multiple types of data-manipulation operations with an array.

Problem

There are 25 prime numbers between 2 and 100. Find a method to generate these prime numbers.

Solution

A prime number is defined as a natural number that is not perfectly divisible by any other number except by 1 and by itself. The following are the prime numbers within 100d:

2, 3, 5, 7, 11, 13, 17, 19, 23, 29, 31, 37, 41, 43,
47, 53, 59, 61, 67, 71, 73, 79, 83, 89 and 97.

In general, 1 is not considered as a prime number.

A simple method to generate prime numbers is to generate an array of natural numbers first. Then, starting from 2, every second number is deleted (replaced by 0). Searching for the next undeleted number and getting 3 would start the next iteration of deleting every third number starting from it. Similarly, deleting would continue with every 5th, 7th, 11th numbers and so on. If 100d is the maximum limit, then searching for undeleted numbers may be terminated at half of it, i.e., at 50d. This method of deletion to generate prime numbers is illustrated schematically in Figs 13.6(a–d). After completion of deletion, the prime numbers may be shifted to generate a compact list. A program may be developed using this technique, which is given below. Note that out of the available 128 bytes of the internal data memory space, we are to leave 100d locations for storing the natural numbers up to 100. That leaves 20 bytes for registers and stacks. Therefore, it is preferable to start the storage from location 1BH and to extend it up to 7EH.

Algorithm

Steps from 1 to 3 are for generating natural numbers up to 100d.

Step 1: Select bank #0 and set SP at 07H. Load R7 by 100d, clear the accumulator and load R0 (natural number storage pointer) by 1BH, which is the first storage location.

Step 2: Increment the accumulator by 1, and store at the location pointed by R0. Then increment R0 by 1 to point next storage location and decrement R7 by 1.

Step 3: If R7 is not 0, then go to Step 2.

Steps from 4 to 9 are for generating prime numbers by method of deletion.

Step 4: Load R0 by 1BH (start of array) and R7 by 32H (50d).

Step 5: Increment R0 by 1, and get the number pointed by R0 in the accumulator. If the accumulator contains 0, then go to Step 9.

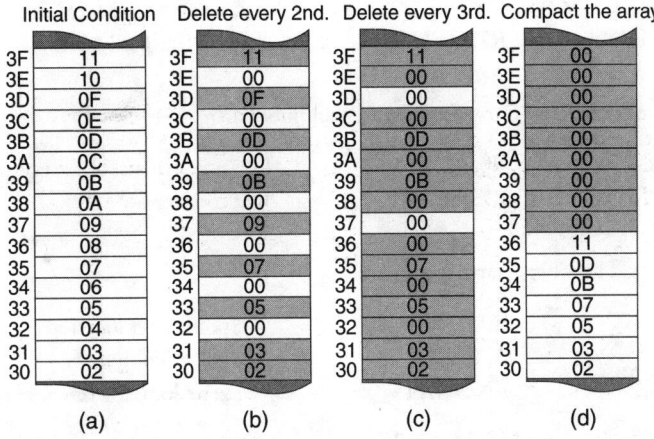

Figure 13.6 Schematic of the deletion method to generate prime numbers

Step 6: Save the number from the accumulator in R6 as offset. Load R1 by the present value of R0.

Step 7: Add offset in R6 with R1. If R1 is the last address of array, then go to Step 9.

Step 8: Clear the location pointer by R1. Then go to Step 7.

Step 9: Decrement R7 by 1. If R7 is not 0, then go to Step 5.

Steps from 10 to 19 are for compacting the array of undeleted numbers (prime numbers).

Step 10: Load R0 by 1BH to point the start of array.

Step 11: If the location pointed by R0 does not contain 0, then go to Step 18.

Step 12: Copy R0 in R1.

Step 13: Increment R1 by 1. If R1 is not pointing to the end of array, go to Step 15.

Step 14: Go to Step 19.

Step 15: If the location pointed by R1 is not containing 0, go to Step 17.

Step 16: Go to Step 13.

Step 17: Copy the prime number pointed by R1 into the location pointed by R0. Then clear the location pointed by R1.

Step 18: Increment R0 by 1. Then go to Step 11.

Step 19: Return.

```
; Program to generate prime numbers upto 100d.
; Register bank used #0. Stack starts from 08H with SP as 07H, allowing 18 bytes.

PRIME:    MOV     PSW, #00H      ; select bank #0
          MOV     SP, #07H       ; stack from 08H onwards
          MOV     R7, #64H       ; counter for 100d
          CLR     A              ; natural number generator
          MOV     R0, #1BH       ; start of storage area for natural numbers

; Generate and store natural numbers up to 100d, starting from location 1BH.

GENAT:    INC     A              ; generate next natural number
          MOV     @R0, A         ; save in its location
```

```
            INC       R0                    ; next storage location
            DJNZ      R7, GENAT             ; continue till 100d
```

; Initialize for search of undeleted numbers (non-zero numbers).

```
            MOV       R0, #1BH              ; point to start of array, storing 01
            MOV       R7, #32H              ; counter for 50d
```

; Searching for undeleted numbers starts from here.

```
NXTNUM:     INC       R0                    ; point next location
            MOV       A, @R0                ; get its content
            JZ        ENDNXT                ; zero, look for the next one
```

; On encountering one prime number, delete all locations which are storing its multiples.

```
            MOV       R6, A                 ; store present offset
            MOV       A, R0
            MOV       R1, A                 ; R1 is pointer for deleting
DELALL:     MOV       A, R1
            ADD       A, R6
            MOV       R1, A
            CJNE      A, #0FEH, NEXT1       ; is it the last address?
NEXT1:      JNC       ENDNXT                ; R1 exceeded 100d
            MOV       @R1, #00H             ; clear this location
            SJMP      DELALL                ; continue deleting
ENDNXT:     DJNZ      R7, NXTNUM            ; continue for all
```

; Now compact the array.

```
            MOV       R0, #1BH              ; point to start of array
COMPAQ:     CJNE      @R0, #00H, GETNXT     ; is the term 0?
            MOV       R1, 00H               ; copy R0 in R1
SRCHNM:     INC       R1                    ; see next higher location
            CJNE      R1, #0FFH, INVEST     ; end of array?
            SJMP      OVER                  ; no more to compact
INVEST:     CJNE      @R1, #00H, SHIFT1     ; it is a prime number
            SJMP      SRCHNM                ; contains 0, look forward
SHIFT1:     MOV       A, @R1                ; copy the prime number
            MOV       @R0, A                ; to lowest blank position
            MOV       @R1, #00H             ; clear copied location
GETNXT:     INC       R0                    ; to find next slot
            SJMP      COMPAQ                ; continue till end of list
OVER:       RET                             ; over
```

 Another method of solution is to take every integer and find whether it is divisible by any other integer except 1 and itself. If this logic is implemented, then the program would be more compact.

SUMMARY

Applications of instructions such as MOVX, MOVC, MUL, DIV, DA A, XCH, XCHD and SWAP were discussed in this chapter through suitable software development examples. Several techniques of algorithms, such as prime number generation by method of deletion, 4-bit left shift of 16-bit operand and counting number of words from an ASCII text file, have also been discussed. Proper usage of flags and bit-addressable data memory area for that purpose have also been demonstrated by suitable example cases.

POINTS TO REMEMBER

- DIV AB instruction leaves the remainder in the B register and the quotient in the accumulator.

- The accumulator must be incremented by 1 before the execution of MOVC A, @A+PC instruction, if the table starts from 0.

REVIEW QUESTIONS

Evaluate Yourself

1. Which of the following instructions performs 16-bit increment?

   (a) INC R0   (b) INC R1

   (c) INC DPTR   (d) None of these

2. What is the accumulator content at return from routine REVRSE of Example 13.3?

   (a) End address   (b) Last term

   (c) Undefined   (d) None of these

3. If in Example 13.5, the second instruction of UPCLK routine is changed from ADD A, #01H to INC A, what would be the resulting change?

   (a) Overflow of minute counter

   (b) DA A instruction cannot be used

   (c) No change at all in performance

   (d) None of these

4. Which of the following instructions communicates with the external data memory?

   (a) MOVC A, @A+DPTR   (b) MOVX A, @R1

   (c) Either one   (d) None of these

5. SWAP instruction exchanges a pair of nibbles

   (a) Within the accumulator

   (b) Between A and an indirectly addressed location

   (c) Between A and a directly addressed location

   (d) None of these

6. One-bit flag bits may be located within

   (a) Bit-addressable area

   (b) Unused bits of PSW and also bit-addressable area

   (c) Any unused data memory location

   (d) None of these

7. Which of the following operations with two prime numbers would always generate another prime number?

   (a) Addition   (b) Multiplication

   (c) Division   (d) None of these

8. In Example 13.7, if the flag bit is set (1) and the current term is a character, then which of the following condition is applicable?

(a) Scanning through space

(b) Scanning through a word

(c) End of a word

(d) None of these

9. If a table look-up procedure is adopted for 8-bit HEX to BCD conversion, then the length of the table would be

(a) 10 bytes   (b) 100 bytes

(c) 250 bytes   (d) None of these

10. The minimum size of the stack for Example 13.2 program must be

(a) 2 bytes   (b) 4 bytes

(c) 8 bytes   (d) None of these

Search for Answers

1. What are the conditions for matching two arrays?

2. Can a subroutine have more than one RET instruction?

3. Is there any difference between a 'load' and a 'copy'?

4. Why is INC A instruction necessary in some, and not all, cases of application of MOVC instruction?

5. What is the difference in reversing an array of odd number of terms and an array of even number of terms?

6. Why were MOVX A, @Ri-type instructions not used in the program of Example 13.1?

7. Is it possible to generate prime numbers from 100 to 199 using 8051?

8. In Example 13.8, both R0 and R1 were used to compact the array. Is it possible to achieve the same thing using only one pointer (either R0 or R1)?

9. What was the purpose of flag bit in Example 13.7?

10. In Example 13.5, the clock display routine used BCD counters for hour and minute. What would be the changes if these counters were in binary?

Think and Solve

1. What happens if we do not place CLR C instruction before executing any SUBB instruction?

2. Is the last CLR C instruction essential for the program of Example 13.1?

3. How is it possible to handle 16-bit entries for the table look-up method using MOVC instruction?

4. What would happen if in Example 13.3 all three XCH instruction of subroutine CHANGE were replaced by XCHD instructions while all other parameters remained identical?

5. Try to rewrite the program of Example 13.6 using RLC A instruction. Then, compare the execution time of both the versions.

6. Write a program to convert any 8-bit binary number to its BCD equivalent.

7. Write a program to update the hour, minute and second counters of a RTC, assuming that all counters are in BCD and the display follows 12-h format.

8. Write a program to calculate the average of an array of unsigned positive integers. The array starts from 31H, and the number of terms in the array is available in location 30H. Store the calculated average in the location 2FH.

9. A random array of integers was generated and stored from location 31H onwards, storing its number of terms at location 30H. However, although the algorithm generally does not permit the repeat of any integer, to check this, develop a program ensuring that there is no repetition of any term. In case of repetition, the program should come out with CY flag as set; otherwise, CY flag should be cleared.

10. Develop a program to generate prime numbers by the method of divisions.

14 EXTERNAL INTERRUPTS

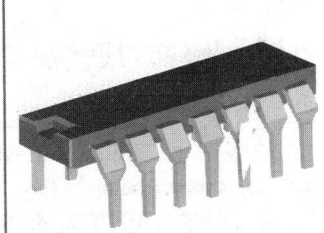

CHAPTER OBJECTIVES

In this chapter, the reader is introduced to 8051 interrupts in general and its external interrupts in particular. After completion of the chapter, the reader should be able to understand

- Overall functioning of interrupts.
- How an external interrupt works.
- Vector addresses for interrupts.
- Priority of interrupts.
- Application of special function registers (SFRs) related to the external interrupts.
- How to develop Interrupt Service Routine (ISR).
- How to generate software interrupts

14.1 | Introduction

As we have completed the discussions on the software instruction set of MCS-51, we may now be introduced to one of its very important features, the interrupts. Its distinctive features allow us to subdivide and present this topic in three parts. This chapter introduces interrupts in general and deals mainly with external interrupt features. The next chapter presents the timer interrupts offered by 8051. In Chapter 16, we will discuss about serial interrupts and serial communications.

14.1.1 | What is an Interrupt?

An interrupt is a signal or a command (indication) which forces the processor to leave its current sequence of activities and jump to another *predefined* set of activities as per the interrupt nature. Let us take an example. Assume that you are reading a book (not my book!) and your mobile phone rings. After noting down the page you are reading or placing a bookmark, you answer the call and complete your discussions. You leave the handset at its place and again pick up the book and start reading from the same portion where you had left to answer the call. This, in short, is what happens during an interrupt.

As we know, a processor is always busy executing instruction after instruction in a sequential manner. This sequence is broken by any interrupt which is generally a signal. However, as this sequence is to be resumed, the processor notes where it was executing and then jumps somewhere else to attend the interrupt; this is known as 'servicing the interrupt'. After completion of this servicing, the previous sequence is resumed.

14.1.2 | Vectored Interrupt

Interrupts are generally multiple in nature, and they are vectored. What is meant by a vectored interrupt? Taking our previous everyday example, assume that instead of the mobile ring tone, you hear the doorbell. In that case, you would proceed towards the door without looking at your handset. This indicates that different external signals force us to take up different assignments at different physical coordinates. Therefore, there is always a change of direction for any particular interrupt. As the vector means a quantity with a direction parameter, the interrupts are known as vectored interrupts as every interrupt has its own branching address.

The 8051 offers five different interrupt sources (six in case of XX52 and upward devices). These six sources, along with their corresponding branching addresses and default priorities, are presented in Table 14.1.

Table 14.1 Interrupts and their vector addresses and priorities of 8051 and 8052

Interrupt source	Vector address	Concerned SFR and bit	Designation of SFR bits	Priority (1 = highest)
External interrupt 0 (INT0)	0003H	TCON.1	IE0	1
Timer 0 interrupt	000BH	TCON.5	TF0	2
External interrupt 1 (INT1)	0013H	TCON.3	IE1	3
Timer 1 interrupt	001BH	TCON.7	TF1	4
Serial interrupt (Transmit and receive, both)	0023H	SCON.0	RI	5
		SCON.1	TI	
Timer 2 interrupt (only for XX52 and above)	002BH	T2CON.7	TF2	6

 Note the pattern of vector addresses in Table 14.1. The least significant digit is either 3 or B, alternately. Also note that the first vector address (for INT0) is at 0003H. Therefore, if this interrupt is used in any system, then the reset initialization routine, starting from 0000H, must place an AJMP or LJMP instruction at starting address 0000H and place the reset initialization routine at a suitable location in the program memory. As the starting addresses of program memory are earmarked for interrupt service routines, it is a good practice to place the reset initialization routine from, say, 0100H onwards.

14.2 | External Interrupts of 8051

In Fig. 2.3 of Chapter 2, we have presented all pin numbers with their nomenclature for 8051. The same diagram is presented here in Fig. 14.1, and it highlights the location of the two external interrupt input pins of 8051.

Note that INT0 is the alternate function of port pin 3.2, while INT1 is the alternate function of I/O port pin 3.3. Certain predefined types of external signals received through these input pins are capable of forcing the process to branch at certain predefined locations as indicated in Table 14.1. To perform these operations, the programmer must preprocess a few bits of some related SFRs of 8051. We will now take a closer look at those concerned SFRs, which are related with external interrupt handling.

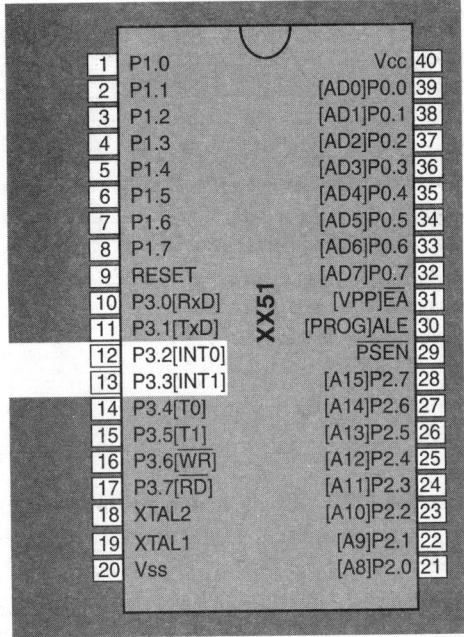

Figure 14.1 External interrupt input pins of 8051

14.3 | SFRs for External Interrupt

As far as external interrupts of 8051 are concerned, there are three SFRs directly related with these. They are TCON, IE and IP. We will discuss only the related bits of these SFRs. Other bit functions of these SFRs would be discussed in the following chapters on Timer interrupt and serial interrupt.

14.3.1 | TCON (bits 0, 1, 2 and 3)

TCON is a bit-addressable SFR with byte address 88H. Its bits may individually be designated as TCON.0, TCON.1 and so on, with bit addresses, like 88H, 89H, etc., as shown in Fig. 14.2. Alternatively, these bits may also be referred to as IT0, IE0 and so on. Out of 8 bits of TCON, we will discuss its lower 4 bits. The upper 4 bits of TCON would be discussed in the next chapter, which is on timer interrupts.

Bits 0 and 1 of TCON are related with external interrupt INT0, whereas bits 2 and 3 are for INT1. In general, the program developer does not handle TCON.1 and TCON.3, as these bits are operated by the processor itself. An external interrupt through INT0 would set TCON.1 (also called IT0) and would be cleared by the processor when the processor branches to the interrupt service routine of INT0 that is located at 0003H of program memory. Similarly, TCON.3 is set and cleared by the processor itself for INT1 external interrupt.

Setting bits IE0 or IE1 of TCON through software commands would generate software interrupt. The branching addresses would be same as those of INT0 or INT1. 8085-oriented reader may be aware of software interrupt generating mnemonics of 8085, like: RST0, RST1, etc. MCS-51 does not offer any such special mnemonic for software interrupts (refer to Section 14.5).

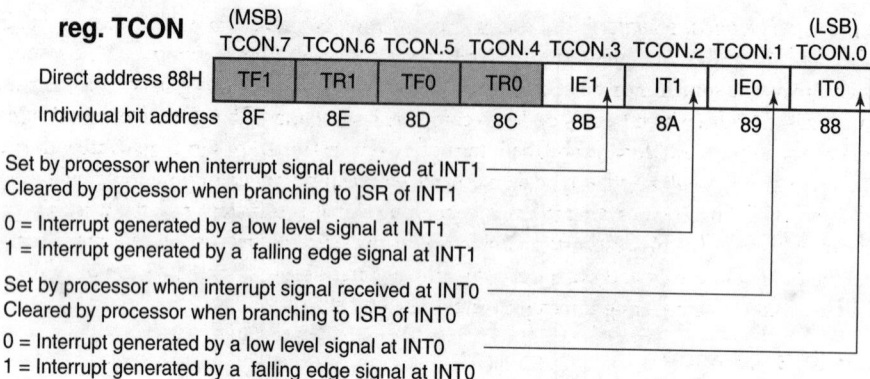

Figure 14.2 External interrupt control bits in TCON SFR

Selection of signal type: Bits TCON.0 and TCON.2 are user programmable and specify the type of signal to be taken as a valid external interrupt-generating signal. If TCON.0 is cleared, then a low-level signal at INT0 (at pin 12 of 8051) would generate the interrupt [Fig. 14.3(b)]. On the other hand, if TCON.0 is set, then a falling-edge (high-to low-level transition) signal would generate an external interrupt through INT0 [Fig. 14.3(a)]. Similarly, the condition of TCON.2 would specify the type of signal capable of generating INT1 interrupt [Figs 14.3(c) and (d)]. It may also be noted that input signal types for INT0 and INT1 are selectable independently.

 What would happen if the interrupting signal is a high level one? Well, in that case the interrupting signal must be inverted through a NOT gate (eg. 74LS14) and the concerned interrupt must be programmed as low level triggered.

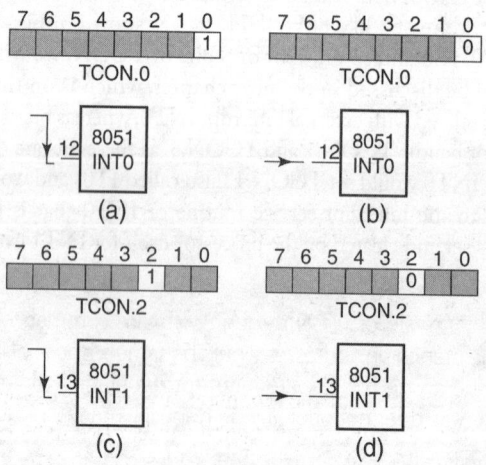

Figure 14.3 Signal-type selection for INT0 and INT1 through TCON

Why is the provision given for selection? The reader may ask why this provision was planned in MCS-51. As a matter of fact, it helps the designer to interface different types of devices through interrupt. A low-level triggered interrupt signal may take some time for its recognition, as the processor may be busy in completing the execution of an instruction, say MUL, which takes 4 cycles to complete its execution. Therefore, a low-level signal must be maintained low for a minimum duration by the interrupting device (interrupt signal generating device).

On the other hand, a falling-edge type signal need not keep on generating multiple falling edges to be recognized. A single falling edge of the interrupt-generating signal is sufficient to latch it within the internal flip-flop of the processor. In this case, after completion of the ongoing instruction, the processor would pay attention to the *trapped* interrupt and branch accordingly. As interrupting devices may not be of the same type, two different provisions of generating interrupts were offered by 8051.

14.3.2 | IE (bits 0, 2 and 7)

The main function of this SFR is to enable or disable different interrupts by setting or clearing its designated bits. IE is a bit-addressable SFR, with individual bit addresses and bit designations, and utilities for only relevant bits are shown in Fig. 14.4. Its most significant bit (bit 7) may be taken as the global interrupt control bit. If this bit is cleared to 0, all 8051 interrupts would be disabled. Setting this bit as 1 permits only those interrupts to be active (able to generate interrupts) whose corresponding bits in IE are set.

For an external interrupt like INT0, its individual enable/disable bit is IE.0, which is the least significant bit of IE SFR. Therefore, to enable (allow) an external interrupt INT0, both IE.7 and IE.0 are to be set. If any one of these two is cleared, INT0 would be disabled. Similarly, to enable INT1, both IE.7 and IE.2 must be set. In this case, IE.2 is the individual enable/disable bit for INT1.

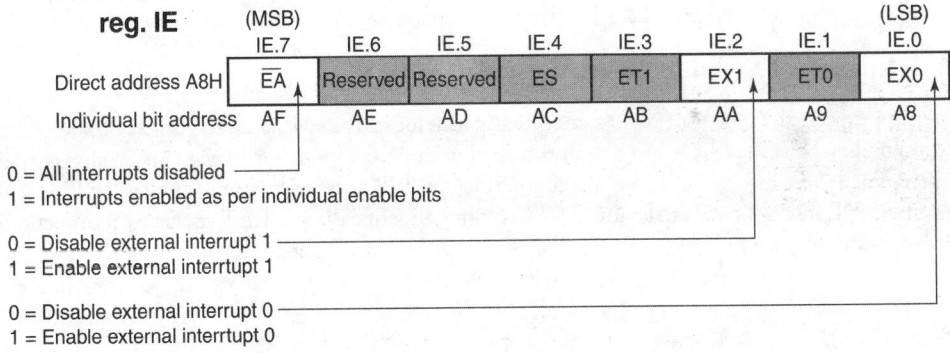

Figure 14.4 External interrupt-enable bits of IE SFR

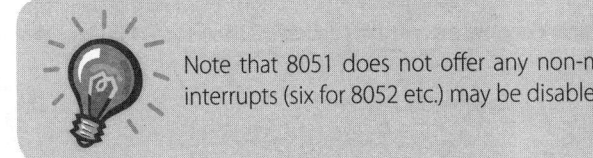 Note that 8051 does not offer any non-maskable interrupt, like TRAP of 8085. All five interrupts (six for 8052 etc.) may be disabled or enabled individually.

14.3.3 | IP (bits 0 and 2)

IP is also a bit-addressable SFR, which helps the programmer to resolve the priority issues of multiple interrupts. This priority issue of interrupts comes into the picture only when there is a concurrency, i.e., more than one interrupt are acknowledged at the same instant. In such a situation, which one would be serviced first is decided by the priority assigned to that interrupt.

In Table 14.1, default priorities of all interrupts are presented in the last column. Here the highest priority is designated as 1 and is assigned to the external interrupt INT0. INT1 is assigned a priority of 3. This indicates that if both INT0 and INT1 are acknowledged simultaneously, first INT0 and then INT1 would be serviced. These default priorities may not be changed by the programmer, unless the IP SFR is used.

SFR IP offers some flexibility over these default priorities. If the programmer wants to allow INT1 higher priority over INT0, then IP.2 (bit 2 of SFR IP) has to be set as 1 and IP.0 has to be cleared by the programmer before activating interrupts. As indicated in Fig. 14.5, any interrupt bit set as 1 would allow higher priority to that interrupt. The reader may want to know that if both IP.0 and IP.2 are set in SFR IP and both INT0 and INT1 interrupts occur simultaneously, which one would be serviced first. Well, in this case, the processor would resolve the issue by using the default priorities (Table 14.1), and INT0 would be serviced first.

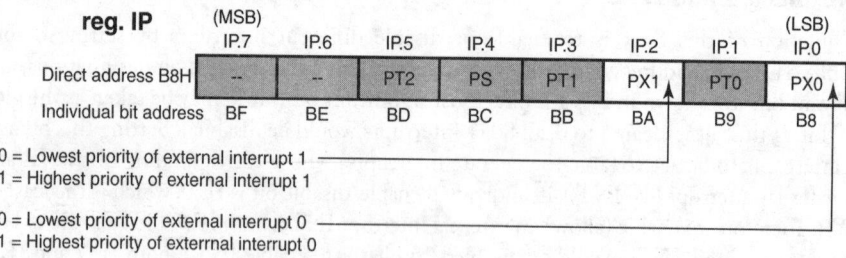

Figure 14.5 External interrupt priority-setting bits of IP SFR

14.4 | Interrupt Service Routine (ISR)

An interrupt expects the processor to do something special for it. The special task(s) is accommodated within the corresponding ISR. The ISR is, just like any other subroutine, composed of relevant instructions to perform some predefined job. However, unlike other subroutines, an ISR must be terminated by a RETI instruction and not by RET instruction. Instruction RETI is almost identical with RET, as far as its functioning is concerned, with only one exception. RETI allows or enables the disabled interrupt. When is the interrupt disabled? Whenever any interrupt is acknowledged and the processor branches to the corresponding vector address, it automatically disables the interrupt. If any ISR is terminated by an RET instruction that interrupt would never be acknowledged again.

As it was indicated earlier, every interrupt of 8051 has its own vector address or branching address as shown in Table 14.1. For INT0, its ISR must start from program memory location 0003H, and for INT1, the corresponding ISR must start from program memory location 0013H. The general practice in these cases is to place the ISR at any suitable location of program memory and incorporate an unconditional jump instruction at the corresponding vector address of that interrupt. We now present two checklists, one for INT0 and the other for INT1, for convenience in software development.

14.4.1 | Checklist for INT0

(i)   Place ISR for INT0, which is terminated by RETI instruction, at program memory location 0003H. If the ISR is long, it may be placed anywhere within the program memory, and an unconditional jump instruction should be placed at 0003H to access the ISR.

(ii) If necessary, set priority through IP.0.

(iii) Select the interrupt signal type by setting TCON.0 (falling-edge triggered) or by clearing it (low-level triggered).

(iv)   Enable INT0 by setting IE.0.
(v)   Enable the global interrupt control bit by setting IE.7.

Sequence of Steps (ii) and (iii) may be interchanged. Step (ii) is optional. It is a good practice to enable the global interrupt control bit at the end of all operations.

14.4.2 | Checklist for INT1

(i)   Place ISR for INT1, which is terminated by RETI instruction, at program memory location 0013H. If the ISR is long, it may be placed anywhere within the program memory, and an unconditional jump instruction should be placed at 0013H to access the ISR.
(ii)   If necessary, set priority through IP.2.
(iii) Select the interrupt signal type by setting TCON.2 (falling-edge triggered) or by clearing it (low-level triggered).
(iv)   Enable INT1 by setting IE.2.
(v)   Enable the global interrupt control bit by setting IE.7.

14.5 | Software Interrupts

Contrary to the hardware interrupts evoked by hardware signals, software interrupts are generated by software commands rather than some external hardware signals, and therefore are programmable.

Other processors, like 8085, offer special instructions, such as RSTn, for the software interrupt generation. The 8051 processor does not offer any such dedicated instruction for the software interrupt generation. However, its interrupt control bits may be manipulated to generate software interrupts through bit-manipulation commands.

In Table 14.1, it was indicated that TCON.1 and TCON.3 are related with INT0 and INT1 external interrupts. Whenever any external signal is received at INT0, the bit TCON.1 is set automatically by the processor. This bit is cleared by the processor itself whenever it branches to the concerned ISR for INT0. The same is also valid for INT1 interrupt and TCON.3 bit.

Using the SETB TCON.1 instruction, TCON.1 may be set to generate the INT0 interrupt without the existence of any physical hardware signal at the INT0 input pin of 8051. If this is implemented, then the processor would take it as a valid INT0 interrupt and immediately branch to INT0 ISR after clearing TCON.1 by itself. Similarly, TCON.3 may also be set to generate another software interrupt related to INT1.

14.6 | Benefits of Interrupt

In general, interrupts release the processor for its other duties rather than continuously watching an input to be true; this is known as 'polling'. For example, if an ADC is interfaced with the processor through one of its ports, say P1.0, and is busy in converting the input signal, the processor is to continuously read the End of Conversion signal from the ADC to correctly record the digital output from it, which is available immediately after the conversion [Fig. 14.6(b)]. Had the processor been interfaced through an interrupt

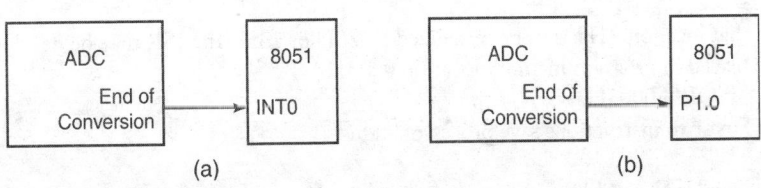

(a)   (b)

Figure 14.6 Schematic for (a) interrupt-driven and (b) polling method for data acquisition

input [Fig. 14.6(a)], the processor would have been allowed to be busy in other duties rather than checking the End of Conversion output continuously. As soon as the processor is interrupted, it would have immediately executed its ISR to acquire the data from the ADC.

 Accurate information of the external interrupt generating signal, specially about its timing features, must be known to the software developer for developing an accurate delay routine to avoid unwanted multiple interrupts generated by the same signal.

14.7 | Solved Examples

Example 14.1

Purpose: How to develop ISR and initialization routine for INT0, programmed for falling edge, and how to enable and disable INT0.

Problem

A system is to read an external sensor continuously, and if its reading exceeds the 7-bit value more than 20d times, then the reading should be terminated. Assume INT0 input is connected with the external circuit in such a way that it generates a high-to-low signal as an indication to read input port P1 and store it in data memory location 30H. If input at P1.7 is high, then memory location 31H should be decremented by 1. If location 31H becomes 0, then INT0 must be disabled. Initialize 31H by 20d before allowing first INT0 interrupt. Write the ISR and initialization routine for this INT0 interrupt.

Solution

A schematic representation of the problem is presented through Fig. 14.7. As it indicates, the procedure may be subdivided into three separate modules as follows.

(i)   Read port P1 at every INT0 interrupt.
(ii) Store this data at location 30H. If P1.7 is equal to 1, then decrement content of location 31H by 1.
(iii) If content of location 31H is 0, then disable INT0.

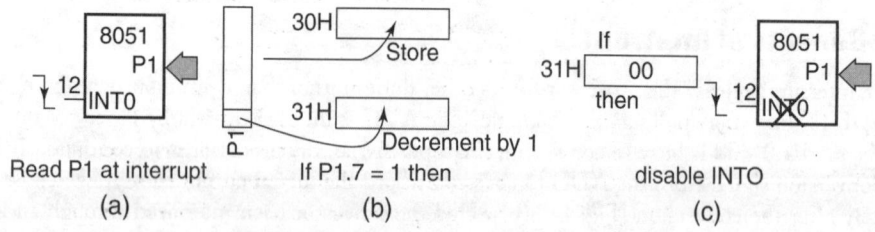

Read P1 at interrupt   If P1.7 = 1 then   disable INTO
(a)   (b)   (c)

Figure 14.7 Schematic representation of Example 14.1

It may be noted that location 31H is to be initialized by 20D or 14H. The ISR may be developed as follows, and it should be placed in program memory location 0003H.

                    ORG 0003H
; ISR for INT0. First two instructions save processor status.

EXTI0:	PUSH	PSW	; save processor status word on stack
	PUSH	ACC	; save accumulator content on stack

; Read port P1 and store in 30H. This is done by following two instructions.

```
                MOV     A, P1        ; read port P1
                MOV     30H, A       ; save reading in location 30H
```

; Check bit 7 (MSB) of the input data. If it is 0, then no action. In that case, restore status and return from
; interrupt.

```
                RLC     A            ; rotate MSB to carry
                JC      HIGHIN       ; continue at HIGHIN if MSB is 1
        LOWIN:  POP     ACC          ; restore accumulator content from stack
                POP     PSW          ; restore processor status word from stack
                RETI                 ; return from interrupt
```

; If bit 7 is not 0, then decrement content of location 31H by 1. Quit, if it is not 0.

```
        HIGHIN: MOV     A, 31H       ; get content of location 31H
                DEC     A            ; decrement it by 1
                MOV     31H, A       ; save it back
                JNZ     OVER         ; return, if it is not 0
```

; content of location 31H is 0. Disable INT0 and return.

```
                CLR     IE.0         ; disable INT0 interrupt
        OVER:   POP     ACC          ; restore accumulator content from stack
                POP     PSW          ; restore processor status word from stack
                RETI                 ; return from interrupt
```

An alternative routine may be developed in the following manner.

```
        EXTI0:  PUSH    PSW          ; save processor status word on stack
                PUSH    ACC          ; save accumulator content on stack
                MOV     30H, P1      ; copy port 1 input to 30H
                MOV     C, P1.7      ; copy port 1 MSB to CY flag
                JNC     OVER         ; return, if no carry
        HIGHIN: DJNZ    31H, OVER    ; decrement location 31H content by 1
                CLR     IE.0         ; disable INT0, as location 31H is cleared
        OVER:   POP     ACC          ; restore accumulator content from stack
                POP     PSW          ; restore processor status word from stack
                RETI                 ; return from interrupt
```

The initialization routine should be placed in such a location that before the first external interrupt INT0 is received, it is executed. A normal practice is to place such an interrupt initialization routine within the reset initialization routine, so that every time a system is reset, the routine is executed.

```
        INITX0: MOV     31H, #14H    ; load location 31H by 20d
                SETB    TCON.0       ; select falling-edge triggered INT0
                SETB    IE.0         ; enable the external interrupt INT0
                SETB    IE.7         ; enable all interrupts (global)
```

C-version

```
#include <regx51.h>
sbit in = P1^7;
register unsigned char *S1, *S2;
void main()
{
        IE = 0x81;
        P1 = 0xFF;
        S1 = 0x30;
        S2 = 0x31;
        *S2 = 20;
        IT0 = 1;
        while(1)
        {
        }                       // infinite loop
}

void ext0() interrupt 0
{
        if (in == 1)
                (*S2)--;
        if (*S2 == 0)
                IE = 0;
        *S1 = P1;
}
```

Example 14.2

Purpose: How to develop ISR for low-level sensitive INT1 and then enable or disable it.

Problem

In a system, INT1 is connected with a hardware circuit, which generates a low-level signal as an interrupt. The signal is active (remains low) for 10 microseconds. The INT1 is to be used to read port 2 and to store it in data memory location from 30H onwards. Whenever this port input becomes 00H, INT1 should be disabled. Develop the initialization routine and also the ISR.

Solution

For storing the port 2 reading from location 30H onwards, it is preferable to use a pointer, which itself may be stored in location 2FH. This must be done at the time of initialization. The following is the initialization routine for INT1.

```
INITX1:     MOV     2FH,#30H        ; initialize pointer at 2FH for 30H to start
            CLR     TCON.2          ; select low-level triggered INT1
            SETB    IE.2            ; enable the external interrupt INT1
            SETB    IE.7            ; enable all interrupts (global)
```

The following is the ISR for INT1 to be placed at location 0013H.

```
            ORG    0013H

EXTI1:      PUSH   PSW        ; save status
            PUSH   ACC
            MOV    R0, 2FH    ; load pointer
            MOV    A, P2      ; read port
            MOV    @R0, A     ; save reading
            INC    R0         ; update pointer for next storage
            MOV    2FH, R0    ; save back pointer
            JNZ    OVER       ; check port reading in A to disable INT1?
            CLR    IE.2       ; disable INT1
OVER:       POP    ACC
            POP    PSW        ; restore status
            RETI
```

The execution time of the ISR is more than 10 microseconds. Therefore, at the time of returning from interrupt, the interrupting signal must be back to high level. No special precaution is necessary in this case of low-level interrupting signal.

Example 14.3

Purpose: To highlight the precautions to be taken for level-sensitive interrupts.

Problem

Assume an interrupting device is interfaced with 8051 through its INT0 input pin. The device generates a low-level triggered interrupt, which lasts for several milliseconds. Its ISR is to store the data available at P0 in the memory location pointed by R0 and then increment R0 by 1 for the next ISR usage. Discuss a suitable mechanism to process this interrupt.

Solution

This is a case where unwanted multiple processing of the same interrupt might take place if proper precautions are not implemented. To start with, it is given that the low-level triggered interrupt remains low for a few milliseconds. Moreover, the duty of the ISR is to read a port, store it, update the pointer and terminate the ISR. These actions would need a few microseconds, which are much less for the termination of the interrupting signal. As the interrupting signal would still be available at INT0 input, the ISR would again be evoked for another set of unwanted storage and so on. This would go on till the interrupting signal remains low (does not go back to its high level).

    Through hardware: One simple hardware method to solve this problem is to set the interrupting signal as high, before executing the RETI instruction, through some flip-flop. The output of the flip-flop should be connected with the INT0 input, and a port pin of 8051 should be connected with the 'clear' input of the flip-flop (Fig. 14.8). Before executing RETI, the ISR would process the port pin so that the input at INT0 goes high.

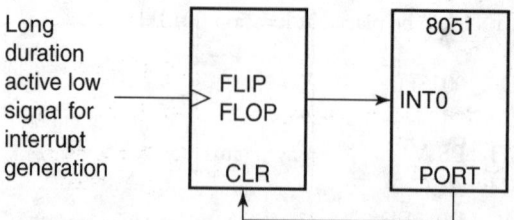

Figure 14.8 Schematic of interrupting signal-controlling hardware

Through software: Another method is to apply a software routine to generate a few milliseconds of delay before executing RETI instruction of the ISR. A delay routine may be developed for this, ensuring the delay is sufficient to guarantee the removal (change of state) of the external interrupt signal. However, in this method, the process itself remains occupied till the interrupting device withdraws the interrupt-generating signal. A sample delay routine is given below.

```
DELAY:      MOV   R7, #00H      ; load counter by max count
DEL1:       DJNZ  R7, DEL1      ; kill time
            RET
```

This delay routine, when called, would consume about a half millisecond's time. Each execution of DJNZ instruction takes 2 cycles. With a 12 MHz external crystal, 1 cycle takes 1 microsecond. Therefore, each execution of DJNZ instruction would take 2 microseconds. This would be executed 256 times to count down the original value in R7 to 0. Therefore, execution of DJNZ alone would consume 512 microseconds. The first instruction (MOV) and the last instruction (RET) of the routine would consume (1+2) 3 cycles resulting in a total delay of 515 microseconds. If the delay is to be increased, another register may be used in cascade with R7.

SUMMARY

In total, 8051 offers five interrupts, namely, two external interrupts (INT0 and INT1), two timer interrupts (Timer 0 and Timer 1) and one serial interrupt. Each one of these five interrupts has its own vector address where its ISR is to be stored. They also have their default priorities, which may be altered through manipulation of different bits of the SFR, IP. To enable or disable all interrupts simultaneously, the most significant bit of SFR IE is to be used. Other bits of this SFR allow enabling or disabling of individual interrupts if IE.7 is set. The lower 4 bits of TCON SFR are related with INT0 and INT1 interrupts. Their input characteristics may be programmed as either falling-edge triggered or low-level sensitive by manipulating these bits. Any ISR must be terminated by a RETI instruction, which enables further interrupts. It is a good programming practice to save the processor status and relevant registers before proceeding with any ISR. However, before returning from the ISR, they must be restored.

POINTS TO REMEMBER

- The software developer must ensure that before execution of RETI in an external interrupt's ISR, the interrupting signal is back to normal.

- To avoid bouncing of the edge-triggered, interrupt-generating signal, a flip-flop may be used before interrupt input.

REVIEW QUESTIONS

Evaluate Yourself

1. If TCON.0 is set, which of the following signals would generate an interrupt through INT0?

   (a)   ⌐___

   (b)   ⌐‾‾‾

   (c)   low level

   (d) None of these

2. If all bits of SFR IP are cleared and INT0 and INT1 interrupts are received simultaneously, which one would be serviced first?

   (a) INT0

   (b) INT1

   (c) Either of the two

   (d) None of these

3. An interrupt, like INT1, is enabled by setting

   (a) IE.7

   (b) IE.2

   (c) both of these

   (d) none of these

4. If any ISR is too long, its vector address must contain a

   (a) CALL instruction

   (b) Jump instruction

   (c) RETI instruction

   (d) none of these

5. Any ISR must have at least

   (a) one RETI instruction

   (b) one RET instruction

   (c) one RETI and one RET instruction

   (d) none of these

6. INT0 input pin of 8051 is assigned to

   (a) no other function

   (b) another function as P3.2

   (c) two more functions as P3.2 and T0

   (d) none of these

7. What is the minimum duration for a low-level signal to be recognized as an interrupt?

   (a) More than 1 cycle

   (b) More than 2 cycles

   (c) More than 3 cycles

   (d) None of these

8. What happens when TCON.1 is set by some software instruction?

   (a) INT0 is acknowledged for a falling edge

   (b) INT0 is disabled

   (c) If it is enabled, then a software interrupt is generated

   (d) None of these

9. Which of the following interrupts is enabled immediately after a system reset?

   (a) INT0

   (b) INT1

   (b) Either of these

   (d) None of these

10. Assuming that initially both interrupts were enabled, what would happen if, during execution of ISR of one interrupt, another interrupt signal interrupts the processor?

    (a) ISR of other interrupt would be executed immediately

    (b) If the later interrupt is of higher priority, then only its ISR would be executed, otherwise not

    (c) The second interrupt ISR would be executed after completion of the RETI instruction of the first ISR and another instruction

    (d) None of these

Search for Answers

1. What is meant by vectored interrupts?

2. What is the purpose of TCON SFR?

3. What is meant by falling-edge triggered?

4. What happens if IE.7 is cleared?

5. What are the characteristics of an ISR?

6. Is it always necessary to assign a priority to an interrupt through IP SFR? Justify your answer.

7. What alternative solution is available for a longer ISR?

8. Why are multiple interrupts not generated by a low-level triggered signal even if it stays low for a longer duration than, say, 10 cycles?

9. Is it possible to change the vector address of an interrupt by software commands or otherwise?

10. What are the benefits of an interrupt?

Think and Solve

1. How would it be possible for 8051 to recognize a high-level interrupt?

2. What are the general characteristics of any interrupt?

3. What is the relation between an interrupt and a subroutine?

4. If all interrupts have their default priorities, what is the purpose of providing IP SFR?

5. How can we generate software interrupts in 8051?

6. Why are accumulator and PSW saved before processing any interrupt?

7. What precautions should we take to develop the ISR for a low-level triggered interrupt?

8. Is there any limitation about the length of an ISR?

9. In which condition may we need some extra hardware interfacing to deal with a low-level external interrupt signal?

10. Under which conditions may an ISR have multiple RETI instructions?

15 TIMER/COUNTER INTERRUPTS

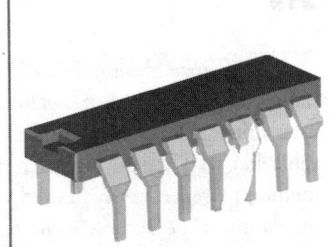

CHAPTER OBJECTIVES

In this chapter, the reader is introduced to Timer/Counter interrupts of 8051 and 8052. After completion of the chapter, the reader would be able to understand

- Overall functioning of Timer interrupts.
- How Timer interrupt works.
- Counters and Counter interrupts.
- Generating time delay by using Timers.
- Functions of different SFRs related to Timer/Counter interrupts.
- How to develop Timer/Counter Interrupt Service Routines.

15.1 | Introduction

In the previous chapter, we discussed about the general features of interrupts. This included their basic characteristics, operational details, vector addresses and priorities. From these discussions, it might be recalled that 8051 offers two Timers/Counters, namely, Timer 0 and Timer 1. Higher versions of 8051, like xx52, offer one extra Timer/Counter, namely, Timer 2. In this chapter, we will concentrate our discussions on interrupts generated by these Timers/Counters.

15.1.1 | What is a Timer?

In general, a Timer may be considered as a simple multi-bit Counter, which produces a signal at the time of its overflow or underflow during counting, depending upon whether it is counting up or down. A clock signal is necessary to activate this counting process. A simplified functioning of a hypothetical 2-bit Timer is shown in Fig. 15.1.

Clock   Counter
pulse → | 0 | 0 |
(1 Hz)

(a)

1
⌐⌐ → | 0 | 1 |

(b)

2
⌐⌐ → | 1 | 0 |

(c)

3
⌐⌐ → | 1 | 1 |

(d)

4
⌐⌐ → | 0 | 0 | → Timer Interrupt

(e)

Figure 15.1 Functioning of a 2-bit count-up Timer

As we may observe from Fig. 15.1, the 2-bit up Counter is initially cleared, and with every clock pulse (falling edge), the Counter value is incremented by 1. Finally, when it overflows from 11B to 00B, the Timer interrupt signal is generated. It may be noted that instead of 00B, the Timer may be initialized by any other value, like 01B or 10B, and then the clock pulses may be allowed to increment the Counter. This is the normal method to generate any desired time delay, estimated between starting the counting of the Timer and receiving an interrupt from it, at the termination of the counting process.

It is needless to mention that for every Timer, there is a *resolution* or *step-duration*, which is dependent upon the input clock frequency. It is not correct that any delay may be generated by a specific timer. For example, a timer running with a 1 Hz clock cannot generate accurately a delay of 2.3 seconds. To produce this accuracy, we are to change the clock frequency to 10 Hz.

15.1.2 | Purpose of a Timer

This leads to the fact that the delay, between start of counting and the generation of the overflow signal (interrupt), may be controlled by loading the Timer with a suitable initial value. In the case of the example 2-bit Timer, if we assume that the clock frequency is 1 Hz, then it would allow a delay of 4 seconds when it is initialized as 00B. On the other hand, if the initial value of the same Counter is 10B, then it would generate a delay of 2 seconds. As a matter of fact, this programmable delay generation, using a clock signal of uniform frequency, is the fundamental purpose of any Timer.

In practice, these Timers are 8 bits or 16 bits and are normally operated through an internal clock of the processor. As and when necessary, its Counter is loaded by the desired count value, and the Timer is started. After starting, the processor may remain busy in other operations, and at the terminal count, the Timer generates an interrupt and thus draws the attention of the processor. It is a normal practice, just like the case of external interrupts, to provide a dedicated interrupt service routine (ISR) for servicing the Timer interrupt.

15.1.3 | Differences Between a Timer and a Counter

Although every Timer must have a Counter, this Counter may be used in either of the two following ways:

(i)   to generate an interrupt after a time delay (designated as Timer) or
(ii)   to generate an interrupt after counting some specific number of external events (designated as Counter).

For the first case, the Counter of the Timer counts the processor clock pulses. In other words, it uses the processor's own oscillatory circuit. If the mechanism is functioning in this manner, it is designated as a Timer, the aim of which is to generate a programmable time delay. In the second case, the same Counter counts the number of pulses of some external signal. Note that this external signal may or may not be periodic in nature. When the unit functions in such a manner, it is called a Counter. This difference is highlighted in Fig. 15.2.

Both Timers of 8051 may function as Counters also. Whether they would function as Timer or Counter, is a factor which is programmable through C/$\overline{T}$ bits of TMOD SFR. These details would be discussed in Section 15.4.1 of this chapter.

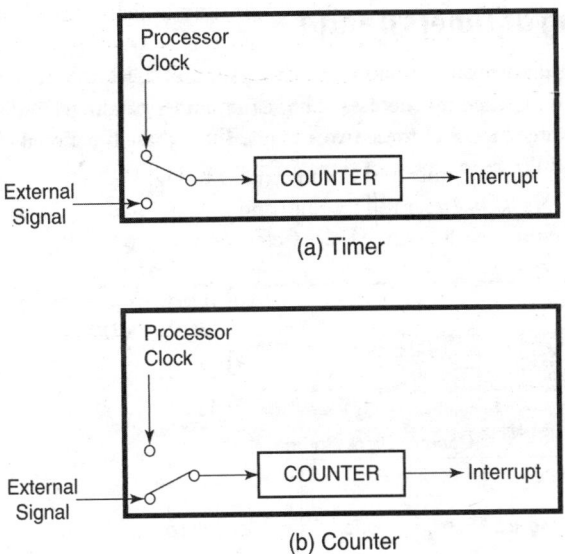

Figure 15.2 Difference between (a) Timer and (b) Counter

15.2 | 8051/52 Timers

As already indicated, 8051 offers two independent 16-bit count-up Timers designated as Timer 0 and Timer 1. The functions of these two Timers are more or less identical with some minor variations. A third Timer, designated as Timer 2, is available in higher versions of 8051, like 8052. The function of Timer 2 is widely different from that of the other two Timers and demands separate discussion. We will take up Timers 0 and 1 at present. After completion of the discussions on these, we will take up Timer 2. However, for a quick reference, the salient features of all three Timers are presented in Table 15.1

Table 15.1 Salient features of 8051/52 Timers

Timers	Vector address	Modes	Related SFRs
Timer 0	000BH	0–13-bit Timer/Counter 1–16-bit Timer/Counter 2–8-bit auto-reload 3–dual 8-bit Timers	TCON–control operations TMOD–Timer mode select IE–interrupt enable IP–interrupt priority setting, TH0, TL0–Timer 0 Counters
Timer 1	001BH	0–13-bit Timer/Counter 1–16-bit Timer/Counter 2–8-bit auto-reload	TCON–control operations TMOD–Timer mode select IE–interrupt enable IP–interrupt priority setting TH1, TL1–Timer 1 Counters
Timer 2	002BH	16-bit auto-reload 16-bit capture Baud-rate generator	T2CON–Timer 2 control RCAP2H, RCAP2L–for capture IE–interrupt enable IP–interrupt priority setting TH2, TL2–Timer 2 Counters

15.3 | Functioning of Timers 0 and 1

Throughout the following discussions pertaining to Timers 0 and 1, we would take Timer 0 as an example case and refer to Timer 1 only when it is necessary. For easier understanding, the discussions are divided into three parts related to the three blocks of these two Timers. These three functional blocks are

  (i)   signal source selection for Timer and Counter (refer to Fig. 15.3),
  (ii) RUN/STOP CONTROL (refer to Fig. 15.4) and
  (iii) counting configurations in different Modes (refer to Figs 15.6–15.9).

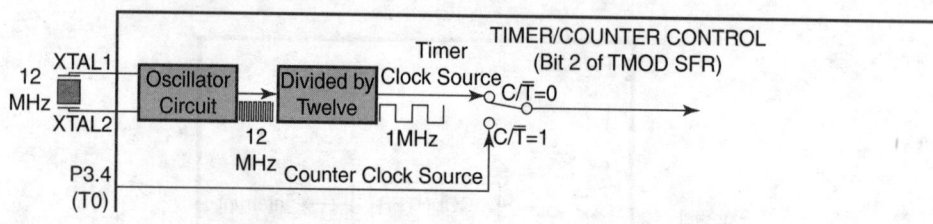

Figure 15.3 Schematic of Timer/Counter sources for Timer 0

 I request the reader not to skip this section as it is essential to understand how these timers function. It would then be easier to develop necessary software using these timer/counters.

15.3.1 | Timer/Counter Select Block

To start with, whether it would function as a Timer or as a Counter is decided by the condition of the $C/\overline{T}$ bit of TMOD SFR (Fig. 15.3). If this $C/\overline{T}$ bit is set, then the function would be as a Counter. If the $C/\overline{T}$ bit is cleared, the function would be as a Timer. Referring Fig. 15.5, we observe that condition of bit 2 of TMOD decides for Timer 0, while condition of bit 6 of TMOD selects the Timer/Counter mode of Timer 1. Note that TMOD is not a bit-addressable SFR.

This TIMER/COUNTER CONTROL block is represented by a SPDT switch in Fig. 15.3. The arrow from this switch indicates the signal flow to RUN/STOP CONTROL block, which we will discuss next. Presently, we will make ourselves familiar with two different sources of this switch, source for Counter and source for Timer, which are located at the left side of the switch.

For Timer 0, the source, when it is programmed to act as a Counter, is P3.4 (pin 14) of the processor, which is marked as T0. For Timer 1, the source for its Counter-mode operation would be P3.5 (pin 15) of 8051, which is marked as T1. A periodic or irregular signal may be fed to these pins to serve as the input for the respective Counter. Note that a falling edge of this external signal would increment the count by 1.

When functioning as a Timer, the sources for both Timer 0 and Timer 1 are identical, and that is the processor clock signal, which is always divided by 12 by the processor itself. As an example case, an external crystal of frequency 12 MHz is shown to be attached with the processor through its XTAL1 and XTAL2 pins in Fig. 15.3. In such a case, the processor clock would be of 12 MHz frequency, and by dividing it by 12, a final frequency of 1 MHz would be obtained. If a 6 MHz crystal is attached, then the frequency of the processor clock would be 6 MHz, and by dividing it by 12, the input for the Timer would be 500 KHz.

Finally, we may observe from Fig. 15.3 that the SPDT switch simply selects one of the two signal sources. We will now discuss the RUN/STOP CONTROL mechanism, which is shown in Fig. 15.4; here the portion, which we have already discussed, is shown without any shade.

15.3.2 | Run/Stop Control Block

The RUN/STOP CONTROL mechanism is a simple on/off switch, which is controlled by TR0 bit of TCON (TR1 bit for Timer 1), Gate bit of TMOD and P3.2 (INT0) input (P3.3 for Timer 1). The first two are software controls, and the last one allows an external hardware signal and therefore may be taken as the hardware control.

As the RUN/STOP CONTROL block shows (rectangular box at top right of Fig. 15.4), the oscillatory signal from the left would pass (RUN) if the output from the AND gate is 1. If this output is 0, then the input oscillatory signal is disconnected and the mechanism enters into the STOP state. To allow the RUN state, TR0 should be set as 1, and the output from the OR gate must be 1. This last condition may be fulfilled by either clearing the Gate bit of TMOD SFR (software control) or pulling up P3.2 (hardware control), if it is possible. The related Truth table is presented in Table 15.2.

Table 15.2 Truth table for Timer 0 Run/Stop control

TR0	Gate	P3.2	Run/Stop control
0	x	x	0 (stop)
1	0	x	1 (run)
1	1	0	0 (stop)
1	1	1	1 (run)

To conclude our discussion on Timer/Counter 0, we note that (refer to Figs 15.3 and 15.4)

(i) $C/\overline{T}$ bit of TMOD SFR decides that it functions as a Timer ($C/\overline{T} = 0$) or as a Counter ($C/\overline{T}=1$). For Timer 0, $C/\overline{T}$ is TMOD.2, and for Timer 1, $C/\overline{T}$ is TMOD.6.

(ii) The Timer clock input is always the processor clock, which is always divided internally by 12 (decimal).

(iii) The Counter input is always an external signal through T0 (for Timer/Counter 0) or T1 (for Timer/Counter 1).

(iv) TR0 bit of TCON SFR, Gate bit of TMOD SFR and P3.2 bit control the RUN/STOP of Timer/Counter 0. The first two are software control, and the last one is hardware control.

Both Timers, 0 and 1, are equipped with 16-bit Counters in the form of two 8-bit SFRs. SFRs TH0 and TL0 are for Timer 0, while SFRs TH1 and TL1 are for Timer 1. In each case, the counting (always up-counting) may be done with 8-bit, 13-bit or 16-bit configuration depending upon the selected mode. For mode selection of Timers 0 and 1, TMOD SFR is provided.

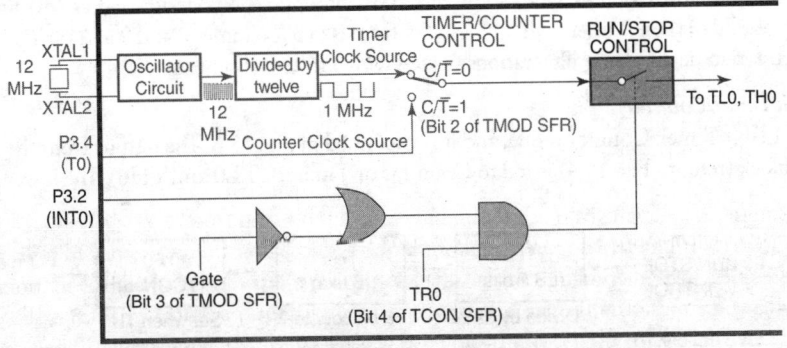

Figure 15.4 Schematic of RUN/STOP CONTROL mechanism for Timer 0

 If we need not control the run/stop operation by external signal, then the Gate bit of the concerned timer (in TMOD) may be cleared and only TR-bit may be used to run/stop the timer.

15.4 | SFRs for Timer 0 and 1 Interrupts

In Table 15.1, the SFRs related with Timers were enlisted. Some of these are common to all interrupts, like IE and IP SFRs. Others are related with some specific Timer interrupt as indicated in the table. We will now discuss the functions of these SFRs in detail.

15.4.1 | TMOD

SFR TMOD is meant for configuring both Timer 0 and Timer 1. As shown in Fig. 15.5, its lower 4 bits (0–3) controls Timer 0, and the upper 4 bits (4–7) controls Timer 1. Bit assignments of both sets are identical, i.e., both bit 0 and bit 4 are designated as M0 and so on.

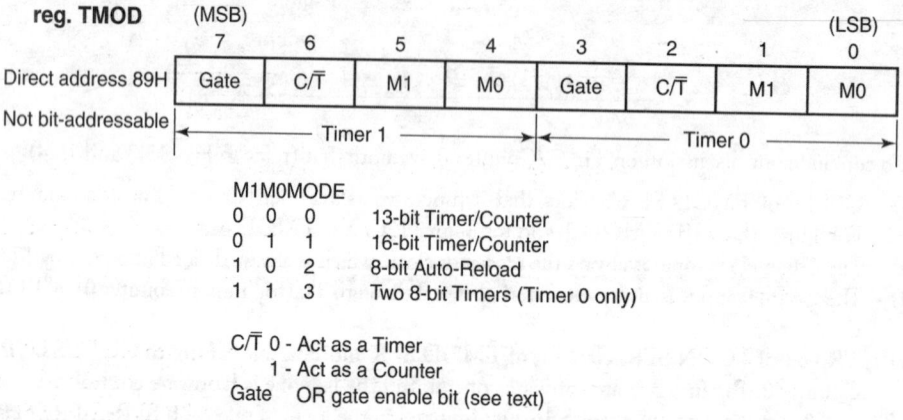

Figure 15.5 Timer interrupt control bits of TMOD SFR

For both Timer 0 and 1, the bits, designated as 'Gate', enable or disable the corresponding OR gate, which we have already discussed (refer to Fig. 15.4). Similarly, the function of C/$\overline{T}$ bit, responsible for selecting the function as either Timer or Counter, was also discussed. In either case, bits designated as M0 and M1 finalize the mode of operation of these Timers, which may vary from 0 to 3 for Timer 0 and 0 to 2 for Timer 1. All these modes are discussed in details using illustrations related with Timer 0 as an example.

Mode 0 (13-bit Timer/Counter)

Mode 0 is the 13-bit Timer/Counter mode, and it is explained in Fig. 15.6. The output from the RUN/STOP CONTROL block (refer to Fig. 15.4) is fed to Counters of Timer 0 (TL0 and TH0). The lower 5 bits of TL0

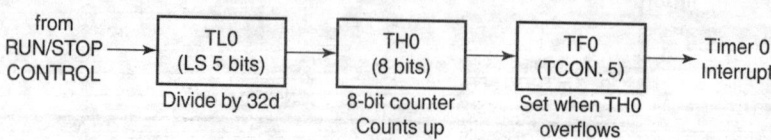

Figure 15.6 Mode 0 (13-bit Timer/Counter)

divide the input signal by 32d, and its output is used by TH0 to count up. When TH0 overflows from FFH to 00H, TF0 (bit 5 of TCON) is set, indicating that Timer 0 interrupt is generated. TF0 is cleared by the processor itself when control branches to 000BH.

The counting does not automatically stop after overflow of Counters and generation of the interrupt. To stop the counting, software commands are necessary through some appropriate routine to modify the control of RUN/STOP CONTROL block.

Mode 1 (16-bit Timer/Counter)

In comparison to Mode 0, Mode 1 differs only by allowing TL0 to be 8 bit instead of 5 bit. In this mode, the output from the RUN/STOP CONTROL block (refer to Fig. 15.4) is also used by TL0 and thereafter by TH0.

A rollover from all 1s to all 0s sets the TF0 bit of TCON SFR, indicating the Timer 0 interrupt (Fig. 15.7). If a 12 MHz crystal is externally interfaced between XTAL1 and XTAL2 pins of 8051, then the frequency of output from RUN/STOP CONTROL block would be 1 MHz. If both TL0 and TH0 are cleared before the start of counting, then the interrupt would occur after 2^{16} microseconds or 65,536 microseconds.

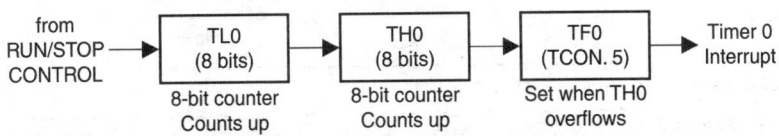

Figure 15.7 Mode 1 (16-bit Timer/Counter)

Just like Mode 0, the Counters keep on counting even after generating the interrupt signal. Some commands from the appropriate routine are necessary to stop the Timer.

Mode 2 (8-bit auto-reload)

In the previous two modes, an overflow from TH0 had set the TF0 flag and generated Timer 0 interrupt. In Mode 2, the overflow from TL0 does this job. TL0 functions as an 8-bit Counter, and whenever its overflow occurs, it sets the TF0 flag and reloads it from the content of TH0 (Fig. 15.8). As before, the output from RUN/STOP CONTROL block activates TL0.

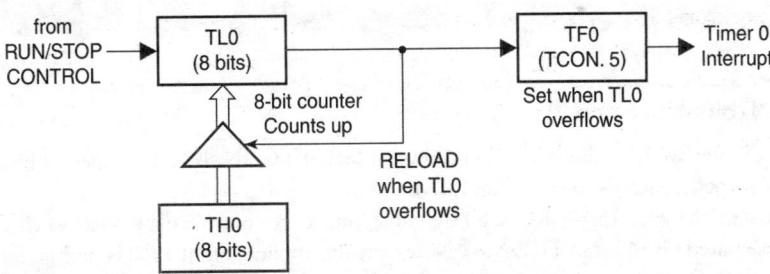

Figure 15.8 Mode 2 (8-bit auto-reload)

Therefore, in this mode, once the Timer Counters are properly initialized and the Timer started, the interrupts would automatically be generated at regular intervals till the Timer is turned off. The delay between any two such interrupts would depend on the value loaded in TH0. Timer run control bit must be used to run/stop the Timer.

Mode 3 (Two 8-bit Timers)

Modes discussed so far are applicable independently to Timer 0 as well as Timer 1. However, Mode 3 is applicable only for Timer 0. *Timer 1 would stop working if it is set for Mode 3.*

As illustrated in Fig. 15.9, Mode 3 converts Timer 0 to two independent 8-bit Timers. TL0 gets its source signal from RUN/STOP CONTROL block and works as Timer or Counter. Its overflow generates Timer 0 interrupt through TF0 bit of TCON. ISR for this interrupt must be placed at the location 000BH.

TH0 gets its source from the oscillator clock divided by 12, which is the signal. It is designated as Timer Clock Source in Fig. 15.3. TH0 works as a Timer, and its run/stop is controlled by the TR1 bit of TCON SFR. Note that in Mode 3, TH0 cannot function as a Counter. During overflow, it sets the TF1 bit of TCON, which generates Timer 1 interrupt. Note that this ISR must be placed at the location 001BH. Although Mode 3 of Timer 0 uses the flag (TF1) and the location of ISR is reserved for Timer 1, Timer 1 still may be used for Modes 0, 1 or 2 for baud-rate generation or for some other purpose, which does not need any explicit Timer 1 ISR.

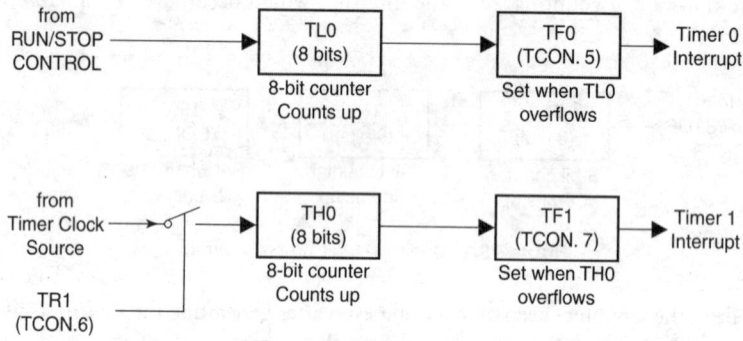

Figure 15.9 Mode 3 (two 8-bit Timers ONLY FOR TIMER 0)

> Mode 2, the 8-bit auto-reload mode is most frequently used. Next is mode 1, the 16-bit counter mode. Mode 0 is rarely used in systems designed around 8051.

15.4.2 | TCON (bits 4, 5, 6 and 7)

In the previous chapter, we have discussed about lower 4 bits of TCON SFR. Its upper 4 bits are earmarked for Timer-control operations as shown in Fig. 15.10.

As indicated in the Fig. 15.10, RUN/STOP of Timer 0 is controlled by TR0 (TCON.4), and for Timer 1, the designated bit is TR1 (TCON.6). Note that the application of TR0 is highlighted in Fig. 15.4 where we have taken Timer 0 as an example case. TR1 has an additional duty if Timer 0 is programmed to operate in Mode 3. In that case, TR1 controls the RUN/STOP operation of the TH0 part of Timer 0 (refer to Fig. 15.9).

Whenever a Timer overflows from all 1s to all 0s, the corresponding TF flag bit of TCON is set by that Timer mechanism. TF0 is set when Timer 0 overflows, and TF1 is set when Timer 1 overflows. Whenever

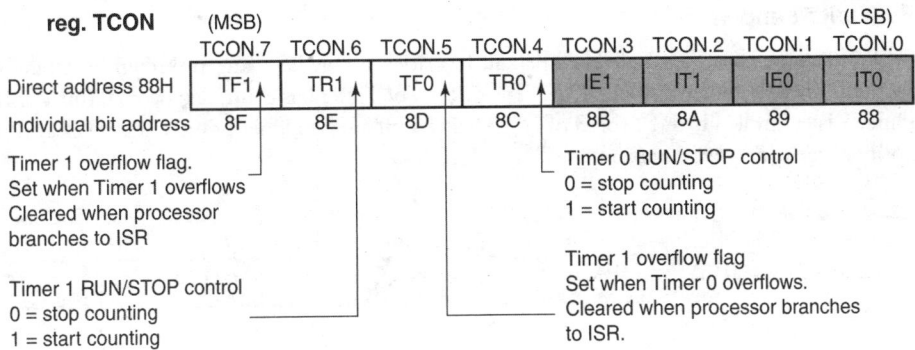

Figure 15.10 Timer interrupt control bits in TCON SFR

the processor branches to the ISR of the interrupting Timer, the corresponding TF flag of that Timer is automatically cleared by the processor itself.

Therefore, for the software developer, TR0 and TR1 are the flag bits to be controlled carefully to start/stop the appropriate Timer. TF bits may be checked to get the indications only.

15.4.3 | IE (bits 1, 3 and 7)

Even when a Timer completes the up-counting and overflows to set the corresponding TF flag, the processor may not acknowledge it if the corresponding Timer's interrupt is not enabled. To enable or disable any Timer interrupt, 8051 provides the IE SFR, which we have already introduced in the last chapter. Relevant bits of this SFR are illustrated in Fig. 15.11. The most significant bit, $\overline{\text{EA}}$, functions as global control for all interrupts by enabling or disabling all interrupts simultaneously. Bits 1 (ET0) and 3 (ET1) enable or disable Timer 0 and Timer 1 interrupts, respectively. Setting any bit enables the corresponding Timer interrupt, and clearing that bit disables it.

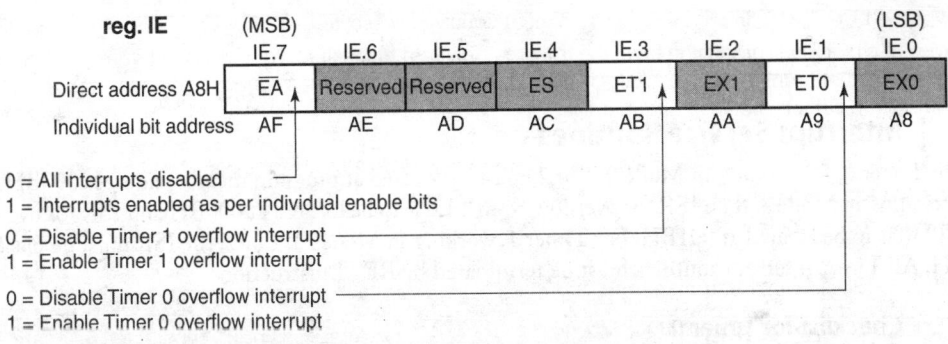

Figure 15.11 Timer interrupt enable bits of IE SFR

15.4.4 | IP (bits 1 and 3)

Although all interrupts have their individual default priorities, it may be changed through software by programming the IP SFR as depicted in Fig. 15.12. The priority of Timer 0 is controlled by PT0 (bit 1 of IP) and that of Timer 1 is controlled by PT1 (bit 3 of IP). Setting a bit allows higher priority and its clearing allows lower priority of the concerned interrupt.

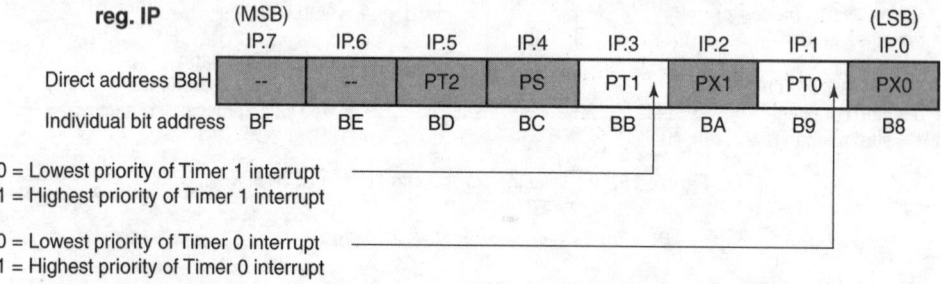

Figure 15.12 Timer interrupt priority setting bits of IP SFR

15.4.5 | TL0, TH0, TL1, TH1

These SFRs serve as Counters of corresponding Timers and are not bit-addressable. TL0 and TH0 belong to Timer 0, while TL1 and TH1 belong to Timer 1. Their direct addresses are presented in Table 15.3.

To make any Timer functional, these Counters must be loaded with appropriate values, remembering the fact that the Counters would count up (increment the SFR values) for each clock pulse (falling edge), and interrupt would be generated when there is an overflow from all 1s to all 0s of the Counters.

Table 15.3 Direct addresses of Timer Counters (SFRs)

SFR	Direct address	Description
TL0	8AH	Timer 0 Counter (lower byte)
TH0	8CH	Timer 0 Counter (higher byte)
TL1	8BH	Timer 1 Counter (lower byte)
TH1	8DH	Timer 1 Counter (higher byte)

15.5 | Interrupt Service Routines

ISR for Timer 0, functioning in Mode 0, 1 or 2, must be located at program memory address 000BH. If it is functioning in Mode 3, then ISR for overflow from TL0 is to be located at 000BH and ISR for overflow from TH0 is to be located at 001BH. For Timer 1, working in Modes 0, 1 or 2, the ISR must be placed at 001BH. All Timer interrupt routines must be terminated by RETI instruction.

15.5.1 | Checklist for Timer 0

The following are the salient points to be taken into account for Timer 0 handling.

(i)   Place ISR for Timer 0, terminated by RETI instruction, at program memory location 000BH. If the ISR is long, it may be placed anywhere within the program memory, and an unconditional jump instruction is to be placed at 000BH to continue at the ISR.

(ii)   Load TL0 and TH0 by the appropriate count values. While calculating these count values, remember that from these values the up-counting would start and interrupt would be generated at the time of rolling from all 1s to all 0s.

(iii)   Select the mode of operation through bits 0 and 1 of TMOD and that of Timer/Counter mode through bit 2 of TMOD.

(iv)   Clear Gate bit (bit 3 of TMOD) so that RUN/STOP may be software controlled by TR0.

(v)   If necessary, set priority through IP.1.

(vi)   Start Timer 0 by setting TR0 of TCON (bit 4).

(vii)   Enable Timer 0 interrupt by setting IE.1.

(viii)   Enable global interrupt control bit by setting IE.7.

(ix)   To stop Timer 0, clear TR0 (TCON.4).

15.5.2 | Checklist for Timer 1

The following are the salient points to be taken into account for Timer 1 handling.

(i)   Place ISR for Timer 1, terminated by RETI instruction, at program memory location 001BH. If the ISR is long, it may be placed anywhere within the program memory, and an unconditional jump instruction is to be placed at 001BH to continue at the ISR.

(ii)   Load TL1 and TH1 by appropriate count values. While calculating these count values remember that from these values the up-counting would start and interrupt would be generated at the time of rolling from all 1s to all 0s.

(iii)   Select the mode of operation through bits 4 and 5 of TMOD and that of Timer/Counter mode by bit 6 of TMOD.

(iv)   Clear Gate bit (bit 7 of TMOD) so that RUN/STOP may be software controlled by TR1.

(v)   If necessary, set priority through IP.3.

(vi)   Start Timer 1 by setting TR1 of TCON (bit 6).

(vii)   Enable Timer 1 interrupt by setting IE.3.

(viii)   Enable global interrupt control bit by setting IE.7.

(ix)   To stop Timer 1, clear TR1 (TCON.6).

15.6 | Timer 2

Timer 2 is a 16-bit Timer/Counter available only in XX52 and its upward versions of MCS-51. Timer 2 is not available in 8051. In comparison to Timer 0 and 1, Timer 2 is similar in certain ways and also different in other parameters. The following are its similarities and differences with the previous two Timers.

15.6.1 | Similarities with Timers 0 and 1

(i)   Functions as either Timer or Counter selectable by C/T2 bit of T2CON.

(ii)   In Timer configuration, the same source clock is used with the same 'divided by twelve' processor clock frequency.

(iii)   Run/Stop is controlled by TR2 bit of T2CON.

(iv)   Uses two 8-bit Counters TL2 and TH2 for up-counting.

(v)   Priority adjustable by IP SFR.

(vi)   Timer 2 interrupt is enabled and disabled by IE SFR.

15.6.2 | Differences with Timers 0 and 1

(i)   Interrupt flag must be cleared by the Timer 2 ISR.

(ii)   Three different modes: 16-bit auto-reload, 16-bit capture and baud-rate generator.

(iii)   Vector address 002BH for placing Timer 2 ISR.

(iv)   Awarded lowest priority among all six interrupts (adjustable by IP).

(v)   Controlled by T2CON SFR (bit-addressable).

(vi)   Separate capture registers are provided.

 As we can observe, only the baud rate generator mode of timer 2 is somewhat similar with that of timer 1. Other two modes of Timer 2 are different. Also note that 16-bit auto-reload mode is very much helpful in certain cases of software development, which is not available in Timers 0 and 1.

15.6.3 | Modes of Timer 2

As indicated above, the modes of Timer 2 are different from the first two Timers. Unlike Timers 0 and 1, the modes of Timer 2 are dependent on RCLK, TCLK, CP/RL2 and TR2 bits of T2CON SFR, which we will discuss shortly. Table 15.4 presents the relation between different modes of Timer 2 and the bits of T2CON.

Table 15.4 Mode setting for Timer 2

RCLK + TCLK	CP/RL2	TR2	MODE
x	X	0	Off
0	0	1	16-bit auto-reload
0	1	1	16-bit capture
1	X	1	Baud-rate generator

As we may deduct, Timer 2 would not perform at all if TR2 bit (bit-2 of T2CON) is cleared. If this bit is set, then Timer 2 would function in 16-bit auto-reload mode provided CP/RL2 bit (bit 0 of T2CON) and both RCLK and TCLK bits of T2CON are cleared. In this mode, TL2 and TH2 are loaded from RCAP2L and RCAP2H (refer to Fig. 15.13) when all bits of TH2 and TL2 roll from all 1s to all 0s, generating Timer 2 interrupt. This mode is useful to generate longer duration continuous interrupts; it is somewhat similar to Mode 2 of Timers 0 and 1, which was 8-bit auto-reload mode. Reloading may also be carried out by the external signal T2EX (P1.1).

The mode is changed to 16-bit capture mode if CP/RL2 bit of T2CON is set, keeping TCLK and RCLK bits cleared. In this mode, the values of Counters TL2 and TH2 may be copied to RCAP2L and RCAP2H, respectively, even if the counting process is ongoing. This may be achieved by the external signal T2EX (P1.1).

Timer 2 may also be used as a baud-rate generator by setting TR2 and any of the bits RCLK or TCLK. The condition of CP/RL2 bit is irrelevant in this case.

15.6.4 | Functioning of Timer 2

Fig. 15.13 illustrates the internal architecture of Timer 2 configured in its 16-bit auto-reload mode. Note that the signal-conditioning part for Timer/Counter is more or less identical with that of Timers 0 or 1, with only one major change. The RUN/STOP CONTROL block is controlled by TR2 bit only, which is

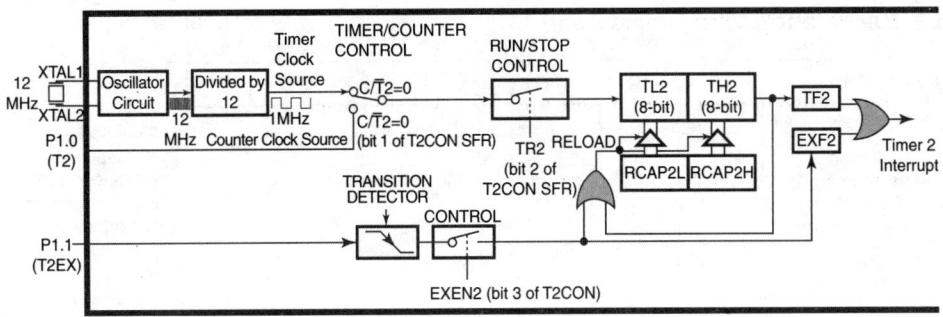

Figure 15.13 Schematic architecture of Timer 2 in auto-reload mode

bit 2 of T2CON SFR. The Counters, TL2 and TH2, are interfaced with RCAP2L and RCAP2H registers. This interfacing is *bi-directional* (only one direction is shown in Fig. 15.13). Timer 2 interrupt is generated by activation of either of the two flags, namely TF2 and EXF2. They are bit 7 and bit 6 of T2CON SFR. Once Timer 2 interrupt is evoked, the flag responsible for generating the interrupt must be cleared by the ISR itself. This is a major difference with the previous two Timers where the interrupt flag was automatically cleared by the processor itself during branching to the ISR.

External control through T2EX pin (P1.1) is also available for capturing the count or reloading TL2 and TH2.

15.6.5 | Timer 2 SFRs

Table 15.5 presents a brief description of only those SFRs which are related with Timer 2. The first one, T2CON, is bit addressable and is used for controlling various features of Timer 2, which we will discuss shortly in detail. TL2 and TH2 are used for counting, and the other two are capture registers.

Table 15.5 Timer 2 SFRs

SFR	Address	Bit-addressable	Function
T2CON	C8H	Yes	Timer 2 control
RCAP2L	CAH	No	Timer 2 capture (low)
RCAP2H	CBH	No	Timer 2 capture (high)
TL2	CCH	No	Timer 2 Counter (low)
TH2	CDH	No	Timer 2 Counter (high)

T2CON

T2CON controls most of the functions of Timer 2 as illustrated in Fig. 15.14. Bit 0 (CP/RL2) controls capture (1) or auto-reload (0) through software commands using this bit. However, when being used for baud-rate generator for serial interface, maintaining either RCLK as 1 or TCLK as 1 overlooks the commands through CP/RL2 bit.

Bit 1 of T2CON (C/T2) indicates whether it would function as a Timer (C/T2 - 0) or a Counter (C/T2 - 1). Bit 2 (TR2) controls the RUN/STOP feature. Setting TR2 would run Timer 2, while clearing it would stop Timer 2. Bit 3 of T2CON is designated as EXEN2. We may observe from Fig. 15.13 that a 0 in this bit would not allow the external signal through T2EX (P1.1) to capture or reload Timer 2 Counters. For this purpose EXEN2 must be set.

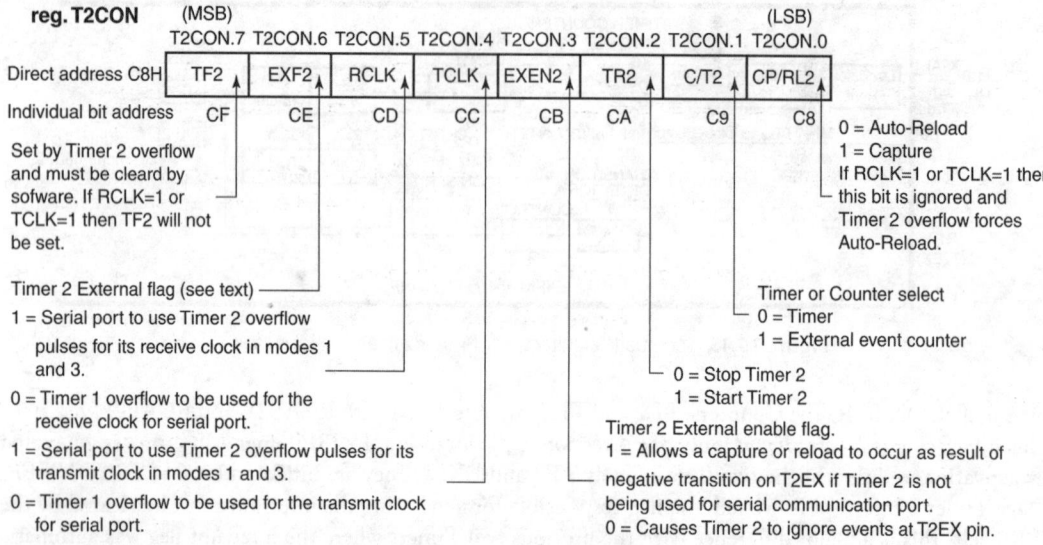

Figure 15.14 T2CON SFR

If Timer 2 is to be used as baud-rate generator, bits designated as RCLK and TCLK should be used. If TCLK is set, then transmit clock of serial transmission would be derived from Timer 2 counting. If RCLK is set, the receiving part of serial interface would use Timer 2. If any one of these 2 bits is 0, the corresponding clock would be obtained from Timer 1. Note that if Timer 2 is to function as 16-bit auto-reload mode or 16-bit capture mode, then both RCLK and TCLK must be 0.

EXF2 is an alternate interrupt-generating flag bit of Timer 2 (T2CON.6), which is activated by external hardware signal (falling edge only) through T2EX pin (P1.1). Timer 2 ISR should read both TF2 and EXF2 to indicate which one has generated the interrupt and service accordingly. The ISR should clear that bit by software command before returning from the ISR.

TF2 is the interrupt-generating flag bit (T2CON.7). Overflow from Counters would set this bit and draw the processor's attention for a Timer 2 interrupt. The ISR for Timer 2 is expected to clear this bit through software commands before returning from the ISR through the RETI instruction. The bit is not automatically cleared by the processor itself at the time of branching to the ISR.

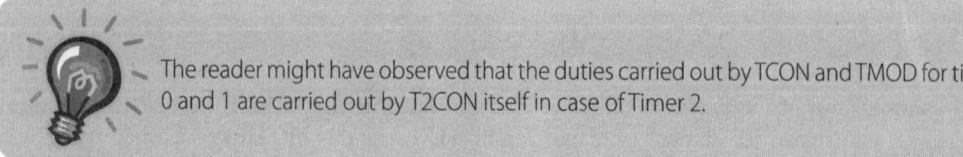

The reader might have observed that the duties carried out by TCON and TMOD for timers 0 and 1 are carried out by T2CON itself in case of Timer 2.

IE (bits 5 and 7)

Bit 5 of IE SFR (ET2) is earmarked for Timer 2 interrupt. Setting this bit as 1 would enable Timer 2 interrupt if the global interrupt enable bit (bit 7, $\overline{\text{EA}}$) is set. As indicated in Fig. 15.15, clearing this bit ET2 would disable Timer 2 interrupt.

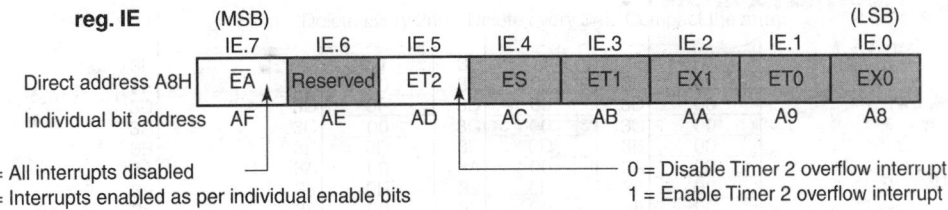

0 = All interrupts disabled
1 = Interrupts enabled as per individual enable bits

0 = Disable Timer 2 overflow interrupt
1 = Enable Timer 2 overflow interrupt

(Note: Remaining bits' functions explained at appropriate places)

Figure 15.15 Register IE

IP (bit 5)

Timer 2 interrupt has, by default, the lowest priority of all six interrupts of XX52. However, setting bit 5 of IP SFR (PT2) would allow it higher priority as illustrated in Fig. 15.16.

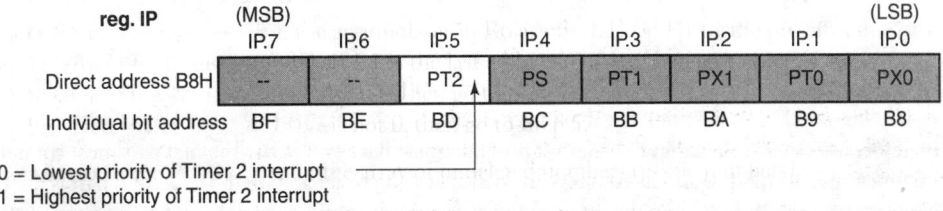

0 = Lowest priority of Timer 2 interrupt
1 = Highest priority of Timer 2 interrupt

(Note: Remaining bits' functions explained at appropriate places)

Figure 15.16 Register IP

15.6.6 | Checklist for Timer 2

The following are the salient points to be taken into account for Timer 2 handling when it operates in 16-bit auto-reload mode.

(i) Place ISR for Timer 2, terminated by RETI instruction, at program memory location 002BH. If the ISR is long, it may be placed anywhere within the program memory, and an unconditional jump instruction is to be placed at 002BH to access the ISR. Note that ISR2 must clear TF2 flag.

(ii) Load TL2 and TH2 by appropriate count values. During calculating these count values remember that from these values the up-counting would start and interrupt would be generated at the time of rolling from all 1s to all 0s. Load RCAP2L and RCAP2H by reload values.

(iii) Select mode of operation through bits 0, 4 and 5 of T2CON and that of Timer/Counter mode by bit 1 of T2CON.

(iv) Clear bit 3 of T2CON so that RUN/STOP may be software controlled by TR2.

(v) Timer 2 has the lowest priority by default. If necessary, set priority through IP.5.

(vi) Start Timer 2 by setting TR2 of T2CON (bit 2).

(vii) Enable Timer 2 interrupt by setting IE.5.

(viii) Enable global interrupt control bit by setting IE.7.

(ix) To stop Timer 2, clear TR2 (T2CON.2).

15.7 | Solved Examples

Example 15.1

Purpose: To develop subroutines to initialize and react through ISR for mode 0 (13-bit Timer/Counter mode) interrupt of Timer 0 so that it is reloaded and restarted.

Problem

Assuming that a 12 MHz crystal is attached with a 8051 microcontroller what is the maximum time delay possible to be generated if its Timer 0 is decided to be used in Mode 0? Develop the initialization routine for this purpose and also the relevant ISR.

Solution

Mode 0 of timer allows us to use any timer (Timer 0 or Timer 1 of 8051) to generate either a single interrupt or multiple interrupts. For the second case, the general procedure is stop-reload-run, which means, after timer interrupt at the terminal count, the timer must be stopped, reloaded and restarted. For the first case, the timer is to be simply loaded once, started and stopped after the terminal-count interrupt. We must remember that the up-counting of any timer of 8051 continues even after its terminal count, till it is stopped by the appropriate software command.

As mode 0 allows us to configure the timer in 13-bit up-counting mode, therefore the maximum delay that might be generated would be 213 or 8192 in decimal. For 12 MHz external crystal, the timer would be running with a frequency of 1 MHz. Therefore, it can generate a maximum delay of 8192 microseconds or 8.192 milliseconds. Note that if Timer 0 is used, then its TL0 to be loaded with xxx0000B and TH0 to be loaded with 00000000B to generate the maximum delay.

Therefore, we can visualize that the relevant program must have two parts. The first part must configure the timer to function in its mode 0 and its counters are loaded and the timer is started. The second part is the ISR for the timer interrupt, which must stop the timer. Following example routines are designed to serve these purposes.

; Initialization routine for Timer 0 in mode 0.

```
INITM0:   MOV    TL0,#0BH
          MOV    TH0,#00H
          MOV    TMOD,#0F0H    ; Timer 0 in mode 0
          SETB   TCON.4        ; start Timer 0
          SETB   IE.1          ; enable Timer 0 interrupt
          SETB   IE.7          ; enable global interrupt
```

; ISR for Timer 0 to be placed at 000BH of program memory.

```
          ORG    000BH
ISRT0:    CLR    TCON.4        ; stop Timer 0
```

; Place other instructions here to perform necessary duties for Timer 0 interrupt.

```
          RETI                 ; return from Timer 0 interrupt
```

Example 15.2

Purpose: To develop initialization routine and ISR for Timer 0 working in mode 1 (16-bit up-counting mode) so that it is always reloaded and restarted by its ISR.

Problem

Assuming that 8051 microcontroller is interfaced with a 12 MHz crystal, develop a program to generate Timer 0 interrupts at every 50 milliseconds continuously.

Solution

In Mode 1 (16-bit counting), the maximum count value is 2^{16} or 65,536. With a 12 MHz external crystal, the Timer Counters would be receiving a frequency of 1 MHz. Therefore, the count value for generating a delay of 50 milliseconds (50,000 microseconds) would be 50,000 in decimal or C350H. As this is to be counted up, and all 1s to all 0s would generate an interrupt, the registers are to be loaded with a value of

$$FFFFH - C350H + 1 = 3CAFH + 1 = 3CB0H.$$

This 16-bit value must be loaded in TL0 and TH0. As 16-bit auto-reload is not possible in Timer 0, the ISR of Timer 0 must stop, reload and restart the Timer to generate continuous interrupts.

The software is to be developed in two parts. The first part would be initialization of Timer 0, which is generally carried out after reset initialization. The second part would be the ISR for Timer 0, placed at program memory location 000BH. The following are the necessary parts of program listings.

; Initialization routine for Timer 0, place after system reset initialization routine.

```
INITM0:   MOV    TL0, #0B0H      ; load TL0 by lower value of count
          MOV    TH0, #3CH       ; load TH0 by higher value of count
          MOV    TMOD, #0F1H     ; Timer 0 Mode 1, Gate cleared
          SETB   TCON.4          ; start Timer 0 by setting TR0
          SETB   IE.1            ; enable Timer 0 interrupt
          SETB   IE.7            ; enable global interrupt
```

; Timer 0 interrupt service routine. Start from location 000BH.

```
          ORG    000BH
ISR0:     CLR    TCON.4          ; stop Timer 0
          MOV    TL0, #0B0H      ; reload TL0
          MOV    TH0, #3CH       ; reload TH0
          SETB   TCON.4          ; restart Timer 0
```

; Complete other duties of ISR here. Save PSW and Accumulator, if necessary. At the end
; restore these values.

```
          RETI                   ; return from Timer 0 interrupt
```

Note that stopping, reloading and restarting the Timer would add up a minor delay in successive interrupts.

C-version

```c
#include <regx51.h>

void timer0(void) interrupt1
{
        TH0 = 0x3C;
        TL0 = 0xB0;
}
```

```
void main()
{
        TMOD = 0x01;
        TH0 = 0x3C;
        TL0 = 0xB0;
        IE = 0x82;
        TR0 = 1;
        While(1)
        {
        }
}
```

Example 15.3

Purpose: To be familiar with the development of a RTC using Timer 1 working in Mode 2 (8-bit auto-reload mode).

Problem

Develop a software for a RTC using Timer 1. Assume that 8051 is running on 12 MHz crystal.

Solution

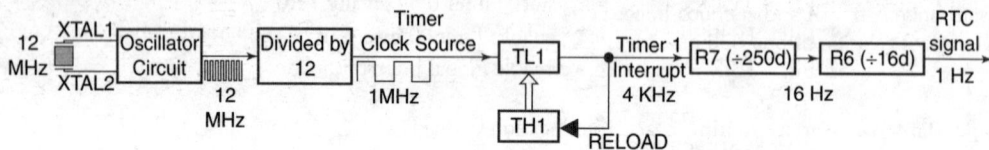

Figure 15.17 Schematic arrangement for 1 Hz signal generation

We assume that if we can generate 1 Hz signal, then any RTC may be implemented thereafter. We may use Timer 1 in Mode 2 (8-bit auto-reload mode). This would generate continuous interrupts, and there would be no need to stop, reload and restart the Timer. For 8-bit Counter with 12 MHz crystal externally, interrupts may be generated at every interval of 250 microseconds. We would use internal registers in cascade to count these interrupts to produce 1 Hz signal. We would need two such registers. The first one would count down from 250 (decimal) to 0. The second one would count down 16 (decimal) to 0. Whenever any register is cleared, it must be reloaded by its reset value. Fig. 15.17 explains the overall strategy. The initialization part and the ISR part are presented below.

```
; Initialization of Timer 1 in Mode 2 (auto-reload mode) to generate continuous interrupts at the intervals of
; 250 microseconds, using 12 MHz external crystal.
; This part should be placed after reset initialization.
; Note that Timer 1 Counters would be counting up and
; R7 and R6 would be counting down.

INITM1:  MOV    R7, #0FAH          ; initialize by 250d to count down
         MOV    R6, #10H           ; initialize by 16d to count down
         MOV    TH1, #06H          ; initialize for 250d up-count (auto-reload)
         MOV    TL1, #06H          ; initialize for 250d up-count at start
         MOV    TMOD, #20H         ; initialize Timer 1 for mode 2, gate cleared
         SETB   TCON.6             ; start Timer 1
         SETB   IE.3               ; enable Timer 1 interrupts
         SETB   IE.7               ; enable global interrupt
```

; Timer 1 ISR. Interrupt at every 250 microseconds.
; This routine must be placed at 001BH.

```
         ORG    001BH
ISR1:    PUSH   ACC            ; save accumulator content
         PUSH   PSW            ; save processor status
         DJNZ   R7, T1OVR      ; count 250 interrupts
         MOV    R7, #0FAH      ; reload by 250d
         DJNZ   R6, T1OVR      ; count 16 times
         MOV    R6, #10H       ; reload by 16d
```

; One second elapsed from previous case of R6 being 0.
; Update real-time clock Counters for second, minute and hour display.
; Then continue at T1OVR.

```
T1OVR:   POP    PSW            ; restore processor status
         POP    ACC            ; restore accumulator content
         RETI                  ; return from Timer 1 interrupt
```

Various parts of a Real Time Clock (RTC) are explained in this book at different appropriate places. Placing these pieces together would give you complete details of an RTC.

C-version

```c
#include <regx51.h>

unsigned int cnt=0;
unsigned char sec=0, min=0,hr=0;

void main(void)
{
        // initialize Timer 1
        TH1 = 0x06;
        TL1 = 0x06;
        TMOD = 0x20;
        TR1 = 1;
        ET1 = 1;
        EA = 1;

        // Terminate here
        while(1);
}

Void timer1Isr(void) interrupt TF1_VECTOR
{
        // would be executed at every 250 us
```

```
        cnt++;

        if(cnt >= 4000) // one second elapsed
        {
            cnt = 0;
            sec++;
            if(sec > 59)
            {
                sec = 0;
                min++;
                if(min > 59)
                {
                    min = 0;
                    hr++;
                    if(hr > 23)
                    {
                        hr = 0;
                    }
                }
            }
        }
}
```

Example 15.4

Purpose: To understand the Mode 3 operation of 8051 Timer 0.

Problem

Develop routines to generate two different waveforms using Timer 0 in Mode 3.

Solution

When configured to operate in mode 3, Timer 0 would operate as two independent 8-bit timers (refer Fig. 15.9). Note that mode 3 is applicable only for Timer 0 and is not applicable for Timer 1 of 8051. When functioning in its mode 3, the terminal count of TL0 register would generate Timer 0 interrupt and terminal count of TH0 register would generate Timer 1 interrupt. Therefore, we may load two different values in these two registers, configure Timer 0 to function in its mode 3 and start the timer.

The stop-reload-restart commands would be necessary for each interrupt to generate continuous waves from each counter (TL0 and TH0). It means the auto-reload is not applicable in this mode (mode 3 of Timer 0). Note that the maximum count value for each case would be limited to 8-bit only. Let us load TL0 by 06H and TH0 by 82H. As all timer counters of 8051 counts up, therefore 06H in TL0 would generate minimum frequency. With a 12 MHz crystal, it would generate interrupt at every 250 micro-seconds, with our stop-reload-restart scheme. TH0 would generate interrupt at every 125 micro-seconds. The relevant software routines would be as follows.

```
; Initialization routine for Timer 0 in Mode 3.
; Timer 0 interrupt @ 4 KHz. Timer 1 interrupt @ 8 KHz.
;
```

```
INITM3:  MOV    TL0, #06H
         MOV    TH0, #82H
         MOV    TMOD, #0F3H        ; Timer 0 in mode 3
         SETB   TCON.4             ; start Timer 0
         SETB   IE.1               ; enable Timer 0 interrupt (for TL0)
         SETB   IE.3               ; enable Timer 1 interrupt (for TH0)
         SETB   IE.7               ; enable global interrupt
```

; Timer 0 interrupt service routine. Start from 000BH.

```
         ORG 000BH
ISRT0:   CLR    TCON.4             ; stop Timer 0.
         MOV    TL0, #06H          ; reload its counter
         SETB   TCON.4             ; restart Timer 0
         RETI
```

; Timer 1 interrupt service routine. Start from 001BH.

```
         ORG 001BH
ISRT1:   CLR    TCON.4             ; stop Timer 0
         MOV    TH0, #82H          ; reload its counter
         SETB   TCON.4             ; restart Timer 0
         RETI
```

Example 15.5

Purpose: To be familiar with Timer 2 working in 16-bit auto-reload mode.

Problem

Using 12 MHz external crystal, develop a software to get continuous interrupts from Timer 2 after every 50 milliseconds.

Solution

This problem is partially similar to Example 15.1 above. To generate 50 milliseconds delay using 12 MHz crystal, the same 16-bit value would be necessary. However, in Example 15.1, this value was reloaded after receiving an interrupt by stopping, reloading and restarting Timer 0. This would not be necessary in Timer 2 in its 16-bit auto-reload mode. Therefore, the ISR would be simpler. However, TF2 must be cleared by the ISR.

; Initialize Timer 2 for continuous interrupt at every 50 milliseconds.

```
INITM2:  MOV    TL2, #0B0H         ; load TL2 by lower value of count
         MOV    TH2, #3CH          ; load TH2 by higher value o count
         MOV    RCAP2L, #0B0H      ; reload value for TL2
         MOV    RCAP2H, #3CH       ; reload value for TH2
         MOV    T2CON, #00H        ; 16-bit auto-reload Timer
         SETB   T2CON.2            ; start Timer 2
         SETB   IE.5               ; enable Timer 2 interrupt
         SETB   IE.7               ; enable global interrupt
```

; Interrupt service routine for Timer 2. This must be located at location 002BH.

```
              ORG     002BH
    ISRT2:    CLR     T2CON.7       ; clear TF2 of T2CON
              PUSH    PSW           ; save status
              PUSH    ACC
```

; Include instructions here as per requirements.

```
              POP     ACC           ; restore status
              POP     PSW
              RETI                  ; return from Timer 2 interrupt
```

C-version

```c
#include <regx51.h>

void main(void)
{
        // initialize Timer 2

        TH2 = 0x3C;
        TL2 = 0xB0;
        RCAP2H = 0x3C;
        RCAP2L = 0XB0;
        T2CON = 0x00;
        TR2 = 1;
        ET2 = 1;
        EA = 1;

        while(1);
}

void timer2Isr(void) interrupt 5
{
        TF2 = 0; // clear interrupt request
}
```

Example 15.6

Purpose: To be familiar with the Counter operation of Timer 1 and the simultaneous usage of multiple Timers. Timer 0 would work in Mode 2 as a Timer, and Timer 1 would work in mode 1 as a Counter.

Problem

An optical encoder attached with a motor is generating square wave pulse with a frequency related to its RPM. Develop a hardware scheme and the related software so that the RPM may be displayed through Port 0 at every interval of 50 milliseconds.

Solution

For this problem, we may use both Timer 0 and Timer 1. Timer 0 may be used in its Mode 2 (auto-reload mode). Assuming 12 MHz external crystal, this may be programmed to generate continuous interrupts at every 250 microseconds. Cascading two general-purpose registers, an interval of 50 milliseconds may be calculated.

Timer 1 may be configured as a Counter with its input from Port 3.5 (T1) (refer to Fig. 15.18), which may receive the signal from the optical encoder. At every 50 milliseconds, the Counters of Timer 1 may be read and displayed through Port 0.

; Initialize Timer 0 in Mode 2 (8-bit auto-reload) and Timer 1 in 16-bit Counter mode.
; Timer 1 would only count and would not generate any interrupt. It would be read before overflow.

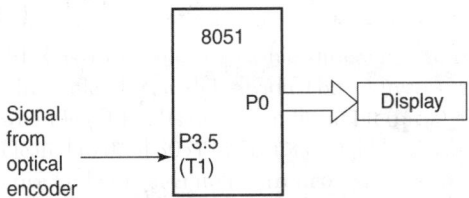

Figure 15.18 Configuration for Timer 1 as Counter for external signal

```
INITM1:MOV    R7, #0FAH      ; initialize by 250d to count down
       MOV    R6, #10H       ; initialize by 16d to count down
       MOV    TH0, #06H      ; initialize for 250d up-count (auto-reload)
       MOV    TL0, #06H      ; initialize for 250d up-count at start
       CLR    A
       MOV    TL1, A         ; clear both Counters of Timer 1
       MOV    TH1, A
       MOV    TMOD, #52H     ; Timer 0 Mode 2 Timer 1 Mode 1 Counter
       SETB   TCON.4         ; start Timer 0
       SETB   TCON.6         ; start Timer 1 as Counter
       SETB   IE.1           ; enable Timer 0 interrupts
       SETB   IE.7           ; enable global interrupt
```

; Interrupt service routine for Timer 0.
; Read and reset Timer 1 Counters. Display through P0.

```
       ORG    000BH
ISR00: DJNZ   R7, T0OVR
       MOV    R7, #0FAH      ; reload by 250d
       DJNZ   R6, T0OVR
       MOV    R6, #10H       ; reload by 16d
       PUSH   ACC            ; save accumulator
       MOV    A, TL1         ; read lower byte of Timer 1 Counter
       MOV    P0, A          ; display
       CLR    A
       MOV    TL1, A         ; clear Timer 1 Counters
       MOV    TH1, A
       POP    ACC            ; restore accumulator
T0OVR: RETI
```

SUMMARY

Microcontroller 8051 offers two 16-bit up-count Timers, namely, Timer 0 and Timer 1. Higher versions, like 8052, offer a third 16-bit up-count Timer, namely, Timer 2. The first two are more or less similar in nature and offer four operational Modes, 0, 1, 2 and 3 (last one is only for Timer 0); they can operate as both Timers and Counters, which are software programmable. The counting may be controlled (frozen or restarted) through both hardware and software provisions. By mode selection, these may be configured as 13- or 16-bit Timers or an

8-bit auto-reload Timer. In Mode 3, Timer 0 borrows some resources of Timer 1 and works as two independent 8-bit timers.

Timer 2 offers three modes, namely, 16-bit auto-reload, 16-bit capture and baud-rate generator. For this purpose, two special capture registers (RCAP2H and RCAP2L) are provided for the Timer. Each of these three Timers has two 8-bit Counters. SFRs, like TMOD, TCON, T2CON, IE and IP, offer various controlling features for these Timers.

POINTS TO REMEMBER

- To get continuous interrupts from Timer 0 or 1 operating in its Mode 1, it must be stopped, reloaded by the 16-bit up-count value and restarted.

- Mode 1 (16-bit count) and Mode 2 (8-bit auto-reload) are the most widely used modes for Timers 0 and Timer 1.

REVIEW QUESTIONS

Evaluate Yourself

1. 8051/8052 Timers are capable of counting

   (a) up   (b) down

   (c) both (programmable)   (d) none of these

2. Timer/Counter configuration is selectable by

   (a) hardware   (b) software

   (c) either way   (d) none of these

3. How many modes are offered by Timer 0?

   (a) Two   (b) Three

   (c) Four   (d) None of these

4. How many modes are offered by Timer 1?

   (a) Two   (b) Three

   (c) Four   (d) None of these

5. Setting TF0 bit of TCON by SETB instruction would

   (a) generate a software interrupt

   (b) nothing

   (c) start reverse counting

   (d) none of these

6. The source for counting of Timer 0 in its Counter mode would be

   (a) P3.2 (INT0)   (b) P1.1

   (c) P3.4 (T0)   (d) none of these

7. To which SFR does the RUN/STOP control bit TR0 for Timer 0 belong?

   (a) TCON   (b) T2CON

   (c) TMOD   (d) None of these

8. If Timer 0 is configured in Mode 3, then overflow from TH0 would set the bit, named

   (a) TF0   (b) TF1

   (c) TL0   (d) none of these

9. The priority of Timer 2 is programmable through the SFR, named

   (a) T2CON   (b) RCAP2

   (c) RCAP2H   (d) none of these

10. The three operating modes of Timer 2 are 16-bit auto-reload mode, 16-bit capture mode and

   (a) PWM mode

   (b) baud-rate generator mode

   (c) dual-timer mode

   (d) none of these

Search for Answers

1. What is the difference between a Timer and a Counter?

2. How is RUN/STOP control achieved in Timers 0 and 1?

3. How is RUN/STOP control achieved in Timer 2?

4. How does Timer 1 function in its Mode 3?

5. What is the relation between external crystal and Timer function as a Timer?

6. What are the functions of RCLK and TCLK bits of T2CON SFR?

7. What are the differences between Timer 0 and Timer 2?

8. What is the role of IE register in Timer operations?

9. How can the priority of a Timer interrupt be changed?

10. At which address should the ISR of Timer 2 be placed?

Think and Solve

1. Which of the processor resources is shared equally by all the three Timers?

2. What would happen if a Timer, say Timer 0 in Mode 1, is not stopped after its terminal counting and interrupt generation?

3. What is the reason for providing multiple interrupt flags in Timer 2?

4. Is it possible to change the priority of a Timer during its run time?

5. Is it possible for the ISR of a Timer, say Timer 0, to start or stop another Timer, say Timer 1 or Timer 2?

6. What would happen if a Timer is started but its interrupt is not enabled?

7. What is the advantage of the capture mode of Timer 2?

8. If all the three Timers are assigned higher priority by setting their corresponding bits in IP register and if all three Timers accidentally generate interrupts at the same time, which order of Timer interrupt servicing would be followed by the processor?

9. Write a program to generate a square wave with 50% duty cycle using Timer 0. Assume the frequency to be 10 KHz with the external crystal frequency of 12 MHz.

10. Assume that both Timer 0 and Timer 1 are operating simultaneously. Develop suitable ISRs for these Timers so that during servicing of Timer 0, Timer 1 should not interrupt it and vice versa.

16 SERIAL COMMUNICATION AND SERIAL INTERRUPTS

CHAPTER OBJECTIVES

In this chapter, the reader is introduced to Serial Interrupts of 8051. After completion of the chapter, the reader should be able to understand

- Types of serial communication.
- Role of serial interrupt in serial communication.
- How 8051 implements serial communication in its different modes.
- SFRs related with serial communications.
- Developing subroutines for serial communication.

16.1 | Introduction

In this third and final chapter on interrupts, we complete the discussions on 8051 interrupts related to serial communications. This type of communication is necessary in multiprocessor environment. Fig. 16.1 shows general classification of communication.

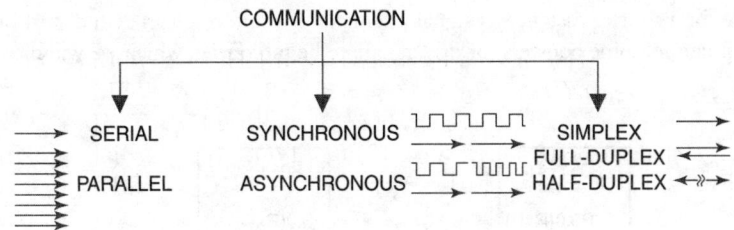

Figure 16.1 General classification of communication

16.1.1 | Serial and Parallel Communications

When the classification is based on number of data bits communicated simultaneously, we find serial and parallel communications. In the first case, only 1 bit is communicated at a time while multiple bits, usually 8 or its multiples, are communicated at the same time in the second case. Therefore, serial communication is a slower process but demands lesser number of signal lines with respect to parallel communication.

16.1.2 | Simplex, Full-Duplex and Half-Duplex Communications

The capabilities of simultaneous data flow directions generate three cases, namely, simplex, full-duplex and half-duplex. In simplex communication, data flow is unidirectional like TV transmission. In the second case, for full-duplex communication, the data flow is bi-directional at the same time, e.g. telephone. In the third case, the data flow is time-shared bi-directional, i.e. unidirectional at any given time. Hewlett Packard Interface Bus (HPIB) is an example of half-duplex communication.

16.1.3 | Synchronous and Asynchronous Communications

Another way to classify is to check the existence of any common timing standard between transmitting and receiving devices. If both are operating at the same frequency, then the communication is synchronous, like the communication between a processor and its internal or external memory device. In this case, data must be communicated within a predefined time-slot agreed by both ends.

On the other hand, if both are being operated at their own frequencies, the communication between them is designated as asynchronous. As there is no predefined time-slot, therefore, in most cases this asynchronous communication needs some handshaking signals. Communication between a PC and its printer may be taken as an example of asynchronous communication.

 Parallel asynchronous communication is also possible to be implemented in a 8051 based system by using its ports. However, in this chapter we concentrate only on serial communications by 8051.

16.2 | Overview of Serial Communication

MCS-51 offers full-duplex asynchronous serial communication feature through its external pins, P3.0 (RxD) and P3.1 (TxD). Here, RxD stands for receiving data and TxD stands for transmitting data. To communicate with another microcontroller, signals from these pins are to be interfaced as shown in Fig. 16.2. Note that this scheme is applicable for a shorter distance as signal attenuation would result in data distortion for a longer distance. In such a situation, some boosters (or drivers) are to be interfaced, which we will discuss in Section 16.5 of this chapter.

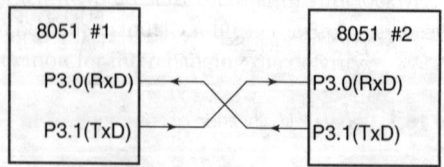

Figure 16.2 Interface to establish serial communication (for very short distance)

16.2.1 | How Serial Communication Works

To get an overview of serial communication, let us imagine (refer to Fig. 16.2) that 8051(#1) is having 10 bytes in its internal RAM location which should be sent to 8051(#2). Note that 8051(#1) must be having its own serial data transmission subroutine while 8051(#2) also must be equipped with its own serial data receiving subroutine. We assume that 8051(#1) knows the starting address of its 10 bytes and 8051(#2) knows where to store the bytes after receiving those. What would happen next is described in an *over-simplified manner* through following steps:

(i)   8051(#1) gets first byte from RAM and places in its transmitting buffer.

(ii) Transmitting buffer sends the byte, 1 bit at a time, to 8051(#2).

(iii) 8051(#2) receives the byte, bit by bit, and stores in its receiving buffer.

(iv) When its receiving buffer is full, 8051(#2) stores the received byte in its own RAM location.

(v)   The procedure is repeated 10 times, and the communication is completed.

The reader may ask that what is the role of interrupt in this communication. Well, every time its transmitting buffer becomes empty (all bits transmitted), 8051(#1) would generate a serial interrupt indicating *transmission complete*. Naturally, its serial ISR would service that interrupt by placing another byte in the transmitting buffer, unless all bytes are transmitted. Similarly, every time the receiving buffer is full, 8051(#2) would generate a serial interrupt indicating *received a byte*. Its serial ISR would service that interrupt by storing the received byte from receiving buffer to the earmarked area in its RAM. Therefore, once initiated, the processor need not look after the communication for either transmitting or receiving. The entire process would be interrupt driven.

However, in our generalized discussion, we have omitted many details, which we would discuss in due course in the remaining sections of this chapter. To start with, we first take up the serial data format.

8085 is also capable of serial communications through its SOD and SID pins. However, the input through SID must be polled by 8085 as its serial communication is not interrupt driven, like 8051.

16.2.2 | Serial Data Format

Although, it is true that every byte is broken into 8 bits and transmitted one by one, some indicators are attached with it to mark the beginning and ending of any such transmitted byte. This is necessary to distinguish a byte from the next one and reduces the chances of errors. Fig. 16.3 illustrates the format of a byte used in serial communication.

As we may observe, every byte is preceded by a *start bit*, which is always low and followed by one (sometimes two) *stop bit(s)*, which is (are) always high. After the start bit, bit 0 (LSB) is transmitted next, followed by bit 1 and so on. Finally, bit 7 (MSB) is transmitted last, followed by the stop bit. These start and stop bits are

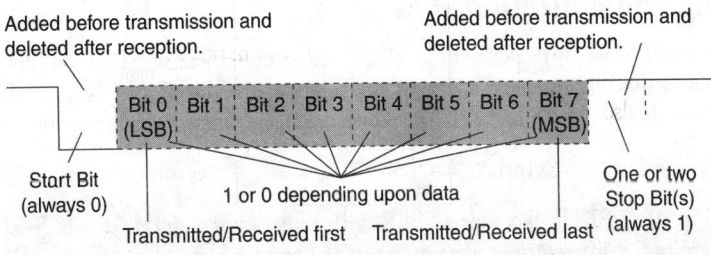

Figure 16.3 Eight-bit serial data communication format

automatically added by the transmitting processor and discarded by the receiving processor, before transmission and after reception of the byte. Therefore, to communicate a byte, it is necessary to physically communicate at least 10 bits of information. We are about to discuss more regarding it during the discussions of baud rate.

Although the format remains same, the logic levels may be represented by either amplitude (Amplitude Modulation or AM) or frequency (Frequency Modulation or FM). Serial communication of 8051 uses amplitude modulation, i.e. logic 0 is represented by 0V and logic 1 by 5V. An idle state (no activity) is also represented by 5V. Duration of this logic level for each bit depends on the selected baud rate, which we would discuss now.

16.2.3 | Baud Rate

Baud rate means the rate of information communication, generally expressed per second. In communication, standard baud rates are available to vary the speed of data transfer. These standard rates are 300, 600, 1200, 2400, 4800, 9600 and 19200 bps. Although in practice, we use the unit bits per second or bps, however, the term *bit* should be replaced by the term *symbol* for a more meaningful and appropriate name. In general, the term bit is associated with byte, and 8 bits make a byte. However, for BPS, the number of bits per second includes the start bit and stop bit, violating the normal correlation between a bit and a byte. To make the matter simpler, we should remember that to get the number of bytes transferred per second for a baud rate of 300 bps, with one stop bit, we should calculate it as 300/10 = 30 bytes per second. We divide by 10 as we include a start bit and a stop bit with every byte of 8 bits.

Before any communication, this baud rate must be finalized and known to both sides of communication. This ensures correct data collection. The receiver side samples the received signal at the middle of any bit as shown in Fig. 16.4.

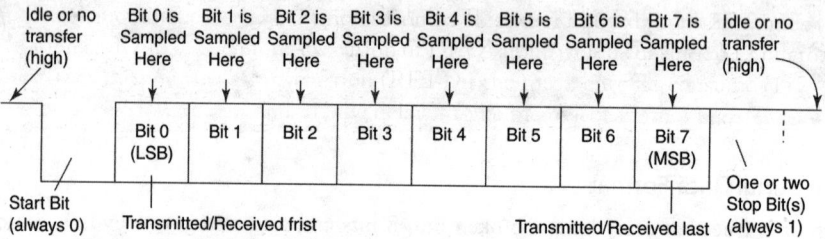

Figure 16.4 Time of sampling of received signal

To implement these timing features, generally the help of a Timer is sought after. Moreover, the frequency of the external crystal is fixed at 11.0592 MHz. We will discuss why such a special frequency of the crystal is necessary if serial communications to be implemented by the system. Presently, we take a look at the Special Function Registers (SFRs) related with serial communications.

16.3 | Serial Communication SFRs

MCS-51 offers three SFRs directly related with serial communication. These are SBUF, SCON and PCON. SBUF is used as both transmitting and receiving buffer. SCON, along with 1 bit of PCON, controls various features of serial communications. It also uses IE and IP SFRs which are already introduced in previous chapters.

16.3.1 | SBUF

As illustrated in Fig. 16.5, SBUF is a pair of SFR with same address, 99H. In MCS-51 architecture, it is the only case where two 8-bit registers were assigned the same address. One of the pair is *write only* and earmarked for data transmission. The byte to be transmitted through TxD is stored in this register by the software. It is then the duty of the processor to add start and stop bits and send all bits one-by-one to the

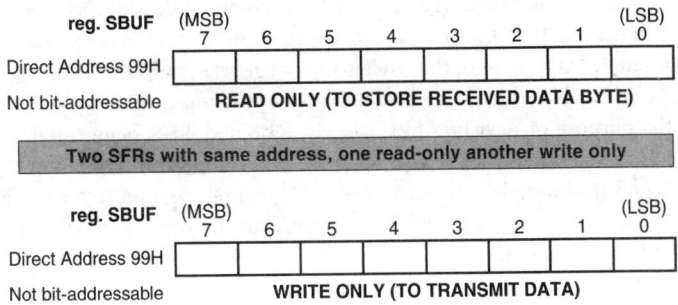

Figure 16.5 Serial communication buffer SBUF

serial output pin, TxD. Once all 8 bits are sent out, a serial interrupt is generated to draw the attention of the controlling software. Note that the *serial transmission is initiated just by writing a byte in SBUF.*

The other one of this pair is *read only* and stores only a complete byte received through RxD. Please note that the data in form of bits, along with start and stop bits, received through RxD input pin, initially stored in a temporary register, bit wise. After completion of collection of all eight data bits, it is immediately copied to SBUF in its read-only part, and a serial interrupt is generated. It is then the duty of the controlling ISR to read the received data from SBUF.

16.3.2 | SCON

SCON is a bit-addressable SFR with address 98H. In the previous chapter, we have observed that timer-interrupt features are controlled by two different SFRs, TMOD and TCON. For serial interrupts, SCON performs all those duties alone (neglecting 1 bit of PCON). Its bit 0, designated as RI, is set by the processor hardware whenever the serial communication interface received a complete byte through RxD. Note the minor variations of exact timing of generation of this interrupt through RI for different Modes from Fig. 16.6. It is the duty of concerned interrupt routine to clear this flag.

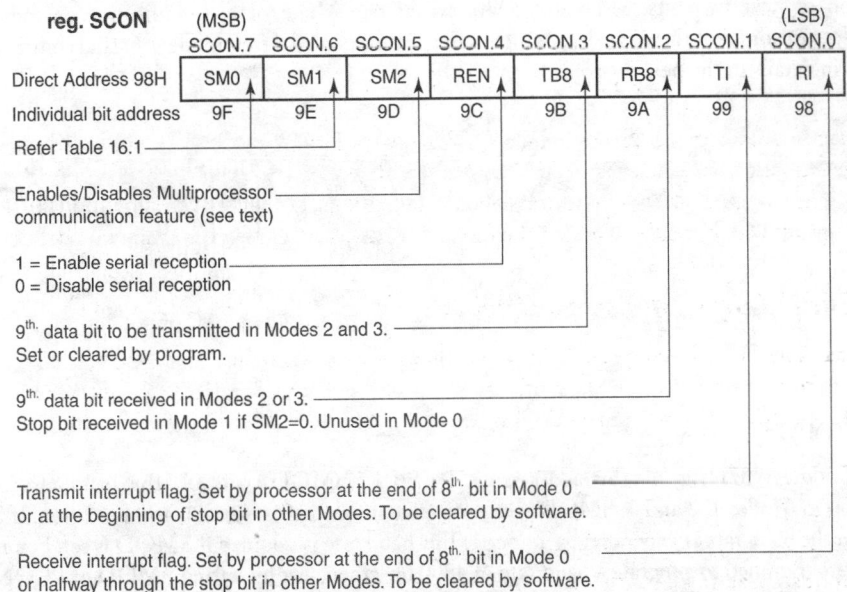

Figure 16.6 Bit assignment of SFR SCON

TI (bit 1 of SCON) is the flag, which is set by the serial communication hardware at the completion of transmission of a byte indicating that it is now free to accept another fresh byte for transmission. In this case also there is a minor difference in timings of the exact instant of generation of the interrupt through TI (refer to Fig. 16.6). This flag also to be cleared by the concerned interrupt service routine.

Understanding the purpose of next two bits, namely, RB8 and TB8, is essential. In some networks, a ninth bit is introduced in a standard byte of 8 bits. Purpose of this ninth bit is to target or address some configuration. Modes 2 and 3 (to be explained shortly) of serial communication (refer to Table 16.1) offer this 9-bit communication feature. In such cases, RB8 would contain the received ninth bit and TB8 should be loaded as the ninth bit to be transmitted.

Table 16.1 Modes of serial communications

Mode	SM0	SM1	Description	Baud rate
0	0	0	Shift register	$F_{osc}/12$
1	0	1	8-bit UART	Variable
2	1	0	9-bit UART	$F_{osc}/64$ or $F_{osc}/32$
3	1	1	9-bit UART	Variable

The next bit, REN, enables or disables the serial data reception (not data transmission). Clearing this bit would disable and setting would enable the reception part of serial communication. *Remember that condition of this bit does not disturb the serial data transmission.*

SM2, bit 5 of SCON, enables multiprocessor communication feature in Modes 2 and 3, if this bit is set. In that case, RI would not be activated (no serial data reception interrupt) if the received ninth bit (RB8) is 0. If SM2 is set in Mode 1, then RI would not be activated till a valid stop bit is received. This bit (SM2) should be cleared to 0 in Mode 0.

Condition of next two bits, SM1 and SM0, selects the Mode of communication as indicated in Table 16.1. Note the order of these two bits, which is unconventional. All four modes of serial communication are explained in details in the next section.

If the network contains only two 8051s, then mode 1 of serial communication is the best solution. The ninth bit need to be incorporated if there are more than two processors in the network.

16.3.3 | PCON

SFR PCON, address 87H, is not bit-addressable. Its bit 7 (SMOD) controls the baud rate of serial communication in Modes 1, 2 and 3. SMOD is not effective in Mode 0 of serial communication. If SMOD is cleared, then the baud rate is generated by Timer 1. This baud rate is doubled if SMOD is set. For example, if Timer 1 is programmed to generate a baud rate of 4800 bps, then, just by setting SMOD of PCON it may be converted to 9600 bps, without any change in Timer 1 counters (Fig. 16.7).

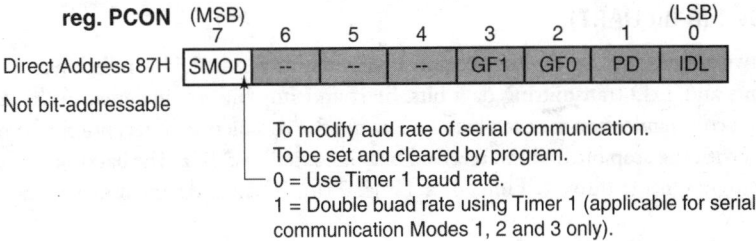

reg. PCON (MSB)

7	6	5	4	3	2	1	(LSB) 0
SMOD	--	--	--	GF1	GF0	PD	IDL

Direct Address 87H

Not bit-addressable

To modify aud rate of serial communication.
To be set and cleared by program.
0 = Use Timer 1 baud rate.
1 = Double buad rate using Timer 1 (applicable for serial communication Modes 1, 2 and 3 only).

Figure 16.7 Power mode control (PCON)

Although PCON is not bit addressable, two one-bit flags for general purpose usage (GF0 and GF1) are provided within PCON. How these flag bits may be used is left to the system developer.

16.4 | Modes of Serial Communication

As presented in Table 16.1, 8051 offers four serial communication modes. They are now explained in details.

16.4.1 | Mode 0 (Shift Register Mode)

In Mode 0, RxD pin is used for serial data communication in half-duplex mode. TxD pin is used as a clock signal, which is always output from the processor as shown in Fig. 16.8. No start bit or stop bit are used in this case. The clock output through TxD is always 1/12 of the oscillator's frequency. Bit 0 is communicated first and bit 7 is communicated last. Mode 0 is provided for very fast data sampling. This mode is never used in multiprocessor communication.

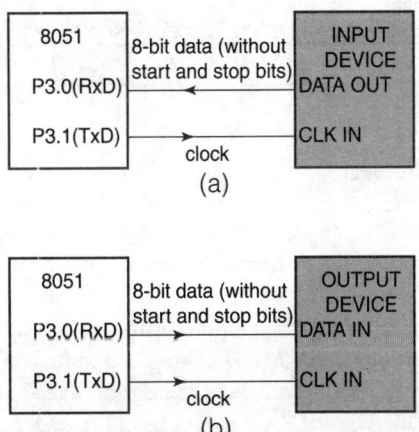

Figure 16.8 Serial communication Mode 0 configuration (a) receive data and (b) transmit data

16.4.2 | Mode 1 (8-bit UART)

This and next two modes are for multiprocessor environment. Mode 1 is a full-duplex mode with RxD receiving data bits and TxD transmitting data bits. Start and stop bits are included in this communication, making it a 10-bit communication for each byte of data. At the completion of reception of every byte, RB8 of SCON is loaded with the stop bit, and the received byte is copied in SBUF. The baud rate of communication is variable and programmable through Timer 1 or Timer 2 (for XX52 and similar devices only).

16.4.3 | Mode 2 (9-bit UART with Fixed Baud Rate)

This and the next mode are designed to communicate 9 bits of data, which includes 8 bits of the communicated byte and a programmable ninth bit. Therefore, total number of bits communicated per byte of data is 11 in this case. This ninth bit may be Parity bit from Program Status Word (PSW) SFR or programmable as 0 or 1. In either case, before transmission, this ninth bit must be available in TB8 of SCON of the transmitting side. At the receiving end, after receiving the data, the ninth bit is moved to RB8 of SCON and eight data bits are copied in SBUF. The stop bit is ignored. The baud rate may be either 1/32 or 1/64 of the oscillator frequency, depending upon the content of bit-7 of PCON SFR.

16.4.4 | Mode 3 (9-bit UART with Variable Baud Rate)

This mode is identical with Mode 2 in all respects except the baud rate being variable and programmable through Timer 1 or Timer 2 (for XX52 and similar devices only).

16.5 | Serial Communication Issues

16.5.1 | Multiprocessor Communication

As indicated in Fig. 16.9 and also in Table 16.1, Modes 2 and 3 of serial communication deals with one extra bit, designated as ninth bit. This bit is especially useful for multiprocessor communication. In this case, multiprocessor means more than two processors. In the previous discussions regarding SFR SCON, it was described how the condition of the bit SM2 (SCON.5) controls the RI generated interrupt. It may be recalled that if SM2 is set then unless the received ninth bit, placed in RB8 of SCON, is 1, no serial reception interrupt would be generated. This feature, as explained through Fig. 16.9, helps in multiprocessor environment.

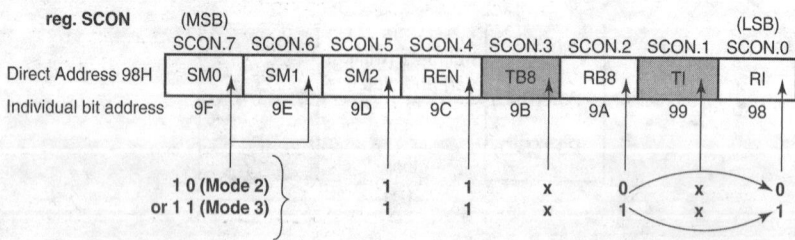

Figure 16.9 Enabling/disabling data reception interrupt by SM2 and ninth bit

For the sake of example, we take up a small multiprocessor network with only three microcontrollers, as illustrated in Fig. 16.10(a). The left one is the *leader* controller and remaining two are *follower* controllers. Let us assume that, for this networking, the address of the rightmost follower controller is 00000001B and for the middle one, it is 00000000B. As usual, TxD output from the leader controller is interfaced with RxD input of both follower controllers. Finally, we assume that SM2 bits within SCON SFR of both follower controllers are initially set as 1.

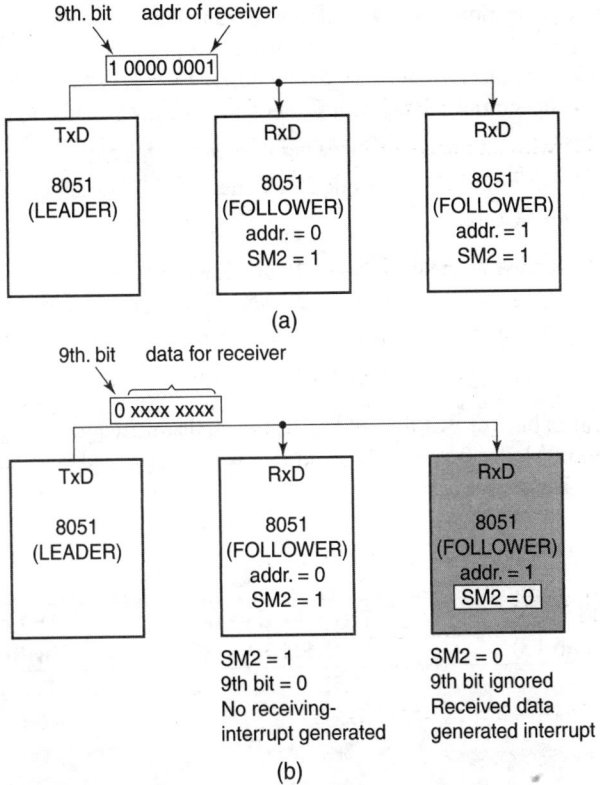

Figure 16.10 Example of multiprocessor communication using ninth bit, (a) send address, (b) send data

In this situation, if the leader controller needs to send some data to the follower with address 00000001B, then it should first target that follower controller by sending its 8-bit address with its ninth bit as 1. As the SM2 bit of both followers is set, therefore, both followers would receive this address [Fig. 16.10(a)]. However, only the follower of the target address would react by clearing its SM2 bit to 0.

In such a situation, if the leader controller sends several bytes of data with its ninth bit as 0, then these data bytes would be ignored by the follower with address 00000000B, as its SM2 bit is still set. However, follower with address 00000001B would accept the data set because it had already cleared its SM2 bit [Fig. 16.10(b)]. The situation would remain identical with more number of followers.

There is no hard and fast rule for assignment of addresses of microcontrollers. Generally higher addresses are assigned for far away processors to maintain uniformity in all the works related with networking.

16.5.2 | Role of Timer in Baud-Rate Generation

For 8051, baud rate is controlled by Timer 1 overflows. For 8052 and similar devices, it is controlled by Timer 1 or by Timer 2, or by both (different baud rates for transmission and reception). For subsequent paragraphs, for easier discussions, we take 8051 with Timer 1 as baud-rate generator, as an example case.

In general, for baud-rate generation Timer 1 is allowed to function in its Mode 2 (auto-reload mode) to produce periodic overflow signals. We should note that for baud-rate generation, Timer 1 need not be enabled to interrupt the processor. That means *there need not be any interrupt service routine placed at location 001BH.* For baud-rate generation, the operations related with Timer 1 initialization are the following.

1. Load TL1 and TH1 with proper count-up values (same value for both).
2. Initialize TMOD to set Timer 1 in its Mode 2 in Timer mode, Gate-bit cleared.
3. Start Timer 1 through TCON.

To get the count value to be loaded in the counters of Timer 1, we must finalize the baud rate. Table 16.2 presents the count values or reload values for some standard baud rates assuming the oscillator frequency to be 11.0592 MHz. These are obtained by solving the following relation:

$$\text{Baud rate} = (2^{\text{SMOD}} \times F_{\text{osc}})/[32 \times 12 \times (256 - \text{TH1})] \tag{16.1}$$

where SMOD is the content of bit 7 of PCON and F_{osc} is the oscillator frequency.

For serial communication Modes 0 and 2, the baud rate is fixed as already indicated in Table 16.1. In these cases, Timer activation is not necessary.

Table 16.2 Reload values in TH1 for 11.0592 MHz crystal

Baud rate (Modes 1 and 3)	TH1 reload value (with SMOD = 0)	TH1 reload value (with SMOD = 1)
300	A0H	40H
600	D0H	A0H
1,200	E8H	D0H
2,400	F4H	E8H
4,800	FAH	F4H
9,600	FDH	FAH
19,200	–	FDH

16.5.3 | Why 11.0592 MHz?

In Section 16.2.3, it was indicated that a crystal having a frequency of 11.0592 MHz is used if MCS-51 is to offer serial communication features. Although in general, in other cases a frequency of 12 MHz is used with MCS-51 devices, there is a reason to select 11.0592 MHz frequency during implementation of serial communications.

Equation 16.1, introduced in the previous section, may be rewritten as

$$\text{TH1} = 256 - 2^{\text{SMOD}} \times F_{\text{osc}}/(32 \times 12 \times \text{Baud Rate}) \tag{16.2}$$

As a matter of fact, entries of Table 16.2 may be easily calculated by using eq. 16.2, where F_{osc} was taken as 11.0592 MHz. However, if we try to calculate any of those TH1 values using F_{osc} as 12 MHz, the result would lead to fraction and should be rounded-off, generating minor errors in initial communications, which would propagate with time. For example, the value of TH1 for a baud rate of 4800 bps using a crystal frequency of 12 MHz, would be

$$\begin{aligned} \text{TH1} &= 256 - 2^0 \times 12000000/(32 \times 12 \times 4800) \\ &= 256 - 6.510416666 \\ &= 249.48958333 \text{ (in decimal)} \end{aligned}$$

whereas under identical condition (SMOD = 0), using 11.0592 MHz crystal, the value would be

$$TH1 = 256 - 2^0 \times 11059200/(32 \times 12 \times 4800)$$
$$= 256 - 6$$
$$= 250 \text{ (in decimal)}$$

To conclude, 11.0592 MHz is the LCM of 32, 12 and all standard baud rates and would, therefore, produce integer results. This is not the case with 12 MHz crystal.

16.5.4 | Initiation of Data Transmission and Reception

Whenever a byte is written in SBUF, the transmission begins automatically. This is valid for all four modes of communication. Therefore, the programmer must ensure that after completion of loading all other registers, the byte to be transmitted is loaded in SBUF.

Reception is initiated in Mode 0 when REN = 1 and RI = 0. For Modes 1, 2 and 3, reception is initiated as soon as a start bit is received through RxD, if REN = 1. In all modes, if REN = 0, then no serial data is received through RxD pin.

16.5.5 | Signal Boosting for Long Distance Transmission

In Fig. 16.2, it was indicated that the depicted hardware connection for serial communication is valid for a very short distance. To communicate through a longer distance, boosting of the signal is necessary. Several ICs are available in the market, MAX232, MC1488 and MC1489 etc. for this purpose.

MAX232 from Texas Instruments offers two drivers and two receivers in a single 16-pin DIP operating with a single power supply of 5V. Motorola (presently ON Semiconductor) MC1488 is a quad transmitter, and MC1489 is a quad receiver. The first one needs +12V and −12V supply, and the second one needs single 5V supply.

16.6 | Serial Communication Routines

As such, the serial communication has two parts, transmission and reception. In each case, instructions are necessary for initialization and interrupt handling phases. Moreover, as there would be more than one microcontroller involved in serial communications, therefore, the serial communication routines must be present in all microcontroller units taking part in communication. We now discuss initialization steps and interrupt handling details one after another. In the following discussions, it is assumed that we are using only 8051 and not 8052, which restricts us with only Timer 1 usage and not Timer 2, for baud-rate generation.

16.6.1 | Initialization Steps for Serial Communication

For baud-rate generation, we initialize Timer 1 in its Mode 2 (auto-reload mode). For this purpose, TMOD to be loaded with 2xH, where x would be the value relate with Timer 0's requirements. TL1 and TH1 must be loaded with suitable values (refer to Table 16.2) for the expected baud rate. Setting TR1 bit of TCON SFR, we start Timer 1. There is no need to enable Timer 1 interrupt through IE SFR.

Bit 7 of PCON (SMOD) is to be set or cleared depending upon the baud-rate requirements (refer to Table 16.2). SFR SCON to be initialized for desired mode, through SM0 and SM1. For multiprocessor communication, SM2 has to be adjusted accordingly. If it is the transmitting side then TB8 should have proper value. The transmission may then be started by loading the first byte, to be sent out, in SBUF. If it is the receiving side then REN must be set to allow the receiving of serial data. For full-duplex communication, both transmitting and receiving features are to be enabled. Bit 4 of IE SFR is to be set along with IE.7 to enable serial interrupt. If necessary, then IP.4 also has to be set for modifying the priorities of serial interrupt.

16.6.2 | Serial Communication Interrupt Service Routine

Although to be handled separately, both transmission and receiving interrupts are having a common vector address for their ISR, which is 0023H. The serial communication ISR, placed at this address, should check both TI (bit 1) and RI (bit 0) of SCON SFR to ensure which one is the source of the present interrupt and service accordingly. If TI bit is set then the processor completed 1-byte data transmission and the next byte (if any) to be written immediately in the SBUF SFR. If RI bit is set then the processor completed the reception of a full data byte, presently available in SBUF, which must be immediately read. Before leaving ISR, the interrupt generating bit (TI or RI) must be cleared by the routine itself to allow the next interrupt. As per normal practice, the interrupt service routine must be terminated by a RETI instruction.

16.7 | Solved Examples

Example 16.1

Purpose: Demonstration of serial communication in Mode 0.

Problem

A small numeric printer is attached with serial port of 8051 for printing numerical data. Design a scheme for interfacing, and discuss about the software for serial communication.

Solution

For such a small printer, serial interface is sufficient. Essentially it would be a simplex communication and Mode 0 (shift register mode) of data transmission may be adopted. Fig. 16.11 shows the schematic. Note that D0 reaches printer first and D7 last for a byte. There is no start bit or stop bit. TxD of 8051 outputs the clock whose frequency is Oscillator frequency divided by 12. If the processor is interfaced with a 12 MHz crystal, then TxD would output a clock of frequency 1 MHz.

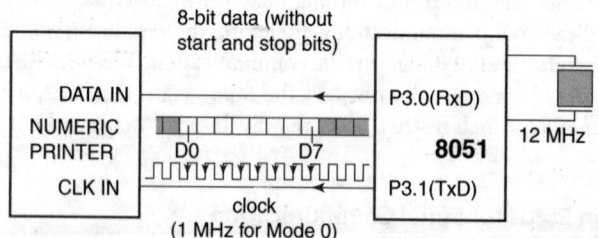

Figure 16.11 Schematic of Mode 0 serial interface with printer

The related software should select Mode 0 through SCON during initialization, set the priority of serial interrupt, if necessary and enable serial interrupt. Whenever a byte to be transferred, it may be loaded in SBUF, which would initiate the data transmission through RxD. At the end of transmission, a serial interrupt through TI of SCON would be generated, which may be used to load another byte in SBUF if necessary.

Example 16.2

Purpose: Using Mode 1 of serial communication, establishing full-duplex communication between two processors with a predefined baud rate.

Problem

Two 8051s are interfaced for full-duplex communication. Assuming the crystal frequency of both to be 11.0592 MHz, develop the software necessary for serial communication with a baud rate of 4800.

Solution

In this case Mode 1 may be selected by both processors, as it is a communication between only two devices. Moreover, same software routines would be applicable for both. For a baud rate of 4800, using Table 16.2, the reload value, with SMOD being 0, would be 0FAH for Timer 1 SFRs TL1 and TH1. The initialization routine may be written as follows:

```
INIT:   MOV    TL1, #0FAH        ; for baud rate 4800, starting
        MOV    TH1, #0FAH        ; for baud rate 4800, reload
        MOV    TMOD, #20H        ; select Mode 2 (auto reload) for Timer 1
        ANL    PCON, #7FH        ; to use Timer 1 baud rate, clear SMOD of PCON
        SETB   IP.4              ; assign higher priority for Serial interrupt
        MOV    SCON, #70H        ; Mode 1, receive enable
        SETB   TCON.6            ; start Timer 1
        SETB   IE.4              ; enable serial interrupt
        SETB   IE.7              ; allow global interrupt
```

; Initiate transmission with first byte. This may be pointed by R0 of bank #1.

```
        PUSH   PSW               ; save status
        MOV    PSW, #08H         ; select bank #1
        MOV    A, @R0            ; get first byte to be transmitted
        MOV    SBUF, A           ; save in serial buffer, this starts transmission
        INC    R0                ; point to next byte
        POP    PSW               ; restore original bank
```

The serial interrupt service routine should start from program memory location 0023H. It is assumed that register bank #1 is reserved for this routine. It is also assumed that registers R0 and R1 of register bank #1 have already been loaded with starting addresses of output buffer and input buffer, respectively. Register R7 contains number of bytes to be transferred out, and R6 contains number of bytes expected from outside.

; Serial interrupt service routine for both transmission and reception.
; Save status, select bank #1.

```
        ORG    0023H
SERISR: PUSH   PSW               ; save processor status
        PUSH   ACC               ; save accumulator
        MOV    PSW, #08H         ; select register bank #1, reserved for this ISR
```

; Check for reception interrupt first.

```
        JBC    SCON.0, GETRXD    ; if received then clear RI and process
```

; Previous byte transmission through TxD is complete. Clear TI. Then check for any more transmission
; through R7. If yes then load next byte in SBUF through R0. Update R0.

```
NXTTXD: CLR    SCON.1            ; clear TI flag of SCON
        DJNZ   R7, DATOUT        ; transmission not over yet
        SJMP   SOVR              ; transmission over
```

```
DATOUT:  MOV   A, @R0        ; get next byte to be sent out
         MOV   SBUF, A       ; load in buffer, transmission starts
         INC   R0            ; point to next byte
         SJMP  SOVR          ; restore and return
```

; One byte received through RxD is waiting in SBUF. Store it through R1 and update R1. Using R6, check for any more pending. If no then clear REN of SCON to terminate reception.

```
GETRXD:  MOV   A, SBUF       ; read data from buffer
         MOV   @R1, A        ; save in its area
         INC   R1            ; next saving address
         DJNZ  R6, SOVR      ; more to be received
         CLR   SCON.4        ; make REN=0 to stop reception
```

; Restore original bank and status with accumulator and return.

```
SOVR:    POP   ACC           ; restore accumulator
         POP   PSW           ; restore PSW and switch to previous bank
         RETI                ; return from interrupt
```

 Note that this is an ideal example for implementation of JBC instruction.

Example 16.3

Purpose: To be familiar with Mode 2 of serial communication.

Problem

Explain how a leader 8051 would communicate with one of the two follower-8051s through serial interface, using a fixed baud rate.

Solution

For multiprocessor communication (refer Fig. 16.10), a ninth bit is essential to identify the processor with which the communication is to be established. Only in Mode 2 and Mode 3 of serial communication of 8051, this ninth bit is available. The only difference between Mode 2 and Mode 3 of serial communication is that in case of Mode 2, the baud rate is fixed and in case of Mode 3 the baud rate is variable.

The fixed baud rate may be either $f_{osc}/32$ or $f_{osc}/64$ in case of Mode 2, where f_{osc} is the external crystal frequency of the system. To implement one of these two baud rates, the most significant bit (bit 7) of the SFR PCON (refer Fig. 16.7), designated as SMOD, must be used. If this bit (SMOD) is cleared then the fixed baud rate would be $f_{osc}/32$. On the other hand, if this bit is set then the fixed baud rate would be $f_{osc}/64$.

Example 16.4

Purpose: To be familiar with Mode 3 of serial communication.

Problem

Explain how a leader 8051 would communicate with one of the two follower-8051s through serial interface, using a variable baud rate.

Solution

Mode 3 is similar to Mode 2, except the baud rate, which is variable (programmable) in case of Mode 3, but not for Mode 2. The range of baud rate varies from 300 to 19200, as indicated in Table 16.2. Just like solved example 16.2, Timer 1 may be utilized to generate necessary baud-rate interrupts. In case of solved example 16.2, the communication was restricted between two systems only. However, in Mode 3, communication between multiple processors is possible, as explained in solved example 16.3.

SUMMARY

8051 supports full-duplex serial communication through its TxD and RxD pins. Eight or nine bits of data are communicated with a leading start bit (always low) and a following stop bit (always high). In case of no communication or idle state, the output remains high. MCS-51 offers three SFRs for this purpose, namely SBUF, SCON and PCON (only MSB). SBUF contains two separate buffers, one read only and another write only, for reception and transmission, respectively.

Four modes of communication are offered from 0 to 3. They are

- 8-bit shift register with fixed baud rate,
- 8-bit UART with variable baud rate,
- 9-bit UART with fixed baud rate and
- 9-bit UART with variable baud rate.

Transmission starts immediately after writing in SBUF, and an interrupt is generated once it is over.

Reception might be enabled or disabled by REN flag of SCON SFR. Completion of reception generates interrupt. Both transmission complete and reception complete interrupt vector are 0023H and TI and RI flags of SCON SFR are to be checked for the source of serial interrupt.

Timer 1, preferably in Mode 2 (auto-reload mode), is to be used to generate the baud rate, if it is a variable one. The fixed baud rate is generated by oscillator clock of the processor. For 8052 and similar devices, Timer 2, either independently or with Timer 1, may be used to generate baud rate. In any case, there is no need to enable the Timer interrupt or place any interrupt routine for the Timer. The overflow of timer counters generates the baud rate. The reload value of the timer counter may be calculated from the relation

$$\text{Baud rate} = (2^{SMOD} \times F_{osc})/[32 \times 12 \times (256 - TH1)]$$

POINTS TO REMEMBER

- RI-flag must be cleared by Serial-data-receive-interrupt-routine to ensure another set of data reception.

- It is a good practice to check for the received byte first followed by the transmitted byte, when a serial interrupt is generated for full-duplex mode.

REVIEW QUESTIONS

Evaluate Yourself

1. Serial communication may be classified as

    (a) simplex, full-duplex and half-duplex

    (b) synchronous and asynchronous

    (c) either of these

    (d) none of these

2. Serial data transmission is initiated by

    (a) placing the data byte in SBUF

    (b) setting TI flag

    (c) enabling Timer 1 interrupt

    (d) none of these

3. If the logic level of a TxD line is high, it indicates

(a) data bit or stop bit

(b) data bit or stop bit or idle state

(c) idle state or stop bit

(d) none of these

4. Using one stop bit, if the baud rate is mentioned as, 9600, then the number of bytes transferred per second would be

(a) 960   (b) 1,067

(c) 1,200   (d) none of these

5. For 9-bit transmission, the ninth bit is used as

(a) stop bit

(b) multiprocessor communication

(c) parity bit

(d) none of these

6. To transmit the ninth bit, it should be placed in

(a) bit 0 of next byte   (b) SMOD of PCON

(c) TB8 of SCON   (d) none of these

7. Which of the following bits, when set, would double the baud rate generated by Timer 1, except in Mode 0 of serial communication?

(a) REN of SCON   (b) SMOD of PCON

(c) TF1 of TCON   (d) None of these

8. Serial data bits being received are initially stored in

(a) temp. buffer   (b) accumulator

(c) SBUF   (d) none of these

9. If five 8051s are serially interfaced, which of the following serial communication mode would yield the best results?

(a) mode 0   (b) mode 1

(c) mode 2   (d) None of these

10. For serial communication, standard frequency of the crystal for 8051 would be

(a) 11.0592 MHz   (b) 11.0952 MHz

(c) 12 MHz   (d) none of these

Search for Answers

1. What is the format of serial communication of a byte of data?

2. Are start and stop bits always included for all modes of serial communication offered by 8051?

3. What is the purpose of REN bit in SCON SFR?

4. What is the difference between Mode 1 and Mode 3 of serial communication?

5. What is the utility of ninth bit?

6. How do we calculate the baud rate?

7. Which Timer is used for serial communication?

8. What is the purpose of Mode 0 of serial communication?

9. What role is played by SBUF in serial communication?

10. How is it possible to stop serial data reception in case of emergency?

Think and Solve

1. How many types of serial communication are possible using 8051?

2. What would happen if serial data reception is going on, but data bytes are not being read from SBUF?

3. Is it possible to generate a baud rate of 75 using Timer 1 and 11.0592 MHz crystal? If yes, then how? (Hint: Use Timer 1 in Mode 1).

4. Develop a method to receive continuous data from a temperature sensor in serial format.

5. Is it possible to use different baud rates for transmission and reception? If yes, then how?

6. A designed system of data acquisition using Mode 0 of serial communication is functioning perfectly. What would happen if the crystal of 8051 is changed from 12 to 11.0592 MHz?

7. What purpose do the boosters serve in serial communication?

8. What is the maximum number of 8051 might be interlinked through serial transmission? Justify your answer.

9. Calculate the reload value for Timer 1 in Mode 2 to generate a baud rate of 2400, if 8051 is interfaced with a 12 MHz crystal.

10. If the MSB of IE SFR ($\overline{EA}$) is not set, would Timer 1 be able to generate the baud rate?

17 INTERFACING: EXTERNAL MEMORY

CHAPTER OBJECTIVES

In this chapter, the reader is introduced to interfacing of external memory devices with 8051 microcontroller. After completion of the chapter, the reader should be able to understand

- How to interface external RAM and ROM with MCS-51.
- Related timing diagrams for external memory access.
- Security features offered by MCS-51 through on-chip program lock bits.

17.1 | Introduction

As we have completed major discussions on internal architecture and instructions of MCS-51, we now may focus on various interfacing issues related to it. We start with external memory interfacing.

17.1.1 | Need of External Memory

At this stage, the reader may ask about the need of external memory interfacing. As a matter of fact, the very purpose of microcontroller is to avoid external interfacing for memory and I/O port (refer to section 1.3 of Chapter 1).

Apparently, this argument is correct. In general, microcontrollers, like 8051, are used as a single chip stand-alone device. However, in certain special cases, some extra interfacing may be helpful. For example, to record temperature or humidity through a long period at shorter intervals may need some extra RAM interfacing for a microcontroller-based temperature or humidity meter. It then becomes a three-chip system as depicted in Fig. 17.1, with 8051, ADC and one extra RAM chip of 8K.

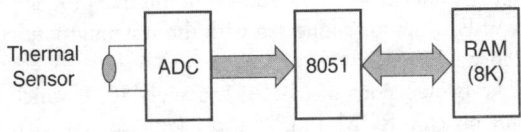

Figure 17.1 A temperature monitoring system with external RAM

 The reason for offering the provision of external memory interfacing in 8051 is the then market trend. The designer tried to make the device as versatile as possible, so that it may cater to most of the customers' demands. The present trend, however, is to offer a larger variety of microcontrollers with different on-chip memory sizes, so that a tailor-made selection is possible.

17.2 | Interfacing Signals

To interface external memory devices, 8051 offers several signals, which we have introduced in Chapter 2. Fig. 17.2 shows only those signals of MCS-51, which are necessary for external memory interfacing. Fig. 17.2(a) presents signals with related pin numbers, and Fig. 17.2(b) illustrates these signals schematically. This also illustrates that 8051 outputs 16-bit address for external memory communication. Lower 8 bits of this address are multiplexed with 8-bit data bus (AD0–AD7). The falling edge of the ALE signal may be used to de-multiplex the lower address bus, using a suitable octal latch, like 74373. To read from external program memory, $\overline{PSEN}$ signal may be used. $\overline{RD}$ and $\overline{WR}$ signals are provided for reading from and writing to external data memory. All these three signals, namely, $\overline{PSEN}$, $\overline{RD}$ and $\overline{WR}$ are active low.

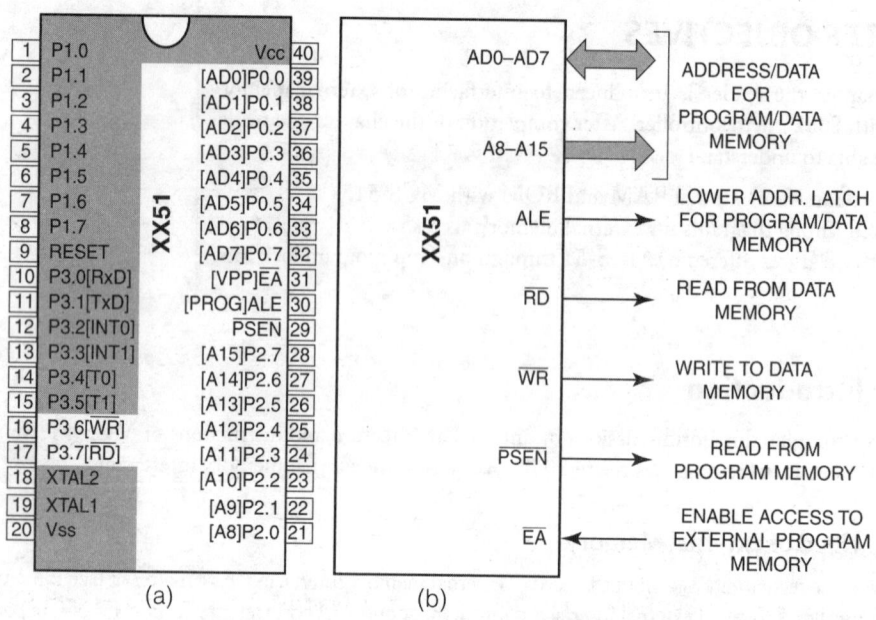

Figure 17.2 8051 signals for external memory interfacing

The input signal, $\overline{EA}$, which is active low, is provided to indicate the processor, which program memory is to be accessed, external or internal? External program memory is accessed under any one of the following conditions:

(i)   whenever input signal $\overline{EA}$ is active, i.e. $\overline{EA} = 0$, or
(ii) whenever the program counter exceeds the address of internal program memory.

To put it in a single sentence, if $\overline{EA}$ input is connected with the system ground, the processor would start its execution from external memory.

To explain this, let us take two versions of MCS-51, namely 8051, which has internal 4K ROM and 8031, which does not have any on-chip ROM. In both cases, let us assume that they are interfaced with 4K of external program memory. Using two different input conditions of $\overline{EA}$, four distinct cases are generated as depicted in Fig. 17.3.

In the first case of 8051 with its $\overline{EA}$ pulled high [Fig. 17.3(a)], processor would first execute from on-chip ROM and when it has reached the internal address 0FFFH, it would seek access to external ROM. In the second case of 8051 with its $\overline{EA}$ connected with the system ground [Fig. 17.3(b)], the processor would ignore its on-chip ROM and start executing directly from the external ROM (note the difference of addresses). In the third case of 8031 with its $\overline{EA}$ pulled high [Fig. 17.3(c)], the processor would seek the first executable byte of

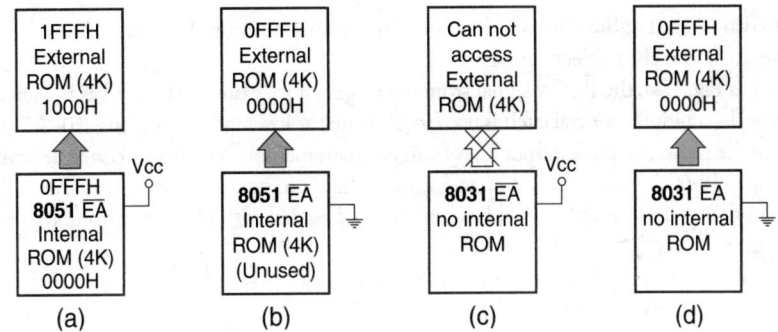

Figure 17.3 Different outcomes of $\overline{EA}$ signal of MCS-51

ROM within the system, as its $\overline{EA}$ = 1. However, as no ROM is available within 8031, the ROM-less version, it would not be able to execute at all. Note that in such a case it cannot execute from external ROM, although it is existing and interfaced with 8031. In the fourth and final case of 8031 with its $\overline{EA}$ input connected with the system ground [Fig. 17.3(d)], execution would start from external ROM.

Therefore, if the internal access is to be completely avoided, $\overline{EA}$ input must be pulled low. In such a case, external ROM must be interfaced. If $\overline{EA}$ = 1, then the processor would execute from external ROM only after completion of execution from its internal ROM.

 Conditions of program memory lock bits also play an important role in external program memory accessing. Interested readers should go through the data sheets of 8051.

17.3 | Program Memory Interfacing

Fig. 17.4 illustrates a schematic of external program memory and its interfacing techniques with 8051. Note that depending upon its size, the program memory will need address lines A0–An as input. Apart from this, a chip-select input and a read input are also necessary. As a normal practice, the chip-select signal is derived

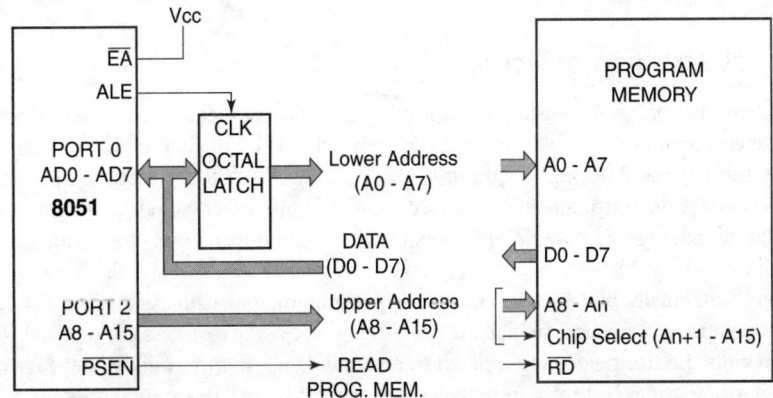

Figure 17.4 Schematic of program memory interfacing with 8051

from a decoder with unused address lines as its input. In the present example, address lines An+1 to A15 may be used for generating the chip-select signal.

As illustrated in Fig. 17.4, the $\overline{\text{PSEN}}$ signal from the processor may directly be used as the read input for program memory. The de-multiplexer 'octal latch' is used to generate the lower address signals, A0–A7, with the help of ALE output from the processor. Data output lines from program memory are directly connected with AD0–AD7 pins of the processor. The $\overline{\text{EA}}$ input is pulled up to ensure external program memory access after completion of accessing internal program memory. Otherwise, $\overline{\text{EA}}$ input should be tied with the system ground to avoid any access to internal program memory.

Fig. 17.5 shows the timing diagram from program memory read operation. The processor clock is depicted at the top (marked as XTAL). Note that each machine cycle of the processor contains six states, and each state has two phases, P1 and P2. Execution of an instruction by the processor may take one or more machine cycles (refer to Appendix A). In each machine cycle, the ALE signal pulses twice. The falling edge of the ALE signal is to be used for latching the lower address A0–A7.

Fetches from external program memory always use 16-bit address. Upper 8 bits of this address are available during states 2, 3 and 4 for the first fetch, and during states 5, 6 and 1 for the second fetch. The data from program memory is sampled at the rising edge of $\overline{\text{PSEN}}$ as shown in Fig. 17.5.

As indicated in Fig. 17.5, Port 0 is used to output lower 8 bits, and Port 2 is used for upper 8 bits of the target address. Port 0 is also used to read data bytes from program memory. During this reading, internal pull-ups are automatically controlled by Port 0 to ensure proper input conditions, so that there is no need to load Port 0 output latch by 1s by the programmer (refer to fig. 3.8 of chapter 3).

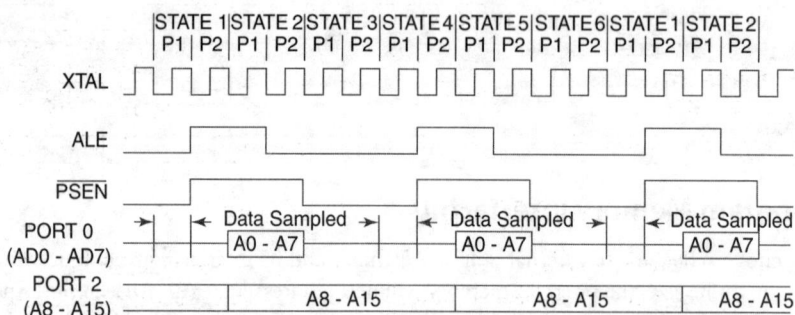

Figure 17.5 Timing diagram for program memory reading

17.4 | Data Memory Interfacing

In contrast with external program memory reading, for external data memory, the address output from the processor may be either 16 bits or 8 bits, as we have already discussed in chapter 12. Sixteen bits of address are produced by the processor during execution of MOVX A, @DPTR instruction. However, during the execution of MOVX A, @Ri instructions (refer to section 12.3), only lower eight bits of address are generated. This helps in page addressing and port P2 SFR is expected to hold upper 8 bits of the target address.

To communicate with external data memory, 8051 provides $\overline{\text{RD}}$ and $\overline{\text{WR}}$ signals through P3.7 and P3.6, respectively (Fig. 17.6). Finally, like the previous case of program memory interfacing, the chip-select signal is to be derived from those address lines, which are not directly necessary to access the external RAM.

Fig. 17.7 presents the timing diagram related to external data memory reading, and Fig. 17.8 illustrates external data memory writing. Note that in both cases slots are reserved for external program memory access. This forces external data memory access to consume two machine cycles.

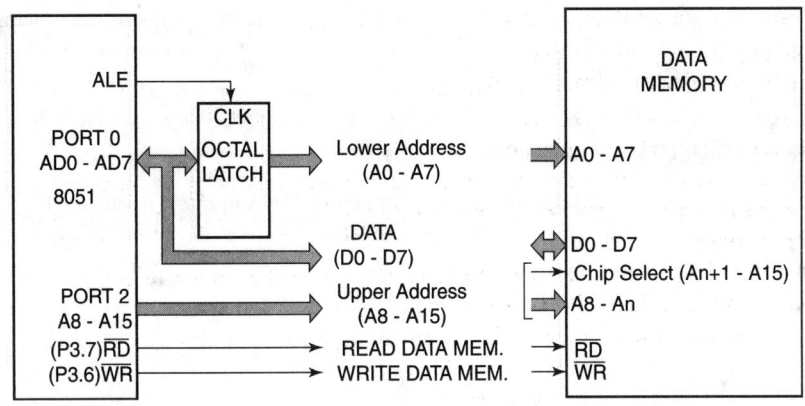

Figure 17.6 Schematic of data memory interfacing with 8051

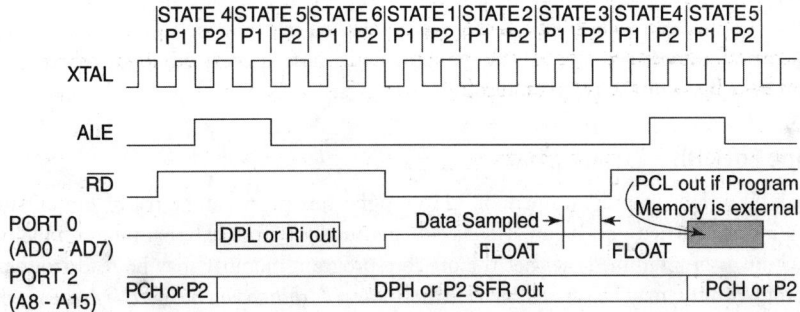

Figure 17.7 Timing diagram for data memory reading

It may readily be observed by comparing Fig. 17.5 with Figs 17.7 and 17.8 that during external program memory access two ALE pulses are generated within one machine cycle, whereas only one ALE pulse is generated for external data memory access. The time of sampling of read-data also varies for external program and data memory access. Both $\overline{RD}$ and $\overline{WR}$ signals are active low, like $\overline{PSEN}$. For external data memory writing (Fig. 17.8), data is valid at the rising edge of $\overline{WR}$ signal. However, for reading from external data memory, data

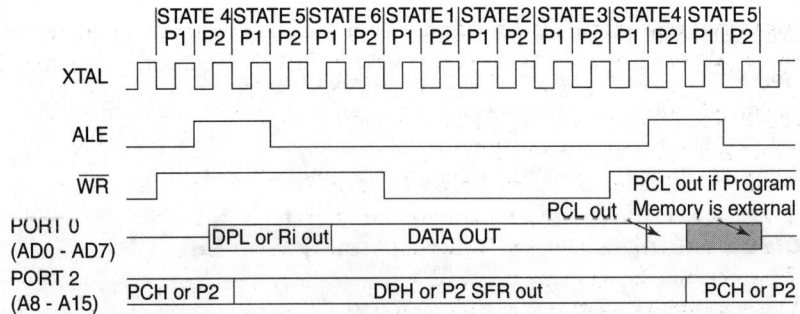

Figure 17.8 Timing diagram for data memory writing

is sampled just before the read strobe is deactivated (goes high). A detailed example of interfacing external program and data memory with MCS-51 would now be taken up after brief discussions about the security features offered in various MCS-51 devices.

17.5 | Software Security Features

To ensure some protection against unauthorized program memory access from external source, MCS-51 offers a program memory locking mechanism. Table 17.1 presents the locking features for various MCS-51 devices.

Table 17.1 MCS-51 program memory lock bits

Device no.	No. of lock bits present
8751H	One
8751BH, 8752BH, 89C2051	Two
87C51, 89C51, 89C52	Three

The lock bits may be programmed by any standard programmer and un-programmed by erasing the whole chip. The program also gets erased in this process. When multiple lock bits are offered, various degrees of on-chip program memory security may be implemented. These are very briefly discussed here. Please refer to the relevant data sheets for detailed features.

17.5.1 | One Lock Bit

With one lock bit, if left un-programmed, on-chip program memory can be read from outside and further programming may be carried out. It can also execute instructions from external program memory. However, when this lock bit is programmed, neither the on-chip program memory may be read from external source nor further programming may be carried out. Furthermore, *the chip would not be able to access external program memory*. All these may be regained by erasing the lock bit along with the content of the program memory. It may be noted that without erasing the program memory, the lock bit cannot be erased.

17.5.2 | Two Lock Bits

With two lock bits, the features offered for the previous case with one lock bit are implemented in two steps. By programming one lock bit and leaving the other un-programmed, the chip may be verified (program memory may be read) but further programming would not be possible. This verification feature may be disabled by programming the second lock bit.

17.5.3 | Three Lock Bits

When three lock bits are offered, programming one lock bit disables the usage of MOVC instruction to access any byte from external program memory. It also samples and latches the input at $\overline{EA}$ during reset and disables further programming. Verifying feature is disabled by programming second lock bit and execution from external program memory is disabled by programming the third lock bit.

17.6 | Solved Example

Example 17.1

Purpose: To understand the details of hardware interfacing for external program and data memory.

Problem

Prepare the interfacing of a MCS-51 based system with 8K of external program memory and 8K of external data memory. Also draw the complete circuit diagram of the interfacing.

Solution

Complete circuit diagram of the interfacing is shown in Fig. 17.9. The system is designed around 8031, the ROM-less version of MCS-51. 2764 and 6264 were used for program and data memory. 2764 is an 8K EPROM and 6264 is an 8K static RAM. Address lines necessary for both of these 8K devices are from A0 to A12.

To generate the lower address bus A0–A7, 74373 octal latch was interfaced with 8031 Port 0. The ALE output of the processor was interfaced to latch the input of 74373. The chip-select signal for 2764 and 6264 was generated by unused address line A15. The read input of program memory was interfaced with $\overline{PSEN}$ output from the processor. For the data memory 6264, the read and write signals were directly obtained from the processor's $\overline{RD}$ and $\overline{WR}$ outputs (P3.7 and P3.6, respectively).

The Vpp and PGM inputs of 2764 were connected with Vcc to input logic 1. CE2 input of 6264 was also connected with Vcc for the same purpose. Finally, the $\overline{EA}$ input of the processor (8031) was connected with system ground to input logic low and thus activating external memory access.

To maintain uniformity and also for easier understanding, some nomenclature of 2764 and 6264 were changed. For example, 2764 has the read input designated as $\overline{OE}$ or output enable. In Fig. 17.9, this has been changed to $\overline{RD}$, for easier understanding.

The reader may note that both 2764 and 6264 are interfaced with A15 address signal working as chip select. However, using Harvard architecture, this would not create any problem as 2764 would be read by $\overline{PSEN}$ signal whereas 6264 would be read by $\overline{RD}$ signal.

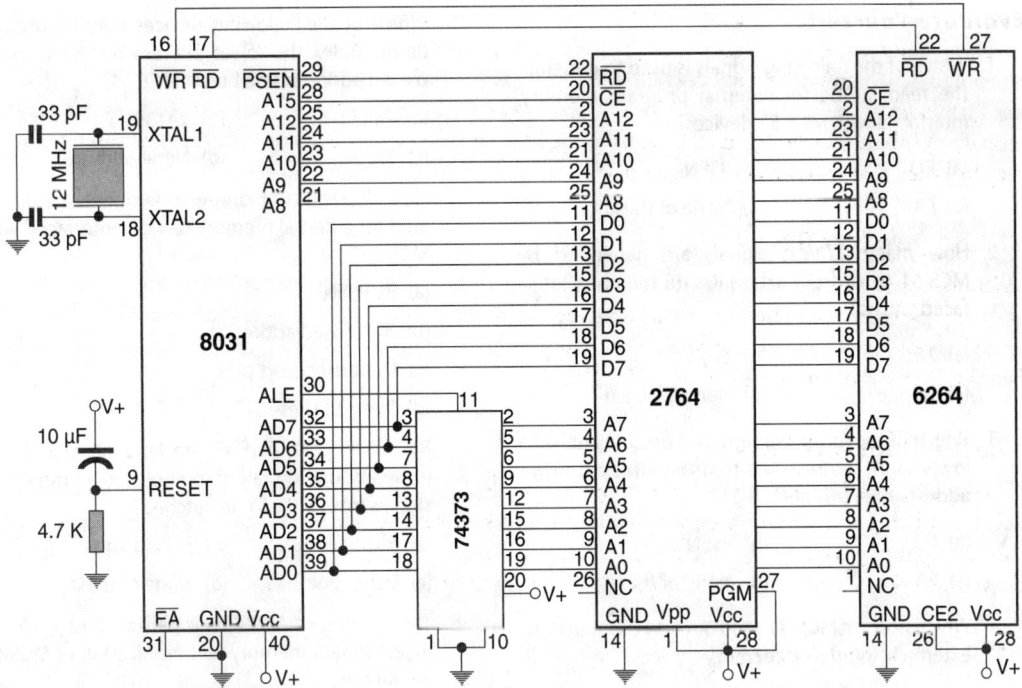

Figure 17.9 Circuit diagram for program and data memory interfacing

SUMMARY

Built around Harvard architecture, MCS-51 supports a maximum of 64K of program memory and 64K of data memory in addition to its internal data memory. The capabilities of on-chip program and data memories may be enhanced by interfacing external program and data memory devices with the processor. To access external program memory, the $\overline{EA}$ input of the processor must be connected with system ground. In such a case, $\overline{PSEN}$ from the processor would generate external program memory read signal. For external data memory reading and writing, both $\overline{RD}$ and $\overline{WR}$ signals are provided by the processor.

MCS-51 generates 16-bit address through its ports 0 and 2 for addressing external program and data memory devices. Lower eight address bits share the same bus with eight data bits. MCS-51 provides the ALE output signal to externally latch the lower address bits, for which an octal latch, like 74373, is necessary. In general, the ALE signal is generated twice within a machine cycle, for program memory addressing. For external data memory addressing, ALE is generated only once in a machine cycle. Instruction MOVX may be used to access external data. If lock bit (s) is (are) programmed then external memory cannot be accessed.

POINTS TO REMEMBER

- If $\overline{EA}$ input is connected with GND, on-chip program memory would never be accessed and execution would start from external program memory only.

- If $\overline{EA}$ input is connected with Vcc, on-chip program memory, if any, would be accessed first and external program memory would be accessed thereafter.

REVIEW QUESTIONS

Evaluate Yourself

1. Which of the following signals is used to generate the read strobe for external program memory, interfaced with MCS-51 devices?

   (a) $\overline{RD}$ (b) $\overline{PSEN}$

   (c) $\overline{EA}$ (d) None of these

2. How many address signals are generated by MCS-51 to address external data memory, interfaced with it?

   (a) 16   (b) 8

   (c) 16 or 8   (d) None of these

3. Which of the following signals is used to latch the lower eight address bits from multiplexed lower address-data bus of MCS-51?

   (a) $\overline{RD}$ (b) $\overline{PSEN}$

   (c) $\overline{EA}$ (d) None of these

4. Which ports of MCS-51 send out address signals to externally interfaced memory devices?

   (a) Ports 0 and 2   (b) Ports 1 and 3

   (c) Ports 0 and 1   (d) None of these

5. Which of the following devices may be used to de-multiplex the lower address bus from multiplexed address-data bus of MCS-51?

   (a) 74373   (b) 74139

   (c) 74138   (d) None of these

6. How chip select or chip-enable signals are generated for external memory devices interfaced with MCS-51?

   (a) By $\overline{PSEN}$

   (b) By unused addresses

   (c) By unused port pins

   (d) None of these

7. By which condition of the latching signal, the lower address signals of multiplexed address-data signals from MCS-51 are latched?

   (a) Rising edge   (b) Falling edge

   (c) Either one   (d) None of these

8. The number of ALE signals per machine cycle for external data memory communication of MCS-51 would be

   (a) one   (b) two

   (c) three   (d) none of these

9. The number of ALE signals per machine cycle for external program memory reading of MCS-51 would be

(a) one   (b) two

(c) three   (d) none of these

10. In MCS-51, if only one lock bit is provided, then programming this lock bit would prevent

(a) further programming

(b) access to on-chip program memory

(c) both of these

(d) none of these

Search for Answers

1. Under which condition it may be necessary to interface external memory with MCS-51 devices?

2. What are the maximum sizes of program memory and data memory that may be interfaced with MCS-51?

3. How many signals are necessary for external memory interfacing with MCS-51? From which ports are they generated? Are all of them alternate functions for these ports?

4. What are the differences between external program memory interfacing and external data memory interfacing with MCS-51?

5. What are the purposes of $\overline{EA}$, $\overline{PSEN}$ and ALE signals of MCS-51?

6. During one machine cycle of opcode-fetch, how many bytes are read from external program memory by MCS-51?

7. How many machine cycles are necessary for MOVX instruction to read a byte from external data memory?

8. What is the purpose of providing lock bits in MCS-51? What is the maximum number of lock bits provided?

9. Explain the functions of the *three lock bits* ystem.

10. Make a list of only those pin numbers w iich function differently in 6264 RAM and 2764 EPROM.

Think and Solve

1. What type of communication takes place between MCS-51 and external program memory, synchronous or asynchronous?

2. Why MCS-51 is equipped with the $\overline{PSEN}$ signal, in spite of the existence of a read signal available from the system?

3. Under which condition 8051 with internal 4K program memory would access external program memory?

4. Why it is not necessary to write 1s in Port 0 latch to read data from external RAM or ROM?

5. What is meant by 'Timing Diagram'?

6. What would happen if, in the circuit of Fig. 17.9, the $\overline{EA}$ input of 8031 is connected with Vcc?

7. What would happen if the chip 8031 is replaced by 8051 in the circuit of Fig. 17.9?

8. Explain the technique of generating the *chip-select* signals for different memory devices.

9. Intel manufactured 8155 with RAM, I/O and Timer and 8755 with EPROM and I/O, to be interfaced with 8085 CPU. If these two devices are to be interfaced with MCS-51, is there any need of 74373? Justify your answer.

10. Design a circuit with 8052 interfaced with an external 6264 RAM in such a manner so that the RAM may be used as data memory as well as program memory.

18 INTERFACING: KEYBOARDS

CHAPTER OBJECTIVES

In this chapter, the reader is introduced to keyboard interfacing, the most widely used input device for any embedded system. After completion of the chapter, the reader should be able to understand

- How does the mechanical contact-type keys function.
- What is the bouncing of keys and how to debounce those.
- How to interface individual keys as well as key-matrix with MCS-51.
- How to develop software routine to read keys and scan a keyboard.

18.1 | Introduction

Keyboard and display are two most widely used user interfaces for any embedded system. In this chapter, we discuss about some important aspects of keyboard interfacing. As a matter of fact, keys may be considered as simplest type of sensors. They are activated by finger pressure or sometimes just by touch. Table 18.1 presents the mechanism used in a few common types of keys.

Table 18.1 Some common types of keys and their mechanism

Key type	Mechanism
Tactile keys (calculator-type keys)	Mechanical contact
Capacitor keys	Capacitance
Membrane keys	Mechanical contact
Hall-effect keys	Magnetic field

18.2 | Contact Type Keys

Mechanical contact type keys are most widely used. Although inexpensive, they are less durable and may not be used in all types of environments. On the other hand, Hall-effect keys are extremely durable and may be used in any ambient conditions including underwater. However, they are of expensive variety. In this chapter, we discuss mostly about mechanical type of keys, which work by mechanical contacts. Functioning of a mechanical contact type key is shown in Fig. 18.1.

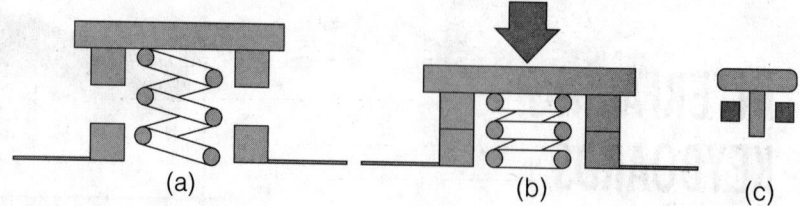

Figure 18.1 Operation of a mechanical type of key : (a) open, (b) closed and (c) symbol

18.3 | Interfacing a Key

Any mechanical key offers two states: open and closed. These mechanical states may easily be converted to corresponding electronic states using the configuration as illustrated in Fig. 18.2. Fig. 18.2(a) shows how one such key may be interfaced with an input port pin using a pull-up resistor. When the key is pressed [Fig. 18.2(b)], the mechanical contact establishes ground potential or logic 'low' at the input pin, through the sinking of current from the pin through the key itself. When the key is released [Fig. 18.2(c)] the input pin experiences logic 'high' as in this case current is sourced to the pin. To interface multiple keys, this scheme would need identical number of port pins. As an example case, we discuss the following problem.

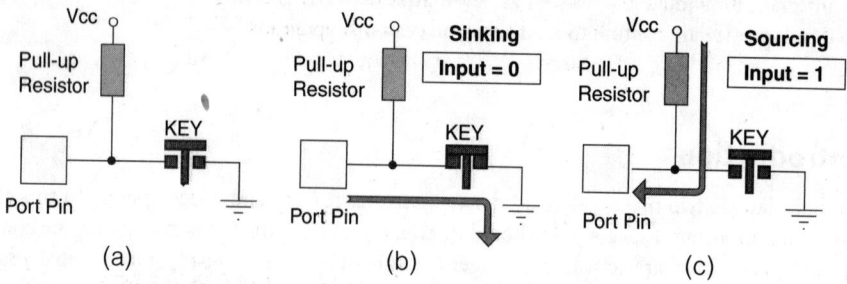

Figure 18.2 (a) Standard key interface with pull-up resistor, (b) inputting 0 and (c) inputting 1

18.4 | Solved Examples

Example 18.1

Purpose: To prepare interface for one-dimensional key array and develop proper branching routine to get access to correct entry points depending upon the key input.

Problem

Four keys to be interfaced with Port 1 of 8051 to offer four arithmetic functions: add, subtract, multiply and divide. Assuming both operands are already available in Accumulator and B-register, design the hardware and develop the software for it.

Solution

For completing the hardware part, four keys (K1, K2, K3 and K4) are interfaced with lowest four port pins of Port 1 as shown in Fig. 18.3. Function offered by each of these keys is marked just above the key. Four pull-up resistors (R1–R4) of 10K each are used in addition to internal pull-ups as shown in the circuit diagram. Note that ground connection (0V) for all keys may be common. However, *each key must be having its own pull-up resistor.*

Assuming that the operands are already present in the registers A and B, a software may be developed, as follows, to read keys and implement its function accordingly.

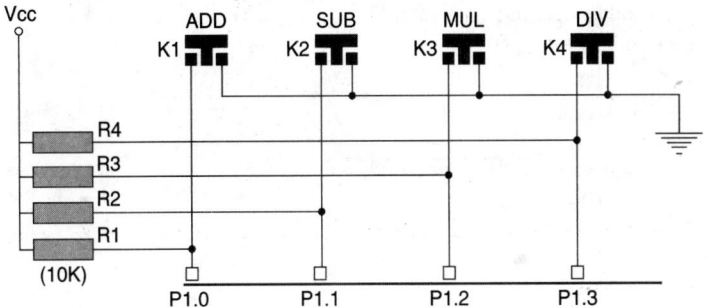

Figure 18.3 Circuit for four function input keys

```
RDKEY4:   ORL     P1, #0FH        ; output 1 for reading P1.0-P1.3
RKLOOP:   JNB     P1.0, ADDIT     ; K1 is pressed
          JNB     P1.1, SUBIT     ; K2 is pressed
          JNB     P1.2, MULIT     ; K3 is pressed
          JNB     P1.3, DIVIT     ; K4 is pressed
          SJMP    RKLOOP          ; wait for a key pressing
ADDIT:    ADD     A, B            ; add both operands
          SJMP    DSPLY           ; display, etc.
SUBIT:    CLR     C
          SUBB    A, B            ; subtract
          SJMP    DSPLY           ; display, etc.
MULIT:    MUL     AB              ; multiply
          SJMP    DSPLY           ; display, etc.
DIVIT:    DIV     AB              ; divide
DSPLY:                            ; place relevant display routine here.
          SJMP    RKLOOP          ; go back and read more keys
```

When executed, the program would be in an infinite loop, which would be waiting for keys and function accordingly.

C-version

```c
#include <regx51.h>

void main(void)
{
        unsigned char  operand1 = 10, operand2 = 20;
        unsigned char result;

        p1 = 0x00;

        // infinite loop

        while(1)
        {
                if(0 == p1_0)
                {
```

```
                result = operand1 + operand2;
                while(0 == p1_0);
        }
        if(0 == p1_1)
        {
                result = operand1 − operand2;
                while(0 == p1_1);
        }
        if(0 == p1_2)
        {
                result = operand1 * operand2;
                while(0 == p1_2);
        }
        if(0 == p1_3)
        {
                result = operand1 / operand2;
                while(0 == p1_3);
        }
    }
}
```

18.5 | Bouncing of Keys

When mechanical switches are pressed or released, the corresponding electrical transitions indicate some undulations just after pressing and releasing the key as illustrated in Fig. 18.4. This is known as bouncing and may be present for a maximum of 40 milliseconds.

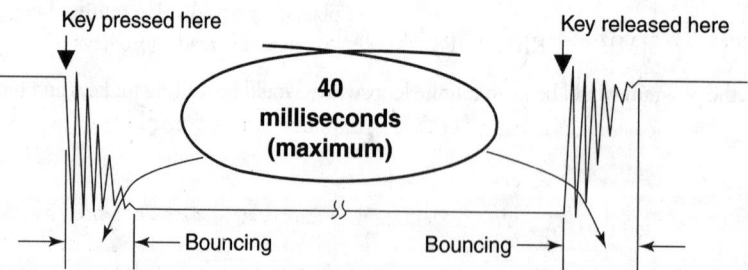

Figure 18.4 Bouncing of mechanical contact-type keys

Bouncing effect is undesirable and generates incorrect key interpretation. Therefore, in general, any key is to be debounced before reading or accepting it as a valid input. This may be achieved by either hardware or software.

18.5.1 | Hardware Debouncing

By incorporating a capacitor and a diode, a simple circuit may be prepared for debouncing the mechanical contact type keys as shown in Fig. 18.5(a). The capacitor may or may not be electrolytic type. However, 1 μFD is a safe value for this capacitor along with a 10K pull-up resistor. The reverse biased diode is to eliminate the negative spikes, which might be generated during contacts. When pressed and released, the resulting waveform would be as shown in Fig. 18.5(b). However, one major disadvantage of hardware debouncing is the increase in overhead expenditure. On the other hand, a software debouncing routine does not add any overhead expenditure, which we would discuss now.

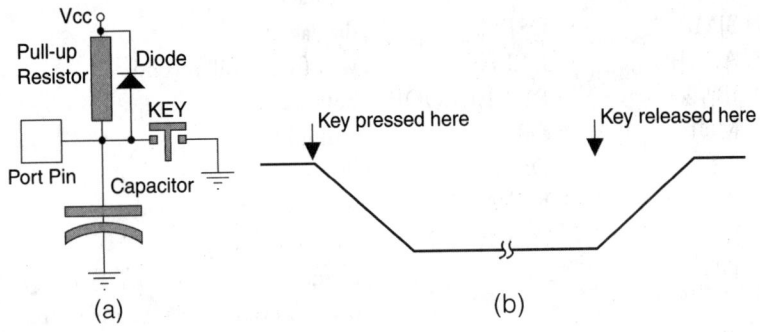

Figure 18.5 Hardware debouncing (a) circuit (b) signal form

18.5.2 | Software Debouncing

Software debouncing is based on the fact that the input signal oscillations, during key pressing and key releasing, do not last for more than 40 milliseconds. Therefore, whenever any key is detected, the input signal is placed under observation for more than 40 milliseconds. If the same signal is present after 40 milliseconds, then, and then only, it is taken as a valid input signal from that key. Following is an example of software debouncing.

Example 18.2

Purpose: To develop a program for software debouncing.

Problem

Modify the software developed for Example 18.1 to debounce the activated key.

Solution

For the problem of Example 18.1, software debouncing may easily be implemented by incorporating a delay routine of 50 milliseconds and calling that delay routine whenever any key is detected. If the same key is still active then it may be taken as a valid one. The modified software is presented below.

```
RDKEY:  ORL       P1, #0FH        ; output 1 for reading P1.0 – P1.3
RKLOOP: JNB       P1.0, ADDIT     ; K1 is pressed
        JNB       P1.1, SUBIT     ; K2 is pressed
        JNB       P1.2, MULIT     ; K3 is pressed
        JNB       P1.3, DIVIT     ; K4 is pressed
        SJMP      RKLOOP          ; wait for a key pressing
ADDIT:  ACALL     DELAY           ; wait for 50 milliseconds
        JB        P1.0, RKLOOP    ; not a valid key command
        ADD       A, B            ; add both operands
        SJMP      DSPLY           ; display, etc.
```

```
SUBIT:    ACALL      DELAY          ; wait for 50 milliseconds
          JB         P1.1, RKLOOP   ; not a valid key
          CLR        C
          SUBB       A, B           ; subtract
          SJMP       DSPLY          ; display, etc.
MULIT:    ACALL      DELAY          ; wait for 50 milliseconds
          JB         P1.2, RKLOOP   ; not a valid key
          MUL        AB             ; multiply
          SJMP       DSPLY          ; display, etc.
DIVIT:    ACALL      DELAY          ; wait for 50 milliseconds
          JB         P1.3, RKLOOP   ; not a valid key
          DIV        AB             ; divide
DSPLY:                              ; place relevant display routine here.
          SJMP       RKLOOP         ; go back and read more keys
```

If 12 MHz crystal is used, the following routine would generate roughly 50 milliseconds delay, when called.

```
DELAY:    MOV        R7, #60H
DELY:     MOV        R6, #00H
DELY1:    DJNZ       R6, DELY1
          DJNZ       R7, DELY
          RET
```

We may observe in this program RDKEY that initially after sensing any key, the program calls the DELAY routine to wait for 50 milliseconds and then checks for the *same key* again. If the same key is still sensed then only it is taken as a valid key, otherwise not. This is one of the many ways of software debouncing.

C-version

```c
#include  <regx51.h>

void Delay50Ms (void);

void main(void)
{
        unsigned char operand1 = 10, operand2 = 20;
        unsigned char result;

        // infinite loop

        while(1)
        {
                if(0 == P1_0)
                {
                        Delay50Ms();
                        if(0 == P1_0)
                        {
```

```
                        result = (operand1 + operand2);
                        while(0==P1_0);
            }
        }
        if(0 == P1_1)
        {
            Delay50Ms();
            if(0 == P1_1)
            {
                        result = (operand1 - operand2);
                        while(0==P1_1);
            }
        }
        if(0 == P1_2)
        {
            Delay50Ms();
            if(0 == P1_2)
            {
                        result = (operand1 * operand2);
                        while(0==P1_2);
            }
        }
        if(0 == P1_3)
        {
            Delay50Ms();
            if(0 == P1_3)
            {
                        result = (operand1 / operand2);
                        while(0==P1_3);
            }
        }
    }
}
void Delay50Ms(void)
{
    volatile unsigned int i=6300;
    while(i>0)
    {
        i--;
    }
}
```

18.6 | Key Matrix

So far we have assumed that each key is sensed by one port pin. This arrangement is adoptable if number of keys are very less. However, for more number of keys, generally a two-dimensional matrix configuration is preferred. This allows us to interface more number of keys through less number of port pins, resulting in economy. As an example case, layouts of Figs 18.6(a) and (b) may be compared.

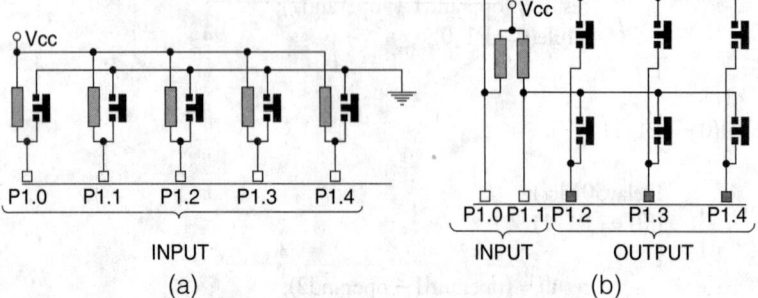

Figure 18.6 (a) 1D and (b) 2D array of keys

What would be the maximum number of keys that might be interfaced with only five port pins? In Fig. 18.6(a), using a linear arrangement, we can interface only five keys. On the other hand, as shown in Fig. 18.6(b), the two-dimensional layout would allow us to interface six keys with five port pins. It may be noted in these two cases that all port pins are in their input mode for the first configuration. In the second configuration, three are output and two are input pins. Similarly, eight port pins would allow us to interface a maximum of 16 keys in 4×4 matrix. The number of keys would come down to eight only, if we interface only one key for each port pin.

 There is no correlation between the physical and electrical layout of keys. That is, for example, the physical layout may be only one array of ten keys, which may be electrically connected as a 2 x 5 array.

18.7 | Scanning Keyboard Matrix

The technique of reading a two-dimensional array of keys would be different from that of a linear array of keys. To explain it, we take the example of the 2×3 matrix, which have just been introduced. To start with, we identify three columns of this matrix driven by P1.4, P1.3 and P1.2. Two rows of this matrix are received by P1.1 and P1.0. A general algorithm for this may be as follows.

Step 1:   We assign a register as key counter and clear it. We assign another register as return code and clear it too. Finally, we assign a third register as column counter and load it by 3, the number of existing columns in the matrix.

Step 2:   Output 0 through one column, say P1.4, and remaining two columns should get the output as 1s as in Fig. 18.7.

Step 3:   We increment the key counter by 1.

Step 4:   Now read one row, say P1.0. This input would be 0 if, and only if, the key K1 is pressed. Pressing no other key at this instant (till P1.4 = 0) would produce a 0 at P1.0. If P1.0 is found to be 0, then we copy the present content of key counter in return code register. No action is taken if P1.0 = 1.

Step 5:   We increment key counter by 1.

Step 6:   Now we read the next row, i.e. P1.1. Just like the previous case, P1.1 would receive 0 if, and only if, K2 is pressed. Otherwise it would show 1. If P1.1 is 0, then we copy the current content of key counter in return code register.

Step 7:   We decrement column counter by 1. As it is not 0, we change the input pattern so that P1.3 outputs 0 and P1.2 and P1.4 output 1s, as in Fig. 18.8.

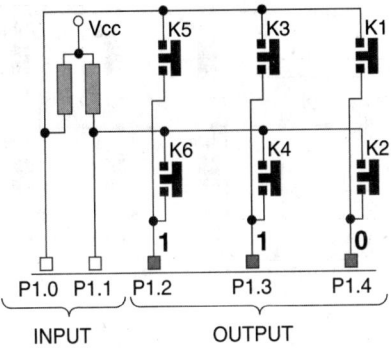

Figure 18.7 Activating first column (to read K1, K2 only)

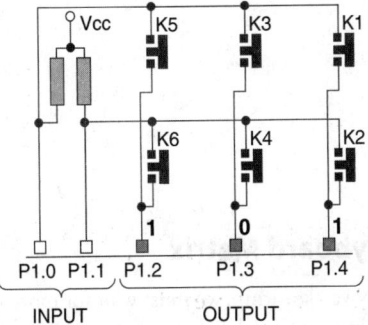

Figure 18.8 Activating second column (to read K3, K4 only)

Step 8: We increment key counter by 1 (same as Step 3).

Step 9: We read row by P1.0, and if it is 0 then copy key counter in return code register. Note that this 0 may only be generated by pressing K3 (same as Step 4).

Step 10: Increment of key counter by 1 (same as Step 5).

Step 11: Read P1.1 and if it is 0 then copy key counter in return code register. This time it is by K4 only (same as Step 6).

Step 12: Decrement column counter by 1. As it is not 0, we change the output pattern as shown in Fig. 18.9 with P1.2 outputting 0 and remaining port pins outputting 1 (same as Step 7).

Step 13: Increment key counter by 1 (same as Step 3).

Step 14: Read P1.0 and if it is 0 (K5 pressed) then copy key counter in return code register (same as Step 4).

Step 15: Increment key counter by 1 (same as Step 5).

Step 16: Read P1.1 and if it is 0 (K6 pressed) then copy key counter in return code register (same as Step 6).

Step 17: We decrement column counter by 1. As it is 0, we terminate the routine. Observe that whichever key is pressed, its code (between 1 and 6) must be there in the return code register. If no key is pressed, return code register would be 0. If more than one key is pressed, although this is not allowed, only the last sensed key would be indicated.

Above description may be apparently a complex and longer one. However, the basic objective is to maintain only one column low at a time and keep on incrementing the key counter before testing each key input. Whenever any input is found low, the value within the key counter is to be copied immediately to the return code register. Softwarewise it is easy to implement. We will develop the complete subroutine after introducing the debouncing part within it.

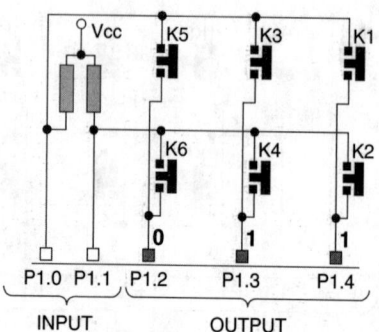

Figure 18.9 Activating third and last column (to read K5 and K6 only)

 In essence, we are reading every key once by assigning its column drive as 0 and reading its row input bit.

18.8 | Debouncing Keyboard Matrix

To debounce any key sensed by above algorithm, we must wait for more than 40 milliseconds to ensure it to be a valid key. Generally, it is achieved by using a debounce counter, which is loaded by the debounce value whenever any new key is detected. After scanning all keys once, this debounce counter is decremented by one and the return code is compared with the previous return code. The procedure is repeated till the debounce counter is 0. In that case the key is completely debounced. Any mismatch with previous return code would force the debounce counter to be reloaded and matching restarted. All these are explained through the following example.

Example 18.3

Purpose: To be familiar with the method of scanning and debouncing a two-dimensional array of keys.

Problem

Sixty-four keys are to be interfaced with 8051 arranged in a 8 × 8 matrix. Give a schematic of the hardware interfacing. Develop a software to generate unique key code for any key pressed. The key code must be fully debounced.

Solution

Hardware interfacing of 64 keys is schematically presented in Fig. 18.10. Port 1 was used as input port and Port 3 as output port. Eight signal lines from Port 3 (scan lines) and eight more signal lines from Port 1 (return lines) form a grid. At each intersection, only one key is accommodated so that the contact between any scan and return lines may be established by pressing the key only at their intersection. Note that if no key is pressed, there would be no connection between any scan line and any return line. For the purpose of stability, all return lines of Port 1 are externally pulled up. The software routine uses register bank #0. Usage of all related registers is explained at the starting of the software routine. This routine, KSCAN, would not return unless it can return a valid debounced key code. If at entry of KSCAN, any key is pressed, KSCAN would wait for the key to be released.

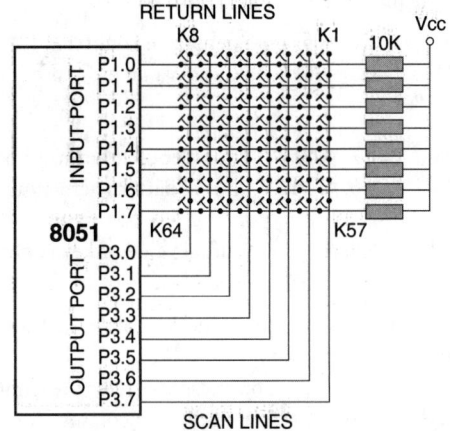

Figure 18.10 Schematic of interfacing 64 keys in 8 × 8 matrix

```
; Name:      KSCAN
; Function:  Scans 64 keys arranged in 8 × 8 matrix and returns the scan code of the active key, if any. If at
;            entry of KSCAN any key is found to be pressed it would wait till the release of that key. KSCAN
;            would not return until a new key is pressed. The returned scan code is fully debounced.
; Input:     none
; Output:    debounced key code in A
; Calls:     SCAN1
; Uses:      none

      KSCAN:    ACALL   SCAN1     ; at return, key code in register A.
                JNZ     KSCAN     ; wait till the key is released
      KSCAN2:   ACALL   SCAN1     ; get fresh key code
                JZ      KSCAN2    ; no key pressed, scan keyboard again
                RET
```

```
; Name:      SCAN1
; Function:  Scans keyboard and returns if no key pressed or with debounced key code.
; Input:     None
; Output:    debounced key code in A. If no key pressed then A = 0.
; Calls:     DELAY
; Uses:      Register bank #0
;            R1 = row counter
;            R2 = debounce counter
;            R3 = final key return code
;            R4 = key code scanned in a cycle
;            R5 = key counter
;            R6 = column activating pattern (only 1 bit would be 0)
;            R7 = column counter
;            A = general purpose and final key code at return
;            CY flag = key bit for testing
```

```
; Initialize for a fresh key debouncing
        SCAN1:  MOV     R2, #05H    ; debounce counter = 5
                MOV     R3, #00H    ; initial value of final return code = 0
; Initialize to scan all 64 keys.
        SCAN:   MOV     R7, #08H    ; number of columns to be scanned
                MOV     R6, #7FH    ; initial pattern activating rightmost column
                CLR     A
                MOV     R5, A       ; initial value of key counter = 0
                MOV     R4, A       ; initial value of fresh return code = 0
```

; Activate one column, update column drive pattern and save it back. Then wait for 1 millisecond and then read ; return lines. Initialize to read returned bits.

```
     NXTCOL: MOV     A, R6       ; get output pattern for present cycle
             MOV     P3, A       ; activate only one column by 0
             RR      A           ; prepare output pattern for next cycle
             MOV     R6, A       ; save pattern back
             ACALL   DELAY       ; wait for 1 millisecond
             MOV     A, P1       ; read return port, all 8 lines
             MOV     R1, #08H    ; counter to read 8 bits
```

; Read returned bits and if any bit is 0 then save key count in fresh return code counter.

```
     NXTROW: INC     R5          ; update key counter
             RRC     A           ; read a bit through carry flag
             JC      ROW1        ; no key pressed in this row for this col
             MOV     R4, 05H     ; copy key counter in fresh return code
     ROW1:   DJNZ    R1, NXTROW  ; read all eight rows
```

; Loop for all eight columns.

```
             DJNZ    R7, NXTCOL  ; complete all eight columns
```

; Fresh return code in R4. If both R4 and R3 are 0 then no key detected.
; Return in that case.

```
             MOV     A, R4       ; fresh return code in A
             ORL     A, R3       ; compare with previous return code
             JNZ     DBIT        ; some key detected, look for debounce
             RET                 ; no key pressed, accumulator = 0, return
```

; Some key detected. If same key and if debouncing over then return with key code in A.

```
     DBIT:   MOV     A, R4       ; to compare with R3
             CJNE    A, 03H, NEW1 ; new key
             DJNZ    R2, SCAN    ; old key, start debouncing
             MOV     A, R3       ; debounced key code in A
             RET                 ; return with debounced key code in A
```

; New key. Reload debounce counter and start fresh scanning.

```
     NEW1:   MOV     R3, A       ; preserve fresh key code in R3
             MOV     R2, #05H    ; reload debounce counter for fresh key
             SJMP    SCAN        ; continue key scanning
```

```
; Name:        DELAY
; Function:    when called, would generate a delay of 1 millisecond.
; Input:       none
; Output:      none
; Calls:       none
; Uses:        R0

   DELAY:    ACALL DELAY1      ; delay for 500 microseconds
   DELAY1:   MOV   R0,#0FAH    ; count of 250d
   DELAY2:   DJNZ  R0,DELAY2   ; loop for 500 microseconds
             RET
```

 Note that how by calling the routine itself, the delay is doubled.

SUMMARY

Keys are most widely used input devices and are of various types such as capacitance type, membrane type and mechanical contact type. In general, they are interfaced with input ports in such a manner so that a key contact generates logic low and a key release generates logic high. Bouncing phenomena, which originates from mechanical properties of the keys, is observed during this pressing and releasing of mechanical contact type keys. They may be debounced by both hardware and software techniques, although the later one is considered to be cost effective. A matrix of keys demand lesser number of port pins for interfacing than a linear array of keys.

POINTS TO REMEMBER

- Any port pin interfaced with a key must output logic 1 to enable the port pin to read the key.

- Bouncing lasts for a maximum of 40 milliseconds and for debouncing, a key must be read for more than 40 milliseconds.

REVIEW QUESTIONS

Evaluate Yourself

1. If decremented by DJNZ instruction, which of the following initial counter values would generate maximum delay?

   (a) #0FFH   (b) #00H

   (c) #60H   (d) None of these

2. Membrane type of keys are activated through

   (a) mechanical contact   (b) capacitance

   (c) induction   (d) none of these

3. Any mechanical contact type key can produce a maximum of

   (a) one state   (b) two states

   (c) three states   (d) none of these

4. To input logic 'low' to any input pin of a port, the relevant circuit should be able to

   (a) sink current   (b) produce 0V

   (c) both of these (d) none of these

278 INTERFACING: KEYBOARDS

5. Bouncing of any mechanical contact type key may last for a maximum of

    (a) 40 microseconds   (b) 4 milliseconds

    (c) 1/4th of a second   (d) none of these

6. For hardware debouncing, the key with pull-up resistor should be attached with

    (a) another resistor   (b) a capacitor

    (c) a diode   (d) none of these

7. For software debouncing, it is to be ensured that the key is

    (a) released

    (b) sensed for more than 40 milliseconds

    (c) sensed twice

    (d) none of these

8. The maximum number of keys may be interfaced with seven Port pins would be

    (a) 7   (b) 10

    (c) 12   (d) none of these

9. To read any two-dimensional keyboard matrix, the output pins should output

    (a) one 0 and others 1

    (b) one 1 and others 0

    (c) either of these

    (d) none of these

10. If more than one key is pressed, then the routine KSCAN would return

    (a) first scanned key   (b) last scanned key

    (c) no key   (d) none of these

Search for Answers

1. How do the Hall-effect keys function?

2. Are springs essential for mechanical contact type keys?

3. What are the essential features of interfacing a mechanical contact type key with a port pin of any microcontroller?

4. What is bouncing? Is it applicable for all types of keys? How can any key be debounced?

5. What is the difference between hardware debouncing and software debouncing?

6. What are the limitations of one-dimensional array of keys?

7. What are the advantages of a two-dimensional array of keys?

8. What is the methodology of scanning a two-dimensional keyboard?

9. What is the role of a debounce counter in any key scanning routine?

10. What would happen if the DELAY routine is not used in the routine SCAN1?

Think and Solve

1. What is the difference between tactile keys and membrane type keys?

2. Assuming that no internal pull-ups available, what would happen if the external pull-up resistor is removed from the interface of Fig. 18.2(a)?

3. When we should interface one key per input pin, and when we should interface multiple keys per input pin?

4. What would happen if all four resistors are replaced by a single resistor in Fig. 18.3?

5. Under which conditions hardware debouncing would be more acceptable than software debouncing? (Hint: RESET)

6. What purpose does a key counter serve in any two-dimensional key scanning routine?

7. How is it possible to generate two different codes for key pressing and key releasing?

8. If a 8051 microcontroller is used to read a keyboard and communicate with host computer through serial interface, what is the maximum number of keys may be interfaced in the keyboard?

9. What would happen if in routine SCAN, register R6 is loaded by #80H instead of #7FH?

10. Is it possible to rewrite SCAN1 routine without using R0?

(**Ans.:** Yes. Use A in place of R0 in delay routine.)

19 INTERFACING: DISPLAY DEVICES

CHAPTER OBJECTIVES

In this chapter, the reader is introduced to different types of widely interfaced display devices with microcontroller-based systems. After completion of the chapter, the reader should be able to understand

- How to interface light-emitting diodes (LEDs) and seven-segment displays.
- How to develop software routines for controlling these devices.

19.1 | Introduction

Most of the microcontroller-based systems are equipped with some types of display devices for displaying output of the system or to generate some feedback signals. LEDs and seven-segment displays are most widely used in different systems as per the demand. Out of these two, LEDs are most inexpensive and easiest to interface and control. In this chapter, we will discuss about both hardware and software aspects of interfacing these display devices. We start our discussions with LED interfacing.

19.2 | LED Interfacing

LEDs are available in various sizes, shapes and colours. Standard sizes are 3, 5 and 8 mm in diameter. Shapes are generally circular. However, rectangular shapes are also available. Red, green, yellow, orange, blue and white are available colours. Bicolour, tricolour, infrared and ultraviolet LEDs are also available. Red LEDs of 5 mm diameter are most widely used as general-purpose indicators. As a normal practice, they are to be interfaced with a current limit of 15 mA (about 7 mA for high-intensity LEDs).

Just like ordinary diodes, LEDs are also having polarities: anode and cathode. Fig. 19.1 shows two views of an LED along with its symbol. For easier recognition of polarity, generally the lead for anode is provided longer than that of the cathode [Fig. 19.1(a)]. When viewed from top, cathode side is seen as a flat surface and not a rounded-one like anode [Fig. 19.1(b)].

LEDs are simplest type of indicators and may be used to display two states, off and on. To interface LEDs, we must remember that

- LEDs glow only when they are forward biased and
- their resistance is almost zero for all practical purpose.

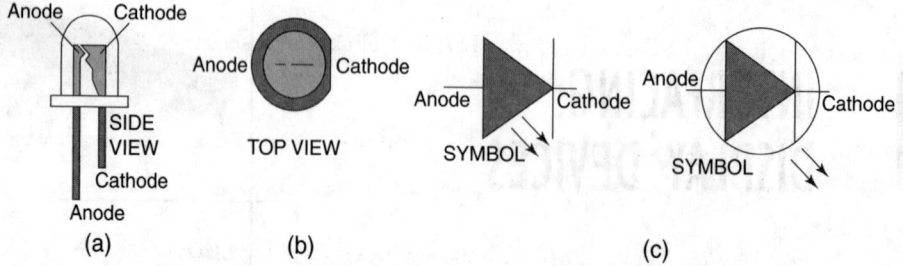

Figure 19.1 LED. (a) Side view, (b) top view and (c) symbols

It means that they must be interfaced with proper polarity and with a resistance in series. This resistance is a current-limiting resistor, and its value is to be calculated as per the applied voltage and current limit of the LED. A simple LED interface is presented in Fig. 19.2, where port pin P1.0 is used to drive the LED. Note that if the LED is to be turned on by logic 0, then it may be directly connected with the port pin [Fig. 19.2(a)], which has the capability to sink sufficient current to ensure proper brightness of the LED. However, if the LED is to be turned on by logic 1 from the port, then the limited capability of sourcing of the port pin becomes a constraint. To eliminate this problem, the port pin may use a buffer, like an inverter gate [Fig. 19.2(b)], which would enhance the current sourcing limit.

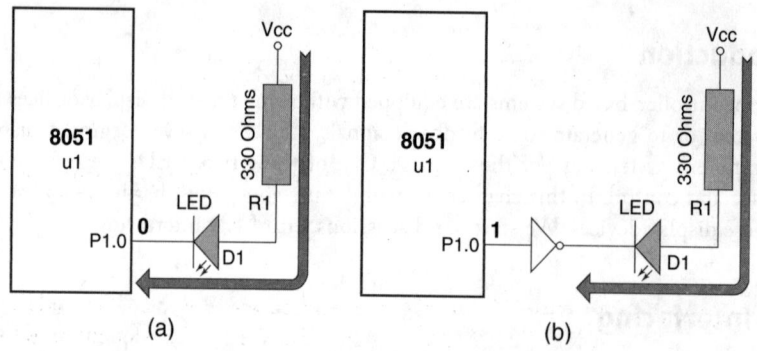

Figure 19.2 Interfacing a LED with 8051 and driving by (a) logic 0 and (b) logic 1

We have not covered surface mounting type (SMD type) LEDs in our discussions as those are used in very limited cases.

19.3 | Solved Examples

In general as indicators, LEDs are either switched on for a long duration as per some key commands or set at the blinking mode. A simple software for LED blinking is presented in the following example. This is followed by another example of using an LED as an indicator for switch operations.

Example 19.1

Purpose: To understand how LED states may be controlled through software.

Problem

Develop a program so that the LED interfaced with Port P1.0 blinks continuously.

Solution

As blinking is nothing but interchanging the on and off states of the LED, therefore, assuming equal durations of on and off states, the following routine may be used.

```
; Blinking LED.
    BLINK:       CPL    P1.0      ; toggle the previous state of the LED
                 ACALL  DELAY     ; wait for some time
                 SJMP   BLINK     ; repeat (infinite loop)
; Delay subroutine
    DELAY:       MOV    R7,#04H
    DELAY1:      MOV    R6,#00H
    DELAY2:      MOV    R5,#00H
    DELAY3:      DJNZ   R5, DELAY3
                 DJNZ   R6, DELAY2
                 DJNZ   R7, DELAY1
                 RET
```

Example 19.2

Purpose: To understand how to coordinate between input and output.

Problem

Two keys and one LED are interfaced with 8051 as shown in Fig. 19.3. Develop a software so that K1 would turn the LED on and K2 would turn it off.

Solution

As it may be observed from Fig. 19.3, keys K1 and K2 are interfaced with Port pins P3.0 and P3.1, respectively, while the LED is interfaced with Port pin P3.7. Moreover, as per the problem definition, K1 would turn on the LED while K2 would turn it off. This may be achieved by the following subroutine.

```
; LED control with two keys.
; First instruction is for initialization.
```

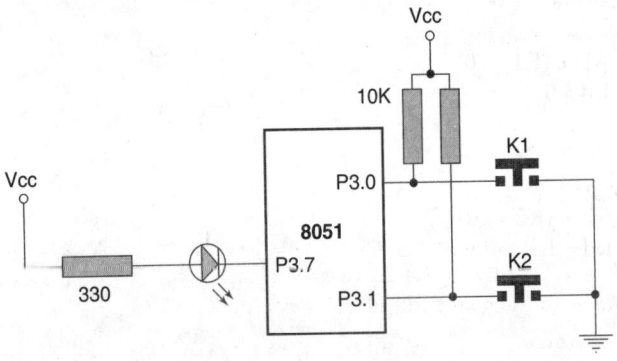

Figure 19.3 LED and keys interfaced with 8051

```
LEDONF:   ORL     P3, #83H      ; turn LED off, set P3.0 and 3.1 as inputs
RDKEY:    MOV     A, P3         ; to read keys
          RRC     A             ; to check K1
          JC      NEXT1         ; K1 not active
          CLR     P3.7          ; K1 active, turn LED on
          SJMP    RDKEY         ; read keys again
NEXT1:    RRC     A             ; to check K2
          JC      RDKEY         ; K2 inactive, go back to read keys
          SETB    P3.7          ; K2 active, turn LED off
          SJMP    RDKEY         ; loop on for keys
```

 A common mistake of students is to interface a LED between Vcc and port pin, without any current limiting resistor is series. Another common mistake is to use a port pin to directly source the necessary current for a LED.

C-version

```c
#include <regx51.h>

sbit K1 = P3^0;
sbit K2 = P3^1;
sbit led = P3^7;

void main()
{
        K1 = 1;
        K2 = 1;
        Led = 1;
        while(1)
        {
            if (K1 == 0)
            {
                while (K1 == 0) { };
                led = 0;
            }
            if (K2 == 0)
            {
                while (K2 == 0) { };
                led = 1;
            }
        }
}
```

19.4 | Seven-Segment Display Interfacing

To display numerical values, seven-segment displays are used widely. They may be seen in clocks and electronic measuring instruments. A few upper and lower case alphabets may also be displayed through it. Seven-segment displays are combination of eight LEDs, either in common anode [Fig. 19.4(b)] or in common cathode [Fig. 19.4(c)] formation. As shown in Fig. 19.4(a), the segments are designated as a, b, c, d, e, f and g and the last or eighth one representing the decimal dot is designated as p. As only seven of its segments are used to display any numeric value, hence the name seven segment.

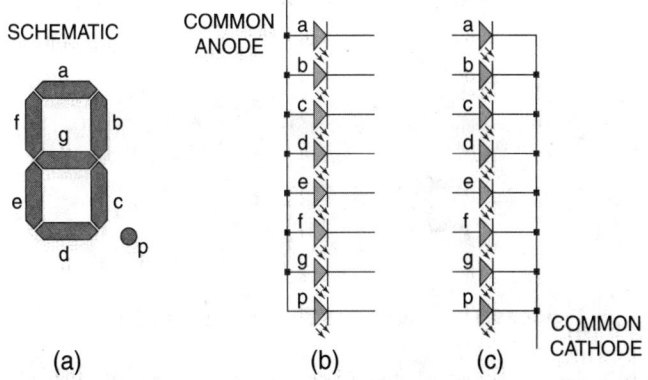

Figure 19.4 Types of seven-segment displays: (a) schematic, (b) common anode and (c) common cathode

To drive these seven-segment display digits, drivers like 7447 are available. These drivers accept 4-bit BCD input and generate the display pattern and also drive the relevant segments. However, a more popular approach is to use multiplexed display scheme, especially when multiple digits are to be driven and power economy is necessary.

Union jack type display and dot-matrix type of display devices are also used in various electronic gadgets. Less power hungry LCD displays are widely used in many battery operated devices.

19.5 | Multiplexed Display

The multiplexed display scheme consumes substantially less amount of current. However, the overall intensity or brightness of the display is lower, although within acceptable limit. The basic technique of multiplexed display scheme is to turn on only one display digit at a time, for a very short duration, say for a few milliseconds. As this is repeated at a faster rate, like 100 times per second, therefore, for the human eyes, it seems that all digits are glowing simultaneously.

For example, if we like to display 21.34, then first the leftmost digit with '2' is turned on for 1 millisecond (refer to Fig. 19.5). After that, the next digit with '1.' is turned on for another millisecond. Next the digit with '3' is turned on for another millisecond and finally the digit showing '4' is turned on for another millisecond.

Figure 19.5 Schematic of four-digit multiplexed display

This completes one cycle of operation, which takes 4 milliseconds. This 4-millisecond cycle is repeated again and again and to human eyes it would appear that the display is showing 21.34 and all digits are glowing simultaneously as indicated at the bottom of Fig. 19.5. Taking the example of four-digit display of a clock, the related hardware interfacing and software are explained through the following example.

Example 19.3

Purpose: To understand the operation of multiplexed seven-segment display.

Problem

Design hardware and software of a four-digit display of a Real Time Clock, to be driven by 8051. The display to be in 24-hour format and should display hour and minute of the day.

Solution

We select common cathode type of display for the system. The circuit diagram is presented in Fig. 19.6. Note that all anodes of any one type of segment are connected together. It means the four anodes of segment 'a' of all four digits are connected together and driven simultaneously by P1.0. Similarly, four anodes of segment 'b' of all the four digits are connected together and driven simultaneously by P1.1 and so on. For driving these segments, a driver, 74245, is interfaced with Port 1 of 8051, which we call as segment driver. Note that logic 1 from Port 1 pin would turn the corresponding segment on.

Each common cathode is activated by a NPN transistor, BC547, and four output pins of Port 3 are used for this purpose. To activate any digit, we are to activate the corresponding transistor, for which related Port 3 pin should output logic 1. External pull-ups are used in Port 3 pins and base of transistors are connected with 2.2K resistors in series. Current-limiting resistors of 10 Ohms were provided at the output of segment driver 74245.

The software is an infinite loop of converting the clock-counter digits into their display pattern and activating one after another four digits with their corresponding display patterns. After activating every digit, a DELAY routine is called to generate 1-millisecond delay. Therefore, in 4 milliseconds each digit is displayed (or refreshed) once, or in other terms, 250 times a second. This rate is fast enough for a flicker-free display and if necessary, may

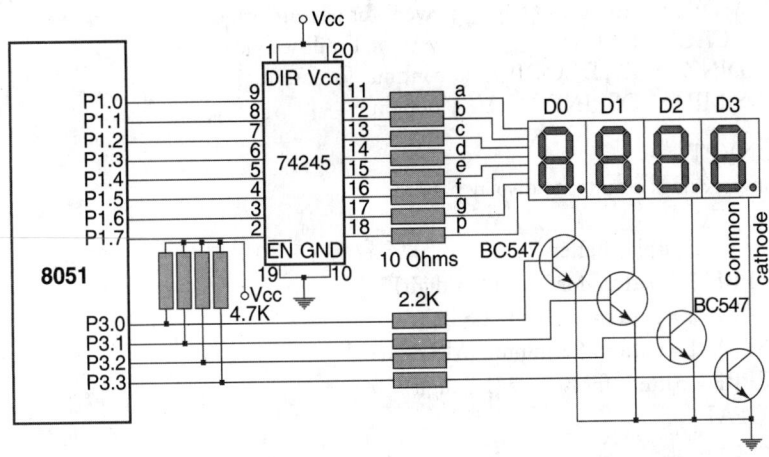

Figure 19.6 Circuit diagram of common four-digit cathode-type seven-segment display

be reduced to 125 times per second by calling the delay routine twice instead of once. The main routine, DSPINF takes the help of three subroutines, namely, CONVPT, SEGPAT and DELAY, all of which are explained below.

```
; Name:    DSPINF
; Function: Converts clock counter to display pattern and continuously refreshes display.
; Input:    CLKBUF 4 bytes to store hour and minute count in unpacked BCD format
;           CLKBUF+3 = Hour (MS digit)
;           CLKBUF+2 = Hour (LS digit)
;           CLKBUF+1 = Minute (MS digit)
;           CLKBUF =Minute (LS digit)
; Output:  see function above
; Calls:   CONVPT, DELAY
; Uses:    register bank #0
;           R1 = Pointer
;           R6 = Digit drive pattern for P3
;           R7 = Digit counter (4 to 0)
;           A = general purpose
;           CY flag
```

```
  DSPINF:  ACALL  CONVPT       ; convert clock count to display pattern
           MOV    R7,#04H      ; count for four digits
           MOV    R6,#01H      ; initial digit drive pattern (leftmost digit)
  DLOOP:   MOV    P3,#00H      ; blank display, all digits off
           MOV    A,R7         ; get current digit number
           ADD    A,#DISPBF-1  ; add with base address
           MOV    R1,A         ; use it as pointer
           MOV    A,@R1        ; get correct segment pattern
           MOV    P1,A         ; output to segment drive port
           MOV    A,R6         ; get digit-drive code
           MOV    P3,A         ; turn on current digit
           RL     A            ; pattern for next digit
```

```
              MOV      R6, A          ; save it for next iteration
              ACALL    DELAY          ; wait for 1 millisecond
              DJNZ     R7, DLOOP      ; continue for four digits
              SJMP     DSPINF         ; loop on
```

; Name: CONVPT
; Function: Converts clock counter to display pattern.
; Input: none
; Output: patterns in display buffer
;   DISPBF+3 = pattern for hour (MS digit)
;   DISPBF+2 = pattern for hour (LS digit)
;   DISPBF+1 = pattern for minute (MS digit)
;   DISPBF = pattern for minute (LS digit)
; Calls: SEGPAT
; Uses: Register bank #0
;   R0 = clock counter pointer
;   R1 = segment pattern buffer pointer
;   R7 = Digit counter (4 to 0)
;   A = general purpose

```
CONVPT:   MOV      R0, #CLKBUF     ; starting address of clock buffer
          MOV      R1, #DISPBF     ; starting address of display buffer
          MOV      R7, #04H        ; counter for four digits' patterns
PTLOOP:   MOV      A, @R0          ; get one number
          ACALL    SEGPAT          ; get its pattern
          MOV      @R1, A          ; save in display buffer
          INC      R0              ; point next number
          INC      R1              ; point next storage area
          DJNZ     R7, PTLOOP      ; continue for four digits
          RET                      ; done
```

; Name: SEGPAT
; Function: Converts a BCD number into its display pattern using table look-up method.
;   Table used is SEGTAB.
; Input: A has the BCD number (0–9) to be converted to display pattern
; Output: A containing display pattern
; Calls: none
; Uses: A

```
SEGPAT:   INC      A
          MOVC     A, @A+PC        ; use table SEGTAB
          RET
SEGTAB:                            ; pgfe dcba = order of segments
          DB       3FH             ; 0011 1111 = pattern for 0
          DB       18H             ; 0001 1000 = pattern for 1
          DB       5BH             ; 0101 1011 = pattern for 2
          DB       4FH             ; 0100 1111 = pattern for 3
          DB       66H             ; 0110 0110 = pattern for 4
          DB       6DH             ; 0110 1101 = pattern for 5
```

```
        DB      7DH         ; 0111 1101 = pattern for 6
        DB      07H         ; 0000 0111 = pattern for 7
        DB      7FH         ; 0111 1111 = pattern for 8
        DB      6FH         ; 0110 1111 = pattern for 9
```

```
; Name:     DELAY
; Function: Generates 1-millisecond delay (approx) when called
; Input:    none
; Output:   nonc
; Calls:    none
; Uses:     R2, R3 of bank #0
```

```
        DELAY:  MOV     R3, #02H
        DEL1:   MOV     R2, #0FAH    ; 250d
        DEL2:   DJNZ    R2, DEL2
                DJNZ    R3, DEL1
                RET
```

When executed, this routine would keep on outputting the segment pattern of a digit and would turn on that digit simultaneously as indicated through Fig. 19.7.

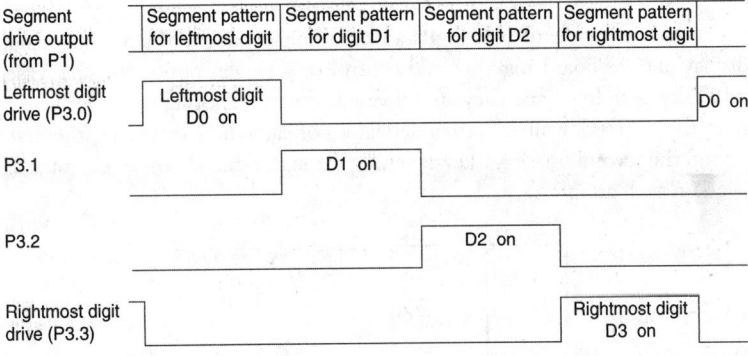

Figure 19.7 Timing diagram for program of Example 19.3

 As sinking larger current is easier than sourcing it, therefore, common cathode type display devices are adopted in more number of designs, than common anode type devices.

C-version

#include <regx51.h>

void Delay50Ms (void);

void main(void)

```
{
        // infinite loop

        while(1)
        {
                P2_0 ^= 0x01;
                Delay50Ms();
        }
}

void Delay50Ms(void)
{
        volatile unsigned int i=6300;
        while(i>0)
        {
                i--;
        }
}
```

19.6 | Multiplexed Keyboard Display Interface

When a system is to have multiple display digits and multiple numbers of keys, the standard practice is to interface both display and keyboard together and control by a single routine. In such cases, the digit-drive lines are also used as key-scan lines, and they are activated one after another (Fig. 19.8), resulting in economy of port pin usage of the microcontroller. During activation of each digit-drive line, the first phase is devoted to display driving and the second phase for key scanning through return lines. In Chapter 25, we will have an elaborate discussion on such an interface.

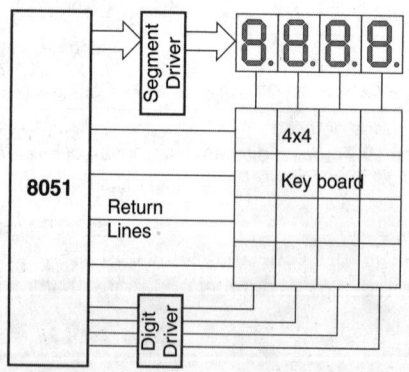

Figure 19.8 Schematic of keyboard display interface

 Intel 8279 is a dedicated device for keyboard display controlling. It has built-in features for key debouncing, display refreshing and many other, like 2-key lock out or n-key rollover mode and so on. All these must be already known to the 8085 or 8086 oriented readers.

SUMMARY

Interfacing of two most widely used display devices, LED and seven-segment display are discussed in this chapter. LEDs are to be interfaced along with a current-limiting resistor in series. It is preferable to drive the LED through logic 0. For a 5 mm red LED generally 10–15 mA of current is allowed. High-intensity LEDs consume less current than this.

Seven-segment displays are common anode or common cathode type and available in various sizes and colours. Generally, multiple numbers of displays are used in cascade form and multiplexing method is used to drive these display digits so that current consumption is substantially less.

POINTS TO REMEMBER

- It is a good practice to place a 330 Ohms resistor between Vcc and anode of any LED, with its cathode connected to port pin.

- Common cathode seven-segment displays are preferable over common anode version as they may easily be interfaced with key-matrix also.

REVIEW QUESTIONS

Evaluate Yourself

1. When viewed from top, the cathode side of a cylindrical LED is
   - (a) circular
   - (b) flat
   - (c) either of these
   - (d) none of these

2. LEDs are capable of displaying
   - (a) one state
   - (b) two states
   - (c) three states
   - (d) none of these

3. LEDs should be used with
   - (a) nothing
   - (b) a resistor in parallel
   - (c) a resistor in series
   - (d) none of these

4. Seven-segment displays are available as
   - (a) common anode
   - (b) common cathode
   - (c) both of these
   - (d) none of these

5. 7447 drivers for seven-segment display accept
   - (a) BCD input
   - (b) hexadecimal input
   - (c) binary input
   - (d) none of these

6. In multiplexed display scheme with seven-segment display,
   - (a) only one digit is driven at a time
   - (b) only one segment is driven at a time
   - (c) only one segment of only one digit is driven at a time
   - (d) none of these

7. Generally, in multiplexed display scheme, all digits are refreshed
   - (a) once in a second
   - (b) 10 times a second
   - (c) 100 times a second
   - (d) none of these

8. To drive a 12-digit multiplexed seven-segment display, the number of port pins for segment drive would be
   - (a) 8
   - (b) 12
   - (c) 16
   - (d) none of these

9. In general, multiplexed seven-segment display scheme is adopted to
   - (a) save space
   - (b) save power
   - (c) brighten the display
   - (d) none of these

10. seven-segment displays are of
    - (a) LED type
    - (b) LCD type
    - (c) both of these
    - (d) none of these

Search for Answers

1. How many varieties of LEDs are commercially available?

2. Why a resistor in series is always necessary to interface any LED?

3. Why a driver is necessary if an LED is to be driven by logic 1?

4. What is the difference between common anode and common cathode type of seven-segment display?

5. Why a digit is to be converted to its display pattern before display?

6. How are seven-segment displays multiplexed?

7. Why a delay routine is necessary for software for refreshing multiplexed display?

8. How interrupts may be used for multiplexing seven-segment display digits?

9. How is economy achieved by interfacing keyboard and display simultaneously?

10. Is it possible to use a decoder, like 74138, as digit driver?

Think and Solve

1. Why LEDs should never be used to replace diodes?

2. Make two lists of electronic devices using LEDs and seven-segment displays.

3. Like seven-segment displays, which types of LEDs are available in common anode and common cathode configurations?

4. Apart from displaying numerical values, can seven-segment displays be used for displaying English alphabets?

5. What would happen if in the Example program 19.1, register R7 is initialized as 00H?

6. In Example program 19.1, does the CPL bit instruction get the previous data from output latch or from input pin?

7. Interface with 8051 one key and one LED. Develop a program so that whenever the key is pressed, the LED would be turned on. The LED may be turned off by releasing the key.

8. In spite of being composed of eight LEDs, why the device is known as seven-segment display?

9. Are the contents of the table SEGTAB of Example 19.2 going to change if the connections between Port 3 pins and the display device inputs are modified?

10. Are the contents of the table SEGTAB of Example 19.2 going to change if the connection between Port 1 pins and the display device inputs is modified?

20 INTERFACING: DAC/ADC

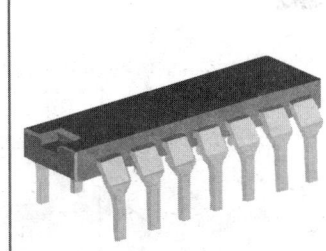

CHAPTER OBJECTIVES

In this chapter, the reader is introduced to interfacing techniques of DAC and ADC with 8051 microcontroller. After completion of the chapter, the reader should be able to understand

- How to interface a DAC with 8051.
- How to interface an ADC with 8051.

20.1 | Introduction

Digital to Analog Converter (DAC) and Analog to Digital Converter (ADC) are widely used in control operations and instrumentation in industries. As their names suggest, a DAC is to convert a digital input to its proportionate analog form. Converting any analog input signal to its proportionate digital form is the job of any ADC. In this chapter, we will discuss how a DAC or ADC can be interfaced with 8051 microcontroller. Because of their wider availability and usage, we will take National 0808DAC and National 0809ADC as our example DAC and ADC.

Audio interfacing is another area where DACs are widely used. Note that DACs can control the volume of the sound. Its tone is controlled by the frequency of the input signal.

20.1.1 | How a DAC Works

Out of several standard methods to convert any digital signal to its proportionate analog form, R-2R ladder network method is most widely used. As shown in Fig. 20.1, each bit of digital input is passed through a set of resistor network proportionate to its weight. The resistors used must be of accurate values. In general, 2R is replaced by a value of 1M and accordingly, R is calculated. Conversion time for this type of DAC is very less, generally a few nanoseconds.

In general, not carbon but metal-film type resistors are used for R-2R ladders for the desired accuracy.

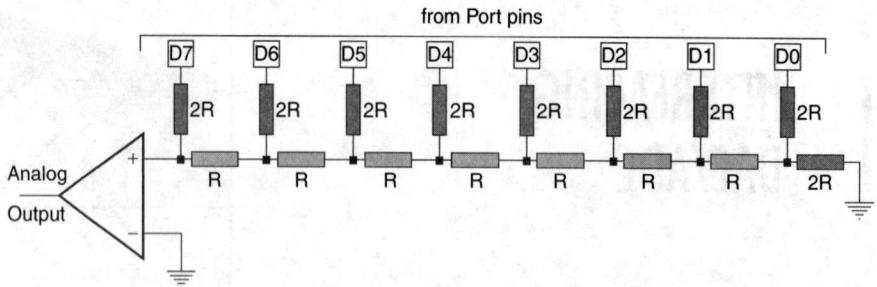

Figure 20.1 Schematic of 8-bit R-2R ladder network DAC

20.1.2 | How an ADC Works

Several methods exist for analog to digital conversion, like: ramp method, successive approximation method, dual slope method and flash conversion method. The basic technique used in ramp method is explained through Fig. 20.2. In this case, the input analog signal is compared with the output of the internal DAC and whenever the output matches, the end-of-conversion signal is generated and data becomes available at the output port. To generate input for the DAC, a register (Counter) is used which keeps on sending digital values generated by continuous increments, starting from 0. Generally, a start conversion signal is necessary to initiate the whole process, and a read signal is necessary to read the data.

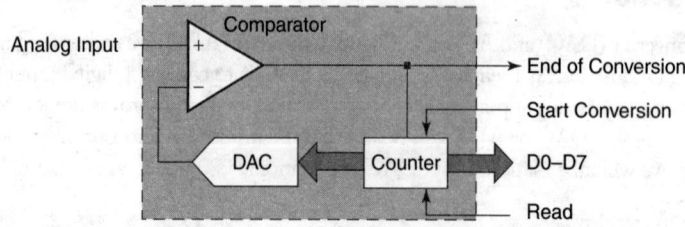

Figure 20.2 Simplified schematic of Ramp type ADC

Such a procedure must depend on sequential logic demanding a clock signal (not shown in Fig. 20.2), which is generally supplied from external source. The whole procedure in this ramp method takes some finite amount of time and not an instantaneous one, like DAC. However, a flash converter type ADC is capable of converting the analog input to its digital form almost instantly, which uses a different technique.

20.1.3 | Sample and Hold

In most of the cases, the conversion from analog to digital takes a finite amount of time and the input analog signal must remain stable during this conversion time. If the signal is of lower frequency then it does not create any noticeable problem. However, for a higher frequency signal, it is essential to maintain the stability of the signal during its conversion. Generally, this is achieved by incorporating a 'Sample and Hold' unit at input, which during its hold state maintains the stability of the input signal through the external capacitor, interfaced with it.

Schematically, a sample and hold unit is illustrated in Fig. 20.3(a) which accepts a digital input as sample and hold signal (hold is active low). Input and output waveforms for different states of sample and hold signal had been shown in Fig. 20.3(b). Note that as long as the hold input is valid (low), the amplitude of the output signal does not change against time in spite of the changes in the input signal. However, during the sample state (high) the output waveform follows the trail of the input waveform.

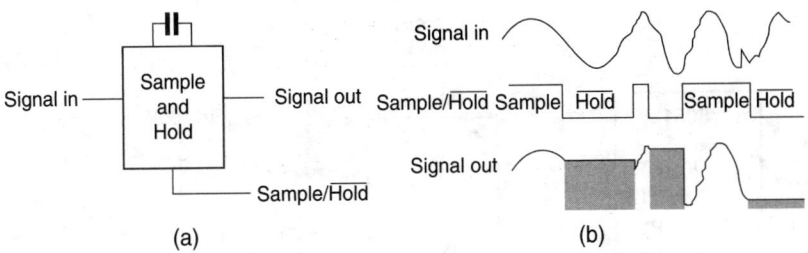

Figure 20.3 Sample and Hold unit: (a) schematic and (b) functional waveforms

 Strictly speaking, a sample and hold unit is unable to hold the input voltage for a long time. The captured value of the voltage decreases slowly. However, most ADCs are fast enough to convert the analog voltage sampled within a reasonable time duration so that the distortion may be neglected for all practical purposes.

20.2 | Interfacing DAC

To interface a DAC with 8051, it is preferable to use any port to provide the digital data input to the DAC. As an example, pin diagram of National 0808DAC is shown in Fig. 20.4 and its interfacing with 8051 is shown in Fig. 20.5.

As we may observe from Fig. 20.4, apart from eight digital input signals, the DAC needs some reference voltage inputs also. Data input lines are designated as A1 (as MSB) through A8 (as LSB). Therefore, at the time of interfacing with Port 1 of 8051, P1.0 is interfaced with A8 and so on (refer to Fig. 20.5). Pin 16 of DAC (COMPENSATION) needs a 0.1 µF disc capacitor between this pin and its Pin 3.

At the output stage, an operational amplifier is necessary to supply sufficient current as indicated in Fig. 20.5. The resistors used should be metal-film resistors of accurate values. It may also be noted that bipolar power supply is necessary in this case as shown in Fig. 20.5.

DACs may be used to generate various types of waveforms such as sine wave, triangular wave, ramps and square waves. Three example software routines based on the interfacing of the circuit of Fig. 20.5 are presented at the end of this chapter. Apart from this, DACs may also be used to drive DC motors and other electrical devices.

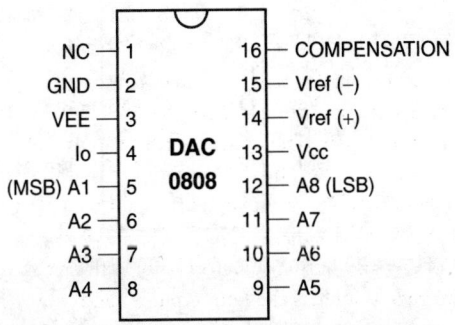

Figure 20.4 Pin diagram of National 0808 DAC

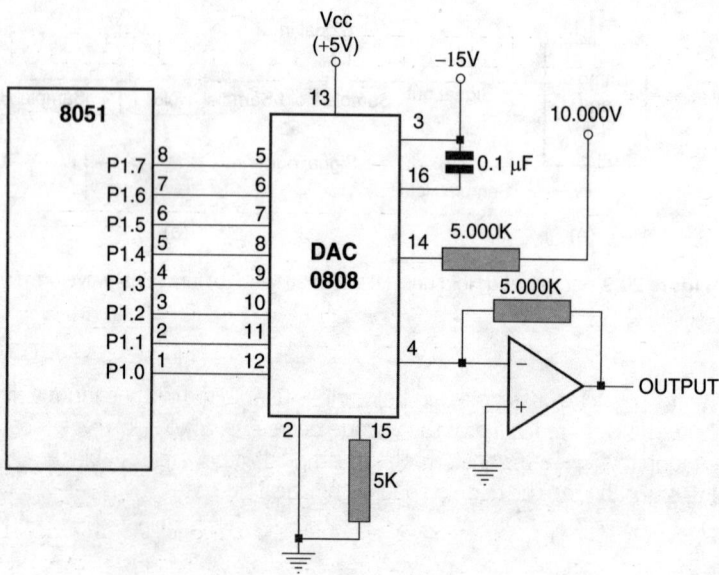

Figure 20.5 Circuit diagram of DAC 0808 interfaced with 8051

20.3 | Interfacing ADC

ADCs may accept single- or multi-channel analog input and the output may be collected either through interrupt or by polling, depending upon the interfacing. Illustrated in Fig. 20.6, ADC 0809 from National semiconductor accepts eight different analog inputs through its eight input channels and converts to the corresponding digital values at a typical rate of 100 microseconds per channel.

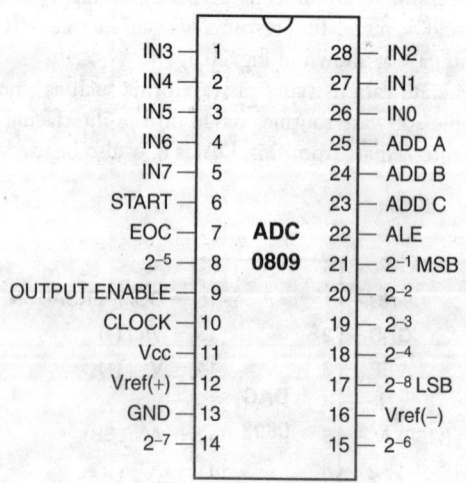

Figure 20.6 Pin diagram of National 0809 ADC

Three address input lines are provided to select any one of eight input channels along with the ALE signal to latch the channel number (refer to Timing Diagram in Fig. 20.7). A start conversion command then

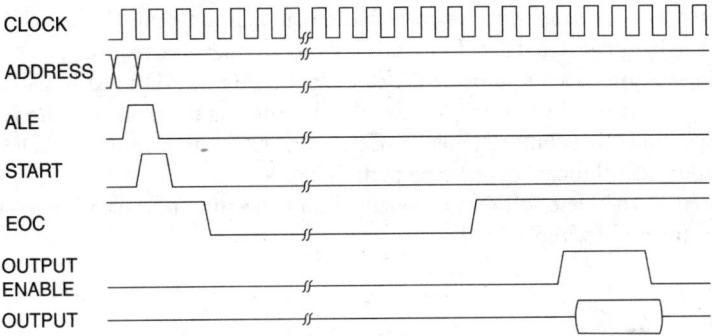

Figure 20.7 Timing diagram of National 0809 ADC

may be applied through the START input. The EOC output remains low throughout the conversion time and goes high after the digital conversion is over and data being available at the output buffer (D0–D7). Generally, the EOC output is interfaced with some interrupt input of the microcontroller. This digital output (D0–D7) may be collected through the output enable (OE) signal and the whole process may then be repeated for another input channel. To convert a fixed channel continuously, the EOC signal may itself be used as the START signal by hardware connections. The conversion time is dependent upon the external clock source necessary at the CLOCK input of the device. The bandwidth is from 10 KHz to 1.28 MHz with a typical value of 640 KHz. This clock may be generated either through a 555 or similar timer or by 8051 itself. For proper functioning, reference voltage inputs are also necessary for the device. A typical interfacing of ADC 0809 with 8051 is shown in Fig. 20.8.

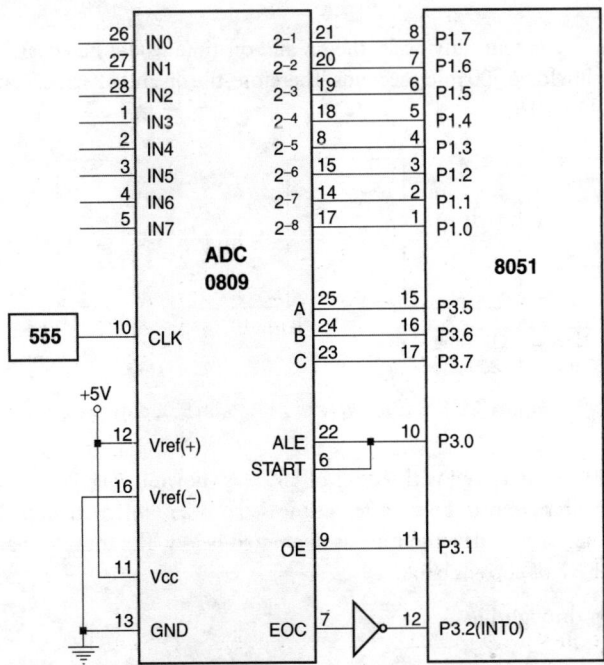

Figure 20.8 Interfacing of 0809 ADC with 8051

This ADC interface is interrupt driven for which external interrupt input INT0 of 8051 was used. Port 1 was used to collect digitally converted data from ADC, and Port 3 signals were used for controlling operation. A separate 555 timer is interfaced to provide the clock input to 0809 ADC. ALE and START inputs are connected together to start the conversion immediately after latching the channel address. It is expected that INT0 is to be programmed as falling-edge sensitive and the EOC output from the ADC is interfaced with INT0 (P3.2) input of 8051 through an inverting buffer, like 7404.

In the next section, complete software is presented for converting the analog signal to its digital value using this interface (refer to Example 20.4).

Note that all signals of 0809 ADC are either active high or falling edge triggered. The reader is suggested to consult data sheets of 0809 ADC for more accurate details about interfacing signals.

20.4 | Solved Examples

Example 20.1

Purpose: Software development for DAC interfacing and understanding some fundamentals of waveform generation.

Problem

Develop a program to generate square waves of 2 KHz frequency with 50 per cent duty cycle using a DAC.

Solution

For a square wave with 50 per cent duty cycle, the on and off time would be equal. To generate 2 KHz frequency, the time period should be 500 microseconds. Therefore, the on and off time would be 250 microseconds in this case as shown in Fig. 20.9.

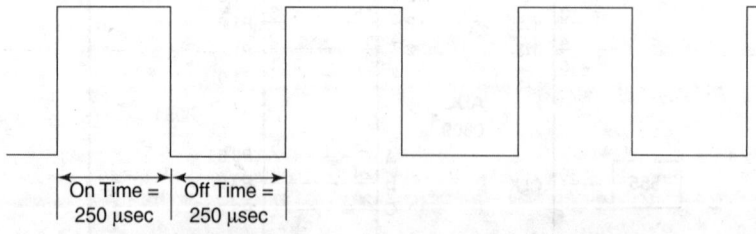

Figure 20.9 Square wave of 2 KHz with 50% duty cycle

Assuming DAC 0808 is interfaced with Port 1 of 8051, as shown in Fig. 20.5, a software may be written to output 00H and FFH alternately to Port 1 after a time delay of 250 microseconds. This delay may either be generated through a Timer or by a delay routine as illustrated below. The routine is an infinite loop and uses accumulator and register R7 of current bank.

; Square wave generation through DAC

```
        SQWAVE:   CLR    A
        SQLOOP:   MOV    P1, A          ; output one state          (1 cy)
```

```
                CPL      A               ; toggle state                      (1 cy)
                NOP                                                          (1 cy)
                MOV      R7, #7AH        ; count for 244 microsecond delay   (1 cy)
      DELAYS:   DJNZ     R7, DELAYS      ; wait                              (2 cy)

                SJMP     SQLOOP          ; continue                          (2 cy)
```

The count value for R7 is calculated as follows. Given in parenthesis against last six instructions of the routine (only those instructions which are repeated) is the execution time in terms of machine cycles necessary. With a 12 MHz crystal, each instruction cycle would consume 1 microsecond. Therefore, excluding the Decrement and Jump if Not Zero (DJNZ) instruction, remaining instructions would consume 6 microseconds. Therefore, DJNZ instruction must consume 244 microseconds. As DJNZ takes 2 microseconds, therefore the count value in R7 should be 244/2 = 122 in decimal. Hexadecimal equivalent of 122 is 7AH, which is the value loaded in R7. Had the no operation (NOP) instruction not been included, R7 would have to be loaded with a value to generate a delay of 245 microseconds, which would have demanded the count value with some fraction and not in integers.

Example 20.2

Purpose: To understand how the DAC output can be controlled through software.

Problem

Develop a program to generate upward ramp wave using a DAC.

Solution

Fig. 20.10 depicts the waveform of upward ramp type signal. It is a linear rise from 0V to the maximum of 5V followed by an immediate drop to 0V. For a downward ramp, it would be a linear decrease from 5V to 0V and then an immediate change to 5V. For saw-tooth type waveform both rise and the fall would be linear without any sudden change.

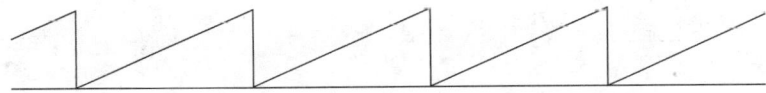

Figure 20.10 Upward ramp-type waveform

Using the DAC interface shown in Fig. 20.5, this waveform may be generated by the following program, assuming the DAC is interfaced with 8051 through Port 1.

```
; Upward ramp wave generation through DAC
      RAMPUP:   CLR      A           ; start with 0V
      UPLOOP:   MOV      P1, A       ; output to DAC
                INC      A           ; overflows to 00H from FFH
                SJMP     UPLOOP      ; continue
```

Note that the DAC interfaced through port P1 is receiving a continuous stream of data from the accumulator at an interval of 4 microseconds (assuming 12 MHz crystal). This is within the bandwidth of the 0808 DAC. After reaching at FFH, the accumulator would overflow for execution of next INC A instruction and would start from 00H again.

Example 20.3

Purpose: To understand how table look-up method may be used to generate complex waveforms.

Problem

Develop a program to generate a sine wave using a DAC.

Solution

To generate a sine wave, the output should be proportionate with ordinates of a sine curve at equal angular intervals. Let us take 15° as the angular interval for which ordinates from 0° to 345° are shown in Table 20.1 and illustrated in Fig. 20.11. For a sine wave, the maximum value or crest would be at 90° and the trough, the lowest value, would be at 270°. For the purpose of scaling, we assume this range as 200 (in decimal) and calculate proportionate hexadecimal values to the nearest integer values. This is presented at the rightmost column of Table 20.1. The third column of the table is obtained from its second column by multiplying it

Table 20.1 Ordinate values in hexadecimal for sine wave generation (Example 20.3)

Angle (in degrees)	Sine value (in decimal)	Scaled equivalent (in decimal)	Hex equivalent
0	0.0	100	64
15	0.2588	126	7E
30	0.5	150	96
45	0.7071	171	AB
60	0.866	187	BB
75	0.9659	197	C5
90	1.000	200	C8
105	0.9659	197	C5
120	0.866	187	BB
135	0.7071	171	AB
150	0.5	150	96
165	0.2588	126	7E
180	0.0	100	64
195	-0.2588	74	4A
210	-0.5	50	32
225	-0.7071	29	1D
240	-0.866	13	0D
255	-0.9659	3	03
270	-1.000	0	00
285	-0.9659	3	03
300	-0.866	13	0D
315	-0.7071	29	1D
330	-0.5	50	32
345	-0.2588	74	4A

with 100d and then adding 100d and finally rounding it off to the nearest integer value. The range of 200d is dependent on the maximum resolution for 8-bit DAC, i.e. 1/255. The software SINEWV is developed with the help of table look-up method, using the table SINTAB. This table SINTAB contains the ordinate values from 0° to 345° at an interval of 15°. We assume that the DAC 0808 is interfaced with 8051 through its port P1 as shown in Fig. 20.5.

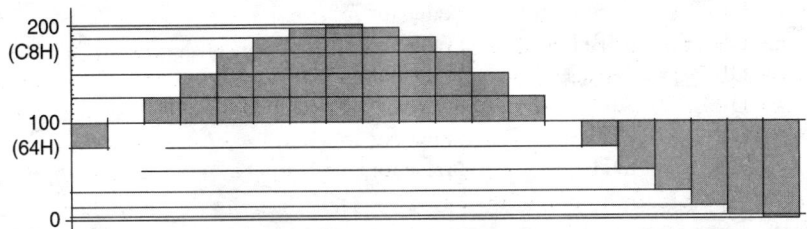

Figure 20.11 Sine wave generated by DAC

Note that the table SINTAB starts with the ordinate value of 345° and proceeds towards 0° in reverse order. This is because the table must be arranged in the way as it would be read by the index values. In this case, the index value is in register R7, which counts from 24 (decimal) to 1. The table SINTAB is not used if R7 is 0. Therefore, the first value for ordinate of 0° is placed at the end of the table SINTAB.

It may also be observed that the routine GETORD does not increment accumulator by one before using the instruction MOVC A, @A+PC, as the table does not use 0 as an index value for table reading.

```
; Sine wave generation through DAC

SINEWV:   MOV    R7, #18H        ; 24d in decimal for ordinates from 0 to 345
SNLOOP:   MOV    A, R7           ; get ordinate number
          ACALL  GETORD          ; get corresponding ordinate through table
          MOV    P1, A           ; output to DAC 0808
          NOP
          NOP                    ; allow some time
          DJNZ   R7, SNLOOP      ; continue for all ordinates
          SJMP   SINEWV          ; repeat all from start, infinite loop

GETORD:   MOVC   A, @A+PC        ; read from table SINTAB
          RET

SINTAB:   DB     4AH             ; value for 345°
          DB     32H             ; value for 330°
          DB     1DH             ; value for 315°
          DB     0DH             ; value for 300°
          DB     03H             ; value for 285°
          DB     00H             ; value for 270°
          DB     03H             ; value for 255°
          DB     0DH             ; value for 240°
          DB     1DH             ; value for 225°
          DB     32H             ; value for 210°
          DB     4AH             ; value for 195°
          DB     64H             ; value for 180°
```

DB	7EH	; value for 165°
DB	96H	; value for 150°
DB	0ABH	; value for 135°
DB	0BBH	; value for 120°
DB	0C5H	; value for 105°
DB	0C8H	; value for 90°
DB	0C5H	; value for 75°
DB	0BBH	; value for 60°
DB	0ABH	; value for 45°
DB	96H	; value for 30°
DB	7EH	; value for 15°
DB	64H	; value for 0°

The output of the software would generate some steps in the profile of sine wave, as shown in figure 20.11. The profile may be smoothened by reducing the interval of ordinates to, say, 5° and calculating more ordinates. The number of ordinates in that case would increase to 72 and would produce a better and well defined sine curve profile.

Example 20.4

Purpose: To understand how interrupt-driven data transfer may be implemented for ADC interfacing.

Problem

Develop a program to convert the analog signal in channel 0 of ADC 0809 and store it in location 30H onwards. The routine should store the value whenever it is called.

Solution

We assume that ADC 0809 is interfaced with 8051 as shown in Fig. 20.8 through Ports 1 and 3. The software would be having two parts: initialization and interrupt service routine for INT0 interrupt, as EOC output from the ADC is interfaced with it. As the program is to convert data as and when called, therefore, the interrupt service routine disables INT0 interrupt before return, which is again enabled by the initialization routine when the data conversion is initiated.

; ADC initialization routine.
; Start conversion for ch. 0 and enable INT0 interrupt as falling edge.

; Analog to digital conversion

```
        INITAD: MOV   R0,#30H    ; initialize pointer for storage area
                MOV   P1,#0FFH   ; make input port to receive data from ADC
                MOV   P3,#00H    ; clear all control lines (outputs)
                SETB  P3.0       ; to latch channel 0 address and start conv.
                NOP
                CLR   P3.0       ; conversion starts
                SETB  TCON.0     ; select falling edge interrupt for INT0
                SETB  IE.0        ; enable INT0 interrupt
                SETB  IE.7        ; enable global interrupt
```

; ISR for INT0 for EOC from ADC.

```
        ORG   0003H

ADC0:   SETB  P3.1      ; read ADC output
        MOV   A, P1     ; read converted data
        MOV   @R0, A    ; save in buffer
        INC   R0        ; next storage area
        CLR   P3.1      ; disable read
        CLR   IE.0      ; disable ADC interrupt
        RETI
```

SUMMARY

DACs and ADCs are widely used in industries for various data acquisition and control operations. In both the cases, some ports are necessary for interfacing. ADCs need external clock source and may be interfaced through interrupt input or polling. As example cases, National 0808 DAC and 0809 ADC pin diagram, signal details and interfacing are explained to generate various types of waveforms and data conversion.

POINTS TO REMEMBER

- To interface any ADC properly, its timing diagram must be thoroughly studied before the software development.

- Unless it is a square-wave generation, 'delay' routines are generally not necessary between successive outputs for waveform generation with a DAC.

REVIEW QUESTIONS

Evaluate Yourself

1. Which of the following techniques is used for DACs?

    (a) SAR
    (b) R-2R network
    (c) Both of these
    (d) None of these

2. Which of the following techniques is used for ADCs?

    (a) SAR
    (b) R-2R network
    (c) Both of these
    (d) None of these

3. Conversion time for DACs are generally in

    (a) milliseconds
    (b) microseconds
    (c) nanoseconds
    (d) none of these

4. For any ADC built around successive approximation technique, which of the following is essential?

    (a) DAC
    (b) Comparator
    (c) Both of these
    (d) None of these

5. Which type of logic is used in ADCs?

    (a) Sequential
    (b) Combinational
    (c) Both of these
    (d) None of these

6. How many 8-bit ports are necessary to interface an 8-bit ADC?

    (a) One
    (b) Two
    (c) Three
    (d) None of these

7. For National 0809 ADC, which of the following signals indicate that converted data is ready and available?

    (a) ALE
    (b) OE
    (c) START
    (d) None of these

8. To convert only one analog channel continuously, which of the following technique may be adopted for 0809 ADC?

    (a) Connect ALE with START
    (b) Connect EOC with START
    (c) Connect OE with START
    (d) None of these

9. What would be the frequency of a square wave if both its on-time and off-time are 25 microseconds?

(a) 200 Hz   (b) 2 KHz

(c) 200 KHz   (d) None of these

10. For Example 20.3, if we add 25 (decimal) with all ordinate values in the Table 20.1, the output would take the shape of a

(a) square wave   (b) triangular wave

(c) ramp   (d) none of these

Search for Answers

1. How many I/O lines are necessary to interface a 12-bit DAC with 8051?

2. How the end of conversion signal is generated for an ADC using successive approximation technique?

3. Why has ADC 0809 a clock input but not DAC 0808?

4. If all three channel select lines (A, B and C) are connected with Vcc, which channel's analog input would be digitized by 0809 ADC?

5. What is the purpose of a Sample and Hold unit? When and where it is to be interfaced?

6. What was the purpose of including the NOP instruction in the SQWAVE program of Example 20.1?

7. How the duty cycle of a square wave may be changed to 75 per cent?

8. What would happen if the instruction INC A is replaced by the instruction DEC A in RAMPUP program of Example 20.2?

9. Why INC A instruction was not incorporated just before the MOVC instruction in subroutine GETORD of Example 20.3?

10. In the software of Example 20.4, why INT0 was disabled in its ISR before RETI instruction?

Think and Solve

1. What value of R would you approve of a R-2R ladder network for a DAC?

2. How does a dual-slope ADC work?

3. How can the flash-converter type ADC convert any analog signal to its digital form almost instantly?

4. How the shape of the sine wave generated by DAC (Example 20.3) may be improved?

5. What do we achieve by connecting START and ALE for ADC 0809?

6. ADC 0809 is capable to convert any analog signal between 0 V and 5 V. How it may be used to convert an analog signal between –1 V and 1 V?

7. What would happen if the converted data is not read till the next conversion-over signal?

8. Which type of waveform is to be input to the DAC of a SAR type ADC if the ADC is to convert an analog signal like a sine wave?

9. Does an ADC take same time to convert analog signals of different magnitudes?

10. In Fig. 20.8 in the interfacing diagram for ADC, Vref (–) and GND are connected together. Is there any special precaution necessary for this connection?

21 INTERFACING: DC MOTOR

CHAPTER OBJECTIVES

In this chapter, the reader is introduced to interfacing of DC motors with the 8051 microcontroller. After completion of the chapter, the reader should be able to understand

- How are DC motors interfaced with 8051 microcontroller.
- How to control speed of revolution of a DC motor using 8051.
- How to control the direction of rotation of a DC motor using 8051.
- How to interface relay and optocoupler for interfacing heavy-duty DC motors.

21.1 | Introduction

Motors are of various types out of which DC motor, Stepper motor and Servomotors are more widely used for different digital-control operations. In this and next two chapters, we will discuss about different interfacing issues related to these motors.

21.1.1 | How DC Motor Works

To recapitulate some fundamental concepts, let us have a quick discussion on the working principle of DC motors. Fleming's left-hand rule is an easier way to remember it. Incidentally, Fleming's right-hand rule is for generators (riGht and Generator both have a 'G': the easier way to remember). As shown in Fig. 21.1, the forefinger indicates the direction of magnetic field (from north to south), the middle finger indicates the direction of current flow and the thumb indicates the direction of force experienced, when current is allowed to pass through a conductor placed perpendicular to any magnetic field.

This force is doubled if the single wire is replaced by a loop as shown in Fig. 21.2. To supply current with proper direction, even during circular rotations, brushes and split-ring commutator are used at the free end of the loop (not shown in Fig. 21.2). In standard DC motors, the number of loops are not one but many, wound around the armature, which increase the torque. Moreover, instead of permanent magnets, electromagnets are used. However, the principle of converting electrical energy to mechanical energy remains unchanged. The rotating part of the motor is designated as *rotor* and the outer static part is known as *stator*. DC motors may also be of a brush-less variety known as brush-less DC motor (BLDM). Two controllable parameters of rotation of any DC motor are its speed and direction. We will now discuss about both of these issues.

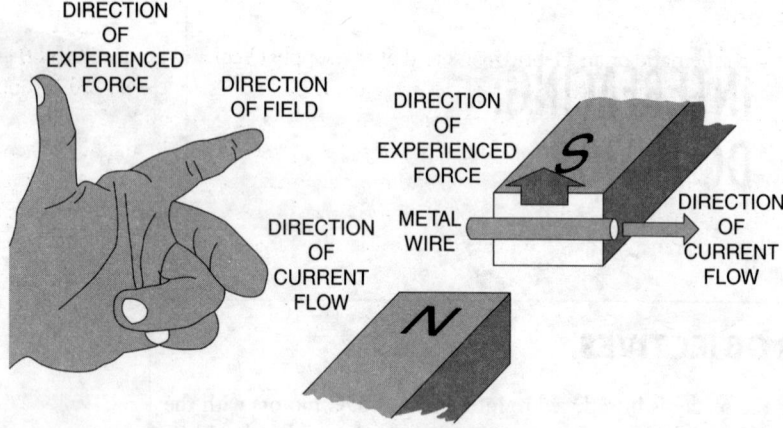

Figure 21.1 Fleming's left-hand rule for motors

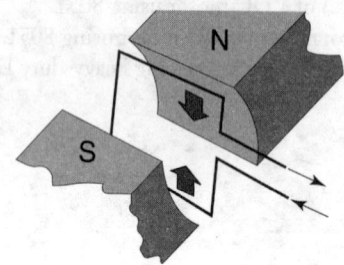

Figure 21.2 How DC motor works

21.2 | Direction Control

Fleming's left-hand rule indicates that the change of direction of current changes the direction of motion. As indicated in Fig. 21.3, the direction of rotation of a DC motor may be reversed by interchanging its input voltage polarity, which reverses the direction of current flow. If we assume that initially it is rotating in clockwise direction [Fig. 21.3(a)], then it would start rotating in anticlockwise direction by the reversed polarity [Fig. 21.3(b)]. When this is to be done manually, a DPDT switch may be used. However, the general practice is to use the H-bridge configuration for interchanging the polarity.

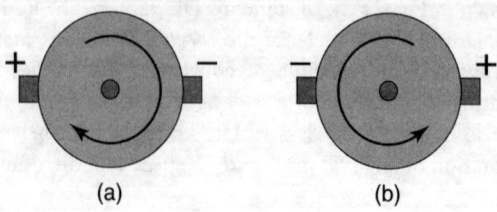

(a)   (b)

Figure 21.3 Direction changing through polarity interchange: (a) clockwise and (b) anticlockwise

21.2.1 | H-Bridge

Fig. 21.4 shows the schematic of an H-bridge. Note that the supply (Vcc) and ground (GND) are available to both pins (inputs) of the motor through four SPST switches. If S1 and S2 are closed and S3 and S4 are open, then the current from Vcc would travel through motor from left to right generating, say, clockwise motion. On the other hand, if S1 and S2 are open and S3 and S4 are closed then the previous input polarity gets interchanged and the current flows from right to left through the motor, generating anticlockwise rotation. It would be interesting for the reader to find out what would happen if S1 and S4 are simultaneously closed. In practice, these switches are replaced by transistors, which may be controlled electronically. This is discussed in the next section.

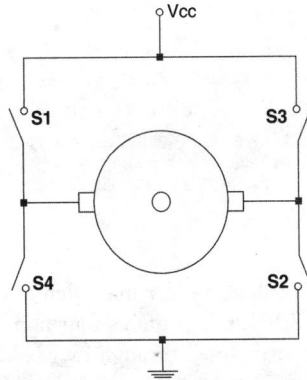

Figure 21.4 Polarity interchanging using H-bridge

21.2.2 | H-Bridge Using Transistors

As shown in Fig. 21.5, four switches of the H-bridge of Fig. 21.4 were replaced by four transistors. Two PNP transistors, TIP127 were used to source the current and two NPN transistors, TIP122 were used to sink the current. Just like the previous case, at any time only two transistors should be on (working) and remaining two should be off. It may be noted that NPN transistors are turned on by logic 1 applied at the base and PNP transistors are turned on by logic 0. Table 21.1 gives the input details of rotation control through these transistors.

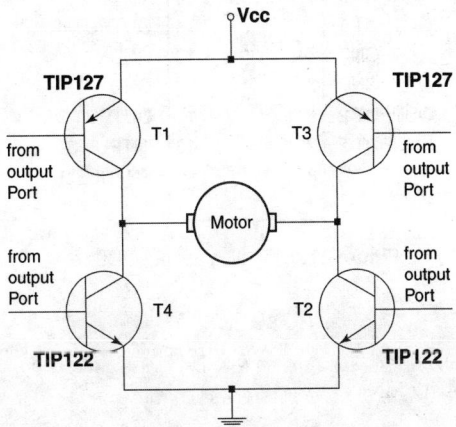

Figure 21.5 H-bridge using transistors

Table 21.1 Direction control through transistors

T1 (PNP)	T2 (NPN)	T3 (PNP)	T4 (NPN)	Remarks
0 (On)	1 (On)	1 (Off)	0 (Off)	Clockwise rotation
1 (Off)	0 (Off)	0 (On)	1 (On)	Anticlockwise rotation
1 (Off)	0 (Off)	1 (Off)	0 (Off)	Stop
0 (On)	0 (Off)	0 (On)	0 (Off)	Stop
0 (On)	X	X	1 (On)	Short circuit
X	1 (On)	0 (On)	X	Short circuit

 In Fig. 21.5, it is indicated that bases of all four transistors are activated by ports of 8051. However, the output from port pins of 8051 may not be strong enough to drive these bases directly and might have to be boosted with another small transistor, like BC547.

21.2.3 | L293D

Dedicated integrated circuits (ICs) are available for driving different types of motors. L293D is one of such ICs, whose pin diagram is shown in Fig. 21.6. It contains four non-inverting drivers placed as two pairs. Each pair may individually be enabled or disabled through its ENABLE input. Each driver is capable of sourcing or sinking 600 milliamperes of current with a peak value of 1.2 amperes. Vs (pin 8) is for motor power supply voltage input, with a minimum limit of 5 V and a maximum of 36 V. Vss (pin 16) is for logic level input voltage with a minimum rating of 5 V.

To drive a DC motor, only one pair of drivers is necessary, which (OUTPUT from L293D) may be connected with poles of the motor directly. Signals marked as INPUT may be driven by port pins of any microcontroller. As there is no heat sink provision, the ground plane in PCB layout to be carefully planned (refer to manufacturer's data sheet).

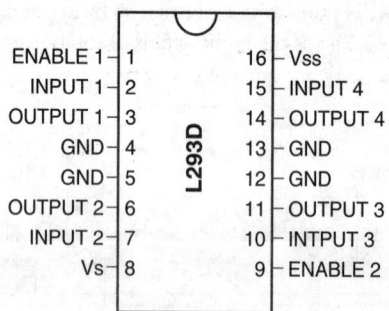

Figure 21.6 Pin diagram of L293D

 Apart from L293D, there are quite a few H-bridge drivers available in the market. Some of them are: L287M, 2295, etc. Interested reader may search in the internet for further details.

21.3 | Speed Control

Speed of rotation of any DC motor may be controlled either by changing its input current or by input voltage. A schematic to implement this may be a variable resistance connected in series with the motor as shown in Fig. 21.7. Note that such a scheme changes both input voltage and input current simultaneously.

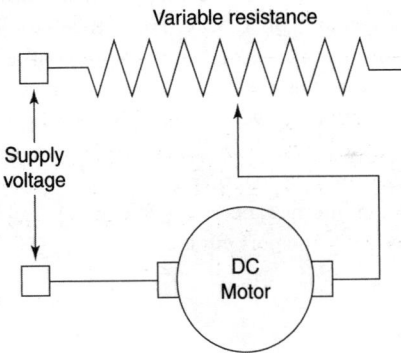

Figure 21.7 Schematic of speed control by varying input voltage/current

21.3.1 | Through DAC

Digital to Analog Converters (DACs) may also be used to control the input voltage of a DC motor. The technique to change the output voltage of any DAC had been discussed in Chapter 20.

21.3.2 | Changing Duty Cycle of Square Wave

Pulse-Width Modulation or PWM is another widely adopted technique to control the input voltage (and, therefore, the speed) of a DC motor. As shown in Fig. 21.8, variation of duty cycle of a square wave changes the value of average output voltage, and the inertial force of the motor keeps it rotating during the off stage of the square wave. Increasing the duty cycle increases the speed of rotation and vice versa. In this case, 'duty cycle' means the on-time with respect to the time taken by one complete on-off cycle.

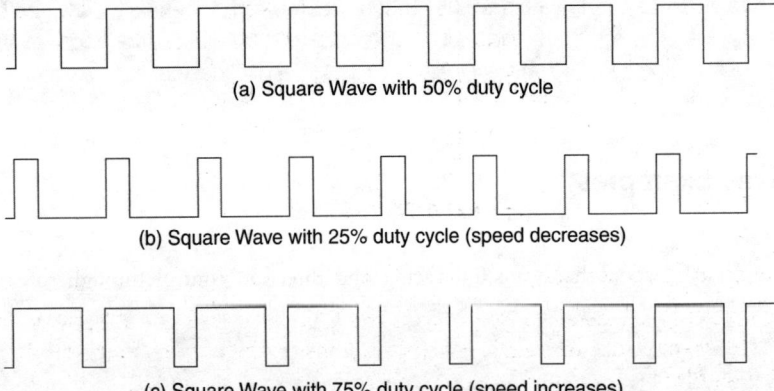

(a) Square Wave with 50% duty cycle

(b) Square Wave with 25% duty cycle (speed decreases)

(c) Square Wave with 75% duty cycle (speed increases)

Figure 21.8 Speed control through duty cycle

21.4 | Relay and Optocoupler

Whatever we have discussed so far was related to small-sized DC motor interfacing. However, to control larger DC motors, which needs larger amount of voltage and current or to operate AC, the normal transistors or L293D would not be suitable. In such conditions we should use a relay. A relay is nothing but an electronic switch, which turns on or off very high voltage, high current devices, driven by digital level signals. In early days, they used to contain mechanical moving parts. However, nowadays Solid State Relay (SSR) is available to replace them.

Fig. 21.9(a) illustrates the schematic of a SSR. Note the built-in optocoupler (also known as optoisolator) at its right side. This optocoupler is necessary to protect the digital circuit from high-voltage transients produced during operation of the relay. We can prepare our own optocoupler by using a LED and a LDR. At the left side of the schematic, we find two MOSFETs connected back to back. Although a single MOSFET is sufficient to handle DC; however, for AC operations two MOSFETs are necessary. Source pins of these two MOSFETs are tied together.

Fig. 21.9(b) shows the application interface, necessary to drive an AC load (a motor or a ceiling fan) by 8051 through a SSR. To activate the relay, the port pin is to output logic 0.

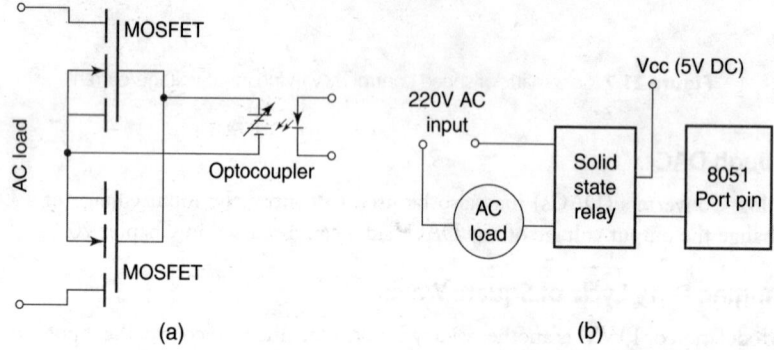

Figure 21.9 (a) Schematic of a SSR with built-in optocoupler and (b) interfacing with 8051 control AC load

As relays are generally driven by 220V AC, therefore, special precaution must be taken to interface a relay with any 8051 based circuit, which is generally powered by 5V DC. It is preferable for a beginner not to attempt any circuit fabrication incorporating a relay. Some experience is essential to construct such an interfacing.

21.5 | Solved Examples

Example 21.1

Purpose: To understand about hardware interfacing and direction control through software for a DC motor.

Problem

Prepare a suitable hardware interface for a DC motor and develop a software routine to change its direction of rotation.

Solution

Using two port pins of 8051, P1.0 and P1.1, a hardware interface for a 12 V DC motor through L293D is presented in Fig. 21.10. To use another DC motor of some other voltage, say 9 V, the Vs input (pin 8) of L293D is to be changed to 9 V DC. Note the presence of a pair of pull-up resistors of 10K each at the Port 1 output. Half of the L293D was disabled by grounding its ENABLE 2 input. Furthermore, its unused input pins 10 and 15 are also connected with system ground. The output through two port pins must be complementary to run the motor. That is, if P1.0 is high, then P1.1 must be low and vice versa. A software routine (DCDIR) to run the DC motor in one direction for about a second and then reversing its direction of rotation for another second is given below. Note that the program is an infinite loop.

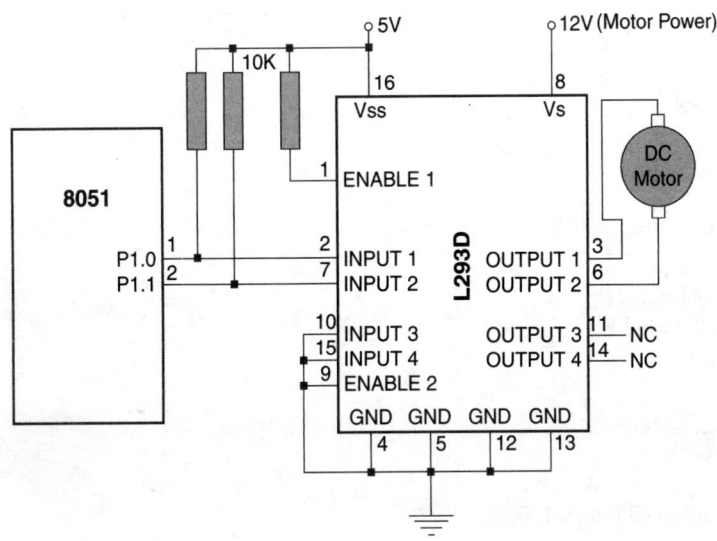

Figure 21.10 Interfacing L293D with 8051 for direction control

; Direction control of a DC motor

```
        DCDIR:    MOV     P1, #0FEH      ; start motor in one direction
        DCLOOP:   ACALL   DELAY          ; wait for about a second
                  XRL     P1, #0FFH      ; change direction by toggling all bits of P1
                  SJMP    DCLOOP         ; continue, infinite loop

        DELAY:    MOV     R7, #08H
        DELAY1:   MOV     R6, #00H
        DELAY2:   MOV     R5, #00H
        DELAY3:   DJNZ    R5, DELAY3
                  DJNZ    R6, DELAY2
                  DJNZ    R7, DELAY1
                  RET
```

To increase or decrease the run time for either direction, the immediate value loaded in R7 in routine DELAY may be adjusted accordingly. Note the usage of XRL instruction to change the direction.

 Note how the XRL instruction was used to toggle the output of port P1.

C-version

```c
#include <regx51.h>

void Delay1Sec(void);

void main(void)
{
        P1 = 0xFE;

        // infinite loop

        while(1)
        {
                Delay1Sec();
                P1 ^= 0xFF;
        }
}

void Delay1Sec(void)
{
        volatile unsigned long i=12000;
        while(i>0)
        {
                i--;
        }
}
```

Example 21.2

Purpose: To understand how the speed of a DC motor may be controlled through software using PWM.

Problem

Develop a subroutine to run a DC motor interfaced with 8051 at alternately faster and slower speeds.

Solution

Assuming the same hardware interface as shown in Fig. 21.10, the speed may be controlled by changing the duty cycle of a square wave output from P1. We interchange the duty cycle between 75 per cent and 25 per cent as shown in Fig. 21.11, the duration of each being ~1 second.

```asm
; Speed control of a DC motor by PWM

        DCFAST:    MOV    R7, #0FFH        ; count for faster speed duration
        FLOOP:     MOV    P1, #0FEH        ; start motor in one direction
                   ACALL  DELAY            ; on-time, faster speed
```

```
              ACALL   DELAY
              ACALL   DELAY         ; 75% duty cycle
              MOV     P1, #0FCH     ; stop the motor
              ACALL   DELAY         ; off-time, faster speed
              DJNZ    R7, FLOOP     ; continue faster speed

DCSLOW:       MOV     R7, #0FFH     ; count for slower speed duration
SLOOP:        MOV     P1, #0FEH     ; switch on for slower speed
              ACALL   DELAY         ; on-time, slower speed
              MOV     P1, #0FCH     ; stop motor
              ACALL   DELAY         ; off-time, slower speed
              ACALL   DELAY
              ACALL   DELAY         ; 25% duty cycle
              DJNZ    R7, SLOOP     ; continue slower speed
              SJMP    DCFAST        ; repeat entire process

; Delay subroutine

DELAY:        MOV     R6, #14H      ; 20 in decimal
DELAY1:       MOV     R5, #0FAH     ; 250 in decimal
DELAY2:       DJNZ    R5, DELAY2    ; generates 500 microseconds delay
              DJNZ    R6, DELAY1    ; 10 milliseconds delay
              RET
```

Note that the motor is started with higher speed to overcome the initial inertia. If the motor does not turn, count value (14H) in R5 of DELAY may be suitably increased. For 75 per cent duty cycle, DELAY is called thrice during on-time while for 25 per cent duty cycle, DELAY is called only once during on-time. The proportion of on-time and off-time fixes the duty cycle.

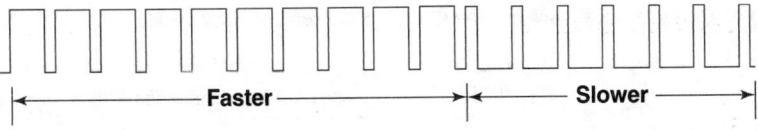

|← ———————— Faster ————————→|← ———— Slower ————→|

Figure 21.11 PWM waveform for speed changing

 A saw-tooth wave form from any DAC may also be used to generate the same effect of the motor speed variations.

Example 21.3

Purpose: To understand how the acceleration of a DC motor may be controlled through software using PWM.

Problem

Develop a routine so that a DC motor accelerates after starting.

Solution

To accelerate a DC motor, its duty cycle to be increased in uniform steps. A schematic representation of the waveform is shown in Fig. 21.12. To implement a noticeable acceleration, duration of each step may be maintained for about a second. The duty cycle may be increased from 30 per cent to 90 per cent in steps of 10 per cent. As indicated in Fig. 21.12, there should be a starting pulse (on-time) of 50 milliseconds duration to overcome the initial inertia. The software is presented below with the same hardware as shown in Fig. 21.10.

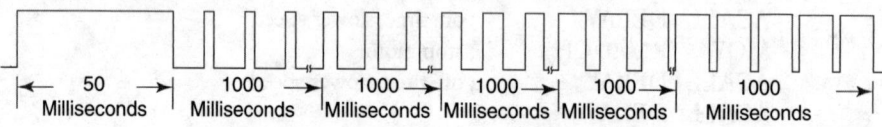

| ← 50 → | 1000 → | 1000 → | 1000→ | 1000→ | 1000 → |
| Milliseconds | Milliseconds | Milliseconds | Milliseconds | Milliseconds | Milliseconds |

Figure 21.12 Waveform to implement acceleration

```
; Acceleration control of a DC motor
; Use register bank #0
; R1 = counter for seven steps from 30% to 90% duty cycle
; R2 = controls delay count starting from 140 and decremented at steps of 20.
; R3 = variable delay count. To be loaded direct from R2 or 200–R2
; R4 = counts 10 cycles of 100 milliseconds each
; R5 = inner delay counter (250–0)
; R6 = outer delay counter, loaded from R3 before start of delay counting
```

```
; Following three instructions generates 50 milliseconds on-time to overcome initial inertia.

        DCACCN:   MOV     P1.#0FDH        ; start motor
                  MOV     R3, #64H        ; 100 in decimal
                  ACALL   VDELAY          ; 50 milliseconds on-time
```

```
; Following two instructions initializes for seven steps of increment duty cycle.

                  MOV     R2, #8CH        ; 140 in decimal for 70 milliseconds delay
                  MOV     R1, #07H        ; seven steps from 30% to 90% duty cycle
```

```
; Loop for incrementing duty cycle starts from here. Each time the loop passes through this point, the speed
; is incremented by 10%.

        ACCNXT:   MOV     R4, #0AH        ; 10 cycles of 100 milliseconds each
```

```
; Start of one cycle of off-time followed by on-time.

        DUTY:     CPL     P1.0            ; stops motor
                  MOV     R3, 02H         ; load R3 from R2 of bank #0
                  ACALL   VDELAY          ; off-time 70 milliseconds
                  CPL     P1.0            ; starts motor
                  MOV     A, #C8H         ; 200 in decimal
                  CLR     C
                  SUBB    A, R2
                  MOV     R3, A           ; R3 = 200 – R2 = on-time
                  ACALL   VDELAY          ; on-time
                  DJNZ    R4, DUTY        ; continue duty cycle for 1 second
```

; Process R2 by subtracting 20 from it for next duty cycle.

MOV	A, R2	; decrease off-time duration for next phase
CLR	C	
SUBB	A, #14H	; 20d
MOV	R2, A	; R2 = R2 – 20d
DJNZ	R1, ACCNXT	; next step of acceleration
RET		; motor runs at full speed

; Variable time delay routine. Variable input from R3 of bank #0.

VDELAY:	MOV	R6, 03H	; load R6 by R3 of bank #0
DELAY1:	MOV	R5, #0FAH	; 250 in decimal
DELAY2:	DJNZ	R5, DELAY2	; generates 500 microseconds delay
	DJNZ	R6, DELAY1	; steps of 20 milliseconds delay
	RET		

 A sine wave form generated by any DAC is also capable of producing acceleration for any DC motor.

SUMMARY

Two major controlling parameters of DC motors, namely, direction control and speed control may be achieved by interfacing through transistor-controlled H-bridge or ICs like L293D. Interchanging the input polarity of any DC motor reverses its direction of rotation. Input voltage change by varying the duty cycle of a square wave, known as PWM, is a widely used technique to control the speed of any DC motor.

POINTS TO REMEMBER

- If a DAC is used to drive a DC motor, then the current requirement of the DC motor and the output current limit of the DAC must be checked.

- Larger pulse width is necessary at start to overcome the initial inertia of any DC motor if it is driven by PWM.

REVIEW QUESTIONS

Evaluate Yourself

1. In Fleming's left-hand rule, the middle finger indicates the direction of

   (a) magnetic field (b) motion

   (c) current flow (d) none of these

2. `Fleming's right-hand rule is for

   (a) generator (b) AC motor

   (c) armature winding (d) none of these

3. The magnetic field is always directed as

   (a) from North to South

   (b) from South to North

   (c) either of these

   (d) none of these

4. To reverse the direction of rotation of any DC motor, we have to

   (a) interchange the magnetic field of the stator

   (b) reverse direction of current

(c) switch to AC input from DC

(d) none of these

5. To make any H-bridge operational (refer to Fig. 21.4), how many switches must remain closed at the same time?

(a) One   (b) Two

(c) Three   (d) None of these

6. If logic 0 is applied at the base of any PNP transistor, it would be

(a) tri-stated   (b) turned off

(c) turned on   (d) none of these

7. Using only one L293D, what is the maximum number of DC motors that could be controlled for direction of rotation?

(a) One   (b) Two

(c) Three   (d) None of these

8. The duty cycle of any square wave signal is its

(a) time period   (b) on-time

(c) off-time   (d) none of these

9. The Vs input of L293D should have the voltage for

(a) motor drive   (b) logic level

(c) double of motor drive (d) none of these

10. If the duty cycle of a square wave to drive a DC motor is changed from 30 per cent to 50 per cent, the speed of the motor would

(a) increase   (b) decrease

(c) remain constant   (d) none of these

Search for Answers

1. What is Fleming's left-hand rule?

2. What is meant by stator and rotor?

3. What is known as BLDM?

4. In how many ways the speed of any DC motor may be controlled?

5. What are the advantages of PWM over other methods?

6. How the present direction of rotation of a DC motor may be reversed?

7. What is an H-bridge? Where are they used for?

8. What purpose does L293D serve?

9. What are input voltage ranges of L293D?

10. How transistors may be used to drive DC motors and control its direction?

Think and Solve

1. What magnetic polarities the following coils windings [Figs 21.13(a) and (b)] would be generating?

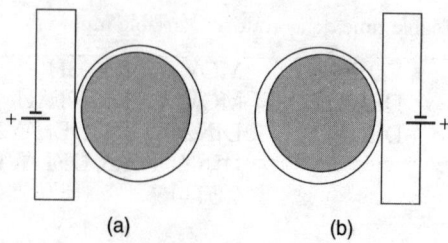

(a)   (b)

2. What is the role of a 'delay' routine in direction control and speed control of any DC motor?

3. Calculate the duty cycle if the on-time is 10 milliseconds and off-time is 90 milliseconds. How would it vary if off-time is decreased to 1 millisecond, while the on-time remains unchanged?

4. Draw the timing diagram of signals from both P1.0 and P1.1, covering direction change of the DC motor. Start the diagram assuming any direction.

5. What is the minimum voltage of any DC motor that may be controlled by L293D?

6. Can only one L293D be capable of controlling four DC motors? Here the controlling is only in terms of starting and stopping.

7. What are the advantages and disadvantages of L239D over a H-bridge prepared from transistors?

8. Assuming that port 1.2 and 1.3 are connected with two keys, K1 and K2, respectively, and all other interfacing of Fig. 21.9 remaining same, develop a program to start the motor when K1 is pressed and stop it when K2 is pressed.

9. Develop a routine to decrease the speed of a DC motor in several steps and finally stop it. Assume that the same L293D interfacing as shown in Fig. 21.9 is used.

10. Modify the circuit of Fig. 21.9, so that it can control two DC motors. Use additional port pins P1.6 and P1.7 for it.

22 INTERFACING: STEPPER MOTOR

CHAPTER OBJECTIVES

In this chapter, the reader is introduced to interfacing of stepper motors with 8051 microcontrollers. After completion of the chapter, the reader should be able to understand

- How do stepper motors work.
- How to design hardware interface for stepper motors.
- How to control speed and direction of rotation of stepper motors.

22.1 | Introduction

While DC motors start rotating immediately after receiving power through its terminals, stepper motors do not rotate in that way. Rotation of stepper motor needs digital pulses, and for every such pulse it would rotate through a fixed angle, generally 1.8° and stop thereafter. Therefore, before discussing the interfacing and controlling techniques, we should have a brief discussion on the working mechanism of the stepper motors.

22.2 | How Stepper Motor Functions

As mentioned above, generally a stepper motor would rotate through 1.8° at application of each pulse. These pulses are generated through a microcontroller to switch on or off the electromagnets (or coils) inside the stepper motor. As a normal practice, the number of these electromagnets is four. Furthermore, this 1.8° rotation is known as full step as there is also a provision of half-step rotation through 0.9°. We will initially investigate the full-step rotation followed by half-step rotation.

22.2.1 | Full-Step Rotation

Fig. 22.1 gives a schematic of fundamental mechanism of a stepper motor. In this case, the stator contains four electromagnets marked as 1 through 4. The rotor is represented by an iron arrow. These electromagnets are activated only one at a time, sequentially. To start with, electromagnet number 1 is activated and the rotor arrow points to it [Fig. 22.1(a)]. Next is the turn of electromagnet number 2, which is turned on after 1 is switched off. This action forces the rotor to rotate towards electromagnet number 2 as shown in Fig. 22.1(b). Note that in this example configuration the rotor would rotate through 90°.

Switching off of electromagnet number 2 and switching on of number 3 occurs next as shown in Fig. 22.1(c) making the rotor to turn further through another step of 90°. Finally, Fig. 22.1(d) shows another rotation of the rotor (arrow) by turning on electromagnet number 4 after switching off electromagnet

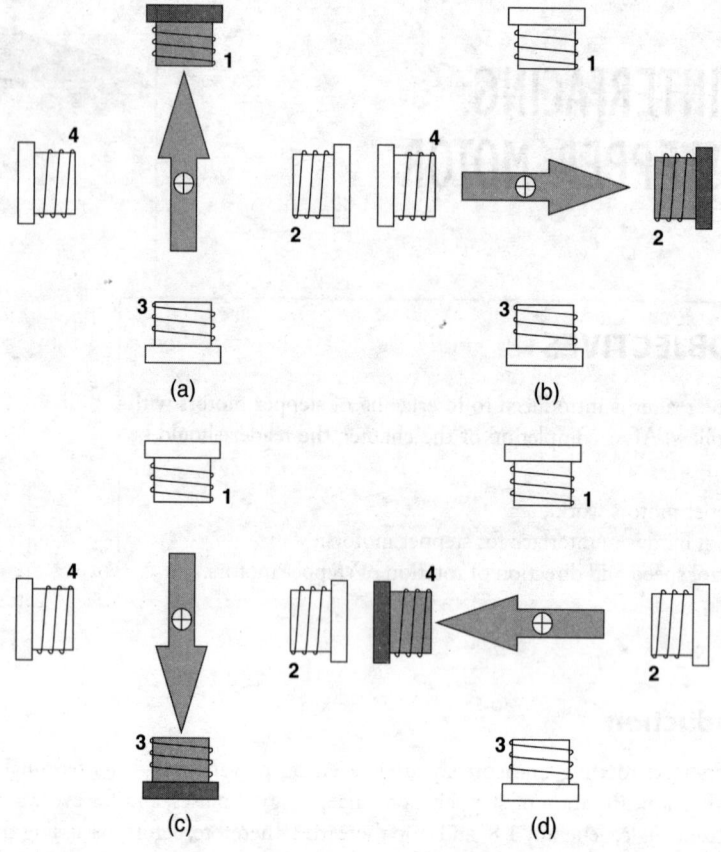

Figure 22.1 Rotation of the stepper motors

number 3. The whole sequence is repeated after this, and the rotor keeps on turning. The reader may note that at any situation the rotation may be terminated not by switching off the electric current but by stopping to send any new pulse. This helps in holding the load by maintaining active torque, which is not possible for any DC motor. However, the step angle for this example would be 90°, which is not a standard step angle. Step angle means the angle through which the rotor rotates after receiving a fresh pulse. The widely accepted step angle is 1.8° and to explain how that is implemented, we take another example for discussions.

22.2.2 | How Step Angle is Controlled

Let us assume that the iron arrow-shaped rotor of Fig. 22.1 is replaced by a five-teeth iron rotor, as shown in Fig. 22.2, although the number and relative positions of electromagnets remain unchanged. In this case, the angle between any two adjacent teeth of the rotor would be 72° as shown in Fig. 22.2(a). Note that five teeth of the rotor are marked as R1 through R5.

Assuming Fig. 22.2(a) as the starting position where coil 1 is active and rotor tooth R1 is attracted by it, next sequence is depicted in Fig. 22.2(b), where coil 2 is activated which attracts rotor tooth marked as R2. Note that coil 2 would not attract R3, placed at an angle of 54°, with greater force than that exerted on R2, resting at an angular distance of only 18°, as R2 is closer than R3. This induces 18° rotation in the rotor. To induce another 18° rotation, coil 3 is to be switched on shown in Fig. 22.2(c). In this case, coil 3 attracts rotor tooth-marked R3, the nearest one.

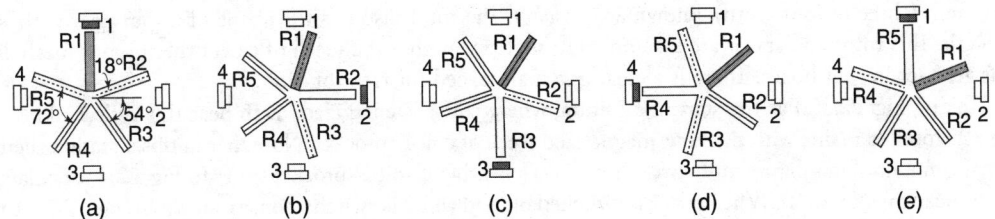

Figure 22.2 Generation of 18° step angle

The clockwise rotation of the rotor would continue in this way by activating coil 4 next and attracting rotor tooth marked R4 [Fig. 22.2(d)]. Finally, Fig. 22.2(e) illustrates how coil 1 is turned on to attract rotor tooth marked R5 and thus completing 72° rotation measured from start.

This example highlights how the change of the shape of rotor helps in bringing down the step angle without any additional electromagnets. The next example illustrates how the step angle of 3.6° is achieved with four electromagnets, with some minor modifications of both stator and rotor.

22.2.3 | Generation of 3.6° Step Angle

Fig. 22.3 shows the sequence of step angles, each one of 3.6°. Note that there are some changes both in stator electromagnets and the rotor iron. Each stator electromagnet has an outer shape in form of

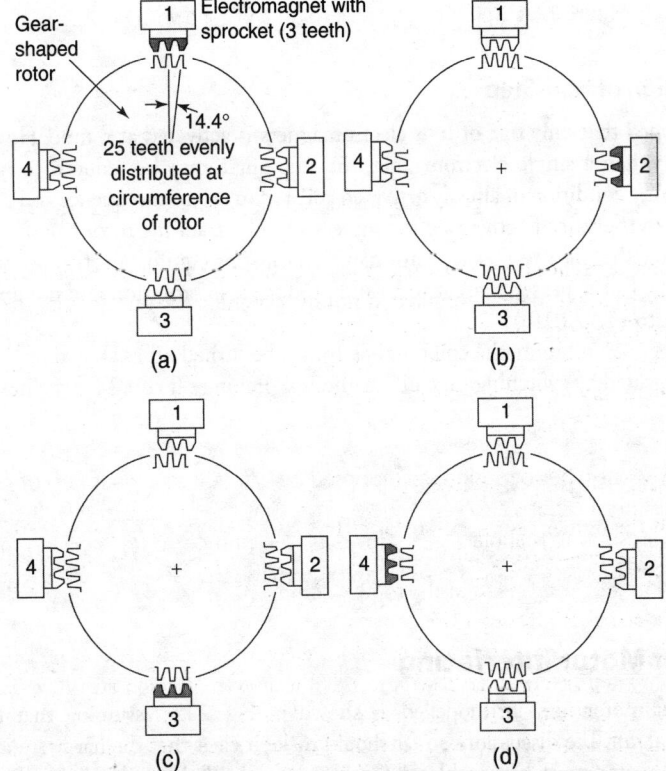

Figure 22.3 Generation of 3.6° step angle

gear teeth (three to four teeth), known as *sprocket*. The rotor also has the shape of a spur gear with, say, 25 teeth. This rotor and stator are assembled in such a way that the teeth of each other do not mesh, like spur gear set in gear boxes. There is a clear gap maintained for free rotation.

Taking Fig. 22.3(a) as the starting position, where coil 1 is energized, teeth near this coil would try to align themselves in line with the now-magnetized teeth of coil 1 sprocket. Note that in this situation there is a small amount of misalignment between rotor teeth nearby of coil 2 sprocket (refer to Fig. 22.4 for enlarged details near coils 1 and 2). When coil 1 is switched off and coil 2 is switched on, as shown in Fig. 22.3(b), the rotor rotates clockwise 3.6° and its teeth are properly aligned with now-active electromagnetic sprocket of coil number 2.

This process is repeated further as shown in Figs 22.3(c) and (d), each time generating a rotation of 3.6°. When coil 1 is turned on next, the rotor completes 14.4° rotation from the illustrated starting position of 22.3(a), which is the angle subtended by one tooth of rotor. It would be clear to the reader now that how by changing the geometry of the rotor, step angle may be controlled.

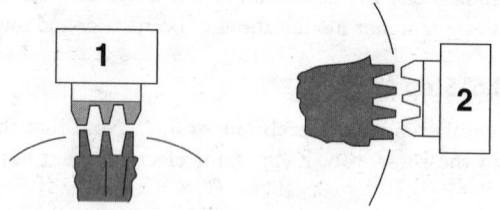

Figure 22.4 Details of rotor near coils 1 and 2 shown in Fig. 22.3(a)

22.2.4 | Generation of Half-Step

So far we have assumed that only one of four electromagnets is activated at a time. However, by alternately activating a pair and then a single electromagnet, the step angle may be reduced to its half. Referring to Fig. 22.5, with a starting condition of already activated coil 1, if in the next sequence coil 2 is activated without turning off coil 1, then the pair of active electromagnets would attract the rotor simult-aneously, with equal force. This would force the rotor to take a position, which must be equidistant from both electromagnets as shown in Fig. 22.5(b). In the present case, this would produce a 45° rotation and not a 90° one, as it was in previous cases [refer to Fig. 22.1(b)].

To induce another 45° rotation, the coil number 1 must be turned off as shown in Fig. 22.5(c). The same technique may be repeated by switching on coil 3 without switching off coil 2 for the next 45° rotation.

As various types of designs are adopted by different manufacturers of stepper motors, their specifications also vary widely. In general, the full step and half-step rotations are uniform for all manufacturers. However, the number of terminal connections varies. Before using any stepper motor, go through the data sheets of the manufacturer.

22.3 | Stepper Motor Interfacing

Electrically, a stepper motor may be modelled as shown in Fig. 22.6, assuming that there are only four electromagnets at its stator. The discussions so far should make it clear that the hardware interfacing necessary to activate a stepper motor must be capable of turning on and off the coils of the electromagnets around it. Generally, these electromagnetic coils are in pair with a common lead for both as shown in Fig. 22.6.

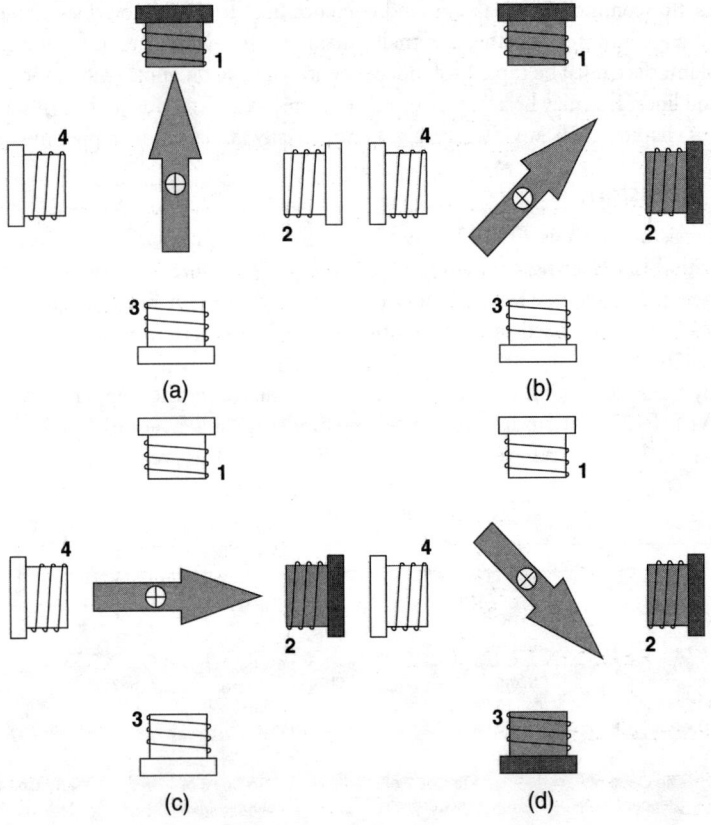

Figure 22.5 Mechanism of half-step generation

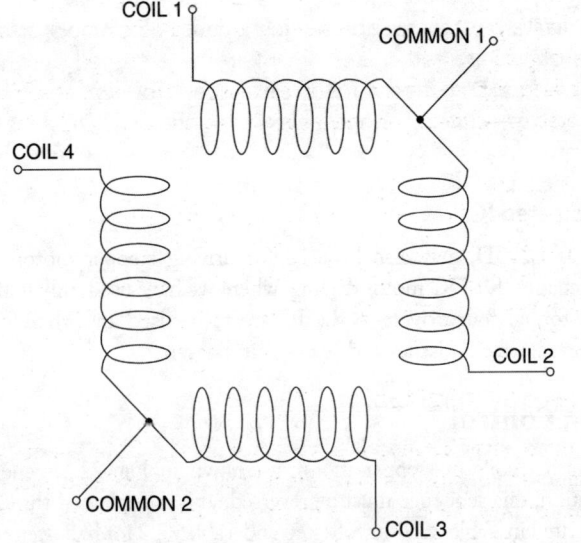

Figure 22.6 Electrical model of a stepper motor

In majority of cases, this common line is the ground reference line. Two common lines, namely, COMMON 1 and COMMON 2 are connected together externally, making a five-wire interface connector for the stepper motor. The external interface must be capable of independently turning on or off each of these four coils through these five connection lines. This may be achieved by using discreet transistors, as we have used for our DC motor drive in the previous chapter. Dedicated ICs are also commercially available for stepper motor driving.

22.3.1 | Using Transistors

Medium power transistors, such as TIP122, may be used for driving coils of small-sized stepper motors. A simple interface with TIP122 transistors is presented in Fig. 22.7. Note that the outputs from emitters of transistors are connected in series with 10 Ohms *10 Watt* resistances for limiting the current flow. Pull-up resistors of 1K (1/4 Watt) were used to drive the base of the transistors from the open collector output of inverting buffers (7416).

To activate any transistor, which would, in turn, activate one coil of the stepper motor, logic 0 from port pin is to be used. As TIP122 contains internal protecting diodes for spikes, additional diodes were not used in the circuit. Example 22.1 may be referred for further discussions on this area.

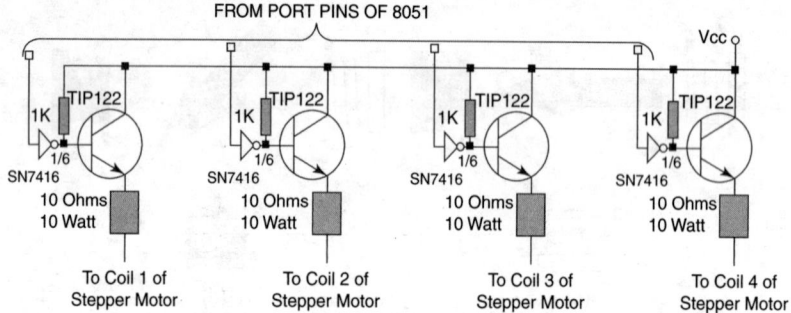

Figure 22.7 Transistor interface for stepper motor drive

 In general any motor requires a higher torque at the time of starting from rest. Once it starts rotating and gathers its momentum, the torque requirement reduces. This is valid specially for DC motors and to some extent, also for stepper motors. Note that we are to supply higher current to generate higher torque.

22.3.2 | Using Dedicated ICs

Dedicated ICs, such as L293D, may also be used for driving stepper motors. L293D has already been introduced in the last chapter for DC motor driving, where we have used only half of it to drive a DC motor. To drive a stepper motor, all four drivers in the IC are to be used and, therefore, enabled. Example 22.2 presents a discussion on interfacing issues and the circuit diagram.

22.4 | Direction Control

The sequence of activating coils of stepper motor, as shown in Fig. 22.1, generates a clockwise rotation. For anticlockwise rotation, this sequence must be reversed. This is explained through the timing diagrams of Fig. 22.8 and also presented in Table 22.1 for full-step and Table 22.2 for half-step rotations. Fig. 22.8(a) shows the sequence for a clockwise rotation, while Fig. 22.8(b) indicates for an anticlockwise one.

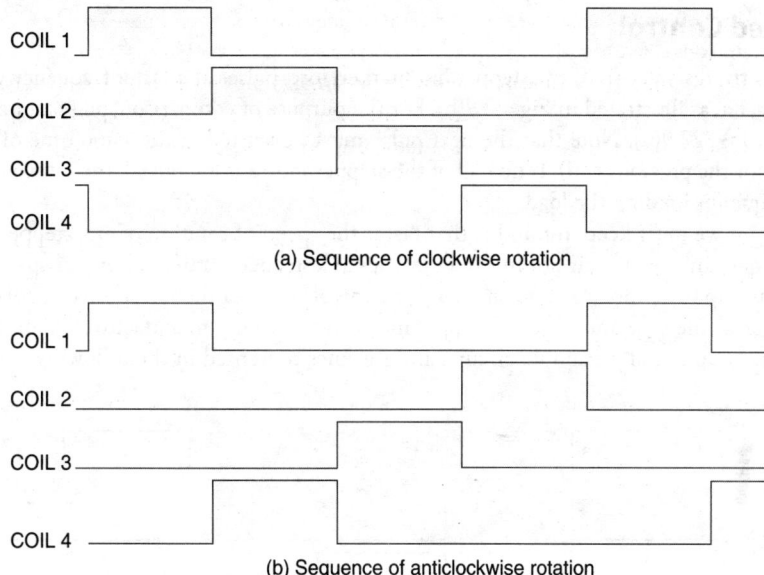

(a) Sequence of clockwise rotation

(b) Sequence of anticlockwise rotation

Figure 22.8 Timing diagram for (a) clockwise and (b) anticlockwise rotations

Table 22.1 Signal sequence for full-step rotation (1.8°)

Coil 1 (A)	Coil 2 (B)	Coil 3 (C)	Coil 4 (D)
1	0	0	0
0	1	0	0
0	0	1	0
0	0	0	1

Table 22.2 Signal sequence for half-step rotation (0.9°)

Coil 1 (A)	Coil 2 (B)	Coil 3 (C)	Coil 4 (D)
1	0	0	0
1	1	0	0
0	1	0	0
0	1	1	0
0	0	1	0
0	0	1	1
0	0	0	1
1	0	0	1

22.5 | Speed Control

As every pulse activates one coil of the stepper motor, therefore, pulses at a faster frequency would run the stepper motor faster, as illustrated in Fig. 22.9(b). For the purpose of comparison, pulses for a slower speed are presented in Fig. 22.9(a). Note that the next pulse must be started at the same time of finishing the previous pulse (for the previous coil). If no coil of the stepper motor is energized, then it would not be able to exert any torque for holding the load.

Another point we must keep in mind is that faster the speed of rotation of any stepper motor, lower would be its torque. This relation is not linear and varies at a steeper gradient at higher speed. As example cases, we now take up two problems, one on direction control and the other on speed control of the stepper motors. However, as the type and make of stepper motors vary widely, manufacturer's datasheets must be consulted before adopting any example circuits and routines presented in the following section.

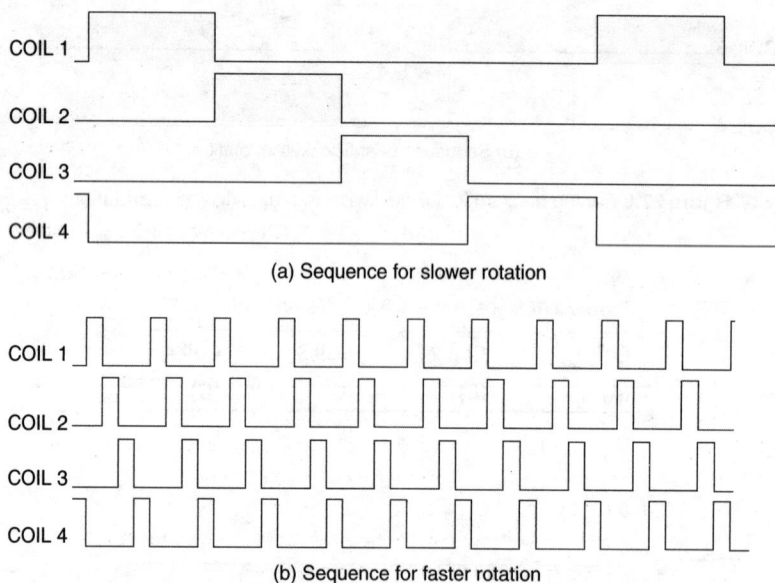

Figure 22.9 Timing diagrams for (a) slower and (b) faster rotation

For every stepper motor, there exists a limit of maximum rotational speed. Even if pulses of higher frequencies are applied beyond this limit, the speed to rotation of the motor would not increase any further. Refer manufacturer's data sheets for these parameters.

22.6 | Solved Examples

Example 22.1

Purpose: To understand hardware interfacing using transistors and software development for direction control of a stepper motor.

Problem

Develop the hardware interface and then develop a subroutine to rotate the interfaced stepper motor as per the content of bit 0 of bit-addressable area of internal RAM of 8051. If the bit is set, then the motor should rotate clockwise, and if the bit is cleared then the motor should rotate anticlockwise.

Solution

Fig. 22.10 shows the necessary hardware for interfacing the stepper motor with 8051, using transistors, controlled by P1.0 through P1.3. We assume that the interfacing is such, so that the sequence presented in Table 22.1 is applicable. The following routine is an infinite loop and continuously checks the bit of address 00H and as per its content, rotates the motor. An interrupt routine would be necessary to read any key and toggle the content of this bit thereafter.

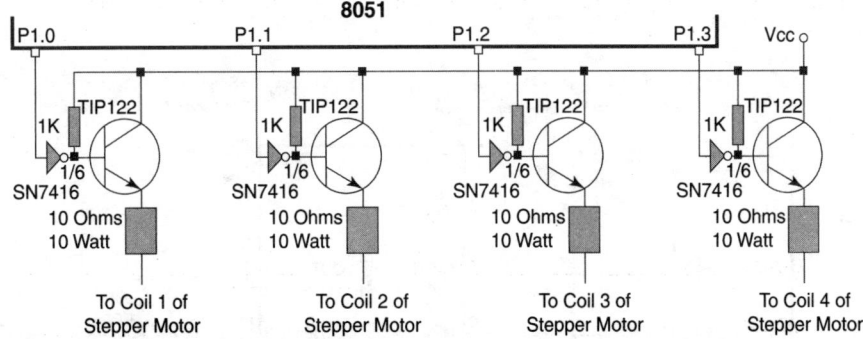

Figure 22.10 Hardware interfacing of stepper motor with 8051 using transistors

```
; Direction control of a stepper motor
    STPDIR: MOV   A, #0EEH      ; initial pattern, check manufacturer's data
            MOV   R7, #04H      ; counter for four pulses
    CHKBIT: JNB   00H, ACLKRT   ; go for anticlockwise rotation
    CLKRT:  MOV   P1, A         ; activate one coil of stepper
            ACALL DELAY         ; wait for some time
            RL    A             ; pattern for next coil, clockwise rotation
            DJNZ  R7, CLKRT     ; continue for four coils
            SJMP  STPDIR        ; check again, etc.
    ACLKRT: MOV   P1, A         ; activate one coil
            ACALL DELAY         ; wait for some time
            RR    A             ; anticlockwise rotation
            DJNZ  R7, ACLKRT    ; continue for four coils
            SJMP  STPDIR        ; start fresh and check bit

    DELAY:  MOV   R6, #0FFH     ; depends upon manufacturer's datasheet
    DELY:   DJNZ  R6, DELY
            RET
```

 It is safer to plan this type of interfacing using port 1 of 8051, which does not offer any alternate functions. Port 3 offers external interrupt input and other signals. Port 0 and 2 offers address and data signals.

Example 22.2

Purpose: Hardware interfacing of a stepper motor using L293D and its speed control by software.

Problem

Prepare a hardware interface using L293D for driving a stepper motor and then develop a routine to run the stepper motor in any one direction at alternately slower and faster speeds.

Solution

Fig. 22.11 shows the hardware interfacing necessary for driving the stepper motor using L293D. Note that both ENABLE 1 and ENABLE 2 were activated by one 10K pull-up resistor. Port pins P1.0 through P1.3 were used for controlling the stepper motor. The following routine is developed to run the stepper motor at alternately fast and slow speeds. Note how the SWAP instruction was used for that purpose.

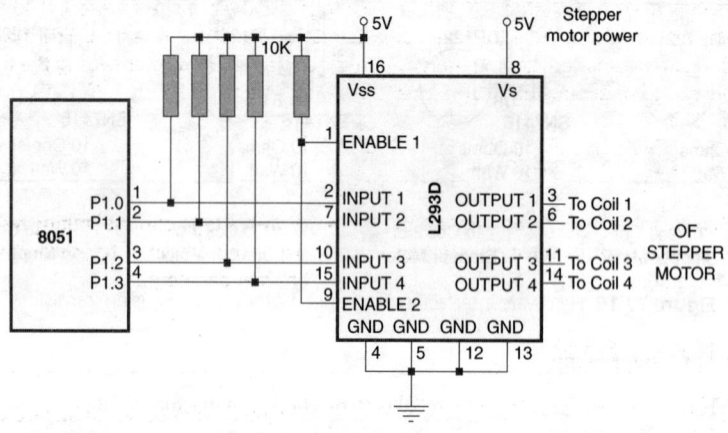

Figure 22.11 Hardware interfacing of stepper motor with 8051 using L293D

```
; Speed control of a stepper motor
; Use register bank #0

STPSLO: MOV   R5, #0FH      ; initial pattern
SPEED:  MOV   R4, #0FFH     ; duration count
START:  MOV   R7, #04H      ; coil number count
        MOV   A, #11H       ; initial pattern
LOOP:   MOV   P1, A         ; activate one coil
        RL    A             ; clockwise rotation
        ACALL DELAY         ; wait
        DJNZ  R7, LOOP       ; complete four coils
        DJNZ  R4, START      ; complete 255 rotations
        MOV   A, R5         ; interchange speed variable
        SWAP  A
        MOV   R5, A         ; to vary the speed
        SJMP  SPEED         ; continue with altered speed

DELAY:  MOV   R6, 05H       ; load R6 from R5 of bank #0
DELAY1: DJNZ  R6, DELAY1
        RET
```

 L293D needs special PCB layout for heat dissipation. It is not possible to attach any heat sink directly with this device. Avoid touching this IC when it is operational, in a working circuit.

SUMMARY

Stepper motors are also called digital motors and used for precision angular movements, like in printer or disk-drive head. Each time it receives a pulse, it rotates through 1.8°, known as full step. The step angle may also be changed to 0.9° known as half step. This is achieved by the placement of four electromagnets at the stator and using a gear-shaped rotor, generally of iron. For full-step rotation, at any time only one electromagnet at stator is energized, forcing the rotor teeth to rotate closest to its sprocket with a few similar teeth just like the rotor. For half-step rotation, alternately a single coil and a pair are energized. To change the direction of rotation, the sequence of coil-activation must be reversed. The speed of rotation might be controlled by changing the frequency of the coil-activation.

POINTS TO REMEMBER

- Manufacturer's data sheet must be considered to finalize the output pattern for driving a stepper motor in any direction.

- High-Wattage current-limiting resistors are essential at the output, if transistors are used to drive any stepper motor.

REVIEW QUESTIONS

Evaluate Yourself

1. Generally, the full-step angle of a stepper motor is taken as
   - (a) 3.6°
   - (b) 1.8°
   - (c) 0.9°
   - (d) none of these

2. Generally, the half-step angle of a stepper motor is taken as
   - (a) 3.6°
   - (b) 1.8°
   - (c) 0.9°
   - (d) none of these

3. The accuracy of rotation of the stepper motor is due to the presence of
   - (a) teeth in rotor
   - (b) electromagnets at stator
   - (c) both of these
   - (d) none of these

4. The unit with electromagnetic teeth at stator is called
   - (a) sprocket
   - (b) spur gear
   - (c) steps
   - (d) none of these

5. The number of pulses to be received by a stepper motor to rotate 360° with full step would be
   - (a) 360
   - (b) 100
   - (c) 60
   - (d) none of these

6. To reverse the direction of rotation of a stepper motor, we are to send the reversed
   - (a) voltage polarity
   - (b) pulse sequence
   - (c) current direction
   - (d) none of these

7. The speed of rotation of stepper motor depends upon the
   - (a) frequency of input pulses
   - (b) time delay between two pulses
   - (c) input voltage of pulses
   - (d) none of these

8. If an imaginary stepper motor has four evenly spaced electromagnets as stator and five teeth in its rotor, then the angle of rotation for full step would be
   - (a) 72°
   - (b) 18°
   - (c) 9°
   - (d) none of these

9. In the circuit of Fig. 22.7, 10-Watt resistors (10 Ohms) are necessary to drive

   (a) more current　　(b) higher voltage

   (c) both of these　　(d) none of these

10. If the speed of rotation is increased, the torque offered by a stepper motor would

   (a) increase　　(b) remain constant

   (c) decrease　　(d) none of these

Search for Answers

1. What are the major differences between DC and stepper motors?

2. How can a stepper motor be electrically modelled?

3. How does a stepper motor rotate?

4. What controls the direction of rotation of a stepper motor?

5. What would happen if the sequence of coil-activation is not maintained?

6. What is the difference between the full-step and half-step?

7. How can the speed of rotation of the stepper motor be controlled?

8. What is the advantage of making the rotor of a stepper motor shaped like a spur gear with multiple teeth?

9. What hardware interface is necessary to control a stepper motor by 8051?

10. Is it possible to use PNP transistors, such as TIP127, to drive the stepper motors?

Think and Solve

1. Why stepper motors are called as 'digital motors'?

2. With respect to the DC motors of the same torque, are stepper motors cheaper in their prices?

3. In general, pulses for all four coils of a stepper motor are of uniform duration. What would happen if the pulse for coil 1 is of double the duration than the pulses for other three coils?

4. What is meant by bi-polar stepper motor?

5. If a DC motor is not rotating, it is unable to offer any torque. How does a stepper motor offer some torque even if it is not rotating?

6. Is a stepper motor capable of offering some torque even if it is switched off?

7. How does the shape of the rotor influence the performance of a stepper motor?

8. Draw the timing diagram for half-step rotation (any direction) of a stepper motor.

9. What are the application areas of the stepper motor?

10. What is meant by *micro-stepping* of the stepper motors?

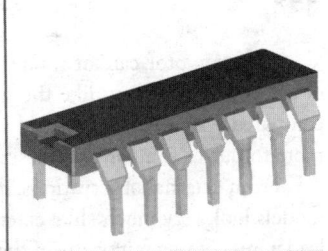

23 INTERFACING: SERVOMOTOR

CHAPTER OBJECTIVES

In this chapter, the reader is introduced to interfacing of servomotors with 8051 microcontrollers. After completion of the chapter, the reader should be able to understand

- How the servomotor is different from a stepper motor or a DC motor.
- How does the servomotor works.
- How to interface a servomotor with 8051.

23.1 | Introduction

Imagine that you are fabricating a robot and presently designing its elbow joint. This joint would be offering only one degree of freedom and rotates through a maximum angle of 180° (Fig. 23.1). If a DC motor or a stepper motor is fitted to perform this duty, it would be underutilized, as they are designed to rotate continuously. In this case, a servomotor is an ideal solution. The same is valid for the knee-joint design (of the robot) also.

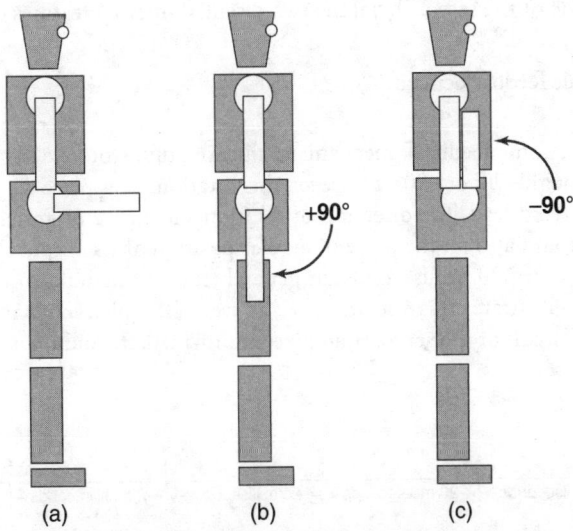

Figure 23.1 Function of elbow-joint of a robot

A servomotor cannot rotate like DC or stepper motors. Generally, they are used where a small amount of rotation is necessary, like the radar of a model aircraft or ventilating fins of air-conditioning machines. As compared with DC motors and stepper motors we have discussed so far, servomotors are capable of offering more torque as compared to its size. However, it is also more expensive than the previous two types of motors.

From external observations, it may initially be difficult to distinguish these three types of motors, as many models look very much alike externally. However, a simpler method to distinguish is to look at the number of lead wires coming out of any of these motors. As illustrated in Fig. 23.2, a DC motor would offer two lead wires, while a stepper motor would show (generally) five (or six) lead wires. A servomotor must be having three lead wires, which is the best way to distinguish it from the previous two. Out of these three lead wires, one is for Vcc, another one is for ground and the third one is for sending pulses. A servomotor does not start rotating if its Vcc and GND leads are connected with power supply, which are the well-known characteristics of any DC motor. Apart from this power supply, accurate electrical pulses are also essential to run any servomotor.

To discuss several aspects of servomotors, we assume some of its parameters, the accuracy and ranges of which should be checked form the relevant datasheets of the manufacturers as they vary considerably.

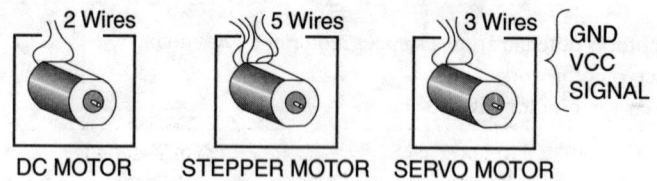

Figure 23.2 How to distinguish DC, stepper and servomotors

23.2 | How a Servomotor Works

Inside a DC or stepper motor, we would see no electrical circuit or PCB. This is not true for a servomotor. Inside, it contains a tiny electronic circuit for pulse sensing and generating a feedback mechanism to rotate its armature (rotor). Apart from these and the standard motor, we would also find a compact gearbox to enhance the output torque of the servo. Therefore, two essential internal features of any servomotor are as follows.

(i)   Internal electronic feedback circuit.
(ii) Built-in gearbox.

In general, due to its electronic feedback mechanism, the armature (rotor) of a servomotor rotates for a fixed angle, typically through +90° or −90°. However, this rotation takes place when the servo receives the *digital pulse* through its third lead. The direction of rotation, clockwise or anticlockwise, depends upon the duration of the pulse and also on the present angular position of its output shaft. This pulse must be applied (or supplied) at every 20 milliseconds, in general (refer to manufacturer's datasheets for exact timing details). Fig. 23.3 illustrates this situation, i.e., at every 20 milliseconds interval, a digital pulse is expected by the internal circuit of the servo. If no pulse is provided, it would not rotate but would simply keep on waiting for a pulse.

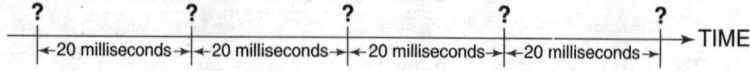

Figure 23.3 A servomotor expects a digital pulse at every 20 milliseconds interval
(refer to manufacturer's datasheet for exact specification)

If the duration of this pulse is more than 1.5 milliseconds [Fig. 23.4(a)], then the armature of the servomotor would rotate clockwise through 90° (or +90°). If the duration of this pulse is less than 1.5 milliseconds [Fig. 23.4(b)], then the armature of the servomotor would rotate −90°. If the duration of this pulse is exactly 1.5 milliseconds [Fig. 23.4(c)], then the shaft of the servomotor would rotate to its neutral position, *which may be either clockwise 90° or anticlockwise 90°, depending upon the present position of its output shaft. However, it would always rotate towards the neutral position of the servo.* Here, we have assumed a hypothetical threshold value of 1.5 milliseconds, which should be verified with manufacturer's datasheet. However, generally this threshold value is close to 1.5 milliseconds. Therefore, we may assume that a pulse of 2 milliseconds would rotate the armature of the servomotor clockwise through a right angle. On the other hand, a pulse of 1 millisecond would rotate it anticlockwise for a right angle.

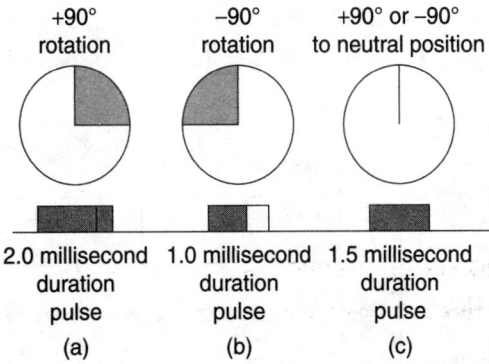

+90° rotation	−90° rotation	+90° or −90° to neutral position
2.0 millisecond duration pulse	1.0 millisecond duration pulse	1.5 millisecond duration pulse
(a)	(b)	(c)

Figure 23.4 (a) Forward, (b) reverse and (c) neutral position-generating pulses (refer to manufacturer's datasheet for exact threshold limit)

These pulses may be just standard TTL signal available directly from the port pin of any microcontroller, like 8051. No extra interfacing, such as buffer or transistor, is necessary to boost it up. However, it is recommended that the port pint output should use an external pull-up resistor of 1K.

The next point we should keep in mind that the servomotor has an internal gear mechanism attached with its motor. This internal gear set not only increases the output torque but also decreases the angular rotational speed. Hence, the 90° rotation (or −90° rotation as the case might be) of the armature would be proportionately *decreased* through the gearbox at the external output shaft of the servomotor. That means after sending a pulse of, say, 2 milliseconds duration, we may externally observe that the output shaft of the servo has rotated only 4.5° because of the gear-set's reduction ratio being 20:1. Total external rotation of a servo is generally limited to 210°, depending upon the manufacturer's datasheet. These extreme limits are controlled by existence of mechanical studs in the gear set as well as internal hardware circuit and its feedback mechanism.

 Note that electrical pulses supplied to a servo are intended for its internal motor's rotation. However, the final or output rotation (at the output shaft) is obtained through the gearbox attached with the motor. The 90° rotation for every pulse is applicable for the internal motor, not for external output shaft. The output shaft rotates through a smaller angle for each pulse, depending upon the reduction ratio of the gearbox.

23.3 | Inside a Servo

To understand the behaviour of a servo, we now take a closer look at its internal mechanism, which is schematically depicted through Fig. 23.5. Here in this schematic we can observe the existence of an electronic circuit (in PCB) at bottom. We can also observe the existence of a stud attached with outer spur gear to limit the rotation of outer shaft. This is achieved by the block (there are two such blocks) attached with the outer casing. Note that at neutral position the stud would remain at equal distance from two blocks.

Whatever we have discussed so far contains following important concepts.

(i)   The rotor of servo rotates through 90° only when it receives a pulse.
(ii)   This rotation may be clockwise or anticlockwise, depending upon the pulse width and present position of the outer shaft with respect to its neutral position.
(iii)   The angular rotation of rotor is reduced at outer shaft because of built-in gearbox.
(iv)   The stud in outer gear does not allow the outer shaft to rotate beyond the block in either direction.

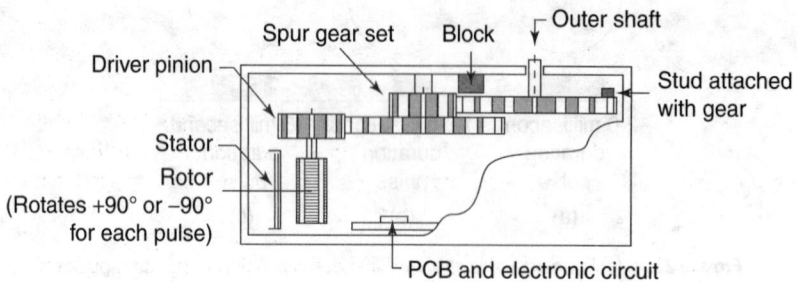

Figure 23.5 Schematic internal details of a servomotor

If the internal studs, meant to block the rotation of the gear, are broken, a servo would also function like a stepper motor and would rotate continuously, rather than a fixed angle.

23.4 | Controlling Direction of Rotation

We have already discussed that the duration of pulse finalizes the direction of rotation of a servo. It was also mentioned that with respect to its threshold value, which we have assumed as 1.5 milliseconds, the pulses might be of three categories, namely, less than, equal to or more than 1.5 milliseconds. We now elaborate these three conditions.

23.4.1 | For Pulses of 2 Milliseconds

In this case, the rotation of the outer shaft would be clockwise, for a small angle, for each pulse. As long as the outer stud fixed at the gear is not obstructed, the outer shaft would rotate against each pulse. However, this angular motion would be stopped if the stud is blocked. Further, similar pulses of 2 milliseconds may damage the servo and are not recommended.

23.4.2 | For Pulses of 1 Millisecond

In this case, the rotation of the outer shaft would be anticlockwise, for a small angle, for each pulse. As long as the outer stud fixed at the gear is not obstructed, the outer shaft would rotate against each pulse. However, this angular motion would be stopped if the stud is blocked. Further, similar pulses of 1 millisecond may damage the servo and are not recommended.

23.4.3 | For Pulses of 1.5 Milliseconds

This pulse is always used to ensure the neutral position of the outer shaft of the servo. There may be three different reactions against this type of pulses depending upon the present position of the outer shaft. The outer shaft may be exactly at the neutral position or away from it, either clockwise or anticlockwise. Reactions for all these three cases are discussed below.

Outer Shaft at Neutral Position: If the outer shaft is exactly at its neutral position then 1.5 milliseconds pulses would not be having any effect on the servo and the outer shaft would not rotate at all. However, in such a condition, repeated 1.5 milliseconds pulses would not damage the servo.

Outer Shaft Away from Neutral Position Clockwise: If the outer shaft is located clockwise away from its neutral position then for every 1.5 milliseconds pulse, the outer shaft would rotate *anticlockwise*, till it reaches the neutral position. Note that the number of 1.5 milliseconds pulses to reach the neutral position depends directly on the initial angular offset of the outer shaft from its neutral position. Once the outer shaft reaches its neutral position, further 1.5 milliseconds pulses would not be having any effect on the servo, and the outer shaft would remain at its neutral position. However, the servo would not be damaged for multiple extra 1.5 milliseconds pulses in such a situation.

Outer Shaft Away from Neutral Position Anticlockwise: If the outer shaft is located anticlockwise away from its neutral position then for every 1.5 milliseconds pulse, the outer shaft would rotate *clockwise*, till it reaches the neutral position. Note that the number of 1.5 milliseconds pulses to reach the neutral position depends directly on the initial angular offset of the outer shaft from its neutral position. Once the outer shaft reaches its neutral position, further 1.5 milliseconds pulses would not be having any effect on the servo, and the outer shaft would remain at its neutral position. However, the servo would not be damaged for multiple extra 1.5 milliseconds pulses in such a situation.

 When pulses of threshold value (1.5 millisecond for our example case) are applied, the direction of rotation at the output shaft would depend on the present position of the shaft. If the shaft is already in its neutral position then it would not rotate at all. Otherwise it may rotate either clockwise or anticlockwise.

23.5 | Interfacing a Servomotor

Hardware interfacing of a servomotor is generally a simple affair as shown in Fig. 23.6. Through the terminal wires attached with the servo, its power and reference ground are to be connected. Generally, the input range varies from 4.8V DC to 6V DC. Manufacturer's datasheet must be checked for the correct voltage and current ratings.

The signal input of the servo, its third terminal, may be directly connected with any port pin of 8051. Although the signal terminal accepts standard TTL voltage levels and corresponding current, it is always safe to use a 1K pull-up resistor as shown in Fig. 23.6.

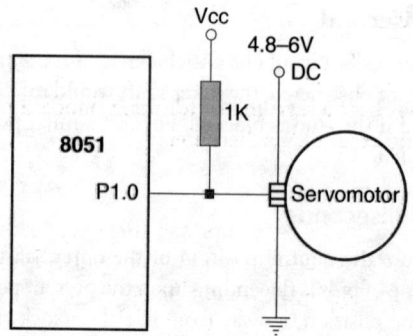

Figure 23.6 Hardware interfacing of servomotor with 8051

23.6 | Home-Position Detection

Although a servomotor has its internal feedback system, for a microcontroller some extra feedback is necessary to correctly control the servo. This is especially necessary at the first power-on condition, a situation where the initial position of the servo-shaft is unknown.

One such home-position detecting set up is illustrated in Fig. 23.7. Three different positions, namely, −90°, Neutral and +90° are sensed by three LDRs. Whenever the 'arm' rotates to a terminal position, the corresponding LDR is blocked to receive the light from its LED, which gives the feedback of the angular position of the motor. The LDRs may be interfaced with any available input port pin of the microcontroller.

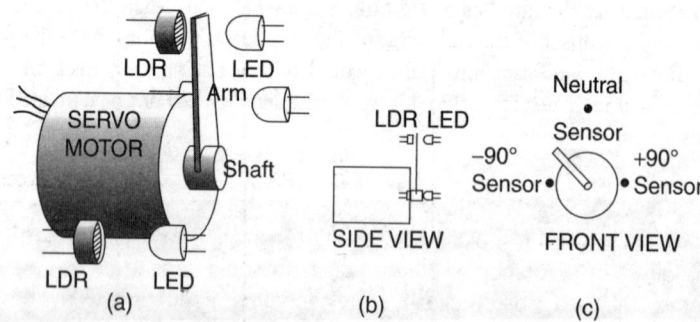

Figure 23.7 Home-position detection with LED and LDR. (a) General view, (b) side view and (c) front view for three positions: −90°, neutral and +90°

Home position detectors are always desirable for a servo as undesirable extra pulses may damage a servo if it is already in its terminal position.

23.7 | Solved Example

Example 23.1

Purpose: To verify the functioning of a servomotor for three standard pulses by designing the related hardware and software and conducting experiments thereafter.

Problem

Prepare a hardware circuit to interface a servomotor and three keys with 8051. Develop a software so that each key, when pressed and released, would generate only one pulse. The durations of pulses should be 1, 1.5 and 2 milliseconds, respectively, for keys K1, K2 and K3.

Solution

A circuit to meet above specifications would consist of three keys, one servomotor properly interfaced and the reset circuit for the 8051-based systems. Using a 20-pin version of 8051, namely, 89C2051, a circuit is prepared as shown in Fig. 23.8. A photograph of the breadboard version of the same circuit is presented in Fig. 23.9. In this case, the servomotor is interfaced with port pin P3.0, and three small keys (K1, K2 and K3) are interfaced with P3.2, P3.3 and P3.4, respectively. Note the 1K pull-up resistor for the servomotor pulse-input line and 10K pull-up resistors for keys. The IC 89C2051 may also be replaced by any other version of 8051 by changing the interfacing pin numbers.

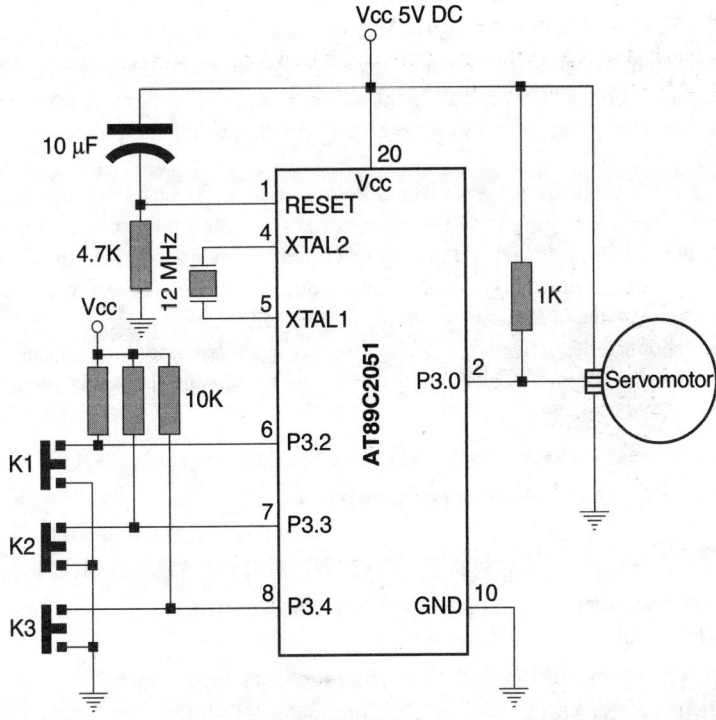

Figure 23.8 A servomotor test circuit

The job of the software would be to generate pulses of accurate real-time duration against each key. Care must be taken so that the pulse is generated after the release of the concerned key. As the duration of pulses

Figure 23.9 Breadboard construction of test circuit

must be 1, 1.5 and 2 milliseconds, a Timer may be used to generate this programmable delay. With 12 MHz external crystal, Timer 0 in its Mode 2 (8-bit auto-reload mode) may be programmed to generate interrupts at every 250-microsecond interval. For generating the pulses, a counter may be used, which would count down the number of interrupts of Timer 0 and stop generating the output for the pulse at its terminal count by turning off Timer 0.

Therefore, the duty of the main program would be to initialize Timer 0 in its mode 2 and starting the timer and enabling Timer 0 interrupt as and when necessary. It would then scan continuously for any key action. If any key is pressed, it would wait till the key is released. Then it would load the counter by an appropriate count value and start the timer. It would then be the duty of the Timer 0 interrupt routine to terminate the pulse after the counter value becomes 0.

To completely debounce the activated key, Timer 0 may again be started for 50 milliseconds and after that the key sensing may be resumed. Following is the complete software listing for the servomotor-testing program.

```
;$$$$$$$$$$$$$$$$$$$$$$$$$$$$$$$$$$$$$$$$$$$$$$$$$$$$$$$$$$$$$$$$$$$$$$$$$$$$$$$$$$$$$$$$$$$$$$$$$
; Program to test a servomotor by generating real-time pulses
      PULSE    EQU      PSW.1                    ; 1 = Timer 0 is busy, 0 = Timer 0 not busy
;$$$$$$$$$$$$$$$$$$$$$$$$$$$$$$$$$$$$$$$$$$$$$$$$$$$$$$$$$$$$$$$$$$$$$$$$$$$$$$$$$$$$$$$$$$$$$$$$$
; Entry point for power-on reset. Reset outputs FFH through all ports.
            ORG 0000H
START:    MOV      P3, #0FEH                     ; motor off, key-input active
          AJMP     START1                        ; continue at START1
;$$$$$$$$$$$$$$$$$$$$$$$$$$$$$$$$$$$$$$$$$$$$$$$$$$$$$$$$$$$$$$$$$$$$$$$$$$$$$$$$$$$$$$$$$$$$$$$$$
; Timer 0 interrupt service routine.
; Duty of Timer 0 routine is to decrement R7 of bank #0 by one and if it is zero then turn off Timer 0 and
; clear port pin P3.0. The timer-on indicator bit (PULSE) also to be cleared.
```

```
                ORG 000BH
TMR0:      DJNZ  R7, TMREND      ; continue pulse
           CLR   TCON.4          ; stop Timer 0
           CLR   P3.0            ; terminate pulse
           CLR   PULSE           ; to indicate duty over to the main loop
TMREND:    RETI
```
; $$$
; Continuation of reset initialization routine.
```
START1:  DJNZ     ACC, START1    ; power-up delay
```
; Initialize Timer 0 in mode 2 for 250 microseconds counting up and allow its interrupt.
; Do not start Timer 0 now.
```
         MOV   TL0, #06H     ; initialize Timer 0 counters
         MOV   TH0, #06H
         MOV   TMOD, #02H    ; Timer 0 in mode 2, gate cleared
         SETB  IE.1          ; enable Timer 0 interrupt
         SETB  IE.7          ; enable global interrupt control
```
; Main loop. Keep reading keys.
```
MAIN:   MOV   A, P3          ; read port 3 for key input
        MOV   R7, #04H       ; assuming K1 pressed for 1 ms pulse
        JB    ACC.2, CHK2    ; K1 not pressed
        SJMP  PROCES         ; K1 is pressed, process further
CHK2:   MOV   R7, #06H       ; assuming K2 pressed for 1.5 ms pulse
        JB    ACC.3, CHK3    ; K2 not pressed
        SJMP  PROCES         ; K2 is pressed, process further
CHK3:   MOV   R7, #08H       ; assuming K3 pressed for 2 ms pulse
        JB    ACC.4, MAIN    ; K3 also not pressed, keep reading keys
```
; Some key is pressed. Wait for its release.
```
PROCES: MOV   A, P3          ; read all keys
        ORL   A, #0E3H       ; set all bits except bits for keys
        CPL   A
        JNZ   PROCES         ; keep reading keys till the key is released
```
; Now start Timer 0 with appropriate value in R7 for a pulse through P3.0.
```
        SETB  PULSE
        SETB  TCON.4         ; start Timer 0
        SETB  P3.0           ; start the pulse
```
; Wait for Timer 0 to complete its job. Then restart Timer 0 for 50 milliseconds for debouncing the
; presently released key. Finally, jump back to main-loop.
```
PLSOFF: JB    PULSE, PLSOFF  ; wait here. Timer 0 ISR would clear PULSE
        MOV   R7, #0C8H      ; for 50 ms debouncing
```

```
            SETB    PULSE
            SETB    TCON.4          ; start Timer 0 again for 50 milliseconds
   DEBIT:   JB      PULSE, DEBIT    ; wait for 50 milliseconds (debouncing)
            AJMP    MAIN            ; loop on for next key command
```
; $$$
end

After completion of the interface, following experiments may be conducted to observe the function of the servomotor against single pulse.

Experiment 1: Power-on the system. Press and release K2 several times till the servo stops turning. Note that at the time of power-on, if the servo is already in its neutral position then it would not rotate at all. However, if it is away from its neutral position then it would rotate either clockwise or anticlockwise depending upon its angular condition before power-on.

Experiment 2: Now keep on pressing and releasing K1. Every time it is pressed and released, it would generate a 1 millisecond pulse forcing the servo to rotate anticlockwise. Observe how much it rotates for each pulse and how many pulses are necessary to rotate it to its terminal position. Stop using K1 once it has reached its terminal position.

Experiment 3: Now use K2 to bring it back to its neutral position. This time K2 would generate clockwise rotation.

Experiment 4: Now use K3 to rotate it further clockwise to its other terminal position. Note the number of pulses required and rotation for each pulse.

Experiment 5: Again use K2 to bring it back to its neutral position. This time K2 would generate anticlockwise rotation.

The best way to understand the working of a servo is to use it. It is insisted that the reader fabricate the circuit and perform indicated experiments with it.

SUMMARY

Servomotors have internal electronic feedback for positions. They are also fitted with gears for producing greater output torque. Generally, they are used for rotation through smaller angles and not for continuous rotations like DC or stepper motors.

Servomotors have three input terminals as compared to two of a DC motor or five of a stepper motor. These three input terminals are for power (4.8–6V DC, in general), ground reference and signal (pulse). This input signal (pulse) is TTL compatible and may be directly obtained from any port pin of a microcontroller. The signal is expected at every 20 milliseconds. The duration of the pulse dictates the direction of rotation of the motor. In general, 1.5 milliseconds pulse-duration indicates rotation to neutral position. A larger duration pulse, such as 2 milliseconds, rotates the motor through 90° (clockwise) while a smaller duration pulse, such as 1 millisecond, rotates the motor through −90° (anticlockwise). These parameters are not exact, and manufacturer's datasheet must be consulted before the interfacing.

POINTS TO REMEMBER

- It is a good practice to send a few 1.5 milliseconds pulses to ensure that the servo rotates to its neutral position after it is powered on.

- 1.5 millisecond pulses may rotate a servo either clockwise or anticlockwise depending upon its present condition with respect to its neutral position

REVIEW QUESTIONS

Evaluate Yourself

1. Numbers of lead wires of a servomotor are

   (a) two

   (b) three

   (c) four

   (d) none of these

2. What are the two special items available inside a servomotor that are not present in the DC or in the stepper motor?

   (a) Rotor and stator

   (b) Coils and shaft

   (c) Gears and electronic circuit

   (d) None of these

3. Input voltage range for servomotors would be from

   (a) 4.8 V to 6 V

   (b) 2.75 V to 6 V

   (c) 3 V to 9 V

   (d) none of these

4. The range of angular rotation of a servomotor varies from

   (a) −180° to +180°

   (b) −90° to +90°

   (c) −45° to +45°

   (d) none of these

5. To rotate the servomotor in anticlockwise direction

   (a) its input voltage polarity has to be reversed

   (b) the input pulse frequency has to be dereased

   (c) input pulse duration has to be decreased

   (d) none of these

6. In general, a servomotor expects a pulse input at every

   (a) 1 millisecond

   (b) 1.5 milliseconds

   (c) 20 milliseconds

   (d) none of these

7. When a servomotor is powered on, its shaft's angular position is

   (a) +90°

   (b) unknown

   (c) −90°

   (d) none of these

8. The input pulse for a servomotor may be generated by a

   (a) port pin

   (b) transistor

   (c) driver

   (d) none of these

9. As compared with its size against other type of motors, servomotors offer

   (a) higher torque

   (b) lower torque

   (c) same torque

   (d) none of these

10. As compared with DC motor and stepper motor, servomotors are

    (a) less expensive

    (b) more expensive

    (c) same

    (d) none of these

Search for Answers

1. How do you quickly distinguish between a servomotor from a DC or a stepper motor?

2. What are the electrical inputs necessary to drive a servomotor?

3. How does a servomotor rotate?

4. How is the direction of rotation controlled in a servomotor?

5. How a servomotor is stopped?

6. What is the purpose of the home-position detector in case of a servomotor?

7. Which part of a servo ensures 90° rotation of its rotor (armature)?

8. What is the purpose of the stud at the output gear of a servo?

9. What is the purpose of the PULSE bit of Example program 23.1?

10. In the set-up of Example 23.1, what would happen if the key K1 is pressed and not released?

Think and Solve

1. If continuous 2 millisecond pulses at 20 millisecond intervals are provided to a servomotor, would it keep on rotating like any DC motor?

2. List a few appropriate usages of servomotors.

3. Why does a servo need a pulse train?

4. Why feedback is necessary for a servomotor?

5. What would be the direction of a servomotor against 1.5 milliseconds pulse input?

6. Assuming that it had not reached its terminal position, for a 10:1 reduction ratio of gearbox, how far the output shaft of a servomotor would turn for a 2 milliseconds pulse?

7. What would happen if the stud at output gear is removed from it?

8. What would happen to a servomotor if it is activated by 1-millisecond duration pulses at every 7 milliseconds?

9. Why the software of Example 23.1 waits for 50 milliseconds after completion of a pulse generation?

10. How software of Example 23.1 may be modified so that just after system reset before starting to sense any key, the servo may be ensured to take its neutral position?

24 POWER MANAGEMENT

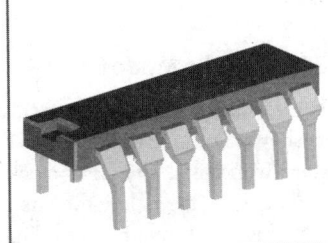

CHAPTER OBJECTIVES

In this chapter, the reader is introduced to the power management features of 8051 microcontroller. After completion of the chapter, the reader should be able to understand some of the details of power management features offered by 8051, like

- What are idle and power-down modes of 8051.
- How are these power-saving modes implemented.
- What are the benefits of each of these modes.

24.1 | Introduction

Power management is the feature available in most of the embedded systems and some other electrical or electronics units, like display of a mobile handset or monitor of a personal computer, which automatically turn off the device or put it in low-powered stand-by mode for power saving. This mechanism not only prolongs the battery life of portable units but also reduces the heat dissipation, which prolongs the life of the machine, reduces the maintenance overhead, as less cooling is necessary, and also protects the environment.

In general, 8051 oriented projects implemented by students avoid the power management features of 8051. However, for any commercial design, power management is one of the important design criteria.

24.2 | Power-Saving Modes

Power management was very briefly introduced in Chapter 1 (Section 1.5), and the related SFR, PCON for 8051 power management was introduced in Chapter 3 (Section 3.9). There it was mentioned that by setting bit 0 or bit 1 of PCON, the processor might be placed in idle mode or power-down mode. In the same chapter, the parameters in the last row of Table 3.3 indicate some benefits of power saving in these two modes. As some of the details of interrupt and Timer are necessary to understand the concepts of idle and power-down modes, detailed discussions on these issues were reserved for this chapter. We will now take a closer look at these two power-saving modes of 8051 and then discuss other related details of the power management issues of 8051.

24.2.1 | Idle Mode

For normal operations, the oscillator clock input is distributed to various parts of the 8051 microcontroller as shown in Fig. 24.1(a). If 8051 is placed in the idle mode by setting bit 0 of PCON as 1, then its CPU stops working as the CPU clock input is frozen. This is illustrated in Fig. 24.1(b) by a dark shade within CPU and by the deletion of its clock connection with the oscillator block. Note that this disconnection is implemented through controlling gates (not shown in Fig. 24.1) engaged for this purpose. Because there is no clock pulse available at the CPU section of the processor, no instruction fetching and execution occur

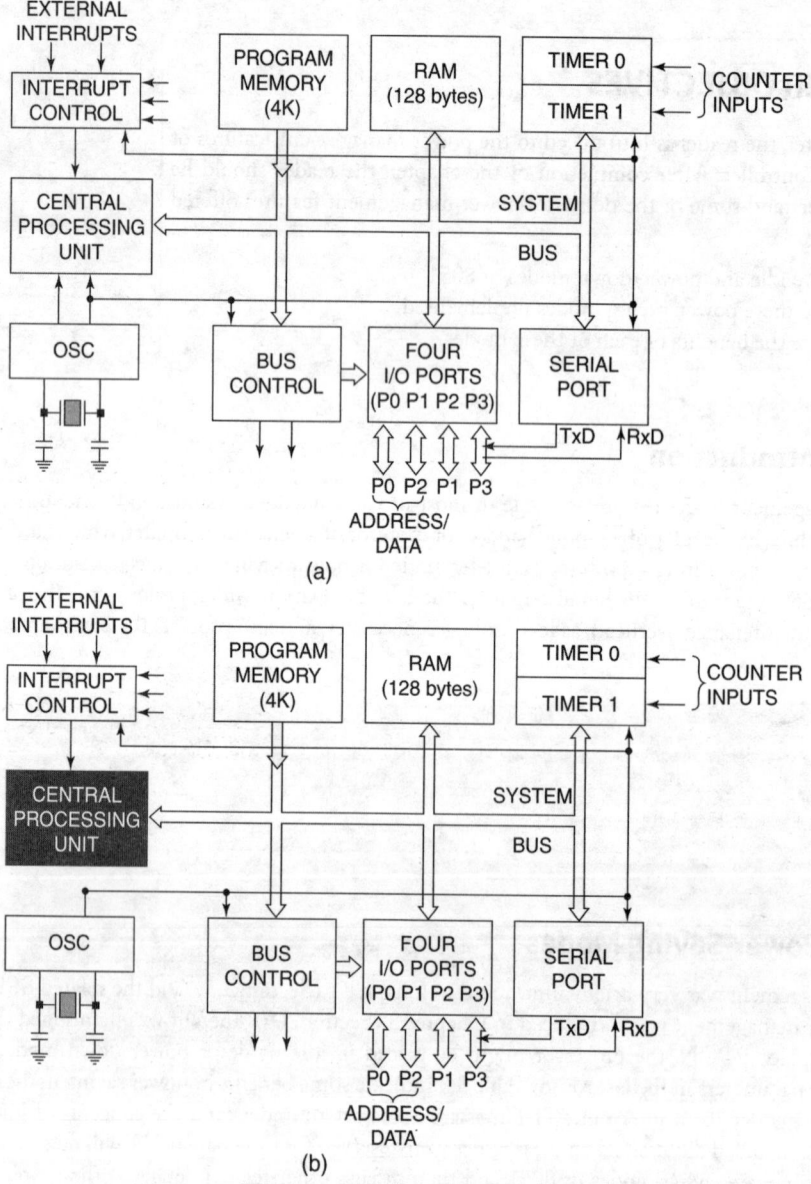

Figure 24.1 (a) Normal functioning and (b) idle mode with CPU frozen

as long as the processor is in the idle mode. The last content of program counter and stack pointer remains unchanged in this frozen condition along with all register contents. However, all other peripheral units, like Timer/Counters, serial ports and interrupts, remain active as they receive the oscillator clock in put in a normal fashion as if nothing has happened. The contents of internal RAM and SFRs remain unc anged. Port pins output data identical to that which they had been outputting at the time of the processor entering in its idle mode. As we may observe from Table 3.3, in idle mode, the 8051 microcontroller consumes 5 mA of current as compared to its demand of 20 mA when it is fully operational.

24.2.2 | Exit from Idle Mode

To come out from the idle mode, either a hardware-reset input or an enabled interrupt signal has to be applied. In either case, the IDL bit (bit 0 of PCON) would be cleared. In case of hardware reset, *the processor would start executing from the very next instruction, which evoked the idle mode.* This may be taken as an exception, as in general, the processor starts working from program memory address 0000H after the hardware reset. For interrupt, the corresponding interrupt service routine would be executed and the RETI instruction at the end of the interrupt service routine would bring back the processor to the instruction left unexecuted at the time of entering in the idle mode. Note that the interrupt may be an external interrupt (INT0 or INT1) or a Timer interrupt or a serial interrupt. It is the duty of the system designer to enable or disable unwanted interrupts for resuming proper execution.

24.2.3 | Power-Down Mode

As described earlier, the processor would enter in power-down mode when bit 1 of PCON is set. In this mode, the on-chip oscillator itself is frozen, and all peripherals, like Timer/Counters, serial and other interrupts including the CPU, become inactive (Fig. 24.2). Content of internal RAM is not disturbed, and port pins continue to output the data as before. Power consumption is reduced to 100 µA, as we may observe from Table 3.3. The Vcc input may be reduced to as low as 2 V during this power-down mode. All interrupts

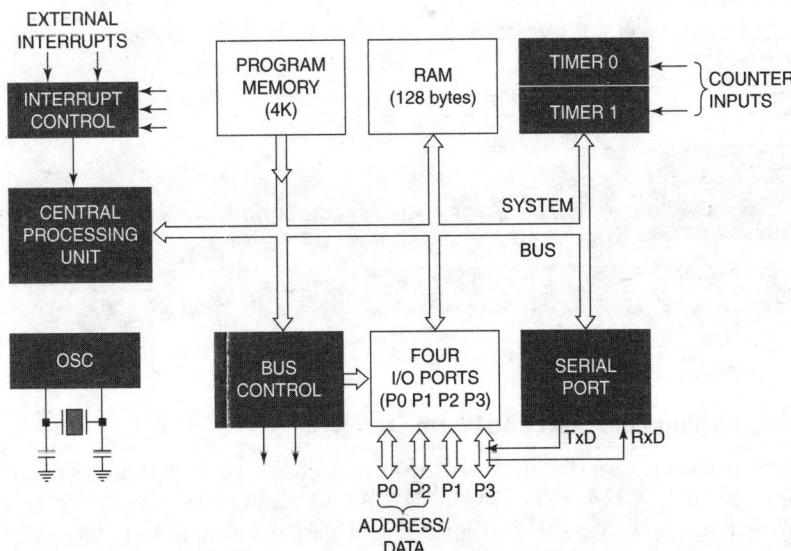

Figure 24.2 Power-down mode with oscillator frozen

become non-functional during power-down mode, and, therefore, unlike the idle mode described in the previous section, it is not possible to bring back the processor to its normal operational mode through some external or internal interrupt.

24.2.4 | Exit from Power-Down Mode

To exit from power-down mode, the hardware-reset input must be applied to the processor. This reset would initialize all SFRs, but the internal RAM content would remain unchanged. The processor would start fetching instruction from program memory address 0000H. The reader must note that different versions of MCS-51 offer marginally different features for the power-down mode and the method to come out from it. The reader must consult the relevant manufacturer's datasheet before implementing these power management modes.

24.2.5 | Difference Between Idle and Power-Down Modes

A comparative study of idle mode and power-down mode is presented in Table 24.1. As we may observe from this table, two major differences are the power consumption and the method of termination. The last parameter, which is the method of termination for power-down mode, may create an initial confusion about how to distinguish a power-on reset and a reset during power-down mode. This is discussed in the next section.

Table 24.1 Comparison between idle and power-down modes

	Idle mode	Power-down mode
Evoked by setting	PCON.0	PCON.1
Terminated by	Any enabled interrupt or hardware reset	By hardware reset only
Power consumption	5 mA	100 μA
Condition of SFRs after exit	Remain unchanged	Initialized by respective reset values
Condition of internal RAM	Unchanged	Unchanged
Condition of ports during sleep	Unchanged	Unchanged
External interrupts, Timer/Counter interrupts and serial interrupt	Remain active	Go inactive
On-chip oscillator	Remains active	Goes inactive

For a beginner, it may be difficult to decide when to apply which mode of power saving. A simpler rule of thumb may be: initially try to use power down mode as it saves maximum amount of current. Remember that a reset is necessary to come out from it and all ports would be initialized. If this application of reset and its effect are undesirable, settle for an idle-mode of power saving.

24.2.6 | Distinguishing Cold Start and Warm Start

In general, after a power-on reset, the internal RAM area of 8051 would contain some random data, as indicated in locations 7BH, 7CH and 7DH of Fig. 24.3. As any hardware reset does not change the internal RAM area, this feature may be used to distinguish a cold start and a warm start. Generally, cold start is designated as the fresh start of any microcontroller when power is applied for the first time. Subsequent hardware-reset signals would generate only warm starts.

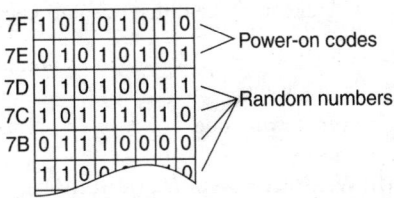

Figure 24.3 Condition of internal RAM with power-on code and random numbers

To distinguish between a cold start and a warm start, 2 bytes of internal RAM (say 7FH and 7EH) may be reserved and used. If during initialization, which follows every cold start, the processor writes a code in these 2 bytes, then during warm start, the code would still be available in unchanged condition as illustrated in Fig. 24.3. Therefore, irrespective of a cold or warm start, the processor must first check these two code bytes. If the power-on code is available, then it should be taken as a warm start, and the processor may proceed accordingly. However, if the power-on code is not available at the designated byte pair, then it must be a power-on reset or a cold start. In that case, the code bytes must be written in their respective locations, and the processor would then proceed accordingly.

As the internal RAM contains random data during power-on condition, the code bytes must be chosen carefully. One simple method is to use AAH and 55H in two successive bytes, say in 7FH and 7EH of internal RAM, as illustrated in Fig. 24.3. In binary form, this would generate a specific pattern, which is not easily available during general power-on condition. It may be noted that this power-on code might be necessary for the power-down mode of 8051, although the same may not be the case for idle mode in general.

24.3 | A Case Study on Power Management

To elaborate the hardware and software details of power management, let us take an example, which we would design by three different methods. In the first method, we would not implement any special power management scheme. In the second method, we would implement the idle mode. The third method would be implemented with power-down mode. In each case, there would be some specialities of hardware and software details which would be highlighted at their appropriate places.

For this case study, we take up a simple problem of random number generation. Let us assume that we need a microcontroller-based unit, which would generate and display any one-digit random number between 0 and 9, in decimal. The system is presented schematically in Fig. 24.4. Essentially, the 8051 microcontroller would need an external key input (RANDOM), which would be interfaced with a one-digit seven-segment display through a 7447 BCD-to-seven-segment decoder cum driver. Whenever the RANDOM key is pressed, a fresh random number between 0 and 9 would be displayed for 5 seconds after which the display would automatically go blank to save the power.

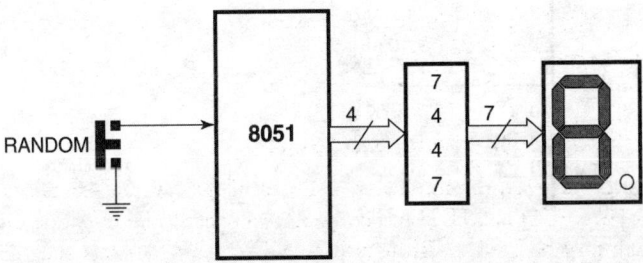

Figure 24.4 Schematic of one-digit random number generator

As we have already planned, we will take up three different solution cases for this system with

(i)   normal design without power management,

(ii)   power management using idle mode and

(iii) power management using power-down mode.

24.3.1 | Case 1: Normal Design Without Power Management

The circuit diagram for the system meant for random number generation is shown in Fig. 24.5. The system has a power-on reset. That means to generate a hardware-reset signal the system has to be switched off and then on.

The key to generate and display a random number (RANDOM) is connected with P3.2. This port pin is selected for interfacing the key as it is also the input for external interrupt INT0 (as an alternate function), which we would utilize for our Case 2 design. The display of the random number would last only for 5 seconds to save power, and thereafter the display would be turned off. To turn the display on and off, we have used the $\overline{BI/RBO}$ input of 7447, which is controlled by P1.7. As long as P1.7 outputs high, the display would remain on. By outputting logic low through P1.7, the display may be turned off.

IC 7447 is a BCD to the seven-segment decoder-cum-driver. Its inputs A0–A3 should receive the decimal digit in BCD form. These four inputs of 7447 are interfaced with the lower four pins of Port P1. Note that 7447 has open collector outputs and therefore needs external pull-ups. Current-limiting resistors are incorporated with the segment-driving lines of 7447. As 7447 outputs are active low, we use a common anode-type seven-segment display.

The software has two modules, reset initialization and the main loop. After reset initialization, which ensures P3.2 to act as an input pin and also blanks the display, the control enters the main loop, which is an infinite loop. The duties of the main loop are continuous key scanning and generation and display of a fresh random number whenever the key is detected. Note the way the random number is generated using the

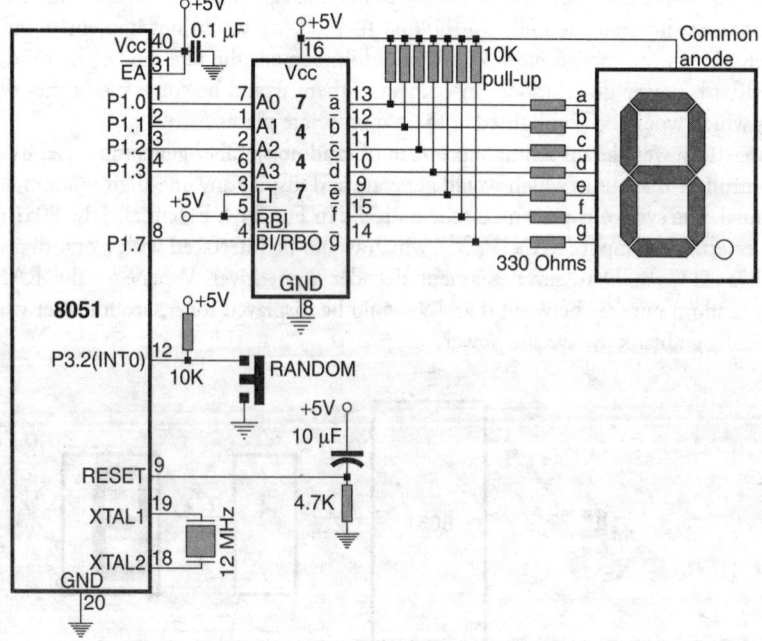

Figure 24.5 Circuit diagram for random number generator (Case 1)

random content of internal RAM location 7CH and 7DH. To turn off the display after 5 seconds, a delay routine is used.

The major part of the job is accomplished by the subroutine GENDSP, which uses two more subroutines, namely, GENRAN and DELAY. GENRAN generates an 8-bit random number, and DELAY takes about 5 seconds to return. The 8-bit random number is truncated to 4 bits by GENDSP so that it does not exceed the BCD limit of 9.

```
; $$$$$$$$$$$$$$$$$$$$$$$$$$$$$$$$$$$$$$$$$$$$$$$$$$$$$$$$$$$$$$$$$$$$$$$$$$
; Program to generate and display a random number (Case 1)
; Program developed by Dr S. Ghoshal, February 2009.
; Whenever RANDOM key is pressed, a random number (0–9) would be
; displayed. The display would automatically be turned off after 5 seconds.

    RNDM1   equ     7DH         ; location of first random number
    RNDM2   equ     7CH         ; location of second random number

; Register bank used #0
; System stack from 08H onwards.
; $$$$$$$$$$$$$$$$$$$$$$$$$$$$$$$$$$$$$$$$$$$$$$$$$$$$$$$$$$$$$$$$$$$$$$$$$$
            ORG     0000H

; Entry point for power-on reset. Accumulator is cleared, and all ports output FFH
; just after power-on reset. Execute a power-up delay and then blank display.

    START:  DJNZ    ACC, START  ; power-up delay
            SETB    P3.2        ; make P3.2 as input
            CLR     P1.7        ; blank display

; Main loop starts from here. Keep reading P3.2 till it goes low (key pressed).

    MAIN:   JB      P3.2, MAIN  ; loop on till key is pressed

; The key is pressed. Generate and display a random digit for 5 seconds. Turn off the
; display after 5 seconds. All these are done by the routine GENDSP.

            ACALL GENDSP        ; generate and display a random number
                                ; for 5 seconds

; Now go back to main loop for key sensing.

            SJMP    MAIN        ; loop on
; $$$$$$$$$$$$$$$$$$$$$$$$$$$$$$$$$$$$$$$$$$$$$$$$$$$$$$$$$$$$$$$$$$$$$$$$$$
; Name:         GENDSP
; Function:     generates a random digit for five seconds and then turns off the display
; Input:        none
; Output:       see function above
; Calls:        GENRAN, DELAY
; Uses:         A, CY

; Generate an 8-bit random number. This is done by the routine GENRAN.

            ACALL GENRAN        ; eight-bit random number in accumulator
```

```
; Now clear higher 4 bits and if the value of lower four bits is more than nine, divide the present number
; by two.

              ANL    A, #0FH          ; clear MS 4 bits
              CJNE   A, #0AH, NEXT    ; 0–9 in A would generate a carry, means OK
    NEXT:     JC     DISPIT           ; generated number is less than ten
              CLR    C                ; generated number is more than nine
              RRC    A                ; therefore, divide it by two

; Display the number through Port 1. Port 1.7 must be made high.

    DISPIT:   ORL    A, #80H          ; set MSB as 1
              MOV    P1, A            ; display random number

; Wait for five seconds. This is done by calling DELAY.

              ACALL DELAY             ; wait for 5 seconds

; Now turn off display by clearing P1.7. Then return.

              CLR    P1.7             ; turn off the display
              RET

; $$$$$$$$$$$$$$$$$$$$$$$$$$$$$$$$$$$$$$$$$$$$$$$$$$$$$$$$$$$$$$$$$$$$$$$$$$$$$$
; Name:     GENRAN
; Function: Generates an 8-bit random number whenever called
; Input:    random content in locations RNDM1 and RNDM2
; Output:   8-bit random number in A
; Calls:    none
; Uses:     A, R0, RNDM1, RNDM2

    GENRAN:   MOV    R0, #RNDM1       ; point to first location for a random number
              MOV    A, @R0           ; load it in A
              RL     A                ; process it
              ADD    A, @R0           ; further process it
              RL     A
              MOV    @R0, A           ; save it back
              DEC    R0               ; point to RNDM2
              XRL    A, @R0
              MOV    @R0, A           ; save it
              RET                     ; 8-bit random number in A

; $$$$$$$$$$$$$$$$$$$$$$$$$$$$$$$$$$$$$$$$$$$$$$$$$$$$$$$$$$$$$$$$$$$$$$$$$$$$$$
; Name:     DELAY
; Function: Generates about 5-second delay whenever called
; Input:    none
; Output:   none
; Calls:    none
; Uses:     R2, R3, R4 of Register bank #0

    DELAY:    MOV    R4, #28H
    DELAY1:   MOV    R3, #00H
```

```
DELAY2:    MOV     R2, #00H
DELAY3:    DJNZ    R2, DELAY3
           DJNZ    R3, DELAY2
           DJNZ    R4, DELAY1
           RET
```

; $$

end

 Note how pseudo-random numbers are generated by the routine GENRAN. Another simple technique is to use the counter of an active timer.

24.3.2 | Case 2: Power Management Using Idle Mode

Without making any change in the hardware circuit shown in Fig. 24.5, the software may incorporate power management using idle mode. Note that from outside, the software functions in identical fashion. That is, whenever the RANDOM key is pressed, the display shows a random number for 5 seconds, after which the display is automatically turned off. However, in this case, the processor enters in idle mode and the RANDOM key has to be activated to bring it out from the idle mode and make it generate the next random number.

As P3.2 is also the input for external interrupt INT0, we will use this interrupt to come out from the idle mode. The complete software listing would be as follows. Note that the routines GENDSP, GENRAN and DELAY are identical in all respects with the previous three routines. Therefore, these three routines are not included here.

; $$
; Program to generate and display a random number (Case 2)
; Program developed by Dr S. Ghoshal, February 2009
; Whenever RANDOM key is pressed, a random number (0–9) would be displayed.
; The display would automatically be turned off after 5 seconds.
; Thereafter processor would enter in idle mode. Press RANDOM again to come out.

```
RNDM1   equ     7DH          ; location of first random number
RNDM2   equ     7CH          ; location of second random number
```

; Register bank used #0
; System stack from 08H onwards.

; $$
```
        ORG     0000H
BEGIN:  AJMP    START         ; entry point for hardware reset
```
; $$
```
        ORG     0003H
```

; Entry point for INT0 external interrupt (by RANDOM key).

```
        EXTIN0:  ACALL  GENDSP      ; generate and display a random digit for 5 seconds
                 RETI
```

; $$
; Reset initialization routine continues.

```
        START:   DJNZ   ACC, START   ; power-up delay
                 CLR    P1.7         ; blank display
```

; Initialize for external interrupt INT0. Then enable it.

```
                 CLR    TCON.0       ; INT0 becomes low-level triggered
                 SETB   IE.0         ; enable INT0 interrupt
                 SETB   IE.7         ; enable global interrupt
```

; $$
; Main loop starts here. Program counter would remain within this main loop whenever the RANDOM key
; (external interrupt INT0) is activated. After executing the ISR of INT0, control would re-enter this main
; loop, which would put the processor back to idle mode.

```
        MAIN:    ORL    PCON, #01H   ; set bit 0 of PCON, processor enters in
                                     ; idle mode
```

; Next two instructions would be executed after pressing RANDOM key, which
; would generate INT0 interrupt thus terminating the idle mode of the processor.
; The processor would execute the ISR for INT0, and after execution of RETI of
; ISR, it would execute the following two instructions.

```
                 NOP
                 SJMP   MAIN         ; loop on, back to idle mode
```

; $$

end

24.3.3 | Case 3: Power Management Using Power-Down Mode

To implement the power-down mode, one minor change is necessary in the hardware circuit of Fig. 24.5. As illustrated in Fig. 24.6, the RANDOM key is disconnected from P3.2 and interfaced in a parallel alignment with the reset capacitor to generate a hardware-reset signal.

In this case too, the software uses the same three subroutines, namely, GENDSP, GENRAN and DELAY, as already used in the first and second case softwares. In this version, the program calls GENDSP to generate and display a random number for 5 seconds after reset initialization. Thereafter bit 1 of PCON is set, putting the processor immediately in the power-down mode. Another hardware-reset signal, produced by activating the RANDOM key, would be necessary to bring the processor back to its normal operational mode.

As some versions of 8051 start executing from 0000H and some other versions start executing from the next instruction after executing power-down mode, the software is developed in such a fashion that it would work properly either way. After imposing power-down mode, the NOP and SJMP instructions bring back the control to its starting point, i.e. 0000H.

; $$
; Program to generate and display a random number (Case 3)
; Program developed by Dr S. Ghoshal, February 2009

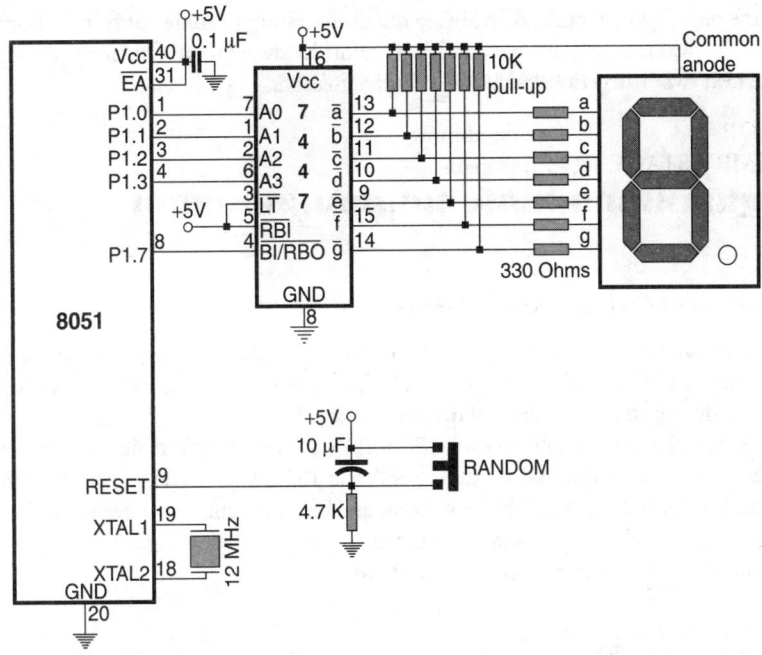

Figure 24.6 Circuit diagram for random number generator (Case 3)

```
; Whenever RANDOM key is pressed, a random number (0–9) would be displayed.
; The display would automatically be turned off after 5 seconds.
; Thereafter the processor would enter in power-down mode. Press RANDOM to
; come out.

    RNDM1  equ    7DH        ; location of first random number
    RNDM2  equ    7CH        ; location of second random number

; Register bank used #0
; System stack from 08H onwards.
; $$$$$$$$$$$$$$$$$$$$$$$$$$$$$$$$$$$$$$$$$$$$$$$$$$$$$$$$$$$$$$$$$$$$$$$$$$$

           ORG    0000H

; Entry point for hardware reset.

    START:  DJNZ  ACC, START  ; power-up delay
            CLR   P1.7         ; blank display

; Generate a random number between 0 and 9, and display for 5 seconds. Then turn off the display. All these
; are executed by the routine GENDSP.

           ACALL GENDSP       ; generate and display random number for 5 seconds

; Next instruction puts the processor in power-down mode. A hardware reset is necessary to come out from it.

           ORL   PCON,#02H   ; set bit 1 of PCON SFR for power-down mode
```

; Processor enters the power-down mode. A hardware reset is necessary to come out from it. Some versions of
; 8051 execute from the next instruction of power down against hardware reset. Other versions start executing
; from 0000H. The next two instructions take care of both situations.

```
NOP                    ; do nothing
SJMP   START           ; loop on
```

; $$

end

24.3.4 | Comparison of Performance in Three Cases

For the purpose of comparison, we assume that in all the three cases, the key to generate the random number
would be activated once in 30 seconds. Therefore, in all the three cases, the maximum power consumption
would take place only during those 5 seconds of display activation.

In the first case, the processor would consume 20 mA of current for the remaining 25 seconds. In the
second and third cases, the processor would consume 5 mA and 100 μA (0.1 mA), respectively. Therefore, with
respect to the first case, power saving would be 75 per cent and 99.5 per cent, respectively. This gives us a rough
idea about power saving through power management, which is represented in Fig. 24.7. Here it was assumed
that the display would consume an average current of 10 mA.

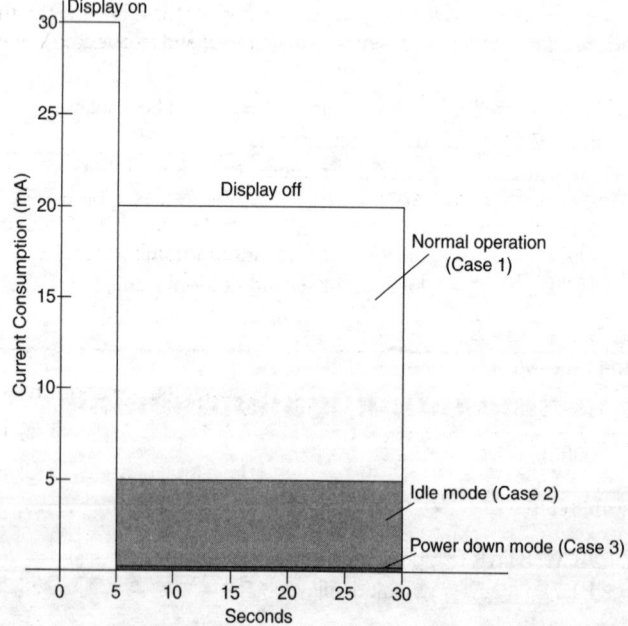

Figure 24.7 Comparison of current consumption for three cases

24.3.5 | Further Improvements

Further power savings may be achieved by placing the processor in its idle mode during 5 seconds of display.
A Timer may be used in this case, generating interrupts at every 50 milliseconds. After 20 interrupts, the
idle mode may be changed to power-down mode. Moreover, the display may be blanked during alternate

50 milliseconds, which would further decrease the power demand. However, the intensity of display would be less than that of the previous case.

 Advanced microcontrollers like AVR or ARM offer better power management features. Interested readers may refer chapter 26 of this book and also manufacturer's data sheets.

SUMMARY

Power management is achieved by freezing a portion or whole processor. Generally this is achieved by preventing the clock input through the activation of some related gates. The 8051 offers two such modes for power management.

The first one, designated as idle mode, decreases the power requirement of the processor by as much as 75 per cent. It halts the CPU of the processor, allowing all other parts to function as they are. A hardware reset or some enabled interrupt input is necessary to bring back the CPU to its functional state.

The second one, designated as power-down mode, decreases the power consumption by 99 per cent or more by freezing the entire processor. A hardware reset is necessary to come out of this mode.

In both of these two modes, the port outputs remain unchanged and so does the internal RAM condition.

POINTS TO REMEMBER

- Check from the manufacturer's datasheets about the address of execution of the processor for power-down mode.

- When set as 1, bit 0 of PCON SFR allows the idle mode and bit 1 of PCON puts the processor in power-down mode.

REVIEW QUESTIONS

Evaluate Yourself

1. How many power-saving modes are offered by 8051?

    (a) One   (b) Two

    (c) Three   (d) None of these

2. Which SFR of 8051 offers the power management facility?

    (a) PCON   (b) TCON

    (c) SCON   (d) None of these

3. How is the idle mode terminated?

    (a) Reset   (b) Interrupt

    (c) Either of these   (d) None of these

4. How is the power-down mode terminated?

    (a) Reset   (b) Interrupt

    (c) Either of these   (d) None of these

5. How much is the power saving by 8051 in idle mode?

    (a) 25 per cent   (b) 50 per cent

    (c) 75 per cent   (d) None of these

6. How much is the power saving by 8051 in power down mode?

    (a) 25 per cent   (b) 50 per cent

    (c) 75 per cent   (d) None of these

7. What happens to the internal RAM locations during power-down mode?

(a) All cleared   (b) No change

(c) Random change   (d) None of these

8. If during idle mode, a hardware-reset signal is applied, from which address would the processor fetch the next instruction?

   (a) 0000H   (b) 000BH

   (c) 0013H   (d) None of these

9. Which input of 7447 may be used to blank display?

   (a) $\overline{BI/RBO}$ (b) A0

   (c) Either of these   (d) None of these

10. What type of output is provided by 7447?

   (a) Active low   (b) Open collector

   (c) Both of these   (d) None of these

Search for Answers

1. What are the benefits of power management?

2. What power management features are offered by 8051?

3. How are these power management features implemented by 8051?

4. What happens to SFRs during power management modes?

5. What are the roles of interrupt and hardware reset in power management?

6. Which power management mode should be implemented if the processor must come out of it when a serial data is received through its serial port?

7. Is it safe to reduce the power-input voltage of the processor during its power-down and idle mode?

8. What is the difference between a cold start and a warm start?

9. How would the processor distinguish between a cold start and a warm start?

10. What are the inputs and outputs of IC 7447?

Think and Solve

1. How maintenance overhead is reduced because of power management?

2. Do we implement power management schemes at the expense of lesser throughput from the processor?

3. Under which condition(s) should the power-down mode be selected?

4. Under which condition(s) should the idle mode be selected?

5. From the datasheets of AT89C51, we observe that it can work for an input voltage range of 4–6 V. Would the power consumption increase with increase of voltage, or would it remain a constant? Justify your answer.

6. Why is the common anode type of seven-segment display required to be used with IC 7447?

7. Which power management mode for an 8051-based system would be best suited if it is to monitor the temperature of a furnace at every second and raise an alarm if the sensed temperature crosses the preset limits? Justify your answer.

8. For the same problem above, if the temperature is to be sensed at every 10 microseconds, which power management scheme would be suitable? Justify your answer.

9. What are the advantages and limitations of IC 7447?

10. How random are the random numbers generated by the algorithm used in the subroutine GENRAN?

25 CASE STUDY: A HOME PROTECTION SYSTEM

CHAPTER OBJECTIVES

In this chapter, the reader is introduced to a full-length design example with 8051 microcontroller as a case study. After completion of the chapter, the reader should be able to understand

- How to plan a system, which is to be designed.
- How are the hardware-interfacing modules prepared.
- How are the software routines developed.
- How to interface a few common sensors with 8051.
- How to implement a multiplexed keyboard-display interface.

25.1 | Initial Planning

As we have completed the study of internal architecture, instruction set and major interfacing techniques related to 8051 microcontroller in the previous chapters, we may now take up a full-scale design example using 8051. Many of the features we have discussed so far would be implemented here. As a case study, let us consider a simple home protection system.

25.1.1 | What is Home Protection System?

By the phrase 'home protection system', we mean the implementation of simple yet effective methods to guard our homes from unwanted incidents such as trespassing, fire hazards, electrical faults or even an empty over-head water tank. To start with, we will plan for two types of security devices. The system should be capable of monitoring all movements in at least two entry/exit points or doors [Fig. 25.1(a)]. We would also implement four touch sensors at any four windows. Any unauthorized opening of these windows would raise an alarm [refer to Fig. 25.1(b)]. An audio output (speaker) to raise the alarm along with a decimal keypad and display system would initially complete the unit.

At a later stage, we can expand the system further by incorporating extra features, like auto-switching of light(s) that depends upon the availability of natural light. The motor for pumping water in the overhead reservoir may be automatically turned on and off.

Now a days, most home protection systems include live CCD cameras which are programmed to transfer images to laptops or mobile handsets. This and many other features of commercial home protection system may be interesting to readers. Here, we have taken up a less ambitious project so that it may be implemented by most of the students.

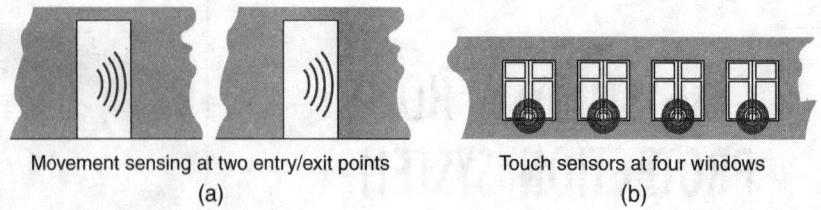

Movement sensing at two entry/exit points   Touch sensors at four windows
(a)   (b)

Figure 25.1 Schematic representation of home-protection system

25.1.2 | Sensor Selection

To sense movement through doors, either LDR or infrared (IR) sensor may be used. We would select the second one as it gives better performance. We would need an IR transmitter and an IR receiver at either side of the door. Windows may be fitted with simple tactile key-type sensors to echo any closed/open position.

 Dedicated transducers for movement detection are available in the market. Even eye-movement detecting sensors are also available. However, to avoid complications in interfacing, these transducers are not adopted for the present project.

25.1.3 | Overall System

Fig. 25.2 shows the schematic of the system. It indicates that the IR sensors (receivers) at doors are interfaced with external interrupt inputs of 8051. One output pin of a port is necessary to generate the signal for both IR transmitters. Signals from sensors at four windows would be interfaced with four input pins of a port. Audio output would be available from one output pin.

For 4-digit 7-segment display and four-by-four keyboard, four scan lines are necessary along with 8-segment-drive lines through a segment driver and four return lines from the keyboard. We generate four scan lines using a digit driver and two output lines of a port.

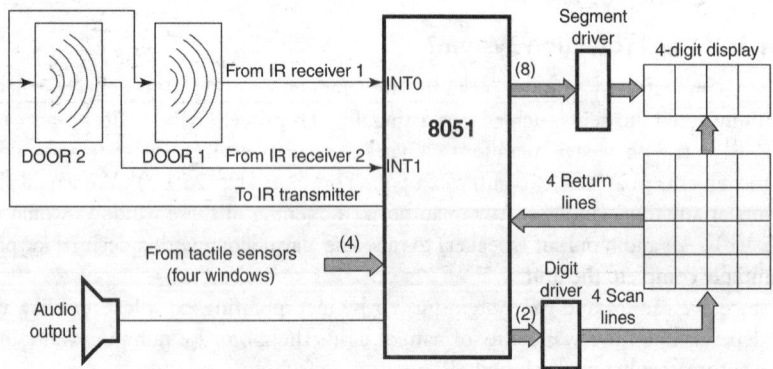

Figure 25.2 Schematic of the system

 Try to plan and assign input and output required for an embedded system right at its early stage of planning. It would be a good practice to set aside a few extra I/O lines for future expansion.

25.1.4 | Software Modules

The software would be divided into four major modules, namely,

- (i) reset initialization,
- (ii) main loop,
- (iii) key servicing and
- (iv) interrupt handling.

The first module would initialize various parameters, including interrupts, and start the Timers. The program then would enter the main loop, which is an infinite loop. If any key pressing is detected, it would be serviced immediately by the key-servicing module. Interrupt service routines would handle all interrupts as and when necessary.

25.2 | Hardware Realization

Fig. 25.3 shows circuit diagram for the controller card of the system. Circuits related with IR sensors are presented in Fig. 25.4. In Fig. 25.3, right side of controller 8051 illustrates the interfacing of keyboard-display module. This contains four 7-segment display digits, which are of high intensity (red colour) and common cathode type. Small size of 1/3rd inch height is sufficient in this case as they consume minimum power. Detailed functions of all sixteen keys are described and also illustrated in software development section.

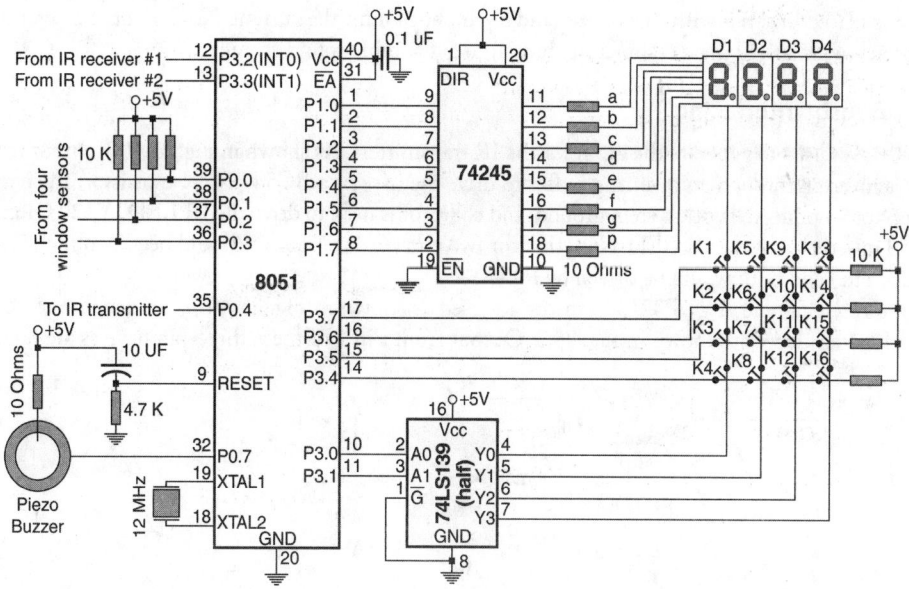

Figure 25.3 Circuit diagram for controller card

25.2.1 | Display Drivers

To drive eight segments of the display unit, one 74245 is interfaced through Port 1 of 8051. Eight resistors of 10 Ohms each are interfaced in series with output lines of 74245 to limit the current at the anodes, which would also provide a limit at the common cathode of the activated digit. As we would be using a multiplexed display-driving scheme, only one digit would be activated at a time. Display would be refreshed from left to right.

One half of a dual two-to-four decoder (74139) was used as the scan-line driver or digit driver. The output of this device is sufficient to drive high-intensity small-sized displays. Note that this 74139 is permanently enabled by connecting its enable input ($\overline{G}$, Pin 1) with system ground. Outputs Y0 through Y3 are connected in such a fashion that digit D1 would be activated if P3.0 and P3.1 both output 0s.

25.2.2 | Key Return Lines and External Pull-Ups

Four-by-four key matrix (K1 through K16) are driven by four scan lines from 74139 and sensed through four return lines interfaced with P3.4 to P3.7. Four external pull-up registers of 10K each are used for a stable input through these pins of Port 3. Note that the keyboard would be scanned from left to right and from top to bottom in the same order as the key numbers shown in Fig. 25.3.

25.2.3 | System Reset and Crystal

In Fig. 25.3, the left side of 8051 shows interfacing of crystal and reset circuit, audio interface and sensor-related interfacing. We use a 12 MHz crystal and prepare a circuit for power-on reset. If the manual reset is considered desirable, then a key may be interfaced between +5V and reset input (Pin 9) across the 10 µF electrolytic capacitor in parallel alignment with the capacitor.

25.2.4 | Audio Interfacing

To generate audio signals, a small piezo-ceramic buzzer is interfaced with P0.7. These buzzers operate with very little current, and the normal sinking capacity of a port pin is sufficient enough to drive it. A 10 Ohms resistance is placed in series with this buzzer and 5V input to limit the current. Note that the buzzer must be properly fixed with a rigid mass (cabinet or chassis) to get audible sound from it.

25.2.5 | Sensor Interfacing

Port pin 0.4 was used to generate the input for the IR transmitters. As shown in Fig. 25.4(a), output from this port pin is inverted through an inverting buffer to drive the base of a BC547 NPN transistor. The emitter of this transistor is connected with system ground, and collector is used to drive the IR LED. A 220 Ohms resistor is used in series with the IR LED. Note that for two entry/exit points, we would need two such interfaces (IR diodes, etc.) and both would be driven by P0.4.

At the receiving end, TSOP32130 sensors are used as shown in Fig. 25.4(b). Pin details and the schematic of this transducer are shown in Fig. 25.5. Output from Pin 1 of these three-pin devices are boosted by

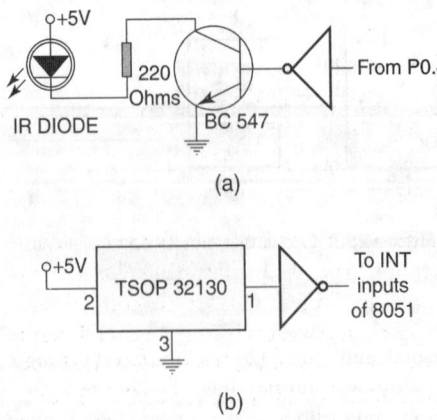

(a)

(b)

Figure 25.4 (a) IR Transmitter and (b) IR Receiver circuits

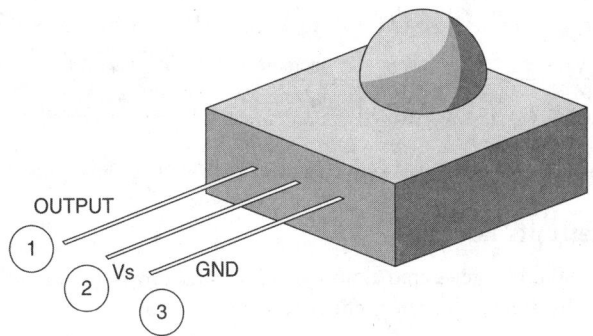

Figure 25.5 TSOP32130-pin diagram

inverter gates and then interfaced with INT0 and INT1 input pins of 8051. Pin 3 is connected with system ground, and Pin 2 is connected with 5V supply. Note that the IR transmitter and receiver are to be so placed that they face each other directly.

Output from four tactile keys from four windows are directly interfaced with P0.0 to P0.3, all four of which are externally pulled up by 10K resistors. One end of each key must be connected with system ground, and the other end must be connected with input port pins.

25.2.6 | Power Supply

The system would need about 500 mA of 5V DC power supply, which may be obtained from a 7805 properly interfaced with a rectifier and a transformer. Suitable battery cells may also be used to get uninterrupted power supply.

 After fabricating any complete system, it is always desirable to check its power requirements, specially if the system is a battery operated one. Set the multimeter in 10 Amp current sensing mode (just to be on the safer side for the protection of the multimeter) and use it in series with the power supply. Note current requirements for display on as well as off conditions.

25.3 | Software Development

Major software modules have already been identified (refer to Section 25.1.4). We will now plan all the details related with these modules.

25.3.1 | IR-Sensing Module

TSOP32130 would be in need of the input frequency of 30 KHz. Timer 0 may be used to generate four cycles of 30 KHz in every 1 millisecond. As per the hardware interface (refer to Fig. 25.3), this would generate interrupts at INT0 and INT1 external interrupt inputs. The concerned interrupt service routines for INT0 and INT1 would increment their respective counters by one for each interrupt. At every 8 milliseconds, this would be verified by Timer 1 interrupt service routine. If any one of these two counters found 0, audio alarm would be set. Otherwise, these two counters would be cleared by the Timer 1 interrupt service routine. This is schematically illustrated in Fig. 25.6.

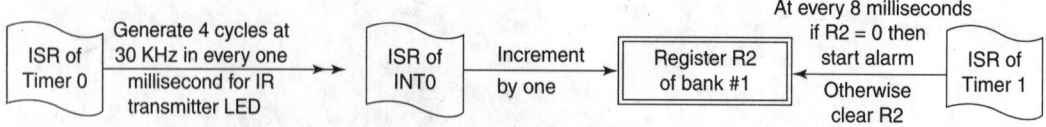

Figure 25.6 Schematic of IR sensor handling for INT0 using Timer 0 and Timer 1

25.3.2 | External Interrupts' Routines

These external interrupts would be generated almost simultaneously by triggering off IRLED through P0.4. These routines are only to increment their respective counters by one and return.

25.3.3 | Timer 0 Interrupt Service Routine

Timer 0 is set in its Mode 2, the 8-bit auto-reload mode, and started by reset initialization routine, just before entering at the main loop. Timer 0 must be programmed to generate interrupts at every 17 microseconds, which in turn generates four cycles at 30 KHz frequency at every millisecond. This is achieved with the help of Timer 1, which sets the IRLED bit at every millisecond. If IRLED bit is set, then Timer 0 ISR toggles the status of P0.4 and decrements its counter by one. Whenever this counter (R7 of bank #0) is cleared, IRLED bit is also cleared by Timer 0 ISR, which eventually turns off the IR transmitter. The flow of logic for Timer 0 ISR is illustrated in Fig. 25.7.

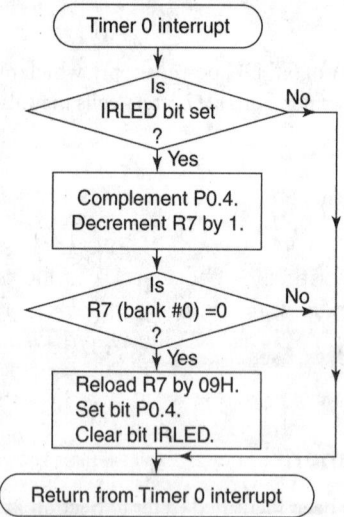

Figure 25.7 Flowchart for Timer 0 interrupt service routine

25.3.4 | Timer 1 Interrupt Service Routine

Timer 1 is also set in its Mode 2, the 8-bit auto-reload mode, to generate continuous interrupts at every 250 microseconds. Timer 1 interrupt service routine performs most of the duties of the program, including those given below.

(i)   At every 250 microseconds, toggle P0.7 (audio output) if BEEP is set.

(ii) At every millisecond, set IRLED bit.

(iii) At every 2 milliseconds, refresh one digit and scan one key column. This leads to refreshing of all four digits once and scanning of all 16 keys in every 8 milliseconds.

(iv) At every 8 milliseconds, read and clear R2 and R3 of bank #1. If any one of these two registers is found zero, then set audio alarm.

(v) If any key is detected, then debounce it for 50 milliseconds.

(vi) For a debounced key, store the key code in SRVKEY and set KFOUND bit to indicate for i ey servicing by the main loop.

25.3.5 | Key Scanning

Fig. 25.8 shows the proposed layout of the keyboard with some of its functions. Although all keys are scanned, only two of these are serviced in the present software and the remaining keys are left for the reader to explore for other possibilities. Scanning of keys starts from the leftmost column and proceeds towards right. Moreover, every column is scanned from top to bottom. The scan code varies from 01H to 10H for 16 keys.

Whenever a fresh key is detected through a complete cycle of 8 milliseconds, it is saved and debounced for 50 milliseconds. Thereafter, the key code is copied to SRVKEY, and KFOUND bit is set to indicate to the main loop that a valid key needs to be serviced. The main loop services the key and waits for about 1 second before clearing KFOUND bit. This (1 second waiting) ensures the key is released for another fresh scanning cycle.

1	5	9	F1
2	6	0	F2
3	7	STOP BEEP	F3
4	8	RE START	F4

Figure 25.8 Keyboard layout

25.3.6 | Display Refreshing

Four bytes from 1CH to 1FH are used as display buffer, and the seven-segment display patterns for four-digit display are stored there as shown in Fig. 25.9. Using R0 of bank #1 as pointer, one of these 4 bytes is accessed in every 2 milliseconds by Timer 1 ISR to refresh the display.

If the display is to be changed, the display buffer must be changed accordingly. Note that the display buffer must contain seven-segment patterns for the display. That means, if we are to display 0 (zero) through digit D1, then simply storing 00H in location 1CH would not serve the purpose. We must store the seven-segment pattern for the present hardware circuit. Referring to Fig. 25.3, the pattern for 0 would be 0011 1111 in binary or 3FH, which would be the correct content in location 1CH to display 0 through D1.

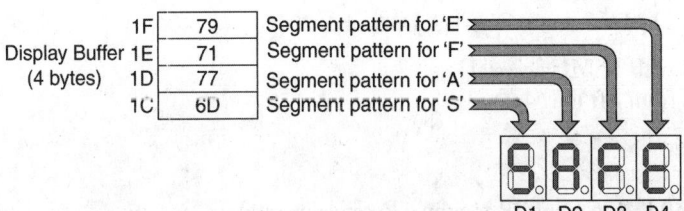

Figure 25.9 Conditions of display buffer and display digits during sign-on message 'SAFE'

25.3.7 | Main Loop

The duty of the main loop is to check for keys continuously and service the key when it is indicated by KFOUND. Audio alarm is immediately started if any key is pressed. After servicing a key, the main loop waits for about a second to ensure the concerned key is released.

25.3.8 | Reset Initialization

Reset initialization clears 128 bytes of internal RAM locations and then performs a self-test by turning on all four digits one by one and finally emitting an audio beep. Ports P0 and P3 are initialized thereafter to read input data. The stack pointer is initialized at 2FH so that system stack starts from 30H onwards. All internal registers and RAM locations, which must have non-zero reset values, are initialized, and the display buffer is loaded with the sign-on message 'SAFE'. The final step is to initialize both Timers and all four interrupts. Control then enters the main loop, which is an infinite loop. Fig. 25.10 shows the usage of all registers, memory locations and addressable bits.

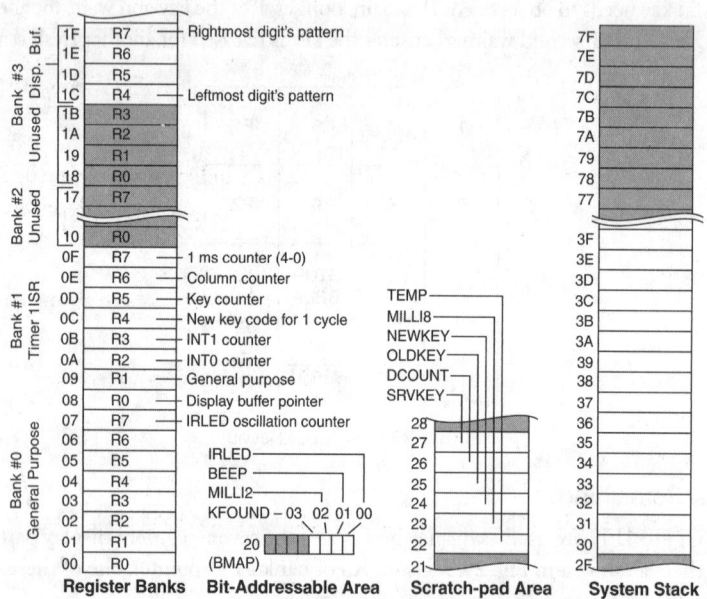

Figure 25.10 Register and memory usage of the software

25.4 | Software Listing

```
; $$$$$$$$$$$$$$$$$$$$$$$$$$$$$$$$$$$$$$$$$$$$$$$$$$$$$$$$$$$$$$$$$$$$$$$$$$$$$$$$$$$$
; Software for Home protection system
; Program developed by Dr Subrata Ghoshal, December 2008

; Program for 8051 with 12 MHz crystal
; Display refreshed from left to right

; Register banks usage
; Bank #0
; R7 used by Timer 0 ISR to count IR toggling. Reset value 09H.

; Bank #1 (used by Timer 1 ISR)
```

```
; R0 display buffer pointer
; R1 for general purpose usage
; R2 used by external interrupt INT0
; R3 used by external interrupt INT1
; R4 new key code (for one cycle of 8 milliseconds)
; R5 key counter
; R6 column counter
; R7 used by Timer 1 interrupt to count 4 interrupts to get 1 millisecond

; 1CH to 1FH for display buffer (display refreshed from left to right)
; 1CH = holds pattern for leftmost digit
; 1FH = holds pattern for rightmost digit

; Bit-addressable area
        BMAP    equ    20H                    ; first byte of bit-addressable area
        IRLED   equ    BMAP.0                 ; 1 = toggle P0.4, 0 = leave it as 1
        BEEP    equ    BMAP.1                 ; 1 = beep, 0 = no beep
        MILLI2  equ    BMAP.2                 ; toggled every millisecond, 0 = display refresh
        KFOUND  equ    BMAP.3                 ; service key in SRVKEY (for main loop)

        TEMP    equ    22H                    ; for temporary storage
        MILLI8  equ    23H                    ; counts 8 milliseconds in step of 2
        NEWKEY  equ    24H                    ; key for last cycle of 8 milliseconds
        OLDKEY  equ    25H                    ; key code, being debounced
        DCOUNT  equ    26H                    ; debounce counter
        SRVKEY  equ    27H                    ; key code to be serviced by main loop

; System stack from 30H onwards
; $$$$$$$$$$$$$$$$$$$$$$$$$$$$$$$$$$$$$$$$$$$$$$$$$$$$$$$$$$$$$$$$$$$$$$$
; Entry point for system reset

                ORG    0000H
        START:  AJMP   RESTRT                 ; continue at RESTRT
; $$$$$$$$$$$$$$$$$$$$$$$$$$$$$$$$$$$$$$$$$$$$$$$$$$$$$$$$$$$$$$$$$$$$$$$
; External interrupt INT0 ISR
; Increment memory location 0AH (R2 of bank #1) by one

                ORG    0003H
        EXINT0: INC    0AH                    ; increment R2 of bank #1 by one
                RETI
; $$$$$$$$$$$$$$$$$$$$$$$$$$$$$$$$$$$$$$$$$$$$$$$$$$$$$$$$$$$$$$$$$$$$$$$
; Entry point for Timer 0 interrupt.

                ORG    000BH
        TMR0:   AJMP   ISRT0                  ; continue at Timer 0 interrupt service routine
; $$$$$$$$$$$$$$$$$$$$$$$$$$$$$$$$$$$$$$$$$$$$$$$$$$$$$$$$$$$$$$$$$$$$$$$
; External interrupt INT1 ISR
; Increment memory location 0BH (R3 of bank #1) by one.

                ORG    0013H
        EXINT1: INC    0BH                    ; increment R3 of bank #1 by one
                RETI
```

```
; $$$$$$$$$$$$$$$$$$$$$$$$$$$$$$$$$$$$$$$$$$$$$$$$$$$$$$$$$$$$$$$$$$$$$$$$
; Timer 1 ISR
; Uses Register Bank #1.
; Timer 1 is programmed in Mode 2 (8-bit auto-reload mode) to generate interrupts at
; every 250 microseconds interval. Duties of Timer 1 are as follows:
;    1. At every 250 microseconds toggle P0.7 (audio output) if BEEP is set.
;    2. At every millisecond, set IRLED bit.
;    3. At every 2 milliseconds, refresh one digit and scan one key column.
;    4. At every 8 milliseconds refresh all four digits once and scan all 16 keys.
;    5. At every 8 milliseconds, read and clear R2 and R3 of bank #1. If 0 is found,
;       then set alarm.
;    6. If any key is detected, then debounce it for 50 milliseconds.
;    7. For a debounced key, set KFOUND bit for main loop for key servicing.
; Note that key scanning is to be skipped if KFOUND bit is set. KFOUND is set
; by Timer 1 ISR when a valid key is scanned and cleared by main loop after servicing that key.

                ORG     001BH
    ISRT1:      PUSH    PSW                 ; save processor status
                PUSH    ACC                 ; save accumulator
                MOV     PSW, #08H           ; select bank #1
                JNB     BEEP, NOBEEP        ; no audio servicing
                CPL     P0.7                ; toggle audio output
    NOBEEP:     DJNZ    R7, OVRT1           ; 1 ms not over yet

; One millisecond over. Reload R7. Set IRLED bit to start IR sensing.

                MOV     R7, #04H            ; reload R7 for next millisecond count
                SETB    IRLED               ; start IR sensor

; Toggle MILLI2 bit. If it is 0, then refresh next display digit, pointed by R0.

                CPL     MILLI2              ; to keep track of 2 milliseconds
                JB      MILLI2, OVRT1       ; check for 2 milliseconds

; Two milliseconds over. Switch to next scan-line column. Read one column of keys.
; Then display pattern for current scan line.

                MOV     P1, #00H            ; blank display
                INC     R0                  ; display pattern (segment) pointer
                INC     R6                  ; scan counter (0 to 3)
                CJNE    R6, #04, STEP3      ; overshoot?
                MOV     R6, #00H            ; reset column counter
                MOV     R0, #1CH            ; reset with leftmost digit pattern
                MOV     R5, #00H            ; reset key counter
                MOV     NEWKEY, R4          ; save key code of this cycle in NEWKEY
                MOV     R4, #00H            ; reset new key code

    STEP3:      MOV     A, R6               ; get current scan or column-drive pattern
                ORL     A, #0FCH            ; set bits from 7 to 2
                MOV     P3, A               ; turn on a fresh scan line (column)
```

; Before sending segment pattern for the digit, read keys of this column.

```
                NOP
                NOP                         ; allow some time to stabilize
                MOV    A, P3                ; bits 4 to 7 are return lines
                MOV    TEMP, A              ; temp save
                MOV    A, @R0               ; get current display pattern (segments)
                MOV    P1, A                ; turn on segments for current digit
```

; Now check for any key. Sensed key return lines were saved in TEMP.
; Skip this part if KFOUND is set.

```
                JB     KFOUND, CHKML8       ; skip if KFOUND is set
                MOV    R1, #04H             ; counter for four return lines checking
                MOV    A, TEMP              ; get back Port 3 input (bits 4 to 7)
    KEYBIT:     INC    R5                   ; update key counter
                RLC    A                    ; MSB shifted to CY flag
                JC     NXTBIT               ; this key not pressed
                MOV    0CH, R5              ; copy key counter value in R4 of bank #1
    NXTBIT:     DJNZ   R1, KEYBIT           ; check all 4 bits for any key pressing
```

; Check for 8 milliseconds

```
    CHKML8:  DJNZ   MILLI8, OVRT1           ; continue for 8 milliseconds
```

; Eight milliseconds over. Reload MILLI8 by 4. Check R2 and R3. If zero, then set alarm. Otherwise clear
; these two.

```
                MOV    MILLI8, #04H         ; reload MILLI8 for next cycle
                MOV    A, R2
                JNZ    CHKR3
                SETB   BEEP                 ; start audio alarm, press STOP BEEP
                                            ; to end
    CHKR3:      MOV    A, R3
                JNZ    CLR2R3
                SETB   BEEP                 ; start audio alarm, press STOP BEEP
                                            ; to end
    CLR2R3:     MOV    R2, #00H
                MOV    R3, #00H
```

; Now check for any valid key. Skip if KFOUND is set.

```
                JB     KFOUND, OVRT1        ; skip key-checking part
                MOV    A, NEWKEY
                JZ     OVRT1                ; no key sensed
                CJNE   A, OLDKEY, NEWDEB    ; new key
```

; Old key, being debounced. See if debouncing is over or not.

```
                DJNZ   DCOUNT, OVRT1        ; yet to be fully debounced
                MOV    SRVKEY, A
                SETB   KFOUND               ; for main loop
                SJMP   OVRT1
```

```
        NEWDEB:   MOV   OLDKEY, A        ; save latest key code to start debouncing
                  MOV   DCOUNT, #05H     ; reload debounce counter by 5
```

; Restore status and return.

```
        OVRT1:    POP   ACC              ; restore accumulator
                  POP   PSW              ; restore processor status
                  RETI                   ; return from Timer 1 interrupt
```
; $$$
; Timer 0 ISR
; Timer 0 is programmed at its Mode 2 (8-bit auto-reload mode) to generate interrupts at every 17 microsec
; onds. It toggles status of P0.4 eight times, and for remaining part of the millisecond, maintains P0.4 output as 1,
; which keeps IR LED switched off. Toggling starts if the IRLED bit is set. If it is cleared, then toggling does
; not take place. Register R7 of bank #0 is used to count number of toggling. Reset value of this register. R7
; of bank #0 is 9.
; The bit variable IRLED is cleared by system reset and set at every millisecond by Timer 1 interrupt routine.
; The IRLED bit is cleared by Timer 0 ISR itself after completion of eight togglings.

```
        ISRT0:    JNB   IRLED, OVRT0     ; exit, no need to toggle P0.4
                  CPL   P0.4             ; toggle IR LED
                  DJNZ  R7, OVRT0        ; toggling IR LED to be continued
                  MOV   R7, #09H         ; Toggling over, reload counter for next turn
                  SETB  P0.4             ; turn off IR LED
                  CLR   IRLED            ; no more toggling
        OVRT0:    RETI                   ; return from Timer 0 interrupt
```
; $$$
; Reset initialization. Following actions are automatically implemented by 8051 during a power-on reset:
;   Program Counter is cleared.
;   Register bank #0 is selected.
;   Stack Pointer is loaded by 07H.
;   All Ports output FFH.
;   Accumulator is cleared.
;   All interrupts are disabled.
; Blank display by clearing Port 1. Clear 128 bytes of internal RAM.

```
        RESTRT:   MOV   P1, #00H         ; blank display, all segments turned off
                  MOV   R0, #7FH         ; point highest internal RAM address
        CLRAM:    MOV   @R0, A           ; clear one byte
                  DJNZ  R0, CLRAM        ; point next lower byte till address 00H
```

; Perform a self-test by turning on all four digits one by one. Then make a beep.

```
                  MOV   R7, #04H         ; to test four digits
                  MOV   P1, #0FFH        ; turn on all segments
        SLFTST:   MOV   A, R7            ; get count value
                  DEC   A                ; scan code ranges from 3 to 0
                  MOV   P3, A            ; turn on one digit
                  ACALL DELAY            ; wait for a second
                  DJNZ  R7, SLFTST       ; continue for four digits
                  MOV   P1, #00H         ; blank display
```

; Now make a beep. Port pin P0.7 controls the buzzer.

```
            MOV    R7,#08H
BEEP1:      MOV    R6,#00H
BEEP2:      CPL    P0.7                 ; toggle buzzer output
            MOV    R5,#80H
BEEP3:      DJNZ   R5, BEEP3            ; wait before next toggling
            DJNZ   R6, BEEP2
            DJNZ   R7, BEEP1
```

; Complete all initialization before entering into the main loop. Start with ports.
; Port 1 had already been blanked for no display.

```
            MOV    P0,#0FFH             ; enable all pins as inputs
            MOV    P3,#0FFH             ; enable all pins as inputs
```

; Initialize system stack from 30H upwards. To achieve this, load SP by 2FH.

```
            MOV    SP,#2FH              ; stack from 30H onwards
```

: All internal RAM locations, including bit-addressable area, had already been cleared.
; Initialize a few bytes, only for non-zero values. Presently active bank is #0.

```
            MOV    R7,#09H              ; Load R7 of bank #0 by 09H
            MOV    PSW,#08H             ; Select bank #1
            MOV    R0,#1FH              ; Load R0 of bank #1 by 1FH (rightmost digit)
            MOV    R6,#03H              ; Load R6 of bank #1 by 03H
            MOV    R7,#04H              ; Load R7 of bank #1 by 04H
            MOV    PSW,#00H             ; Select bank #0
            MOV    MILLI8,#04H
            MOV    DCOUNT,#05H          ; debounce counter
```

; Load display buffer by sign on message 'SAFE'.

```
            MOV    1CH,#6DH             ; segment pattern for 'S'
            MOV    1DH,#77H             ; segment pattern for 'A'
            MOV    1EH,#71H             ; segment pattern for 'F'
            MOV    1FH,#79H             ; segment pattern for 'E'
```

; Initialize both external interrupts as falling edge sensitive. Both Timers in Mode 2.
; Initialize both Timers and start. Also initialize all interrupts.

```
            MOV    TMOD,#22H            ; both Timers in Mode 2, Gates cleared
            MOV    TL0,#0EEH            ; Timer 0 interrupts at every 17 microseconds
            MOV    TH0,#0EEH            ; reload value for Timer 0
            MOV    A,#05H               ; for next four instructions
            MOV    TL1, A               ; Timer 1 interrupts at every 250 microseconds
            MOV    TH1, A               ; reload value for Timer 1
            MOV    TCON, A              ; both INT0 and INT1 falling-edge sensitive
            MOV    IP, A                ; higher priority to both external interrupts
            MOV    IE,#0FH              ; enable INT0, INT1, Timer 0 & Timer 1 interrupts
            SETB   TCON.4               ; start Timer 0
            SETB   TCON.6               ; start Timer 1
            SETB   IE.7                 ; enable all interrupts (global)
```

```
; $$$$$$$$$$$$$$$$$$$$$$$$$$$$$$$$$$$$$$$$$$$$$$$$$$$$$$$$$$$$$$$$$$$$$$$$$$$
;              MAIN LOOP
; Duty of main loop is to poll four inputs from P0.0 to P0.3 and start audio alarm as and when necessary.
; Main loop must also service any key scanned by Timer 1 interrupt service routine. After servicing the key,
; KFOUND bit must be cleared by main loop.

    MAIN:       MOV   SP, #2FH           ; reset system stack
                MOV   PSW, #00H          ; select bank #0
                MOV   A, P0              ; to read window keys inputs
                ORL   A, #0F0H           ; set all bits of higher nibble
                CPL   A                  ; if no key pressed, all bits would be 0s
                JZ    KSRV               ; no window key activated
                SETB  BEEP               ; start audio alarm
    KSRV:       JNB   KFOUND, MAIN       ; no keyboard key found

; One key of keyboard is pressed. SRVKEY contains key code. Branch to proper entry point as per key code.
; In this software, only two keys are serviced, namely, STOP BEEP and RESTART. Suitable routines for
; other keys may be developed accordingly.

                MOV   A, SRVKEY          ; get key code scanned
                CJNE  A, #0BH, CHK2      ; is it STOP BEEP key?
                CLR   BEEP               ; stop audio alarm
                SJMP  KWAIT              ; continue at KWAIT
    CHK2:       CJNE  A, #0CH, KWAIT     ; is it RESTART key?
                LJMP  START              ; restart from reset initialization

; Wait for a second to get the key released.

    KWAIT:      ACALL DELAY              ; wait for key release
                CLR   KFOUND             ; allow fresh key sensing
                AJMP  MAIN               ; continue at main loop

; $$$$$$$$$$$$$$$$$$$$$$$$$$$$$$$$$$$$$$$$$$$$$$$$$$$$$$$$$$$$$$$$$$$$$$$$$$$
;    SERVICE ROUTINE
; $$$$$$$$$$$$$$$$$$$$$$$$$$$$$$$$$$$$$$$$$$$$$$$$$$$$$$$$$$$$$$$$$$$$$$$$$$$
; Name:      DELAY
; Function:  Generates about one second delay whenever called.
; Input:     none
; Output:    none
; Calls:     none
; Uses:      R2, R3, R4 of Register bank #0

    DELAY:   MOV   R4, #08H
    DELAY1:  MOV   R3, #00H
    DELAY2:  MOV   R2, #00H
    DELAY3:  DJNZ  R2, DELAY3
             DJNZ  R3, DELAY2
             DJNZ  R4, DELAY1
             RET
; $$$$$$$$$$$$$$$$$$$$$$$$$$$$$$$$$$$$$$$$$$$$$$$$$$$$$$$$$$$$$$$$$$$$$$$$$$$
end
```

SUMMARY

The construction details of a simple home-protection system are discussed in this chapter. The system is composed of two IR sensors at two doors and four touch-type sensors at four windows. These are interfaced with an 8051 microcontroller, which monitors these sensors continuously. The system also contains a 16-key keyboard and a four-digit display along with an audio output to generate audio alarm.

Timer 1 interrupt service routine oversees all major operations of the system. Timer 0 generates the required frequency for IR transmitter. Two IR receivers generate continuous interrupts through INT0 and INT1 external interrupts. The absence of any one of these two interrupts indicates obstruction between IR transmitter and its receiver, which is used to trigger the audio alarm.

The system performs a self-test and then displays its sign-on message 'SAFE'. An infinite loop monitors all 16 keys and refreshes the display continuously. In the present software, only two keys are serviced, and the remaining key service routines are left for the reader to develop.

POINTS TO REMEMBER

- TSOP-type IR receivers work properly only when its IR transmitter generates the identical carrier frequency of the TSOP-type IR receiver.

- Resistance of any LDR drops to a few Ohms when exposed to light and increases to Mega-Ohms or more in darkness.

REVIEW QUESTIONS

Evaluate Yourself

1. How many pins are there in the TSOP-type IR sensor?

   (a) 2   (b) 3

   (c) 4   (d) None of these

2. What is the expected amount of current required to drive the designed system?

   (a) 500 mA   (b) 1A

   (c) 5A   (d) None of these

3. Which of the following frequencies are used by the selected IR sensor?

   (a) 300 Hz   (b) 3 KHz

   (c) 30 KHz   (d) None of these

4. How many interrupts are involved for each IR-sensing module?

   (a) Two   (b) Three

   (c) Four   (d) None of these

5. If the audio alarm is active, which of the following keys may be used to stop it?

   (a) STOP BEEP   (b) RESTART

   (c) Either of these   (d) None of these

6. For the present hardware, what should be the content of the display buffer 1CH if we want to display '1' through the digit D1?

   (a) 01H   (b) 06H

   (c) Either of these   (d) None of these

7. In the present software, how much time is spent to debounce a key?

   (a) 40 milliseconds   (b) 50 milliseconds

   (c) 100 milliseconds   (d) None of these

8. Which interrupt receives higher priority in the present system?

   (a) Timer 0   (b) Timer 1

   (c) INT0 and INT1   (d) None of these

9. What is the frequency of the audio alarm?

   (a) 3 KHz   (b) 30 KHz

   (c) 300 KHz   (d) None of these

10. Sign-on message is the display during

   (a) system reset   (b) IR blocking

   (c) audio output   (d) none of these

Search for Answers

1. What extra features may later be incorporated within the system?

2. Which port is left for that purpose?

3. Is any register bank left for this purpose?

4. Why are IR sensors used in doors instead of LDRs?

5. What is the purpose of 10 Ohms resistors interfaced with 74245?

6. What are the duties of external interrupt service routines?

7. What is the purpose of waiting for a second after servicing any key?

8. What is the duty of the main loop in the present software?

9. What is the purpose of the self-test module?

10. Are the frequencies of audio alarm and self-test beep identical?

Think and Solve

1. What changes are necessary in the hardware circuit to incorporate a manual reset key?

2. What is the advantage of clearing the entire internal RAM during reset initialization?

3. Why are IR sensors not used in windows?

4. What is the purpose of a digit driver?

5. In the present circuit, only half of 74139 is used. Find a suitable usage for the other half of this device.

6. What is the advantage of using a piezo buzzer over a moving coil-type speaker?

7. What would happen if the IR LED is continuously turned on?

8. How does Timer 1 interrupt service routine control Timer 0 interrupts in the present software?

9. Can you identify the scan codes of each of the 16 keys of the system?

10. Why are SP and PSW reinitialized at the starting of the main loop?

26 ADVANCED MICROCONTROLLERS

CHAPTER OBJECTIVES

In this chapter, the reader is introduced to a few advanced microcontrollers. After completion of the chapter, the reader should be able to understand some of the features of:

- 8-bit microcontroller: Atmel's AVR (Atmega8),
- 16-bit microcontroller: Intel's MCS-96 and
- 32-bit microcontroller: ARM microcontrollers

26.1 | Introduction

After commercial introduction of microcontrollers in the world market, system designers and developers started realizing that it would provide a cost-effective solution to various programmable digital systems in contrast to the systems designed with microprocessors. This rise of demand of microcontrollers has led to the introduction of many varieties of microcontrollers in the IC market. Today, against every microprocessor-based system, there are at least 20 microcontroller-based systems, which may even increase to as many as 50 in the near future as visualized by the experts. Out of these, we will have some brief discussions on three advanced microcontrollers, which have been developed after 8051. However, the information presented in subsequent sections is introductory in nature, and interested readers may consult the relevant manufacturer's datasheets for detailed features, accurate specifications and functioning.

Are these microcontrollers going to completely replace all microprocessors in near future? Till date, the answer is 'no'. Microprocessors have their own advantages also. They are more suitable for a larger system, without any power consumption restrictions. However, who knows what is stored as the future of microprocessors?

26.2 | AVR Microcontrollers

AVR series microcontrollers were developed by Atmel around 1996. Atmel has been the pioneer in introducing *Flash memory* for program storage, available in AVR. These advanced RISC microcontrollers are built around the Harvard architecture. Two popular versions of these microcontrollers are ATmega8 and ATmega16. TINY and XMEGA versions of AVR are also available to cater to various application demands. In this section, we will take a closer look at ATmega8, and finally, we will identify the differences in ATmega16 against it.

26.2.1 | Important Features

The following are some of the important features of ATmega8.

(i) 8-bit advanced RISC architecture
(ii) Register-to-register architecture without any accumulator
(iii) 130 instructions including signed and unsigned multiplication
(iv) 32 general-purpose 8-bit registers, all capable of storing the result
(v) Fully static to 16 MHz operational frequency
(vi) 8K bytes on-chip Flash program memory
(vii) 1K bytes on-chip SRAM for data storage
(viii) 512 bytes on-chip EEPROM
(ix) Programming lock for software security
(x) 23 programmable I/O lines
(xi) Two 8-bit and one 16-bit Timer/Counters.
(xii) 6 channel 10-bit on-chip ADC (8 channel for TQFP package)
(xiii) Programmable serial USART
(xiv) SPI serial interface
(xv) Watchdog Timer with separate on-chip oscillator
(xvi) On-chip analog comparator
(xvii) External and internal interrupt sources
(xviii) Five sleep modes for power management

 Register to register architecture is a feature which totally eliminates the need of any accumulator. In early days of processors, all results were available in the accumulator only. Later the situation improved making any register to hold the result. In 8051, we got a glimpse of it when we observe that results of some logical operations with bytes are also available in some directly addressed memory locations.

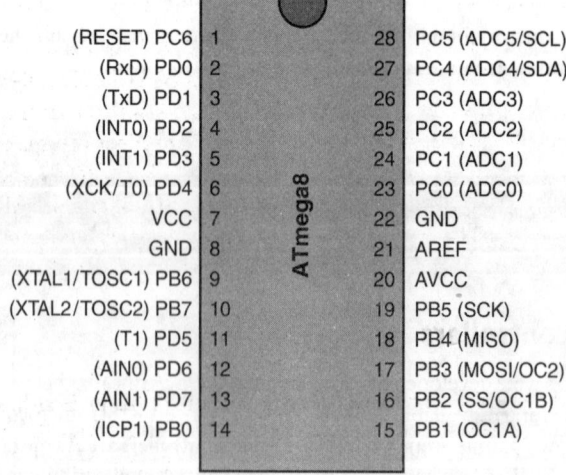

(RESET) PC6	1
(RxD) PD0	2
(TxD) PD1	3
(INT0) PD2	4
(INT1) PD3	5
(XCK/T0) PD4	6
VCC	7
GND	8
(XTAL1/TOSC1) PB6	9
(XTAL2/TOSC2) PB7	10
(T1) PD5	11
(AIN0) PD6	12
(AIN1) PD7	13
(ICP1) PB0	14

ATmega8

28	PC5 (ADC5/SCL)
27	PC4 (ADC4/SDA)
26	PC3 (ADC3)
25	PC2 (ADC2)
24	PC1 (ADC1)
23	PC0 (ADC0)
22	GND
21	AREF
20	AVCC
19	PB5 (SCK)
18	PB4 (MISO)
17	PB3 (MOSI/OC2)
16	PB2 (SS/OC1B)
15	PB1 (OC1A)

Figure 26.1 Pins and signals of ATmega8 (28-pin PDIP)

26.2.2 | Pins and Signals

Fig. 26.1 shows all signals against their respective pin numbers for a 28-pin PDIP of ATmega8. Note that all 23 I/O lines have alternative functions. Only five input lines, reserved for power supply, are not assigned any alternative functions.

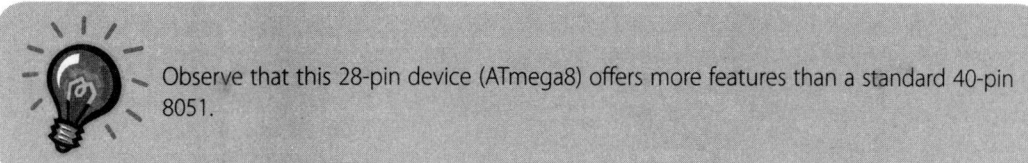

Observe that this 28-pin device (ATmega8) offers more features than a standard 40-pin 8051.

26.2.3 | External Reset and Crystal Inputs (Optional)

An initial observation of Fig. 26.1 would reveal that even the system reset input or external crystal inputs are optional. This is because the device contains a built-in RC oscillator circuit capable of generating any one of the four standard software-selectable frequencies of 1.0, 2.0, 4.0 and 8.0 MHz. Alternatetively, an external crystal may also be connected with Pins 9 and 10 for any desired frequency up to 16 MHz (8 MHz for ATmega8L). There is an internal power-on reset facility to initialize various parameters in absence of any external reset input, which is optional.

26.2.4 | Operating Voltages and Power Consumption

As most of the embedded systems designed with microcontrollers are battery-powered, power supply and its consumption play an important role in all advanced microcontrollers. Operating voltage of ATmega8 is from 4.5 V to 5.5 V, while that of ATmega8L is between 2.7 V and 5.5 V. For an operational speed of 4 MHz with 3V power source, it consumes 3.6 mA when it is active; this is reduced to 1 mA in its idle mode and further reduced to 0.5 μA in power-down mode. Each port pin of ATmega8L is capable of sourcing or sinking 20 mA of current at 5 V and 10 mA of current at 3 V. Note that the *sourcing and sinking limits are identical for this device.*

26.2.5 | Internal Architecture

A simplified internal architecture of ATmega8 is presented in Fig. 26.2. As it may be readily observed, its central feature is the AVR core (the portion not shaded), which consists ALU, 32 general-purpose registers including three index registers, the processor status register or flag register (SREG), program counter (PC), stack pointer and the instruction register with the instruction decoder generating various control lines.

The remaining portion contains program and data memory (Flash, SRAM and EEPROM), ADC with multiplexing unit, I/O ports (designated as B, C and D), Timer/Counters, USART, Interrupt unit, analog comparator interface, oscillator circuits and serial interfaces. The reader should note the absence of any external memory interfacing provision in AVR. The AVR works on the principle that for larger memory demand, some other AVR version needs to be selected.

26.2.6 | Program Memory (Flash)

As all instructions' opcodes for ATmega8 are either 16-bit or 32-bit in length, the on-chip 8K bytes of program memory are arranged as 4K words, which are accessed by a 12-bit PC. For software security, the Flash is divided into two sections, namely, Boot section and Application Program section. The Boot section occupies the higher part of the Flash and the lower portion is left for application programs as shown in Fig. 26.3. Note that the size of the Boot section is configurable through fuses.

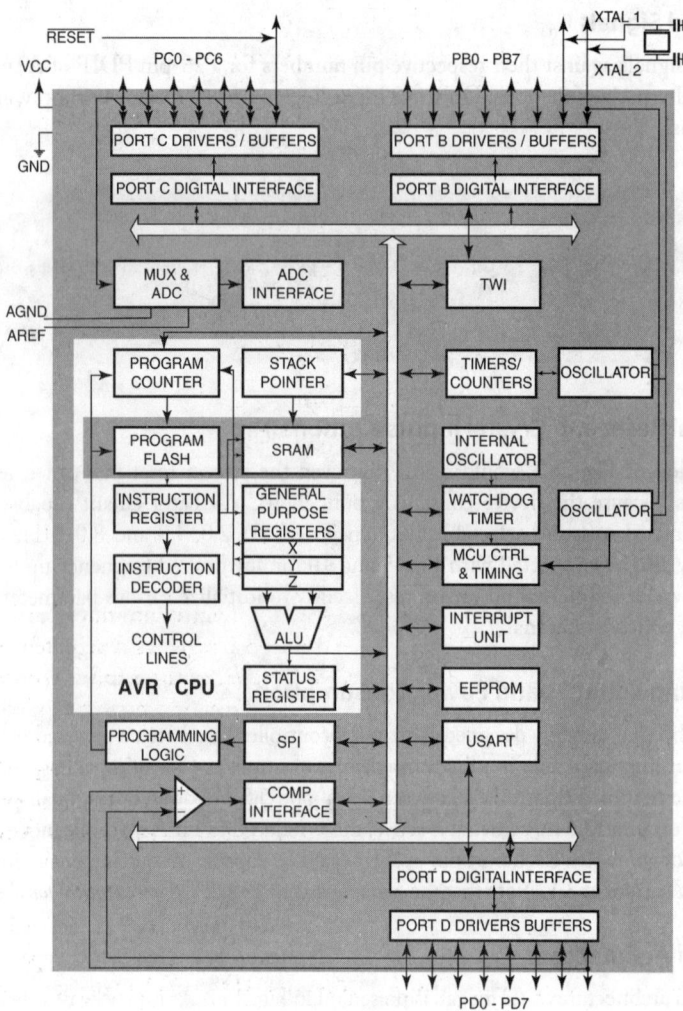

(Courtesy: Atmel Corporation)

Figure 26.2 Simplified internal architecture of Atmega8

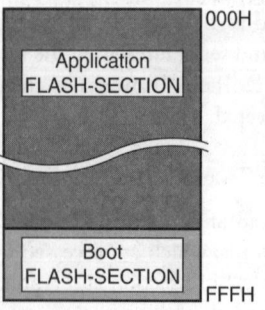

Figure 26.3 Map of program memory (Flash)

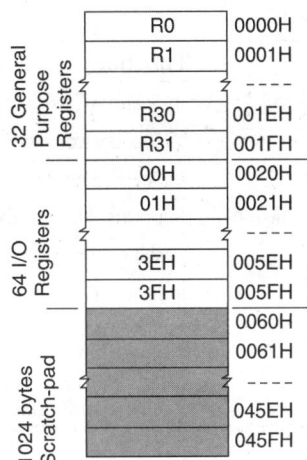

Figure 26.4 Organization of 1,120 bytes of SRAM data memory

26.2.7 | Data Memory (SRAM)

ATmega8 offers 1,120 bytes of data memory (SRAM), which is divided into three parts, as illustrated in Fig. 26.4. The lowest 32 bytes are occupied by 32 general-purpose registers designated as R0, R1 and so on upto R31. Register pairs from R26 to R31 are also designated as three index registers, X, Y and Z. Registers may be addressed by their names (R0, R1, etc.) or by direct or other types of addressing modes. Direct addresses of registers vary from 0000H to 001FH

ATmega8 does not have any accumulator for storing the result of arithmetical or logical operations. The destination register indicated in the opcode of the instruction holds the result. This is known as 'register-to-register architecture'. It would be the designated register pair that holds the 16-bit results. Any one or a pair of 32 general-purpose registers may be used as the destination register.

Sixty-four I/O registers are located after general-purpose registers. In 8051, we have come across Special Function Registers or SFRs. In ATmega8, these I/O registers serve the identical purpose as that of SFRs for controlling various activities of processor operations. Note that two types of addresses are applicable for I/O registers: absolute address (from 0020H to 005FH) and I/O register address (from 00H to 3FH). The remaining 1,024 bytes of SRAM may be used as a scratch pad for data storage.

26.2.8 | Status Register

Located as the highest addressed register of I/O registers (absolute address 005FH), this 8-bit register contains eight flags as shown in Fig. 26.5. Table 26.1 presents the names of these flags along with their functions.

26.2.9 | Stack and Stack Pointer

ATmega8 allows stack anywhere within 1,024 bytes of data SRAM. The stack must avoid the lower 96 bytes of SRAM area containing general-purpose registers and I/O registers. Stack is always pointed

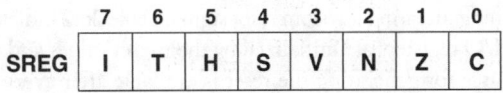

Figure 26.5 Details of SREG of ATmega8

Table 26.1 Descriptions of ATmega8 flags

SREG bit	Name	Designation	Function
0	C	Carry flag	Indicates a Carry after any arithmetical or logical operation
1	Z	Zero flag	Indicates zero result after any arithmetical or logical operation
2	N	Negative flag	Indicates a negative result in arithmetical or logical operation
3	V	Two's complement Overflow flag	Supports two's complement arithmetic
4	S	Sign bit	Result of exclusive OR of N and V fags of SREG
5	H	Half Carry flag	Indicates eventual half Carry in arithmetical operations
6	T	Bit copy storage	Serves as a source or destination for bit-copy operation
7	I	Global interrupt enable	Disables all interrupts if cleared
			If set, then interrupts are enabled by their individual control bits

by the 16-bit stack pointer, which is composed of two 8-bit registers SPH and SPL. Note that a PUSH command decreases the stack pointer and a POP command increases it. That is, the stack grows from higher to lower address, and the highest assigned address of the stack area is to be used to initialize the stack pointer. For example, if 0100H to 01FFH of SRAM is earmarked for stack, then the stack pointer must be initialized by 01FFH with SPH containing 01H and SPL containing FFH. To store the return address for subroutine calls, the SP uses a post-decrement addressing scheme.

26.2.10 | Power Management and Sleep Modes

ATmega8 offers five sleep modes for efficient power management. They are

- Idle mode
- ADC noise reduction mode
- Power-down mode
- Power-save mode
- Standby mode

The SLEEP instruction must be executed with MSB of MCUCR register set and proper sleep mode selected through SM2, SM1 and SM0 bits of MCUCR. A reset or interrupt wakes up the CPU, and after the start-up time and an additional four cycles of halting, the processor starts executing from the next instruction after the SLEEP instruction if it is awakened by interrupts. If it is awakened by reset, then the execution starts from the reset vector.

Compare these power saving modes with those offered by 8051.

26.2.11 | System Reset

ATmega8 offers four sources for generating a system reset signal. Table 26.2 indicates these sources with a brief description. During reset, all I/O registers are initialized by their reset values and the program starts executing from the reset vector. The source for generating the reset is available after every reset in the least significant 4 bits of the MCUCSR register.

Table 26.2 Reset sources for ATmega8

Reset type	Executed when
Power-on reset	The supply voltage is below the power-on reset threshold
External reset	A low-level is applied at external RESET input for longer that the minimum pulse length
Watchdog reset	Watchdog is enabled, and the Watchdog Timer period expires
Brown-out reset	The supply voltage is below the brown-out reset threshold, and brown-out detector is enabled

26.2.12 | Watchdog Timer

Any Timer generates an interrupt at its terminal count, and Watchdog Timer is no exception. However, the specialty of the Watchdog Timer is that instead of generating a normal interrupt and drawing the attention of the processor, it triggers a system reset. This is very useful if the system is in an infinite loop or just 'hanging'. To avoid any unwanted system reset, the Watchdog Timer must be periodically serviced by the system software.

ATmega8 offers a built-in Watchdog Timer, which may be enabled or disabled by the software commands through its register, which is designated as WDTCR. The Watchdog Timer is operated by a separate internal oscillator at 1 MHz frequency.

26.2.13 | ATmega8 Instruction Set

ATmega8 has a rich set of 130 instructions. They are divided into five groups, namely

   (i)   Arithmetical and logical instructions,
  (ii)   Branch instructions,
 (iii)   Data transfer instructions,
 (iv)   Bit and bit-test instructions and
  (v)   MCU control instructions.

Group-wise descriptions of all instructions are presented from Table 26.3 to Table 26.7 along with some brief introductory comments.

Arithmetical and logical instructions: All conventional arithmetical (like add, subtract, etc.) and logical (like AND, OR, XOR, etc.) along with increment by one and decrement by one instructions are available in this instruction group. In general, these instructions deal with 8-bit operations. Both unsigned and signed multiplication instructions are available for 8-bit numbers, including fractional format, producing 16-bit result. Table 26.3 presents a summary of all arithmetical and logical instructions offered in ATmega8.

Branch instructions : Program branching plays an important role for any software, and a wide variety of branching instructions is offered in ATmega8 as presented in Table 26.4. Branching instructions dependent on SREG flag bits are related with PC within the range between −63 and +64 bytes. The Z register (16-bit) plays an important role in some jump and call instructions by providing the 16-bit branching address to be loaded in PC.

Data-transfer instructions: A large number of data transfer instructions are available in ATmega8 with several addressing modes. Some instructions offer post-increment or pre-decrement options through index registers for easier access to arrays. In case of the post-increment option, the data is first copied from the address available in the index register and then the index register content is incremented by one. On the other hand, for the pre-decrement instructions, the content of the index register is first decremented by one and then the data pointed by it is copied to the destination register.

Table 26.3 ATmega8 Arithmetical and logical instructions

Mnemonics	Description	Mnemonics	Descriptions
ADD	Add two registers	NEG	Two's complement
ADC	Add with Carry two registers	SBR	Set bit(s) in register
ADIW	Add immediate to word	CBR	Clear bit(s) in register
SUB	Subtract two registers	INC	Increment
SUBI	Subtract constant from register	DEC	Decrement
SBC	Subtract with Carry two registers	TST	Test for zero or minus
SBCI	Subtract with Carry constant from register	CLR	Clear register
SBIW	Subtract immediate from word	SER	Set register
AND	Logical AND registers	MUL	Multiply unsigned
ANDI	Logical AND register and constant	MULS	Multiply signed
OR	Logical OR registers	MULSU	Multiply signed with unsigned
ORI	Logical OR register and constant	FMUL	Fractional multiply unsigned
EOR	Exclusive OR registers	FMULS	Fractional multiply signed
COM	One's complement	FMULSU	Fractional multiply signed with unsigned

Table 26.4 ATmega8 branch instructions

Mnemonics	Description	Mnemonics	Description
RJMP	Relative jump	BRNE	Branch if not equal
IJMP	Indirect jump to (Z)	BRCS	Branch if Carry set
RCALL	Relative subroutine call	BRCC	Branch if Carry cleared
ICALL	Indirect call to (Z)	BRSH	Branch if same or higher
RET	Subroutine return	BRLO	Branch if lower
RETI	Interrupt return	BRMI	Branch if minus
CPSE	Compare, skip if equal	BRPL	Branch if plus
CP	Compare	BRGE	Branch if greater or equal, signed
CPC	Compare with Carry	BRLT	Branch if less than zero, signed
CPI	Compare register with immediate	BRHS	Branch if half Carry flag set
SBRC	Skip if bit in register cleared	BRHC	Branch if half Carry flag cleared
SBRS	Skip if bit in register is set	BRTS	Branch if T flag set
SBIC	Skip if bit in I/O register cleared	BRTC	Branch if T flag cleared
SBIS	Skip if bit in I/O register is set	BRVS	Branch if overflow flag is set
BRBS	Branch if status flag set	BRVC	Branch if overflow flag is cleared
BRBC	Branch if status flag cleared	BRIE	Branch if interrupt enabled
BREQ	Branch if equal	BRID	Branch if interrupt disabled

Table 26.5 ATmega8 data transfer instructions

Instruction	Description	Instruction	Description
MOV Rd, Rr	Move between registers	ST Y, Rr	Store indirect
MOVW Rd, Rr	Copy register word	ST Y+, Rr	Store indirect and post-increment
LDI Rd, K	Load immediate	ST –Y, Rr	Store indirect and pre-decrement
LD Rd, X	Load indirect	STD Y+q, Rr	Store indirect with displacement
LD Rd, X+	Load indirect and post-increment	ST Z, Rr	Store indirect
LD Rd, –X	Load indirect and pre-decrement	ST Z+, Rr	Store indirect and post-increment
LD Rd, Y	Load indirect	ST –Z, Rr	Store indirect and pre-decrement
LD Rd, Y+	Load indirect and post-increment	STD Z+q, Rr	Store indirect with displacement
LD Rd, –Y	Load indirect and pre-decrement	STS k, Rr	Store direct to SRAM
LDD Rd, Y+q	Load indirect with displacement	LPM	Load program memory
LD Rd, Z	Load indirect	LPM Rd, Z	Load program memory
LD Rd, Z+	Load indirect and post-increment	LPM Rd, Z+	Load program memory and post-increment
LD Rd, –Z	Load indirect and Pre-decrement	SPM	Store program memory
LDD Rd, Z+q	Load indirect with displacement	IN Rd, P	In port
LDS Rd, k	Load direct from SRAM	OUT P, Rr	Out port
ST X, Rr	Store indirect	PUSH Rr	Push register on stack
ST X+, Rr	Store indirect and post-increment	POP Rd	Pop register from stack
ST –X, Rr	Store indirect and pre-decrement	—	—

This post-increment and pre-decrement operations are always limited to 256 bytes as only the lower part of the 16-bit register is changed. Table 26.5 presents the summary of all data transfer instructions offered by ATmega8.

Bit and bit-test instructions: ATmega8 allows transactions of bit-wise information for all 32 general-purpose registers and lower 32 I/O registers. Bits of upper 32 I/O registers cannot be accessed in this fashion. The source or destination of the bit-information transfer must be the T flag bit of SREG, the status register. Bits are identified by their respective register's name and the concerned bit number (0 to 7).

Bit-oriented instructions allow the setting and clearing of all flag bits of SREG and other addressable bits of general-purpose registers and I/O registers. Instruction for swapping 4 bits (nibble) at a time is also offered and is applicable for all 32 general-purpose registers. Table 26.6 presents a summary of these instructions.

MCU control instructions: Out of three instructions of this group (shown in Table 26.7), SLEEP allows power management as discussed earlier. The Watchdog Timer may be reset by the WDR instruction.

Table 26.6 ATmega8 bit and bit-test instructions

Mnemonics	Description	Mnemonics	Description
SBI	Set bit in I/O register	SEN	Set negative flag
CBI	Clear bit in I/O register	CLN	Clear negative flag
LSL	Logical shift left	SEZ	Set zero flag
LSR	Logical shift right	CLZ	Clear zero flag
ROL	Rotate left through Carry	SEI	Global interrupt enable
ROR	Rotate right through Carry	CLI	Global interrupt disable
ASR	Arithmetic shift right	SES	Set signed test flag
SWAP	Swap nibbles	CLS	Clear signed test flag
BSET	Flag set	SEV	Set two's complement overflow
BCLR	Flag clear	CLV	Clear two's complement overflow
BST	Bit store from register to T	SET	Set T in SREG
BLD	Bit load from T to register	CLT	Clear T in SREG
SEC	Set Carry	SEH	Set half Carry flag in SREG
CLC	Clear Carry	CLH	Clear half Carry flag in SREG

Table 26.7 ATmega8 MCU control instructions

Mnemonic	Description
NOP	No operation
SLEEP	Sleep
WDR	Watchdog reset

26.2.14 | Comparison Between ATmega8 and ATmega16

An advanced version of ATmega8 is ATmega16, which offers more program memory and I/O lines. Available in a 40-pin PDIP (also in other packages), ATmega16 offers all instructions available in ATmega8 (plus one extra) and allows Joint Test Action Group (JTAG) interface for debugging, which is explained in the next section.

It also offers one extra PWM channel as compared with ATmega8 and one more sleep mode for better power management. The major differences between ATmega8 and ATmega16 are presented in Table 26.8. For more details about ATmega8 and ATmega16, interested readers may consult the manufacturer's datasheets.

Table 26.8 Differences between ATmega8 and ATmega16

	ATmega8	ATmega16
Number of instructions	130	131
Size of program memory (Flash)	8K	16K
Number of I/O lines	23	32
PDIP	28-pin	40-pin
PWM channels	3	4
Sleep modes	5	6
JTAG interface	Not available	Available

26.2.15 | JTAG Interface

JTAG is a special 4/5-signal interface to test functional characteristics of any circuit board or chip. Originally designed to test circuit boards, JTAG has now been accepted as the IEEE 1149.1 standard and is known as Standard Test Access Port and Boundary-Scan Architecture. Presently JTAG is extensively used for IC testing.

Generally, JTAG connectors are 10-pin or 20-pin and all pins of one side are connected with GND. Any JTAG port of a device offers the following signals.

  (i)   TCK (test clock)
  (ii) TMS (test mode select)
  (iii) TDI (test data in)
  (iv) TDO (test data out)
  (v)   TRST (test reset) [optional]

With only one data line available, JTAG adopted serial protocol for testing. Multiple devices with JTAG port may be daisy chained for purpose of testing.

26.3 | MCS-96 Microcontrollers

MCS-96 represents a family of 16-bit microcontrollers launched by Intel during 1982. It is often referred to as the 8XC196 family or simply 80196. It is widely used in printers, modems and hard-disk drives. These microcontrollers have different operating frequencies and are offered in various packages. For the purpose of initial understanding, we take 8XC196 as a sample member of this family of MCS-96 microcontrollers.

26.3.1 | Important Features

The following are some of the important features of 8XC196.

  (i)   Register-to-register architecture
  (ii)   16K on-chip ROM/OTPROM
  (iii) 488 bytes register RAM
  (iv)   Five 8-bit I/O ports
  (v)   High speed I/O subsystem
  (vi)   Full-Duplex serial port
  (vii) 16-bit multiply and divide instructions
  (viii) 8 or 10-bit A/D converter with sample and hold
  (ix)   Four 16-bit software Timers
  (x)   16-bit up/down Counter with capture
  (xi)   Dynamically configurable 16-bit or 8-bit bus-width
  (xii) Three PWM outputs
  (xiii) 16-bit Watchdog Timer
  (xiv) 28 interrupt sources, 16 vectors
  (xv) $\overline{\text{HOLD}}/\overline{\text{HLDA}}$ bus protocol
  (xvi) Power-down and idle modes

26.3.2 | Pins and Signals

Fig. 26.6 illustrates pin and signal details of 8XC196 as in a 68-pin PLCC. Note that it is also available in 80-pin QFP and 80-pin SQFP. A brief introduction of all major signals is given in Table 26.9. Apart from ton-chip memory, MCS-96 offers the scope for interfacing external memory, if required, for the system. Complete 16-bit address and data lines along with necessary control signals are available for this interfacing.

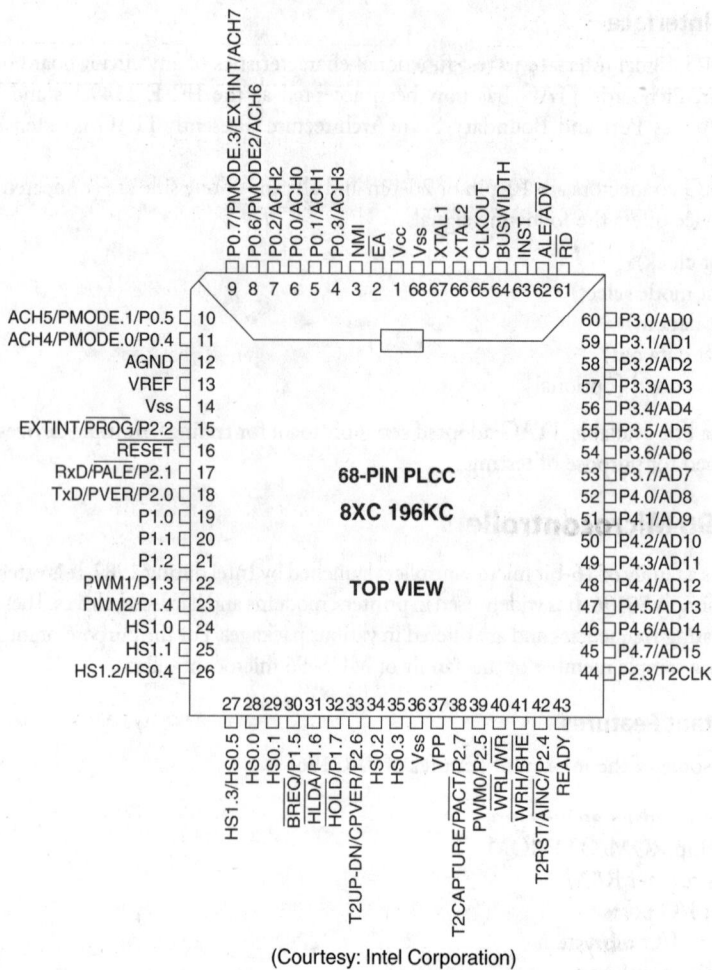

(Courtesy: Intel Corporation)

Figure 26.6 Pins and signals of 8XC196 (68-pin PLCC)

26.3.3 | Internal Architecture

MCS-96 offers a special interface for high-speed I/O devices. HS0 and HS1 pins are reserved for this operation. Out of five ports, one port (Port 0) is an input-only port and others are bi-directional. Port 0 is meant to be used for analog input for ADC channels. Otherwise it can also receive digital inputs.

Various other features are available in MCS-96, some of which are presented in the simplified internal architecture diagram in Fig. 26.7. The most important feature of MCS-96 is its register-to-register architecture. This avoids the conventional accumulator to store the result of all ALU operations. In MCS-96, any register can serve the purpose of the accumulator, and the ALU is for the register set; that is the reason for designating it as *RALU*. This architecture improves the I/O operation speed, as any register content may directly be moved out through the port or received without the need of passing it through an accumulator.

MCS-96 decodes the instructions through micro-coding, as indicated in Fig. 26.7. This micro-coding engine along with RALU, 24 bytes CPU SFR and 488 bytes of register RAM complete the CPU of the device or its core. The system bus width is programmable as either 8-bit or 16-bit through the BUSWIDTH input pin and the programmable condition of the CCR bit.

Table 26.9 Signal description of 8XC196

Signal	Description/Function
Vcc	Main supply voltage (5V).
Vss	Digital circuit ground (0V). There are multiple Vss pins, all of which must be connected.
VREF	Reference voltage for ADC (5V). It is also the supply voltage to the analog portion of ADC and the logic used to read Port 0. It must be connected if these are to function.
ANGND	Reference ground for ADC.
VPP	Timing pin for the return from power-down circuit. It also supplies programming voltage for the EPROM device.
XTAL1	Input of the oscillator inverter and of the internal clock generator.
XTAL2	Output of the oscillator inverter.
CLKOUT	Output of internal clock generator. Frequency is half of the oscillator frequency.
$\overline{\text{RESET}}$	Active-low system reset input.
BUSWIDTH	Input for bus width selection. If CCR bit is 0, bus is always 8 bit. If 1, then a high input allows 16-bit bus and a low input 8-bit bus.
NMI	Non-muskable interrupt input. A positive edge causes a vector through 203EH.
INST	Valid only for external memory accesses. It is high for instruction fetch and low for data reading.
$\overline{\text{EA}}$	Input for external memory access.
ALE/ADV	Address latch enable active only during external memory access.
$\overline{\text{RD}}$	Output only for external memory reading.
$\overline{\text{WR}}$/$\overline{\text{WRL}}$	Output only for external memory writing.
$\overline{\text{BHE}}$/$\overline{\text{WRH}}$	Activated only during external memory write cycles.
READY	Input to lengthen external memory cycles.
HSI	Inputs to high-speed input unit (HIS.2 and HIS.3).
HSO	Outputs from high-speed output unit (six HSO pins are available out of which two are shared).
Port 0	8-bit input-only port for analog or digital inputs.
Port 1	8-bit quasi-bi-directional I/O port.
Port 2	8-bit multi-functional port.
Ports 3 and 4	8-bit bi-directional I/O port shared with multiplexed address-data bus.
$\overline{\text{HOLD}}$	Input requesting control of the system bus.
$\overline{\text{HLDA}}$	HOLD acknowledge output, indicating release of the bus.
$\overline{\text{BREQ}}$	Bus request output for pending external memory cycle.
PMODE	Selects EPROM programming mode.
PACT	Signal indicating programming is in progress or complete.
CPVER	Cumulative program output verification.
PALE	Falling edge indicates Ports 3 and 4 contain valid programming information.
PROG	Falling edge indicates Ports 3 and 4 contain valid programming data.
PVER	Programmed byte verification.
AINC	Auto address input during programming.

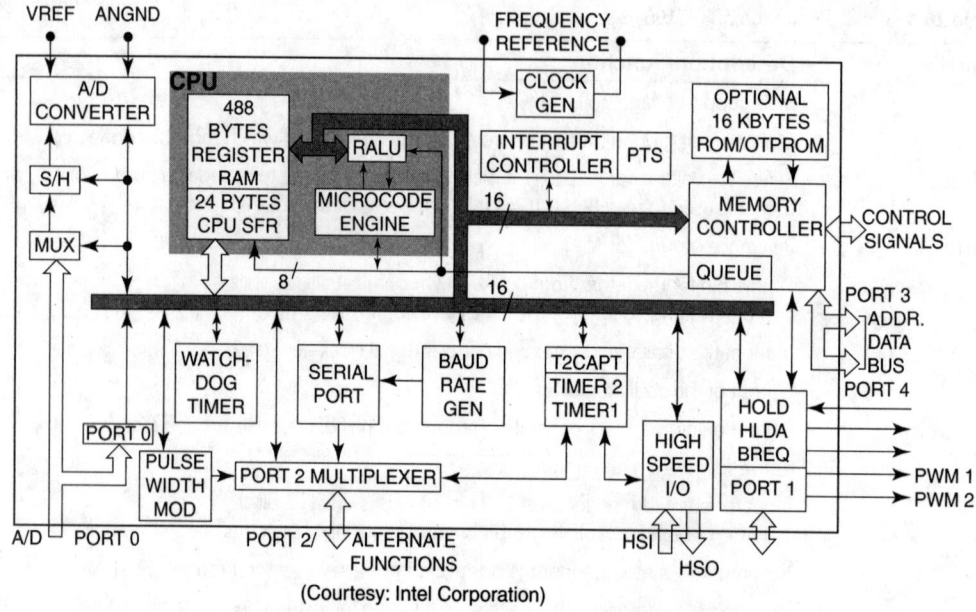

Figure 26.7 Simplified internal architecture of 8XC196

26.3.4 | Memory Map

Unlike 8051 family, MCS-96 is designed around the Princeton architecture, allowing program and data memory to share the same memory space. Total addressable memory space is 64K, which is divided in external and internal parts, as shown in Fig. 26.8. The lower address space of the memory area is reserved for internal on-chip RAM, which also accommodates SFRs and other CPU registers. Interrupts have their own vector address, which is divided in two parts, upper and lower.

26.3.5 | Addressing Modes

MCS-96 offers a rich instruction set and uses the following five addressing modes.

  (i)   Direct
  (ii)   Immediate
  (iii) Indirect
  (iv)   Indirect with auto-increment
  (v)   Indexed

For indirect with auto-increment addressing modes, the 16-bit target address, available in a register pair, is used to transact either 8-bit or 16-bit data, and after this data transaction, the target address, available in the register pair, is incremented by one for 8 bit and by two for 16-bit operand.

   The processor offers a 0 register at address 0000H, which contains 0000H and is available for the purpose of comparison. This 0 register may also be used as a base register for indexed addressing mode, which effectively would generate a direct address anywhere within 64K memory.

26.3.6 | Instruction Set

Instructions offered for MCS-96 may be divided into 11 groups as follows.

  (i)   Arithmetical instructions
  (ii)   Logical instructions

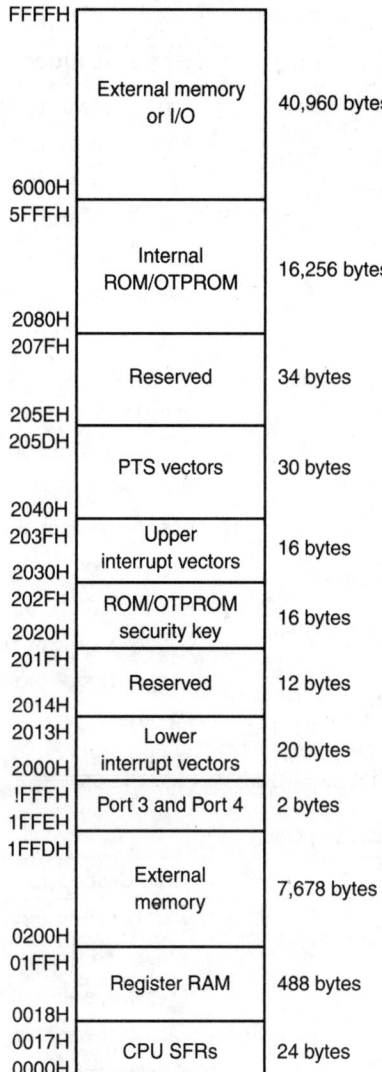

Address	Region	Size
FFFFH	External memory or I/O	40,960 bytes
6000H 5FFFH	Internal ROM/OTPROM	16,256 bytes
2080H 207FH	Reserved	34 bytes
205EH 205DH	PTS vectors	30 bytes
2040H 203FH 2030H	Upper interrupt vectors	16 bytes
202FH 2020H	ROM/OTPROM security key	16 bytes
201FH	Reserved	12 bytes
2014H 2013H 2000H	Lower interrupt vectors	20 bytes
!FFFH 1FFEH	Port 3 and Port 4	2 bytes
1FFDH 0200H	External memory	7,678 bytes
01FFH 0018H	Register RAM	488 bytes
0017H 0000H	CPU SFRs	24 bytes

Figure 26.8 System memory map

(iii) Data transfer instructions
(iv) Stack operations
(v) Jump and call instructions
(vi) Conditional jumps
(vii) Jump on bit conditions
(viii) Loop control
(ix) Single register instructions
(x) Shift instructions
(xi) Special control instructions

These instructions are briefly described through Table 26.10–26.20. Interested readers should consult the manufacturer's datasheets for further details.

Table26.10 Arithmetical instructions

Mnemonics	Brief description
ADD	Add two operands (word)
ADDB	Add two operands (byte)
ADDC	Add with Carry (word)
ADDCB	Add with Carry (byte)
SUB	Subtract (word)
SUBB	Subtract (byte)
SUBC	Subtract with Carry (word)
SUBCB	Subtract with Carry (byte)
CMP	Compare two words
CMPB	Compare two bytes
MULU	Unsigned multiply (word)
MULUB	Unsigned multiply (byte)
MUL	Multiply (word)
MULB	Multiply (byte)
DIVU	Unsigned divide (word)
DIVUB	Unsigned divide (byte)
DIV	Divide (word)
DIVB	Divide (byte)

Table 26.11 Logical instructions

Mnemonics	Brief description
AND	AND two operands (word)
ANDB	AND two operands (byte)
OR	OR two operands (word)
ORB	OR two operands (byte)
XOR	XOR two operands (word)
XORB	XOR two operands (byte)

Table 26.12 Data-transfer instructions

Mnemonics	Brief description
LD	Load a word
LDB	Load a byte
ST	Store a word
STB	Store a byte
LDBSE	Load a byte as word with sign unchanged
LDBZE	Load a byte as word with sign positive (0)

Table 26.13 Stack operations

Mnemonics	Brief description
PUSH	Save on stack
POP	Load from stack
PUSHF	Save PSW on stack
POPF	Load PSW from stack

Table 26.14 Jump and call instructions

Mnemonics	Brief description
LJMP	Unconditional long jump
SJMP	Unconditional short jump
LCALL	Unconditional long call
SCALL	Unconditional short call
RET	Return from subroutine

Table 26.15 Conditional jumps

Mnemonics	Brief description
JC	Jump if Carry is set
JNC	Jump if Carry is not set
JH	Jump if higher
JNH	Jump if not higher
JE	Jump if equal
JNE	Jump if not equal
JV	Jump if $V = 1$
JNV	Jump if $V = 0$
JGE	Jump if greater or equal
JLT	Jump if less
JVT	Jump if $VT = 1$, clear VT
JNVT	Jump if $VT = 0$
JGT	Jump if greater
JLE	Jump if less or equal
JST	Jump if $ST = 1$
JNST	Jump if $ST = 0$

Table 26.16 Jump on bit conditions

Mnemonics	Brief description
JBC	Jump if bit is clear
JBS	Jump if bit is set

Table 26.17 Loop control

Mnemonics	Brief description
DJNZ	Decrement and jump if not 0

Table 26.18 Single register instructions

Mnemonics	Brief description
DEC	Decrement a word by 1
DECB	Decrement a byte by 1
NEG	Two's complement a word
NEGB	Two's complement a byte
INC	Increment a word by 1
INCB	Increment a byte by 1
EXT	Extract the sign of a word
EXTB	Extract the sign of a byte
NOT	Complement a word
NOTB	Complement a byte
CLR	Clear a word
CLRB	Clear a byte

Table 26.19 Shift instructions

Mnemonics	Brief description
SHL	Shift left a bit (word)
SHLB	Shift left a bit (byte)
SHLL	Shift left a bit (long word)
SHR	Shift right a bit (word)
SHRB	Shift right a bit (byte)
SHRL	Shift right a bit (long word)
SHRA	Shift right arithmetic (word)
SHRAB	Shift right arithmetic (byte)
SHRAL	Shift right arithmetic (long word)

Table 26.20 Special control instructions

Mnemonics	Brief description
SETC	Set Carry flag
CLRC	Clear Carry flag
CLRVT	Clear VT flag
DI	Disable all interrupts
EI	Enable all interrupts
NOP	No operation
SKIP	Skip next instruction (2 bytes)

26.4 | ARM Microcontrollers

ARM is a 32-bit microcontroller whose core is designed by ARM Limited and is offered to other manufacturing organizations for adding necessary peripherals, fabrication and sales. Various versions of ARM are available, and it has undergone substantial changes till its inception during the early 1980s. As they are extremely efficient in power saving and operate with very low power consumption, ARM microcontrollers are widely used in modern handsets for mobile communications. ARM processors are also used in various other embedded systems, like iPODs, hand-held gaming units, disk drives and so on. Built with 32-bit architecture, ARM offers several advanced features for embedded applications.

26.4.1 | ARM Core Architecture

The success of ARM largely depended on the simplicity of its architecture. There was no micro-coding used, and this reduced the number of transistors used. A simplified schematic of the core of ARM processor is illustrated in Fig. 26.9.

One important feature in ARM architecture is its barrel shifter. This barrel shifter can perform the necessary preprocessing of register values before it enters the ALU. This allows easier calculations of wider ranges of expressions and addresses.

The register bank contains an array (file) of 32-bit registers holding signed and unsigned values. Signed 8-bit and 16-bit numbers are converted to their signed 32-bit equivalents before being loaded in these registers through hardware conversions. ARM adopts a register-to-register architecture and therefore does not contain any accumulator. The Incrementer updates the address register before the core writes or reads the next data set.

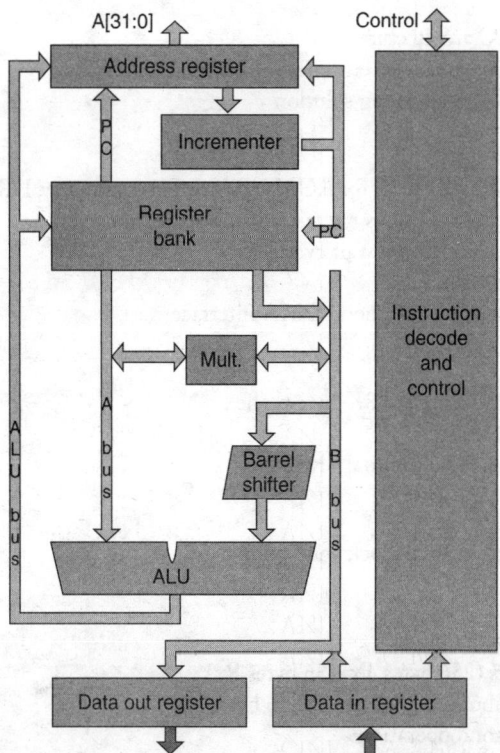

Figure 26.9 Simplified schematic of ARM processor core

26.4.2 | Versions of ARM

The design of ARM was initiated around 1983, and by 1985, the first samples, designated as ARM1, were completed. ARM2 was available in 1986. The improvement continued further, resulting in the initiation of ARM3, ARM6, ARM7, ARM9, ARM11, ARM13 and so on. As it has been already stated at the beginning of this section, ARM Limited designed only the ARM core, which was adopted with necessary peripherals by different manufacturing companies, like Atmel, Samsung and Philips. This has created a very large family of ARM processors with a substantial number of versions.

26.4.3 | Important Features

The complexity of the ARM processor may easily be understood if we glance through its basic features. Out of a very wide variety of ARM processor models, we will take a look at Samsung's S3C4510B, which is fabricated around ARM7TDMI core. Some of the important features of this processor are as follows.

Package type

    (i)   208-pin QFP

Operating frequency

    (i)   Up to 50 MHz

Operating voltage

    (i)   3.3V +/– 5%

Architecture

    (i)   Fully 16/32-bit RISC architecture
    (ii)   Integrated system for embedded ethernet applications
    (iii) Cost-effective JTAG-based debug solution

System manager

    (i)   8/16/32-bit external bus support for ROM/SRAM, Flash memory, DRAM and external I/O
    (ii)   One external bus master with bus request/acknowledge pins
    (iii) Programmable access cycle (0–7 wait cycles)
    (iv) Four-word depth write buffer
    (v)   Cost-effective memory-to-peripheral DMA interface

Unified instruction/data cache

    (i)   Two-way set-associative unified 8K-byte cache
    (ii)   Support for LRU protocol
    (iii) Cache is configurable as an internal SRAM

I²C serial interface

    (i)   Baud-rate generator for serial clock generation

Ethernet controller

    (i)   DMA engine with burst mode
    (ii)   DMA Tx/Rx buffers (256 bytes Tx, 256 bytes Rx)
    (iii) MAC Tx/Rx FIFO buffers (80 bytes Tx, 16 bytes Rx)
    (iv) 100/10 Mbit per second operation
    (v)   Station-management signaling

(vi)   On-chip CAM (up to 21 destination addresses)
(vii)   Full-duplex mode with PAUSE feature
(viii) Long/short packet modes

High-level Data Link Controls (HDLCs)

(i)   HDLC protocol features
(ii)   Address search mode (expandable to 4 bytes)
(iii)   Selectable CRC or No CRC mode
(iv)   Automatic CRC generator preset
(v)   Loop-back and auto-echo modes
(vi)   Modem interface
(vii)   Up to 10 Mbps operation
(viii) Two-channel DMA buffer descriptor for Tx/Rx on each HDLC

DMA controller

(i)   Two-channel general DMA for memory-to-memory, memory-to-UART, UART-to-memory data transfers without CPU intervention
(ii)   Initiated by a software or external DMA request
(iii)   Increments a source or destination address in 8-bit, 16-bit or 32-bit data transfers
(iv)   Four data burst mode

UARTs

(i)   Two UART (serial I/O) blocks with DMA-based or interrupt-based operation
(ii)   Support for 5-bit, 6-bit, 7-bit or 8-bit serial data transmit or receive
(iii)   Programmable baud rates
(iv)   One or two stop bits
(v)   Odd or even parity
(vi)   Break generation and detection
(vii)   Parity, overrun and framing error detection
(viii) x16 clock mode
(ix)   Infra-red (IR) Tx/Rx support (IrDA)

Timers

(i)   Two programmable 32-bit Timers
(ii)   Interval mode or toggle mode operation

Programmable I/O

(i)   Eighteen programmable I/O ports
(ii)   Pins individually configurable to input, output or I/O mode for dedicated signals

Interrupt controller

(i)   Twenty-one interrupt sources, including 4 external interrupt sources
(ii)   Normal or fast interrupt mode (IRQ, FIQ)
(iii)   Prioritized interrupt handling

Phase Lock Loop (PLL)

(i)   The external clock can be multiplied by on-chip PLL to provide high frequency system clock
(ii)   The input frequency range is 10–40 MHz
(iii)   The output frequency is 5 times of input clock. To get 50 MHz, input clock frequency should be 10 MHz

As it is beyond the scope of the present work to discuss even a few salient features from the above list, let us discuss very briefly the recent improvement of the power-management feature, which is offered by ARM controllers and developed in collaboration with National Semiconductor.

26.4.4 | Intelligent Energy Manager (IEM)

A major usage of ARM controller is in handsets. Although primarily they were developed for wireless communications, the coverage has rapidly grown to other domains like, gaming, FM receiver, video player, digital camera and so on. Essentially, all these extra features depend on the functioning of various components of the processor simultaneously, which demands extra current. To conserve the battery life in such situations, ARM introduced Intelligent Energy Manager (IEM) on the chip, which contains the following modules:

(i) ARM Intelligent Energy Controller (IEC)
(ii) National Semiconductor's *Advanced Power Controller* (APCI)
(iii) A custom, SoC specific, Dynamic Clock Generator (DCG)

By controlling the input voltage and also the clock input for different modules, the power demand of the controller is always optimized by the IEM. Please refer to the manufacturer's datasheets for technical details.

26.5 | Renesas Microcontrollers

Renesas Technology Corporation, a joint venture of Hitachi Ltd. and Mitsubishi Electric, founded in 2003, offers a very wide range of efficient microcontrollers of various sizes and architecture to suit practically all demands of different embedded-system designers. These microcontrollers range from 8 bits to 32 bits, having on-chip Flash and SRAM of various sizes, along with I/O ports, Timers, Watchdog Timer, UART and many other features.

The range of the Renesas family of microcontrollers along with some of their application areas is illustrated schematically in Fig. 26.10. As we may observe, the device size varies from 20-pin to about 500-pin. The operating speed ranges from almost static to near about 300 MHz. Interested readers may consult the datasheets of the manufacturer.

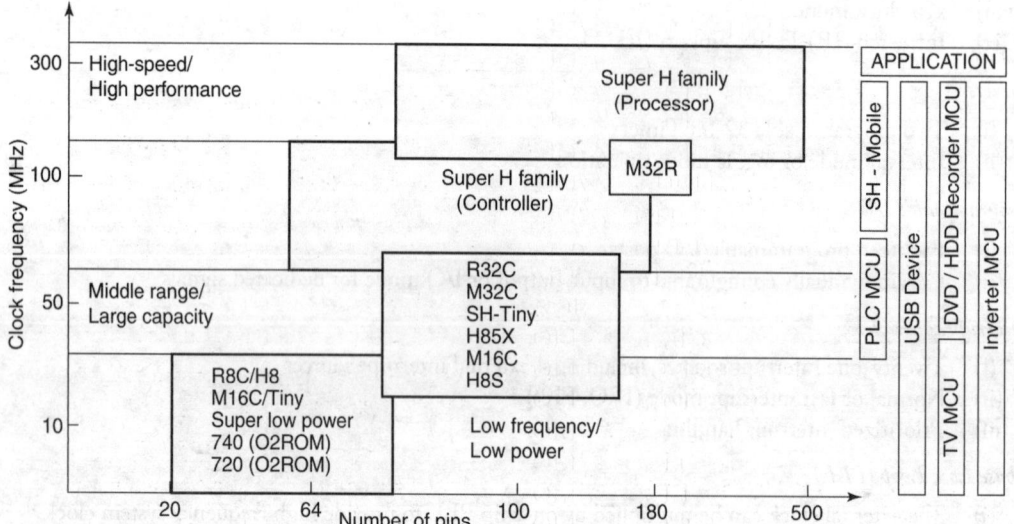

Figure 26.10 Renesas family of microcontrollers

In Japan, Renesas microcontrollers are very popular and extensively used. Their popularity is for a very wide range of microcontroller versions.

SUMMARY

ATmega8, one of the 8-bit AVR microcontrollers from Atmel, offers several advanced features like register-to-register architecture, which eliminates the concept of accumulator for result storage. This 28-pin device available as PDIP offers 8K program Flash, 512 bytes EEPROM and 1K SRAM with 32 general-purpose registers, 23 programmable I/O lines, three Timers, ADC, USART and Watchdog Timer. Built on advanced RISC architecture, it offers 130 instructions with various addressing modes.

Intel's MCS-96 family of microcontrollers are 16-bit devices adopting register-to-register architecture.

Available in 68-pin PLCC, MCS-96 offers a rich instruction set and many improved features like on-chip ADC and high-speed I/O interface.

ARM microcontrollers, built around 32-bit ARM processor core by different manufacturers, are used in most of the handsets of the world. ARM core provides a barrel shifter for faster computing of data and an IEM for power management, which has been developed in collaboration with National Semiconductor.

Recently introduced Renesas microcontrollers offer a wide variety of choice for the embedded-system designers.

POINTS TO REMEMBER

- On-chip memory and I/O of latest microcontrollers vary widely, and in practice, external memory or I/O interfacing with microcontroller is avoided by embedded-system design experts.

- Modern microcontrollers offer more power-saving features to extend the battery-recharging frequency for the system designed with them.

REVIEW QUESTIONS

Evaluate Yourself

1. Introduction of Flash memory for program storage was first introduced by

   (a) Intel   (b) Atmel

   (c) Hitachi   (d) none of these

2. The size of on-chip SRAM of ATmega8 microcontroller is limited to

   (a) 8K bytes   (b) 512 bytes

   (c) 1K bytes   (d) none of these

3. The maximum operational frequency of ATmega8 microcontroller is

   (a) 16 MHz   (b) 8 MHz

   (c) 4 MHz   (d) none of these

4. ATmega8 microcontroller is available in

   (a) 28-pin PDIP   (b) 40-pin PDIP

   (c) 48-pin PDIP   (d) none of these

5. Number of general-purpose 8-bit registers within ATmega8 microcontroller is

   (a) eight   (b) sixteen

   (c) thirty-two   (d) none of these

6. Total number of processor status flags available in SREG of ATmega8 is

   (a) five   (b) six

   (c) seven   (d) none of these

7. The arithmetic-logic unit of MCS-96 is known as

   (a) RALU   (b) ALU

   (c) ALU-96   (d) none of these

8. 8XC196 is available in

   (a) 68-pin PLCC   (b) 40-pin PDIP

   (c) 432-pin QFP   (d) none of these

9. The bus-width of MCS-96 is

   (a) fixed as 16 bits   (b) programmable

   (c) fixed as 32 bits   (d) none of these

10. The barrel shifter of ARM processor performs

    (a) bank switching   (b) bit shifting

    (c) preprocessing   (d) none of these

Search for Answers

1. What are the preset frequencies available from the built-in RC oscillator of ATmega8?

2. How many I/O lines are available in ATmega8?

3. How many index registers are offered in ATmega8 microcontroller? Name them.

4. After arithmetical or logical operations, where are the results stored by ATmega8?

5. What is the purpose of I/O registers in ATmega8 microcontroller?

6. How does the system stack grow in ATmega8 microcontroller?

7. How many signals are necessary for JTAG interface?

8. How many PWM channels are available in Intel MCS-96?

9. Which port of MCS-96 is reserved for ADC input?

10. Who manufactures RENESAS microcontrollers?

Think and Solve

1. Why does AVR provide a built-in oscillator?

2. What is the purpose of providing some EEPROM in ATmega8 microcontroller?

3. Why is the program Flash divided into two sections in ATmega8 microcontroller?

4. How can all interrupts be disabled in ATmega8 microcontroller?

5. What is the purpose of T flag of SREG register in ATmega8 microcontroller?

6. In how many ways can ATmega8 be reset?

7. What is the purpose of a Watchdog Timer?

8. How can the bus width be controlled in Intel MCS-96?

9. What is the speciality of ARM controller with respect to MCS-96 and AVR microcontrollers?

10. What is the purpose of IEM in ARM controllers?

27 INTERFACING INTEL 8255 PPI WITH 8051

CHAPTER OBJECTIVES

The number of I/O pins of Intel 8051 microcontroller is limited to 32 only. In some special design cases these many I/O pins might not be sufficient for the system designer. This chapter introduces a simple method to enhance the number of I/O pins by interfacing 8255, a general purpose port from Intel, with 8051. After completion of this chapter the reader would be able to:

- Understand internal architecture and modes of 8255.
- Interface 8255 with 8051.
- Use 8255 in different modes through 8051.

27.1 | Introduction

From our previous discussions in Chapter 2, we know that 8051 offers four I/O ports, each of 8-bit size. This allows us to avail 32 I/O lines for external interfacing. In most of the cases these 32 I/O lines are sufficient to cater the system designer's need. However, in certain exceptional cases the designer might need more I/O lines. For example, we may assume that an embedded system with 8051 microcontroller needs an external 8-bit ADC (demanding 12 I/O lines), an 8-bit DAC (needing 8 output lines), 8 digit multiplexed 7-segment display and a standard 64-key keyboard (requiring 24 I/O lines together). It may be easily visualized that 32 I/O lines are not sufficient to cater for this I/O demand of the system (we need 44 I/O lines together). In such cases an external port, like Intel 8255 PPI (programmable peripheral interface), may be interfaced with 8051 to get additional I/O lines for the system. In following sections we shall discuss about hardware interfacing and software initialization issues to put these two devices together.

The reader may note that Intel 8255 PPI was originally developed and marketed around 1974 to serve for 8-bit microprocessors from Intel, like 8085. Intel 8051 was introduced in the commercial market during 1980.

27.2 | Overview of 8255

Intel 8255 PPI is available in 40-pin DIP and offers, apart from necessary interfacing signals, 24 I/O lines through three I/O ports, namely: Port A, Port B and Port C, each of 8-bit size. Moreover, Port C may be programmed as two independent 4-bit ports, Port-C *lower* and Port-C *upper*, as illustrated in Fig. 27.1. The reader may note that ports of 8255 are grouped as A and B with port A and port C upper falls in group A while port B and port C lower are designated as group B. Certain functions of 8255 are dependent on these groups which we shall indicate at appropriate places in following sections.

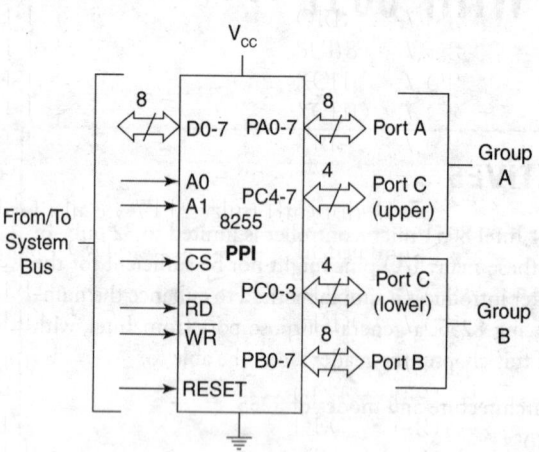

Figure 27.1 Designer's model of 8255

Ports of 8255 are not only programmable as either input or output, but also may function in one of three modes, namely, Mode 0, Mode 1 and Mode 2. Mode 0 is designated as *Basic Input/Output* Mode, Mode 1 as *strobed input/output* mode and Mode 2, which is applicable only for group-A, as *bidirectional bus* mode. Moreover, when configured as output, each bit of port C may be individually set or reset through appropriate software command. To implement all these features, 8255 offers a 8-bit Control Word Register or CWR apart from three 8-bit port registers. To output any data set through any port, the data to be written in the concerned port-register. Similarly, to read a port, the concerned register of the port to be read by the processor or microcontroller. Individual addresses of these internal registers of 8255 are presented through Table 27.1.

Table 27.1 Addresses of internal registers of 8255

Address input		Select
A1	**A0**	
0	0	Port A
0	1	Port B
1	0	Port C
1	1	CWR

The reader may note that both read as well as write operations are allowed for any three port registers. However, the control word register may only be used for writing command words and can never be read. That means CWR is a *write only* register of 8255.

27.3 | Control Words for 8255

Before using 8255 ports, the controller (8051) must initialize 8255 by sending (writing) an appropriate control word in its CWR with both A0 and A1 as 1 (refer Table 27.1). Various bit-details of the control word are explained in Fig. 27.2. Note that when MSB of the control word is set, the remaining part of the control word would assign modes of different ports [Fig. 27.2(a)]. On the other hand, if the MSB of the control word is reset, the remaining part of the control word would set/reset the port C bit [Fig. 27.2(b)]. Let us now take a closer look at different operational modes of 8255.

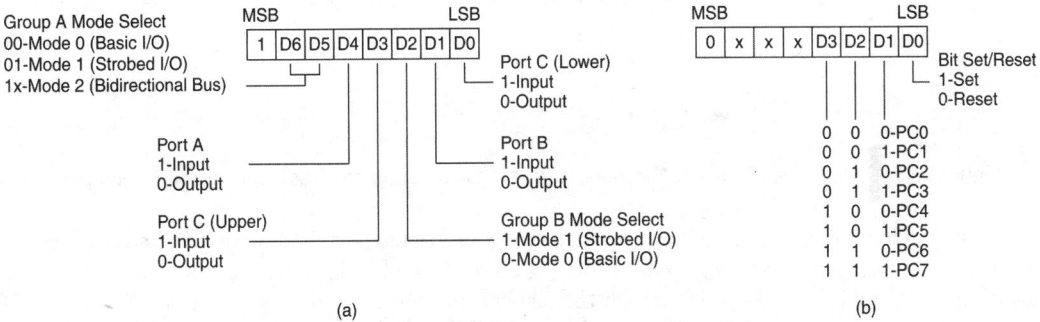

Figure 27.2 Control words for 8255, (a) mode selection, (b) port C bit set/reset

 The reader may note that 8051 ports do not need any control word for initialization. However, they must output logic 1 to serve as input. Moreover all port pins of 8051 are bit-wise programmable as either input or output, which is not available for any port of 8255. Only individual bit set/reset is possible for port C pins while serving as output.

27.4 | Mode 0 of 8255 (Basic Input/Output Mode)

As indicated in Fig. 27.2(a), either group-A ports or group-B ports or both group-A ports and group-B ports may be configured to function in mode 0. Each port of each group may be individually programmed as either input port or output port. Therefore a total of sixteen different possible combinations are allowed by 8255, while all ports are functioning in mode 0.

After configuring any port to function in mode 0 output, the output data must be written in the concerned port register (refer Table 27.1). This data would then be latched within the port. When functioning as input (in mode 0) there is no provision of latching the input data and therefore the input data must remain stable till it is read by the controller. To eliminate this problem, 8255 also offers mode 1 input output scheme, which is described in the following section.

27.5 | Mode 1 of 8255 (Strobed Input/Output Mode)

In strobed input/output mode an external device can initiate data transfer (either input or output) with a port of 8255 (either port A or port B) through appropriate handshaking signals, in asynchronous form. When programmed to function in mode 1, each group of 8255 offers some handshaking signals to allow asynchronous data transfer through the 8-bit port. This 8-bit port may be either port A or port B or both. However, few handshaking signals are offered through relevant port C pins, as shown in Fig. 27.3.

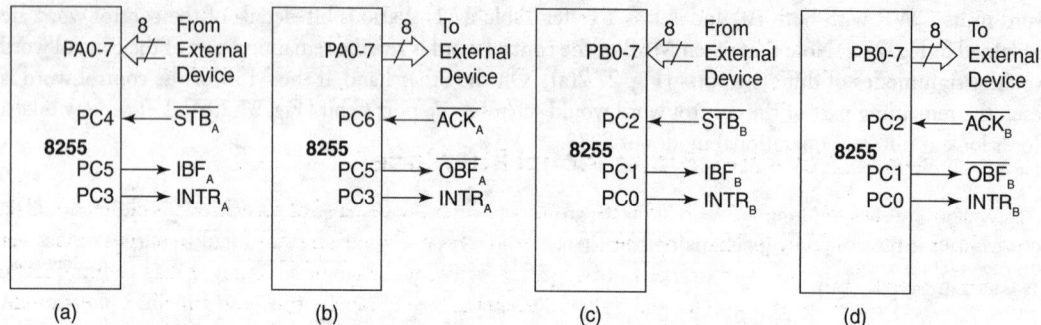

Figure 27.3 Mode 1 of 8255, (a) port A as input, (b) port A as output, (c) port B as input, (d) port B as output

In strobed input mode, an external device may send 8-bit data to port A (or to port B) and indicate that by activating the concerned STB signal low (refer Fig. 27.4). This STB signal is necessary to draw the attention of the port, and is acknowledged through IBF signal from the port to the external device. Once the IBF is received by the external device, it changes the status of its STB output to its original status. Meanwhile 8255 generates an interrupt signal (INTR) to draw the attention of the controller.

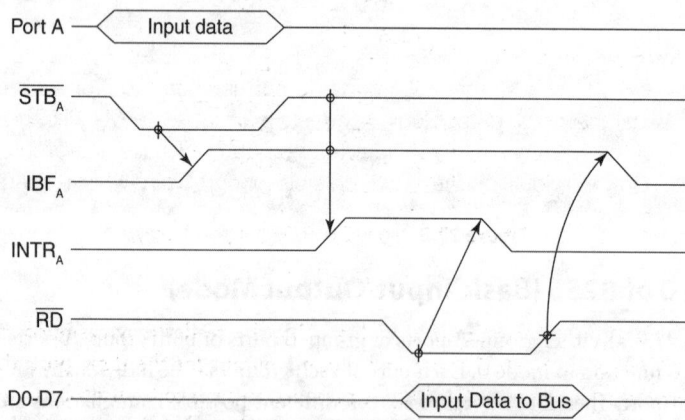

Figure 27.4 Timing diagram for mode 1 input (for port A)

The relevant interrupt service routine reads the port (Port A in this example case) and places the data within the system data bus. The IBF signal is now changed to its original state indicating to the external device that the input buffer is no longer full and another byte may be transferred if necessary.

In strobed output mode, data transaction is more or less similar in nature as that of strobed input mode. However, in this case IBF is replaced by OBF and STB by ACK. INTR signal is available in this case also to serve the same purpose, that is drawing attention of the controller.

It may be noted that in certain cases all pins of port C may not be necessary for handshaking purpose. In these cases remaining pins of port C may be configured to function in mode 0 (input or output) and bit set/ reset command would be applicable if configured as output.

As an example we may assume that port A is configured as mode 1 input and port B is configured in mode 1 output. In that case PC6 and PC7 remain unused for handshaking and may be used in mode 0 either as input or as output.

27.6 | Mode 2 of 8255 (Bidirectional Bus Mode)

One major limitation of mode 1 of 8255 is its inability to cater the demand of asynchronous bidirectional data transfer. It is possible only by changing the concerned port's configuration from input to output or vice versa through a fresh command word to the control register for every change of direction of data communication. This may not be a happy solution for the system designer and to eliminate this problem, 8255 offers mode 2, which is capable of bidirectional asynchronous data transfer with any external device. However, as we have already indicated, mode 2 is applicable only for group A of 8255, which is only for port A. The relevant handshaking signals are catered by five pins of port C, as shown in Fig. 27.5.

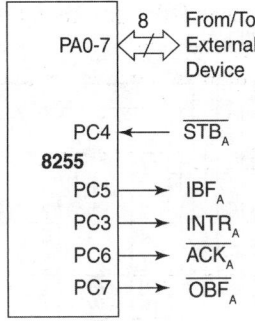

Figure 27.5 Mode 2 of 8255 (for port A only)

Observing handshaking signals depicted in Fig. 27.5, the reader may easily conclude that mode 2 is nothing but a combination of mode 1 input and mode 1 output for the same port, that is port A. In this case only one interrupt-evoking signal ($INTR_A$) is provided to cater to both strobed input as well as strobed output. The concerned interrupt service routine is to react against $INTR_A$ after checking the strobing input STB_A. If it is low then it must be a case of reading from port A. Otherwise it is a case of writing in port A.

27.7 | Bit Set/Reset Mode

As already illustrated in Fig. 27.2, bit set/reset mode is applicable only for port C of 8255. This is possible by sending the proper control word to the control register of 8255 with its most significant bit as 0. The bit position is directed by D1, D2 and D3 of the control word. If its LSB is set (1) then the concerned bit of port C would be set and otherwise it would be reset. Note that using this command to the control register, we can change only one bit of port C at a time, without altering other bits of port C. The reader may note that bit set/reset command would be applicable for port C if its pins are configured as output pins in mode 0. If configured as mode 0 input, then bit set/reset command is not applicable for port C pins.

27.8 | Hardware Interfacing

To interface 8255 with 8051, we are to take care of all signals depicted at the left side of 8255 in Fig. 27.1. They are:

- Data lines D0-D7
- Address lines A0, A1 and chip select signal CE
- Control signals RD and WR
- System RESET.

All of these signals are available either directly or indirectly, from 8051, as illustrated in Fig. 27.6.

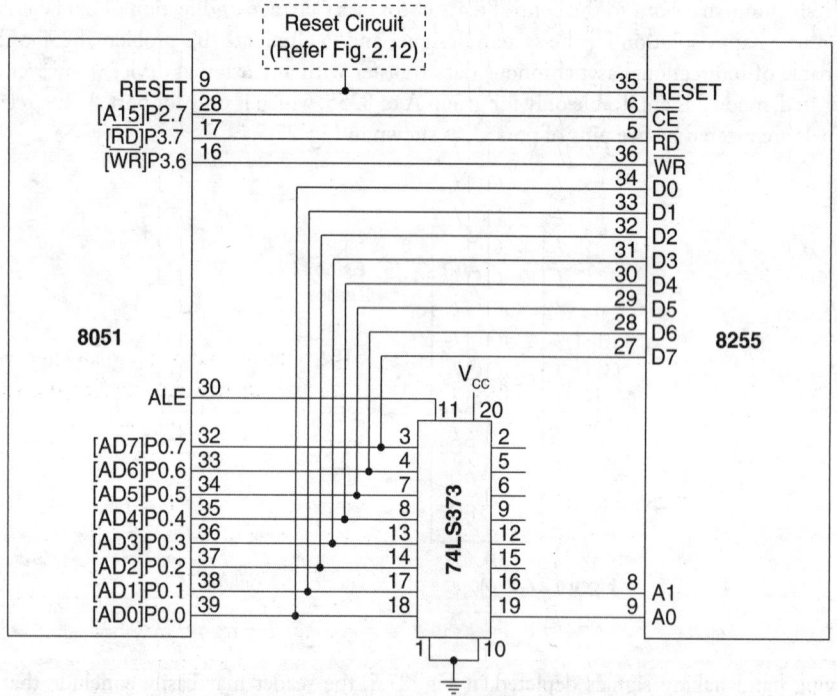

Figure 27.6 Interfacing 8255 with 8051

 The current sourcing capabilities of port B and port C pins of 8255 are higher than that of port A. Any eight pins selected randomly from these two ports can source 1 milli-ampere current at 1.5 volts.

To derive the address signals A0 and A1 from multiplexed address-data signals of 8051, we are to interface 74373 between 8051 and 8255, as illustrated in Fig. 27.6. As a matter of fact the same device (74373) we have already used for the identical purpose in case of external memory interfacing circuit depicted in Fig. 17.9. Data lines D0 – D7 of 8255 may be directly connected with port 0 pins of 8051. The read (RD) and write (WR) control input signals of 8255 may be obtained directly from port pins P3.7 and P3.6

respectively. The reset input for 8255, which is active high, may be obtained from the system reset input as illustrated in Fig. 27.6. Details of this reset circuit is already presented in Fig. 2.12. Finally, for the chip enable (CE) input of 8255, we may directly connect its pin 6 with the highest address output from 8051, that is A15 from P2.7.

As the CE input of 8255 is available directly from A15, therefore internal registers of 8255 would need 16-bit addresses for communication. These addresses are presented in Table 27.2. By sending appropriate command word to the CWR, 8255 may be initialized to function in any mode as intended by the system designer.

Table 27.2 Port addresses for circuit in Fig. 27.6

Internal registers	16-bit address
Port A of 8255	0000H
Port B of 8255	0001H
Port C of 8255	0002H
CWR of 8255	0003H

27.9 | Solved Examples

27.9.1 | Example 1

Eight keys and eight bi-color LEDs are to be interfaced through 8255 so that pressing any key will turn the corresponding LED as green. Releasing the key would immediately change the color of the LED to red. Design the circuit and also develop the necessary software for it.

Solution
In this case we can use 8255 in its mode 0 that is the basic input/output mode. Hardware interfacing details of 8255 with 8051 has already been explained in section 27.8 and illustrated in Fig. 27.6. The same interfacing may be adopted in the present case also. The addresses of internal registers of 8255, as indicated in Table 27.2, would remain valid in this case. To interface eight keys, we may use port C of 8255 and eight bicolor LEDs may be interfaced through ports A and B as illustrated in Fig. 27.7. The reader may note that we have used common anode type bicolor LEDs. Port A pins are used for driving red and port B for green part of each LED. One 330 ohms resistor must be placed in series between each common anode and the Vcc input of

Bi-color LEDs are generally a combination of red and green LEDs in a single package and available as either common anode or common cathode configuration.

each bicolor LED, as shown in Fig. 27.7. As both red and green would never be turned on simultaneously, therefore one resistor may be considered sufficient in this case.

For all port C pins, 1K pull-up resistors are incorporated forcing each input as logic 1 if the corresponding key is not pressed. Pressing any key would force the corresponding input to logic 0. As per the requirement

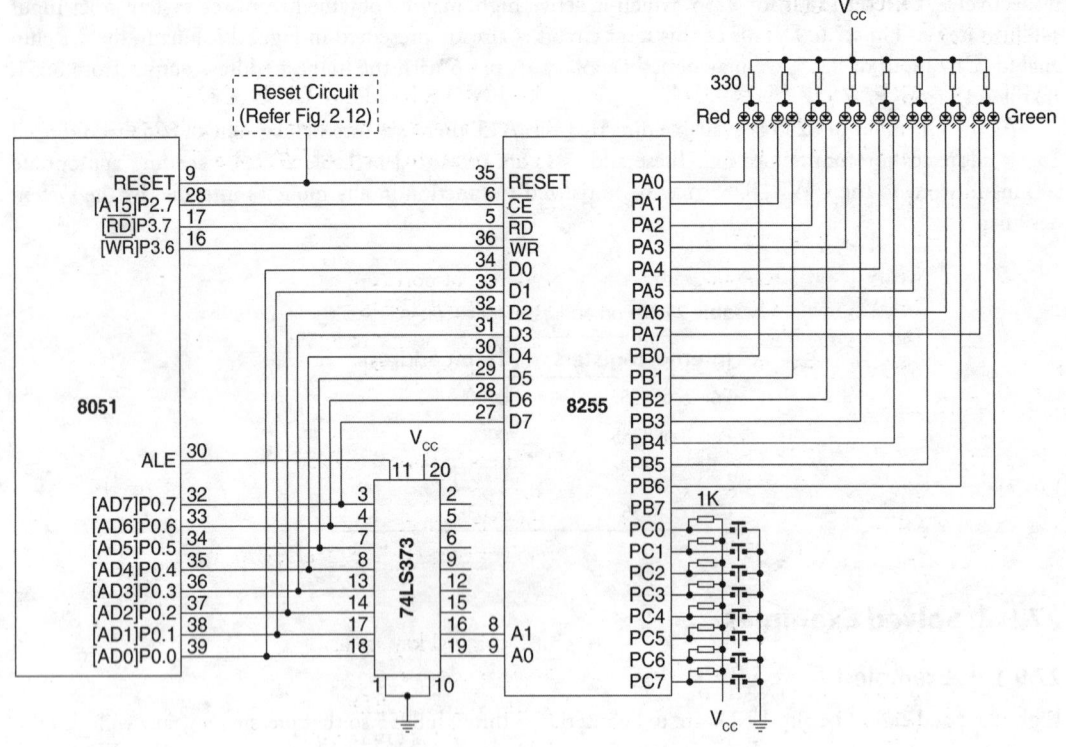

Figure 27.7 Circuit diagram for example 1

of the circuit, initially all bicolor LEDs are to be turned off (port A outputs and port B outputs inactive) and the microcontroller is to keep on reading (polling) the input port C. Whenever any key is sensed to be pressed (corresponding input bit changes from 1 to 0) the similar port pin of port B to be activated (green LED on). The polling of keys to be continued and the release of the key should be reflected by turning off the green LED and turning on the red LED.

At the software side, for faster polling operation the controller, instead of checking all eight bits of port C one by one, should complement port C after reading it and then check if all bits are zero or not. If all bits of complemented value of port C input are zero then no key is pressed. Otherwise some key must have been pressed. The remaining part of the software must service LEDs as per the input status.

```
;$$$$$$$$$$$$$$$$$$$$$$$$$$$$$$$$$$$$$$$$$$$$$$$$$$$$$$$$$$$$$$$$$$$$$$$$$$$$$$$$$$$
; Software for controlling bicolor LED array
; Software developed by Dr. Subrata Ghoshal, August, 2012.
; Intel 8255 PPI is interfaced with 8051.
; 8255 in mode 0, port A and B output, port C input.
; port A controls red LEDs and port B controls green LEDs.
; Keys are scanned by polling.
; If any key is pressed, corresponding bicolor LED turns green.
; When the same key is released the corresponding bicolor LED becomes red.
; Registers used: A, R7, DPTR
;$$$$$$$$$$$$$$$$$$$$$$$$$$$$$$$$$$$$$$$$$$$$$$$$$$$$$$$$$$$$$$$$$$$$$$$$$$$$$$$$$$$
```

```
                  ORG    0000
; initialize 8255

START:     MOV   DPTR, #0003H        ; load DPTR by addr of CWR of 8255
           MOV   A, #89H             ; port A and B output, port C input, all mode 0
           MOVX @DPTR, A             ; send comman word in CWR of 8255

; turn off all LEDs

           MOV   A, #0FFH
           MOV   DPTR, #0000H        ; address of port A of 8255
           MOVX @DPTR, A             ; turn off red LEDS
           INC   DPTR                ; address of port B of 8255
           MOVX @DPTR, A             ; turn off green LEDs

; look for key pressing

ANYKEY:    MOV   DPTR, #0002H        ; address of port C of 8255
KLOOP:     MOVX A, @DPTR             ; read port C
           CPL   A                   ; complement reading
           JZ    KLOOP               ; no key pressed. Keep looking for it.

; some key pressed

           CPL   A                   ; get original key status
           MOV   DPTR, #0001H        ; port B address
           MOVX @DPTR, A             ; turn on green LED
           MOV   R7, A               ; temp save for future use

; wait for key release

           MOV   DPTR, #0002H        ; port C address
KREL:      MOVX A, @DPTR             ; read keys
           CPL   A
           JNZ   KREL                ; the key is still pressed

; key released, turn off green, turn on red.

           MOV   DPTR, #0001H        ; port B address
           MOV   A, #0FFH
           MOVX @DPTR, A             ; turn off green LED
           MOV   DPTR, #0000H        ; port A address
           MOV   A, R7               ; get back old setting
           MOVX @DPTR, A             ; turn on red LED

; loop back

           SJMP  ANYKEY              ; loop on
```

27.9.2 | Example 2

Modify the circuit of previous example so that polling of keys are not necessary.

Solution

To change the key-input configuration to an interrupt driven circuit, where polling of keys are not necessary,
8255 may be used in its mode 1, that is: strobed input mode. All eight keys may be interfaced with port B,

which would function in strobed input mode (Mode 1). To generate the strobing signal all eight outputs from keys must be logically ANDed. Output from this AND gate would generate STB_B signal, which would be received by PC2 of 8255 which is reserved to accept this STB_B signal for group-B. The reader may note that PC1 would generate IBF_B signal (not used in the present case) and the $INTR_B$ (interrupt output signal) would be available from PC0. This interrupt from 8255 may be received at INT0 of 8051 microcontroller and serviced by its ISR.

Port A may be configured as mode 0 output to drive eight common cathode bicolor LEDs. Two pins from port C may be used to control red LED and green LED anodes, as shown in Fig. 27.8.

The reader may note that to turn on any one of eight red LEDs, PC6 must output logic 1. This would make the corresponding BC547 NPN transistor active and Vcc would be allowed at anodes of all red LEDs. Similarly, to turn on any one of eight green LEDS, PC7 must be high. Furthermore, port A output would decide which one of eight LEDs would be turned on. A logic low through any port A pin would activate an LED attached through it. Note that although any port A pin is capable to drive a pair of LEDs (one red and one green) however, depending upon the conditions of PC6 and PC7 outputs, any one of this pair of LEDs may be activated.

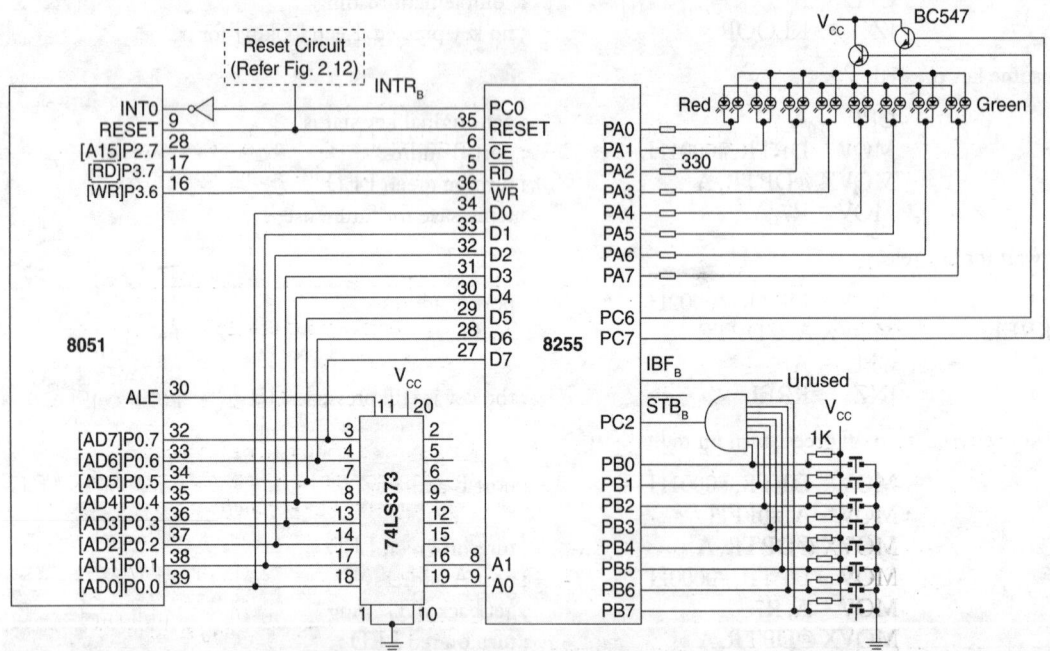

Figure 27.8 Circuit diagram for example 2

A better alternative to enhance the I/O handling capability of 8051 is to interface another 8051 with it. It would allow us to have another 32 I/O lines as compared to only 24 I/O lines of 8255. TxD and RxD signals of 8051 would be sufficient to communicate between two 8051s. Furthermore, the system would have an additional memory and timers, not possible to obtain from an interfaced 8255.

SUMMARY

Intel 8051 offers 32 I/O lines which might be enhanced by interfacing Intel 8255 PPI providing three additional 8-bit ports, namely A, B and C. Moreover three different programmable modes, available by sending appropriate command word to 8255's control register, allow greater flexibility to the system design engineers. These three modes of 8255 are:

- Mode 0: Basic input/output mode,
- Mode 1: Strobed input/output mode, and
- Mode 2: Bidirectional bus mode.

In Mode 0, output is latched but input is not latched and to be scanned by polling. In this mode port C is divided in two 4-bit ports and each one may be individually programmed as either input or output, depending upon the system requirements.

In Mode 1, a strobing signal is available for both input as well as output which may be activated by the external device to trigger the desired action. This triggering of STB signal would result in asynchronous interrupt-driven data transfer. This interrupt (INTR) is automatically generated by 8255 in its mode 1 apart from generating an acknowledgement signal, which may be sent to the external device initiating the data transfer. Both port A and port B may function individually as either input or output. Some of port C pins are required to cater to handshaking signal generation.

Mode 2 is applicable only for port A and is a combination of mode 1 input and mode 1 output. An external device may initiate data reading and data writing in this mode using port A of 8255.

POINTS TO REMEMBER

- Port C pins, which are not used for handshaking signals, may be utilized in mode 0 as input or output.

- After reset, all 8255 ports are assigned as mode 0 output and all pins output logic high.

REVIEW QUESTIONS

Evaluate Yourself

1. How many I/O lines are offered by 8255?
   - (a) 24
   - (b) 32
   - (c) 40
   - (d) None of these

2. Width of 8255 ports are
   - (a) 4-bit
   - (b) 8-bit
   - (c) Either of these
   - (d) None of these

3. Immediately after a system reset, ports of 8255 would function as
   - (a) Mode 0 input
   - (b) Mode 1 output
   - (c) Either of these
   - (d) None of these

4. How many registers are available inside 8255?
   - (a) 3
   - (b) 4
   - (c) 5
   - (d) None of these

5. Apart from address, data, RD, WR and Chip select signals, which one of the following input signals is necessary for 8255 interfacing?
   - (a) Clock in
   - (b) Mode select
   - (c) Reset in
   - (d) None of these

6. Bit set/reset mode of 8255 is applicable for
   - (a) Port C only
   - (b) Ports B and C
   - (c) Ports A, B and C
   - (d) None of these

7. In strobed output mode for group B, which of the following signals is received by the external device, getting the output from port B of 8255?
   - (a) IBF_B
   - (b) OBF_B
   - (c) ACK_B
   - (d) None of these

8. Which one of the following statements is *true* for mode 0 of 8255?
   - (a) Input is latched but output is not latched.
   - (b) Output is latched but input is not latched.
   - (c) Both input and output are latched.
   - (d) None of these

9. Which one of the following ports of 8255 is capable of functioning in mode 2?
   - (a) Port A
   - (b) Port B
   - (c) Port C
   - (d) None of these

10. Address of the control word register of 8255 is

    (a) 00   (b) 01

    (c) 10   (d) None of these

Search for Answers

1. How many pins are there for the DIP of 8255?

2. Which part of port C of 8255 serves group A and which part serves group B?

3. Which purpose is served by the most significant bit (bit 7) of the control word sent to 8255?

4. How many different combinations of inputs/outputs of various ports are possible in mode 0 of 8255?

5. In mode 1 input for 8255, what must be the condition to make INTR change from low to high?

6. What is the major limitation of mode 1 of 8255?

7. What is the special purpose of mode 2 of 8255?

8. How many bits of port C of 8255 may be simultaneously changed using bit set/reset command?

9. For interfacing purpose, how do we generate A0 and A1 for 8255 using available signals from 8051?

10. What is the electrical specialty of port B and port C of 8255?

Think and Solve

1. Why Intel 8255 is designated as *Programmable Peripheral Interface*?

2. Why only port C of 8255 was designed as bit-programmable?

3. If both port A and port B of 8255 are configured to function in mode 1, should the remaining two pins of port C remain unused?

4. Is there any precaution necessary to interface $INTR_A$ or $INTR_B$ of 8255 in mode 1 with INT0 or INT1 of 8051?

5. Referring Fig. 27.3, we find that for group B handshaking signal generation, identical pins of port C are used for both input and output. However, for group A, separate port C pins are used in case of two handshaking signals. What might be the reason behind this?

6. Why mode 2 is applicable only for one group, that is group A? Why it is not offered through both A and B groups simultaneously?

7. How many types of bi-color LEDs are commercially available?

8. Is there any modification necessary in the circuit of solved example 1 (Fig. 27.7) if both red and green LEDs are to be turned on simultaneously? Justify your answer.

9. What are the major differences between 8051 ports and 8255 ports?

10. Using only 8051 ports (without using 8255) is it possible to implement the strobed input scheme as shown in Fig. 27.8?

ANSWERS FOR EVALUATE YOURSELF

Chapter 1
1 (a)   2 (c)   3 (b)   4 (c)   5 (d)   6 (c)   7 (a)   8 (b)   9 (d)   10 (a)

Chapter 2
1 (b)   2 (d)   3 (a)   4 (a)   5 (b)   6 (c)   7 (d)   8 (d)   9 (c)   10 (a)

Chapter 3
1 (b)   2 (a)   3 (c)   4 (d)   5 (c)   6 (a)   7 (d)   8 (a)   9 (c)   10 (b)

Chapter 4
1 (a)   2 (d)   3 (c)   4 (b)   5 (a)   6 (c)   7 (a)   8 (d)   9 (b)   10 (a)

Chapter 5
1 (a)   2 (b)   3 (c)   4 (d)   5 (c)   6 (a)   7 (d)   8 (c)   9 (b)   10 (b)

Chapter 6
1 (c)   2 (a)   3 (b)   4 (c)   5 (a)   6 (b)   7 (b)   8 (c)   9 (d)   10 (b)

Chapter 7
1 (b)   2 (c)   3 (c)   4 (c)   5 (c)   6 (b)   7 (a)   8 (c)   9 (a)   10 (d)

Chapter 8
1 (a)   2 (c)   3 (c)   4 (b)   5 (a)   6 (b)   7 (c)   8 (d)   9 (b)   10 (a)

Chapter 9
1 (b)   2 (a)   3 (c)   4 (c)   5 (d)   6 (a)   7 (b)   8 (c)   9 (b)   10 (a)

Chapter 10
1 (d)   2 (a)   3 (b)   4 (b)   5 (c)   6 (c)   7 (c)   8 (a)   9 (b)   10 (a)

Chapter 11
1 (a)   2 (b)   3 (c)   4 (c)   5 (b)   6 (c)   7 (a)   8 (b)   9 (a)   10 (c)

Chapter 12
1 (b)   2 (a)   3 (a)   4 (b)   5 (a)   6 (c)   7 (d)   8 (c)   9 (b)   10 (a)

Chapter 13
1 (c)   2 (a)   3 (b)   4 (b)   5 (a)   6 (b)   7 (d)   8 (b)   9 (d)   10 (a)

Chapter 14

| 1 (b) | 2 (a) | 3 (c) | 4 (b) | 5 (a) | 6 (b) | 7 (d) | 8 (c) | 9 (d) | 10 (b) |

Chapter 15

| 1 (a) | 2 (b) | 3 (c) | 4 (b) | 5 (a) | 6 (c) | 7 (a) | 8 (b) | 9 (d) | 10 (b) |

Chapter 16

| 1 (c) | 2 (a) | 3 (b) | 4 (a) | 5 (b) | 6 (c) | 7 (b) | 8 (a) | 9 (c) | 10 (a) |

Chapter 17

| 1 (b) | 2 (c) | 3 (d) | 4 (a) | 5 (a) | 6 (b) | 7 (b) | 8 (a) | 9 (b) | 10 (c) |

Chapter 18

| 1 (b) | 2 (a) | 3 (b) | 4 (c) | 5 (d) | 6 (b) | 7 (b) | 8 (c) | 9 (a) | 10 (b) |

Chapter 19

| 1 (b) | 2 (b) | 3 (c) | 4 (c) | 5 (a) | 6 (a) | 7 (c) | 8 (a) | 9 (b) | 10 (c) |

Chapter 20

| 1 (b) | 2 (a) | 3 (c) | 4 (c) | 5 (a) | 6 (b) | 7 (d) | 8 (b) | 9 (d) | 10 (d) |

Chapter 21

| 1 (c) | 2 (a) | 3 (a) | 4 (b) | 5 (b) | 6 (c) | 7 (b) | 8 (d) | 9 (a) | 10 (a) |

Chapter 22

| 1 (b) | 2 (c) | 3 (c) | 4 (a) | 5 (d) | 6 (b) | 7 (a) | 8 (b) | 9 (a) | 10 (c) |

Chapter 23

| 1 (b) | 2 (c) | 3 (a) | 4 (b) | 5 (c) | 6 (c) | 7 (b) | 8 (a) | 9 (a) | 10 (b) |

Chapter 24

| 1 (b) | 2 (a) | 3 (c) | 4 (a) | 5 (c) | 6 (d) | 7 (b) | 8 (d) | 9 (a) | 10 (c) |

Chapter 25

| 1 (b) | 2 (a) | 3 (c) | 4 (b) | 5 (c) | 6 (b) | 7 (b) | 8 (c) | 9 (d) | 10 (a) |

Chapter 26

| 1 (b) | 2 (c) | 3 (a) | 4 (a) | 5 (c) | 6 (d) | 7 (a) | 8 (a) | 9 (b) | 10 (c) |

Chapter 27

| 1 (a) | 2 (c) | 3 (d) | 4 (b) | 5 (c) | 6 (a) | 7 (b) | 8 (b) | 9 (a) | 10 (d) |

APPENDIX A

MCS-51 Instruction Set

(All mnemonics are copyright of Intel Corporation 1980)

The following abbreviations are used in instruction-set listing:

adr 11   11-bit address
adr 16   16-bit address
adr 8   8-bit address
bit   8-bit bit address
rel   8-bit relative offset address
dat 8   8-bit data
dat 16   16-bit data

A.1 | Instructions Arranged by Functional Order

A.1.1 | Arithmetic Operations

Opcode	Bytes/Cycles	Instruction		Flags affected
25	2-1	ADD	A, adr 8	C OV AC
26	1-1	ADD	A, @R0	C OV AC
27	1-1	ADD	A, @R1	C OV AC
24	2-1	ADD	A, #dat 8	C OV AC
28	1-1	ADD	A, R0	C OV AC
29	1-1	ADD	A, R1	C OV AC
2A	1-1	ADD	A, R2	C OV AC
2B	1-1	ADD	A, R3	C OV AC
2C	1-1	ADD	A, R4	C OV AC
2D	1-1	ADD	A, R5	C OV AC
2E	1-1	ADD	A, R6	C OV AC
2F	1-1	ADD	A, R7	C OV AC
35	2-1	ADDC	A, adr 8	C OV AC
36	1-1	ADDC	A, @R0	C OV AC
37	1-1	ADDC	A, @R1	C OV AC
34	2-1	ADDC	A, #dat 8	C OV AC
38	1-1	ADDC	A, R0	C OV AC
39	1-1	ADDC	A, R1	C OV AC
3A	1-1	ADDC	A, R2	C OV AC
3B	1-1	ADDC	A, R3	C OV AC

(Continued)

Opcode	Bytes/Cycles	Instruction		Flags affected
3C	1-1	ADDC	A, R4	C OV AC
3D	1-1	ADDC	A, R5	C OV AC
3E	1-1	ADDC	A, R6	C OV AC
3F	1-1	ADDC	A, R7	C OV AC
D4	1-1	DA	A	C
14	1-1	DEC	A	
15	2-1	DEC	adr 8	
16	1-1	DEC	@R0	
17	1-1	DEC	@R1	
18	1-1	DEC	R0	
19	1-1	DEC	R1	
1A	1-1	DEC	R2	
1B	1-1	DEC	R3	
1C	1-1	DEC	R4	
1D	1-1	DEC	R5	
1E	1-1	DEC	R6	
1F	1-1	DEC	R7	
84	1-4	DIV	AB	C OV
04	1-1	INC	A	
05	2-1	INC	adr 8	
A3	1-2	INC	DPTR	
06	1-1	INC	@R0	
07	1-1	INC	@R1	
08	1-1	INC	R0	
09	1-1	INC	R1	
0A	1-1	INC	R2	
0B	1-1	INC	R3	
0C	1-1	INC	R4	
0D	1-1	INC	R5	
0E	1-1	INC	R6	
0F	1-1	INC	R7	
A4	1-4	MUL	AB	C OV
95	2-1	SUBB	A, adr 8	C OV AC
96	1-1	SUBB	A, @R0	C OV AC
97	1-1	SUBB	A, @R1	C OV AC
94	2-1	SUBB	A, #dat 8	C OV AC
98	1-1	SUBB	A, R0	C OV AC
99	1-1	SUBB	A, R1	C OV AC
9A	1-1	SUBB	A, R2	C OV AC
9B	1-1	SUBB	A, R3	C OV AC
9C	1-1	SUBB	A, R4	C OV AC
9D	1-1	SUBB	A, R5	C OV AC
9E	1-1	SUBB	A, R6	C OV AC
9F	1-1	SUBB	A, R7	C OV AC

A.1.2 | Logical Operations

Opcode	Bytes/Cycles	Instruction		Flags affected
55	2-1	ANL	A, adr 8	
56	1-1	ANL	A, @R0	
57	1-1	ANL	A, @R1	
54	2-1	ANL	A, #dat 8	
58	1-1	ANL	A, R0	
59	1-1	ANL	A, R1	
5A	1-1	ANL	A, R2	
5B	1-1	ANL	A, R3	
5C	1-1	ANL	A, R4	
5D	1-1	ANL	A, R5	
5E	1-1	ANL	A, R6	
5F	1-1	ANL	A, R7	
52	2-1	ANL	adr 8, A	
53	3-2	ANL	adr 8, #dat 8	
E4	1-1	CLR	A	
F4	1-1	CPL	A	
45	2-1	ORL	A, adr 8	
46	1-1	ORL	A, @R0	
47	1-1	ORL	A, @R1	
44	2-1	ORL	A, #dat 8	
48	1-1	ORL	A, R0	
49	1-1	ORL	A, R1	
4A	1-1	ORL	A, R2	
4B	1-1	ORL	A, R3	
4C	1-1	ORL	A, R4	
4D	1-1	ORL	A, R5	
4E	1-1	ORL	A, R6	
4F	1-1	ORL	A, R7	
42	2-1	ORL	adr 8, A	
43	3-2	ORL	adr 8, #dat 8	
23	1-1	RL	A	
33	1-1	RLC	A	C
03	1-1	RR	A	
13	1-1	RRC	A	C
C4	1-1	SWAP	A	
65	2-1	XRL	A, adr 8	
66	1-1	XRL	A, @R0	
67	1-1	XRL	A, @R1	
64	2-1	XRL	A, #dat 8	
68	1-1	XRL	A, R0	
69	1-1	XRL	A, R1	
6A	1-1	XRL	A, R2	
6B	1-1	XRL	A, R3	
6C	1-1	XRL	A, R4	
6D	1-1	XRL	A, R5	

(Continued)

Opcode	Bytes/Cycles	Instruction		Flags affected
6E	1-1	XRL	A, R6	
6F	1-1	XRL	A, R7	
62	2-1	XRL	adr 8, A	
63	3-2	XRL	adr 8, #dat 8	

A.1.3 | Data Movement Operation

Opcode	Bytes/Cycles	Instruction		Flags affected
E5	2-1	MOV	A, adr 8	
E6	1-1	MOV	A, @R0	
E7	1-1	MOV	A, @R1	
74	2-1	MOV	A, #dat 8	
E8	1-1	MOV	A, R0	
E9	1-1	MOV	A, R1	
EA	1-1	MOV	A, R2	
EB	1-1	MOV	A, R3	
EC	1-1	MOV	A, R4	
ED	1-1	MOV	A, R5	
EE	1-1	MOV	A, R6	
EF	1-1	MOV	A, R7	
F5	2-1	MOV	adr 8, A	
85	3-2	MOV	adr 8, adr 8	
86	2-2	MOV	adr 8, @R0	
87	2-2	MOV	adr 8, @R1	
75	3-2	MOV	adr 8, #dat 8	
88	2-2	MOV	adr 8, R0	
89	2-2	MOV	adr 8, R1	
8A	2-2	MOV	adr 8, R2	
8B	2-2	MOV	adr 8, R3	
8C	2-2	MOV	adr 8, R4	
8D	2-2	MOV	adr 8, R5	
8E	2-2	MOV	adr 8, R6	
8F	2-2	MOV	adr 8, R7	
F6	1-1	MOV	@R0, A	
F7	1-1	MOV	@R1, A	
A6	2-2	MOV	@R0, adr 8	
A7	2-2	MOV	@R1, adr 8	
76	2-1	MOV	@R0, #dat 8	
77	2-1	MOV	@R1, #dat 8	
90	3-2	MOV	DPTR, #dat 16	
F8	1-1	MOV	R0, A	
F9	1-1	MOV	R1, A	
FA	1-1	MOV	R2, A	
FB	1-1	MOV	R3, A	
FC	1-1	MOV	R4, A	
FD	1-1	MOV	R5, A	

(Continued)

Opcode	Bytes/Cycles	Instruction		Flags affected
FE	1-1	MOV	R6, A	
FF	1-1	MOV	R7, A	
A8	2-2	MOV	R0, adr 8	
A9	2-2	MOV	R1, adr 8	
AA	2-2	MOV	R2, adr 8	
AB	2-2	MOV	R3, adr 8	
AC	2-2	MOV	R4, adr 8	
AD	2-2	MOV	R5, adr 8	
AE	2-2	MOV	R6, adr 8	
AF	2-2	MOV	R7, adr 8	
78	2-1	MOV	R0, #dat 8	
79	2-1	MOV	R1, #dat 8	
7A	2-1	MOV	R2, #dat 8	
7B	2-1	MOV	R3, #dat 8	
7C	2-1	MOV	R4, #dat 8	
7D	2-1	MOV	R5, #dat 8	
7E	2-1	MOV	R6, #dat 8	
7F	2-1	MOV	R7, #dat 8	
93	1-2	MOVC	A, @A + DPTR	
83	1-2	MOVC	A, @A + PC	
E0	1-2	MOVX	A, @DPTR	
E2	1-2	MOVX	A, @R0	
E3	1-2	MOVX	A, @R1	
F0	1-2	MOVX	@DPTR, A	
F2	1-2	MOVX	@R0, A	
F3	1-2	MOVX	@R1, A	
D0	2-2	POP	adr 8	
C0	2-2	PUSH	adr 8	
C5	2-1	XCH	A, adr 8	
C6	1-1	XCH	A, @R0	
C7	1-1	XCH	A, @R1	
C8	1-1	XCH	A, R0	
C9	1-1	XCH	A, R1	
CA	1-1	XCH	A, R2	
CB	1-1	XCH	A, R3	
CC	1-1	XCH	A, R4	
CD	1-1	XCH	A, R5	
CE	1-1	XCH	A, R6	
CF	1-1	XCH	A, R7	
D6	1-1	XCHD	A, @R0	
D7	1-1	XCHD	A, @R1	

A.1.4 | Boolean Variable Operations

Opcode	Bytes/Cycles	Instruction		Flags affected
82	2-2	ANL	C, bit	C
B0	2-2	ANL	C, /bit	C

(Continued)

Opcode	Bytes/Cycles	Instruction		Flags affected
C2	2-1	CLR	bit	
C3	1-1	CLR	C	C
B2	2-1	CPL	bit	
B3	1-1	CPL	C	C
72	2-2	ORL	C, bit	C
A0	2-2	ORL	C, /bit	C
92	2-2	MOV	bit, C	
A2	2-1	MOV	C, bit	C
D2	2-1	SETB	bit	
D3	1-1	SETB	C	C

A.1.5 | Call and Jump Operations

Opcode	Bytes/Cycles	Instruction		Flags affected
*	2-2	ACALL	adr 11	
**	2-2	AJMP	adr 11	
B5	3-2	CJNE	A, adr 8, rel	C
B4	3-2	CJNE	A, #dat 8, rel	C
B6	3-2	CJNE	@R0, #dat 8, rel	C
B7	3-2	CJNE	@R1, #dat 8, rel	C
B8	3-2	CJNE	R0, #dat 8, rel	C
B9	3-2	CJNE	R1, #dat 8, rel	C
BA	3-2	CJNE	R2, #dat 8, rel	C
BB	3-2	CJNE	R3, #dat 8, rel	C
BC	3-2	CJNE	R4, #dat 8, rel	C
BD	3-2	CJNE	R5, #dat 8, rel	C
BE	3-2	CJNE	R6, #dat 8, rel	C
BF	3-2	CJNE	R7, #dat 8, rel	C
D5	3-2	DJNZ	adr 8, rel	
D8	2-2	DJNZ	R0, rel	
D9	2-2	DJNZ	R1, rel	
DA	2-2	DJNZ	R2, rel	
DB	2-2	DJNZ	R3, rel	
DC	2-2	DJNZ	R4, rel	
DD	2-2	DJNZ	R5, rel	
DE	2-2	DJNZ	R6, rel	
DF	2-2	DJNZ	R7, rel	
20	3-2	JB	bt, rela	
10	3-2	JBC	bit, rel	
40	2-2	JC	rel	
73	1-2	JMP	@A + DPTR	
30	3-2	JNB	bit, rel	
50	2-2	JNC	rel	
70	2-2	JNZ	rel	
60	2-2	JZ	rel	
12	3-2	LCALL	adr 16	

(Continued)

Opcode	Bytes/Cycles	Instruction		Flags affected
02	3-2	LJMP	adr 16	
00	1-1	NOP		
22	1-2	RET		
32	1-2	RETI		
80	2-2	SJMP	rel	

* Depending upon the call address, ACALL may have the following opcodes:
11H, 31H, 51H, 71H, 91H, B1H, D1 and F1H.

** Depending upon the jump address, AJMP may have the following opcodes:
01H, 21H, 41H, 61H, 81H, A1H, C1H and E1H.

A.2 | Instructions Arranged by their Opcodes

Opcode	Bytes/Cycles	Instruction		Flags affected
00	1-1	NOP		
01	2-2	AJMP	adr 11	
02	3-2	LJMP	adr 16	
03	1-1	RR	A	
04	1-1	INC	A	
05	2-1	INC	adr 8	
06	1-1	INC	@R0	
07	1-1	INC	@R1	
08	1-1	INC	R0	
09	1-1	INC	R1	
0A	1-1	INC	R2	
0B	1-1	INC	R3	
0C	1-1	INC	R4	
0D	1-1	INC	R5	
0E	1-1	INC	R6	
0F	1-1	INC	R7	
10	3-2	JBC	bit, rel	
11	2-2	ACALL	adr 11	
12	3-2	LCALL	adr 16	
13	1-1	RRC	A	C
14	1-1	DEC	A	
15	2-1	DEC	adr 8	
16	1-1	DEC	@R0	
17	1-1	DEC	@R1	
18	1-1	DEC	R0	
19	1-1	DEC	R1	
1A	1-1	DEC	R2	
1B	1-1	DEC	R3	
1C	1-1	DEC	R4	
1D	1-1	DEC	R5	
1E	1-1	DEC	R6	
1F	1-1	DEC	R7	

(Continued)

Opcode	Bytes/Cycles	Instruction		Flags affected
20	3-2	JB	bit, rela	
21	2-2	AJMP	adr 11	
22	1-2	RET		
23	1-1	RL	A	
24	2-1	ADD	A, #dat 8	C OV AC
25	2-1	ADD	A, adr 8	C OV AC
26	1-1	ADD	A, @R0	C OV AC
27	1-1	ADD	A, @R1	C OV AC
28	1-1	ADD	A, R0	C OV AC
29	1-1	ADD	A, R1	C OV AC
2A	1-1	ADD	A, R2	C OV AC
2B	1-1	ADD	A, R3	C OV AC
2C	1-1	ADD	A, R4	C OV AC
2D	1-1	ADD	A, R5	C OV AC
2E	1-1	ADD	A, R6	C OV AC
2F	1-1	ADD	A, R7	C OV AC
30	3-2	JNB	bit, rel	
31	2-2	ACALL	adr 11	
32	1-2	RETI		
33	1-1	RLC	A	C
34	2-1	ADDC	A, #dat 8	C OV AC
35	2-1	ADDC	A, adr 8	C OV AC
36	1-1	ADDC	A, @R0	C OV AC
37	1-1	ADDC	A, @R1	C OV AC
38	1-1	ADDC	A, R0	C OV AC
39	1-1	ADDC	A, R1	C OV AC
3A	1-1	ADDC	A, R2	C OV AC
3B	1-1	ADDC	A, R3	C OV AC
3C	1-1	ADDC	A, R4	C OV AC
3D	1-1	ADDC	A, R5	C OV AC
3E	1-1	ADDC	A, R6	C OV AC
3F	1-1	ADDC	A, R7	C OV AC
40	2-2	JC	rel	
41	2-2	AJMP	adr 11	
42	2-1	ORL	adr 8, A	
43	3-2	ORL	adr 8, #dat 8	
44	2-1	ORL	A, #dat 8	
45	2-1	ORL	A, adr 8	
46	1-1	ORL	A, @R0	
47	1-1	ORL	A, @R1	
48	1-1	ORL	A, R0	
49	1-1	ORL	A, R1	
4A	1-1	ORL	A, R2	
4B	1-1	ORL	A, R3	
4C	1-1	ORL	A, R4	

(Continued)

Opcode	Bytes/Cycles	Instruction		Flags affected
4D	1-1	ORL	A, R5	
4E	1-1	ORL	A, R6	
4F	1-1	ORL	A, R7	
50	2-2	JNC	rel	
51	2-2	ACALL	adr 11	
52	2-1	ANL	adr 8, A	
53	3-2	ANL	adr 8, #dat 8	
54	2-1	ANL	A, #dat 8	
55	2-1	ANL	A, adr 8	
56	1-1	ANL	A, @R0	
57	1-1	ANL	A, @R1	
58	1-1	ANL	A, R0	
59	1-1	ANL	A, R1	
5A	1-1	ANL	A, R2	
5B	1-1	ANL	A, R3	
5C	1-1	ANL	A, R4	
5D	1-1	ANL	A, R5	
5E	1-1	ANL	A, R6	
5F	1-1	ANL	A, R7	
60	2-2	JZ	rel	
61	2-2	AJMP	adr 11	
62	2-1	XRL	adr 8, A	
63	3-2	XRL	adr 8, #dat 8	
64	2-1	XRL	A, #dat 8	
65	2-1	XRL	A, adr 8	
66	1-1	XRL	A, @R0	
67	1-1	XRL	A, @R1	
68	1-1	XRL	A, R0	
69	1-1	XRL	A, R1	
6A	1-1	XRL	A, R2	
6B	1-1	XRL	A, R3	
6C	1-1	XRL	A, R4	
6D	1-1	XRL	A, R5	
6E	1-1	XRL	A, R6	
6F	1-1	XRL	A, R7	
70	2-2	JNZ	rel	
71	2-2	ACALL	adr 11	
72	2-2	ORL	C, bit	C
73	1-2	JMP	@A + DPTR	
74	2-1	MOV	A, #dat 8	
75	3-2	MOV	adr 8, #dat 8	
76	2-1	MOV	@R0, #dat 8	
77	2-1	MOV	@R1, #dat 8	
78	2-1	MOV	R0, #dat 8	
79	2-1	MOV	R1, #dat 8	

(Continued)

Opcode	Bytes/Cycles	Instruction		Flags affected
7A	2-1	MOV	R2, #dat 8	
7B	2-1	MOV	R3, #dat 8	
7C	2-1	MOV	R4, #dat 8	
7D	2-1	MOV	R5, #dat 8	
7E	2-1	MOV	R6, #dat 8	
7F	2-1	MOV	R7, #dat 8	
80	2-2	SJMP	rel	
81	2-2	AJMP	adr 11	
82	2-2	ANL	C, bit	C
83	1-2	MOVC	A, @A + PC	
84	1-4	DIV	AB	C OV
85	3-2	MOV	adr 8, adr 8	
86	2-2	MOV	adr 8, @R0	
87	2-2	MOV	adr 8, @R1	
88	2-2	MOV	adr 8, R0	
89	2-2	MOV	adr 8, R1	
8A	2-2	MOV	adr 8, R2	
8B	2-2	MOV	adr 8, R3	
8C	2-2	MOV	adr 8, R4	
8D	2-2	MOV	adr 8, R5	
8E	2-2	MOV	adr 8, R6	
8F	2-2	MOV	adr 8, R7	
90	3-2	MOV	DPTR, #dat 16	
91	2-2	ACALL	adr 11	
92	2-2	MOV	bit, C	
93	1-2	MOVC	A, @A + DPTR	
94	2-1	SUBB	A, #dat 8	C OV AC
95	2-1	SUBB	A, adr 8	C OV AC
96	1-1	SUBB	A, @R0	C OV AC
97	1-1	SUBB	A, @R1	C OV AC
98	1-1	SUBB	A, R0	C OV AC
99	1-1	SUBB	A, R1	C OV AC
9A	1-1	SUBB	A, R2	C OV AC
9B	1-1	SUBB	A, R3	C OV AC
9C	1-1	SUBB	A, R4	C OV AC
9D	1-1	SUBB	A, R5	C OV AC
9E	1-1	SUBB	A, R6	C OV AC
9F	1-1	SUBB	A, R7	C OV AC
A0	2-2	ORL	C, /bit	C
A1	2-2	AJMP	adr 11	
A2	2-1	MOV	C, bit	C
A3	1-2	INC	DPTR	
A4	1-4	MUL	AB	C OV
A5	unused			
A6	2-2	MOV	@R0, adr 8	
A7	2-2	MOV	@R1, adr 8	

(Continued)

Opcode	Bytes/Cycles	Instruction		Flags affected
A8	2-2	MOV	R0, adr 8	
A9	2-2	MOV	R1, adr 8	
AA	2-2	MOV	R2, adr 8	
AB	2-2	MOV	R3, adr 8	
AC	2-2	MOV	R4, adr 8	
AD	2-2	MOV	R5, adr 8	
AE	2-2	MOV	R6, adr 8	
AF	2-2	MOV	R7, adr 8	
B0	2-2	ANL	C, /bit	C
B1	2-2	ACALL	adr 11	
B2	2-1	CPL	bit	
B3	1-1	CPL	C	C
B4	3-2	CJNE	A, #dat 8, rel	C
B5	3-2	CJNE	A, adr 8, rel	C
B6	3-2	CJNE	@R0, #dat 8, rel	C
B7	3-2	CJNE	@R1, #dat 8, rel	C
B8	3-2	CJNE	R0, #dat 8, rel	C
B9	3-2	CJNE	R1, #dat 8, rel	C
BA	3-2	CJNE	R2, #dat 8, rel	C
BB	3-2	CJNE	R3, #dat 8, rel	C
BC	3-2	CJNE	R4, #dat 8, rel	C
BD	3-2	CJNE	R5, #dat 8, rel	C
BE	3-2	CJNE	R6, #dat 8, rel	C
BF	3-2	CJNE	R7, #dat 8, rel	C
C0	2-2	PUSH	adr 8	
C1	2-2	AJMP	adr 11	
C2	2-1	CLR	bit	
C3	1-1	CLR	C	C
C4	1-1	SWAP	A	
C5	2-1	XCH	A, adr 8	
C6	1-1	XCH	A, @R0	
C7	1-1	XCH	A, @R1	
C8	1-1	XCH	A, R0	
C9	1-1	XCH	A, R1	
CA	1-1	XCH	A, R2	
CB	1-1	XCH	A, R3	
CC	1-1	XCH	A, R4	
CD	1-1	XCH	A, R5	
CE	1-1	XCH	A, R6	
CF	1-1	XCH	A, R7	
D0	2-2	POP	adr 8	
D1	2-2	ACALL	adr 11	
D2	2-1	SETB	bit	
D3	1-1	SETB	C	C
D4	1-1	DA	A	C

(Continued)

Opcode	Bytes/Cycles	Instruction		Flags affected
D5	3-2	DJNZ	adr 8, rel	
D6	1-1	XCHD	A, @R0	
D7	1-1	XCHD	A, @R1	
D8	2-2	DJNZ	R0, rel	
D9	2-2	DJNZ	R1, rel	
DA	2-2	DJNZ	R2, rel	
DB	2-2	DJNZ	R3, rel	
DC	2-2	DJNZ	R4, rel	
DD	2-2	DJNZ	R5, rel	
DE	2-2	DJNZ	R6, rel	
DF	2-2	DJNZ	R7, rel	
E0	1-2	MOVX	A, @DPTR	
E1	2-2	AJMP	adr 11	
E2	1-2	MOVX	A, @R0	
E3	1-2	MOVX	A, @R1	
E4	1-1	CLR	A	
E5	2-1	MOV	A, adr 8	
E6	1-1	MOV	A, @R0	
E7	1-1	MOV	A, @R1	
E8	1-1	MOV	A, R0	
E9	1-1	MOV	A, R1	
EA	1-1	MOV	A, R2	
EB	1-1	MOV	A, R3	
EC	1-1	MOV	A, R4	
ED	1-1	MOV	A, R5	
EE	1-1	MOV	A, R6	
EF	1-1	MOV	A, R7	
F0	1-2	MOVX	@DPTR, A	
F1	2-2	ACALL	adr 11	
F2	1-2	MOVX	@R0, A	
F3	1-2	MOVX	@R1, A	
F4	1-1	CPL	A	
F5	2-1	MOV	adr 8, A	
F6	1-1	MOV	@R0, A	
F7	1-1	MOV	@R1, A	
F8	1-1	MOV	R0, A	
F9	1-1	MOV	R1, A	
FA	1-1	MOV	R2, A	
FB	1-1	MOV	R3, A	
FC	1-1	MOV	R4, A	
FD	1-1	MOV	R5, A	
FE	1-1	MOV	R6, A	
FF	1-1	MOV	R7, A	

A.3 | Instructions Arranged by Alphabetical Order

Opcode	Bytes/Cycles	Instruction		Flags affected
*	2-2	ACALL	adr 11	
25	2-1	ADD	A, adr 8	C OV AC
26	1-1	ADD	A, @R0	C OV AC
27	1-1	ADD	A, @R1	C OV AC
24	2-1	ADD	A, #dat 8	C OV AC
28	1-1	ADD	A, R0	C OV AC
29	1-1	ADD	A, R1	C OV AC
2A	1-1	ADD	A, R2	C OV AC
2B	1-1	ADD	A, R3	C OV AC
2C	1-1	ADD	A, R4	C OV AC
2D	1-1	ADD	A, R5	C OV AC
2E	1-1	ADD	A, R6	C OV AC
2F	1-1	ADD	A, R7	C OV AC
35	2-1	ADDC	A, adr 8	C OV AC
36	1-1	ADDC	A, @R0]	C OV AC
37	1-1	ADDC	A, @R1	C OV AC
34	2-1	ADDC	A, #dat 8	C OV AC
38	1-1	ADDC	A, R0	C OV AC
39	1-1	ADDC	A, R1	C OV AC
3A	1-1	ADDC	A, R2	C OV AC
3B	1-1	ADDC	A, R3	C OV AC
3C	1-1	ADDC	A, R4	C OV AC
3D	1-1	ADDC	A, R5	C OV AC
3E	1-1	ADDC	A, R6	C OV AC
3F	1-1	ADDC	A, R7	C OV AC
**	2-2	AJMP	adr 11	
55	2-1	ANL	A, adr 8	
56	1-1	ANL	A, @R0	
57	1-1	ANL	A, @R1	
54	2-1	ANL	A, #dat 8	
58	1-1	ANL	A, R0	
59	1-1	ANL	A, R1	
5A	1-1	ANL	A, R2	
5B	1-1	ANL	A, R3	
5C	1-1	ANL	A, R4	
5D	1-1	ANL	A, R5	
5E	1-1	ANL	A, R6	
5F	1-1	ANL	A, R7	
52	2-1	ANL	adr 8, A	
53	3-2	ANL	adr 8, #dat 8	
82	2-2	ANL	C, bit	C
B0	2-2	ANL	C, /bit	C
B5	3-2	CJNE	A, adr 8,	C
B4	3-2	CJNE	A, #dat 8, rel	C

(Continued)

Opcode	Bytes/Cycles	Instruction		Flags affected
B6	3-2	CJNE	@R0, #dat 8, rel	C
B7	3-2	CJNE	@R1, #dat 8, rel	C
B8	3-2	CJNE	R0, #dat 8, rel	C
B9	3-2	CJNE	R1, #dat 8, rel	C
BA	3-2	CJNE	R2, #dat 8, rel	C
BB	3-2	CJNE	R3, #dat 8, rel	C
BC	3-2	CJNE	R4, #dat 8, rel	C
BD	3-2	CJNE	R5, #dat 8, rel	C
BE	3-2	CJNE	R6, #dat 8, rel	C
BF	3-2	CJNE	R7, #dat 8, rel	C
E4	1-1	CLR	A	
C2	2-1	CLR	bit	
C3	1-1	CLR	C	C
F4	1-1	CPL	A	
B2	2-1	CPL	bit	
B3	1-1	CPL	C	C
D4	1-1	DA	A	C
14	1-1	DEC	A	
15	2-1	DEC	adr 8	
16	1-1	DEC	@R0	
17	1-1	DEC	@R1	
18	1-1	DEC	R0	
19	1-1	DEC	R1	
1A	1-1	DEC	R2	
1B	1-1	DEC	R3	
1C	1-1	DEC	R4	
1D	1-1	DEC	R5	
1E	1-1	DEC	R6	
1F	1-1	DEC	R7	
84	1-4	DIV	AB	C OV
D5	3-2	DJNZ	adr 8, rel	
D8	2-2	DJNZ	R0, rel	
D9	2-2	DJNZ	R1, rel	
DA	2-2	DJNZ	R2, rel	
DB	2-2	DJNZ	R3, rel	
DC	2-2	DJNZ	R4, rel	
DD	2-2	DJNZ	R5, rel	
DE	2-2	DJNZ	R6, rel	
DF	2-2	DJNZ	R7, rel	
04	1-1	INC	A	
05	2-1	INC	adr 8	
A3	1-2	INC	DPTR	
06	1-1	INC	@R0	
07	1-1	INC	@R1	
08	1-1	INC	R0	
09	1-1	INC	R1	

(Continued)

Opcode	Bytes/Cycles	Instruction		Flags affected
0A	1-1	INC	R2	
0B	1-1	INC	R3	
0C	1-1	INC	R4	
0D	1-1	INC	R5	
0E	1-1	INC	R6	
0F	1-1	INC	R7	
20	3-2	JB	bit, rel	
10	3-2	JBC	bit, rel	
40	2-2	JC	rel	
73	1-2	JMP	@A + DPTR	
30	3-2	JNB	bit, rel	
50	2-2	JNC	rel	
70	2-2	JNZ	rel	
60	2-2	JZ	rel	
12	3-2	LCALL	adr 16	
02	3-2	LJMP	adr 16	
E5	2-1	MOV	A, adr 8	
E6	1-1	MOV	A, @R0	
E7	1-1	MOV	A, @R1	
74	2-1	MOV	A, #dat 8	
E8	1-1	MOV	A, R0	
E9	1-1	MOV	A, R1	
EA	1-1	MOV	A, R2	
EB	1-1	MOV	A, R3	
EC	1-1	MOV	A, R4	
ED	1-1	MOV	A, R5	
EE	1-1	MOV	A, R6	
EF	1-1	MOV	A, R7	
F5	2-1	MOV	adr 8, A	
85	3-2	MOV	adr 8, adr 8	
86	2-2	MOV	adr 8, @R0	
87	2-2	MOV	adr 8, @R1	
75	3-2	MOV	adr 8, #dat 8	
88	2-2	MOV	adr 8, R0	
89	2-2	MOV	adr 8, R1	
8A	2-2	MOV	adr 8, R2	
8B	2-2	MOV	adr 8, R3	
8C	2-2	MOV	adr 8, R4	
8D	2-2	MOV	adr 8, R5	
8E	2-2	MOV	adr 8, R6	
8F	2-2	MOV	adr 8, R7	
92	2-2	MOV	bit, C	
A2	2-1	MOV	C, bit	C
F6	1-1	MOV	@R0, A	
F7	1-1	MOV	@R1, A	
A6	2-2	MOV	@R0, adr 8	

(Continued)

Opcode	Bytes/Cycles	Instruction		Flags affected
A7	2-2	MOV	@R1, adr 8	
76	2-1	MOV	@R0, #dat 8	
77	2-1	MOV	@R1, #dat 8	
90	3-2	MOV	DPTR, #dat 16	
F8	1-1	MOV	R0, A	
F9	1-1	MOV	R1, A	
FA	1-1	MOV	R2, A	
FB	1-1	MOV	R3, A	
FC	1-1	MOV	R4, A	
FD	1-1	MOV	R5, A	
FE	1-1	MOV	R6, A	
FF	1-1	MOV	R7, A	
A8	2-2	MOV	R0, adr 8	
A9	2-2	MOV	R1, adr 8	
AA	2-2	MOV	R2, adr 8	
AB	2-2	MOV	R3, adr 8	
AC	2-2	MOV	R4, adr 8	
AD	2-2	MOV	R5, adr 8	
AE	2-2	MOV	R6, adr 8	
AF	2-2	MOV	R7, adr 8	
78	2-1	MOV	R0, #dat 8	
79	2-1	MOV	R1, #dat 8	
7A	2-1	MOV	R2, #dat 8	
7B	2-1	MOV	R3, #dat 8	
7C	2-1	MOV	R4, #dat 8	
7D	2-1	MOV	R5, #dat 8	
7E	2-1	MOV	R6, #dat 8	
7F	2-1	MOV	R7, #dat 8	
93	1-2	MOVC	A, @A + DPTR	
83	1-2	MOVC	A, @A + PC	
E0	1-2	MOVX	A, @DPTR	
E2	1-2	MOVX	A, @R0	
E3	1-2	MOVX	A, @R1	
F0	1-2	MOVX	@DPTR, A	
F2	1-2	MOVX	@R0, A	
F3	1-2	MOVX	@R1, A	
A4	1-4	MUL	AB	C OV
00	1-1	NOP		
45	2-1	ORL	A, adr 8	
46	1-1	ORL	A, @R0	
47	1-1	ORL	A, @R1	
44	2-1	ORL	A, #dat 8	
48	1-1	ORL	A, R0	
49	1-1	ORL	A, R1	
4A	1-1	ORL	A, R2	
4B	1-1	ORL	A, R3	

(Continued)

Opcode	Bytes/Cycles	Instruction		Flags affected
4C	1-1	ORL	A, R4	
4D	1-1	ORL	A, R5	
4E	1-1	ORL	A, R6	
4F	1-1	ORL	A, R7	
42	2-1	ORL	adr 8, A	
43	3-2	ORL	adr 8, #dat 8	
72	2-2	ORL	C, bit	C
A0	2-2	ORL	C, /bit	C
D0	2-2	POP	adr 8	
C0	2-2	PUSH	adr 8	
22	1-2	RET		
32	1-2	RETI		
23	1-1	RL	A	
33	1-1	RLC	A	C
03	1-1	RR	A	
13	1-1	RRC	A	C
D2	2-1	SETB	bit	
D3	1-1	SETB	C	C
80	2-2	SJMP	rel	
95	2-1	SUBB	A, adr 8	C OV AC
96	1-1	SUBB	A, @R0	C OV AC
97	1-1	SUBB	A, @R1	C OV AC
94	2-1	SUBB	A, #dat 8	C OV AC
98	1-1	SUBB	A, R0	C OV AC
99	1-1	SUBB	A, R1	C OV AC
9A	1-1	SUBB	A, R2	C OV AC
9B	1-1	SUBB	A, R3	C OV AC
9C	1-1	SUBB	A, R4	C OV AC
9D	1-1	SUBB	A, R5	C OV AC
9E	1-1	SUBB	A, R6	C OV AC
9F	1-1	SUBB	A, R7	C OV AC
C4	1-1	SWAP	A	
C5	2-1	XCH	A, adr 8	
C6	1-1	XCH	A, @R0	
C7	1-1	XCH	A, @R1	
C8	1-1	XCH	A, R0	
C9	1-1	XCH	A, R1	
CA	1-1	XCH	A, R2	
CB	1-1	XCH	A, R3	
CC	1-1	XCH	A, R4	
CD	1-1	XCH	A, R5	
CE	1-1	XCH	A, R6	
CF	1-1	XCH	A, R7	
D6	1-1	XCHD	A, @R0	
D7	1-1	XCHD	A, @R1	
65	2-1	XRL	A, adr 8	

(Continued)

Opcode	Bytes/Cycles	Instruction		Flags affected
66	1-1	XRL	A, @R0	
67	1-1	XRL	A, @R1	
64	2-1	XRL	A, #dat 8	
68	1-1	XRL	A, R0	
69	1-1	XRL	A, R1	
6A	1-1	XRL	A, R2	
6B	1-1	XRL	A, R3	
6C	1-1	XRL	A, R4	
6D	1-1	XRL	A, R5	
6E	1-1	XRL	A, R6	
6F	1-1	XRL	A, R7	
62	2-1	XRL	adr 8, A	
63	3-2	XRL	adr 8, #dat 8	

* Depending upon the call address, ACALL may have the following opcodes:
11H, 31H, 51H, 71H, 91H, B1H, D1 and F1H.

** Depending upon the jump address, AJMP may have the following opcodes:
01H, 21H, 41H, 61H, 81H, A1H, C1H and E1H.

A.4 | Instructions Arranged by Operand Types

A.4.1 | Instructions with Bit Operands

Opcode	Bytes/Cycles	Instruction		Flags affected
82	2-2	ANL	C, bit	C
B0	2-2	ANL	C, /bit	C
C2	2-1	CLR	bit	
B2	2-1	CPL	bit	
20	3-2	JB	bit, rel	
10	3-2	JBC	bit, rel	
30	3-2	JNB	bit, rel	
92	2-2	MOV	bit, C	
A2	2-1	MOV	C, bit	C
72	2-2	ORL	C, bit	C
A0	2-2	ORL	C, /bit	C
D2	2-1	SETB	bit	

A.4.2 | Instructions with Direct Addressing

Opcode	Bytes/Cycles	Instruction		Flags affected
25	2-1	ADD	A, adr 8	C OV AC
35	2-1	ADDC	A, adr 8	C OV AC
55	2-1	ANL	A, adr 8	
52	2-1	ANL	adr 8, A	
53	3-2	ANL	adr 8, #dat 8	
B5	3-2	CJNE	A, adr 8,	C
15	2-1	DEC	adr 8	
D5	3-2	DJNZ	adr 8, rel	
05	2-1	INC	adr 8	

(Continued)

Opcode	Bytes/Cycles	Instruction		Flags affected
E5	2-1	MOV	A, adr 8	
F5	2-1	MOV	adr 8, A	
85	3-2	MOV	adr 8, adr 8	
86	2-2	MOV	adr 8, @R0	
87	2-2	MOV	adr 8, @R1	
75	3-2	MOV	adr 8, #dat 8	
88	2-2	MOV	adr 8, R0	
89	2-2	MOV	adr 8, R1	
8A	2-2	MOV	adr 8, R2	
8B	2-2	MOV	adr 8, R3	
8C	2-2	MOV	adr 8, R4	
8D	2-2	MOV	adr 8, R5	
8E	2-2	MOV	adr 8, R6	
8F	2-2	MOV	adr 8, R7	
A6	2-2	MOV	@R0, adr 8	
A7	2-2	MOV	@R1, adr 8	
A8	2-2	MOV	R0, adr 8	
A9	2-2	MOV	R1, adr 8	
AA	2-2	MOV	R2, adr 8	
AB	2-2	MOV	R3, adr 8	
AC	2-2	MOV	R4, adr 8	
AD	2-2	MOV	R5, adr 8	
AE	2-2	MOV	R6, adr 8	
AF	2-2	MOV	R7, adr 8	
45	2-1	ORL	A, adr 8	
42	2-1	ORL	adr 8, A	
43	3-2	ORL	adr 8, #dat 8	
D0	2-2	POP	adr 8	
C0	2-2	PUSH	adr 8	
95	2-1	SUBB	A, adr 8	C OV AC
C5	2-1	XCH	A, adr 8	
65	2-1	XRL	A, adr 8	
62	2-1	XRL	adr 8, A	
63	3-2	XRL	adr 8, #dat 8	

A.4.3 | Instructions with Indirect Addressing Through R0

Opcode	Bytes/Cycles	Instruction		Flags affected
26	1-1	ADD	A, @R0	C OV AC
36	1-1	ADDC	A, @R0]	C OV AC
56	1-1	ANL	A, @R0	
B6	3-2	CJNE	@R0, #dat 8, rel	C
16	1-1	DEC	@R0	
06	1-1	INC	@R0	
E6	1-1	MOV	A, @R0	
86	2-2	MOV	adr 8, @R0	
F6	1-1	MOV	@R0, A	

(Continued)

Opcode	Bytes/Cycles	Instruction		Flags affected
A6	2-2	MOV	@R0, adr 8	
76	2-1	MOV	@R0, #dat 8	
E2	1-2	MOVX	A, @R0	
F2	1-2	MOVX	@R0, A	
46	1-1	ORL	A, @R0	
96	1-1	SUBB	A, @R0	C OV AC
C6	1-1	XCH	A, @R0	
D6	1-1	XCHD	A, @R0	
66	1-1	XRL	A, @R0	

A.4.4 | Instructions with Indirect Addressing Through R1

Opcode	Bytes/Cycles	Instruction		Flags affected
27	1-1	ADD	A, @R1	C OV AC
37	1-1	ADDC	A, @R1	C OV AC
57	1-1	ANL	A, @R1	
B7	3-2	CJNE	@R1, #dat 8, rel	C
17	1-1	DEC	@R1	
07	1-1	INC	@R1	
E7	1-1	MOV	A, @R1	
87	2-2	MOV	adr 8, @R1	
F7	1-1	MOV	@R1, A	
A7	2-2	MOV	@R1, adr 8	
77	2-1	MOV	@R1, #dat 8	
E3	1-2	MOVX	A, @R1	
F3	1-2	MOVX	@R1, A	
47	1-1	ORL	A, @R1	
97	1-1	SUBB	A, @R1	C OV AC
C7	1-1	XCH	A, @R1	
D7	1-1	XCHD	A, @R1	
67	1-1	XRL	A, @R1	

A.4.5 | Instructions with Register Addressing Through R0

Opcode	Bytes/Cycles	Instruction		Flags affected
28	1-1	ADD	A, R0	C OV AC
38	1-1	ADDC	A, R0	C OV AC
58	1-1	ANL	A, R0	
B8	3-2	CJNE	R0, #dat 8, rel	C
18	1-1	DEC	R0	
D8	2-2	DJNZ	R0, rel	
08	1-1	INC	R0	
E8	1-1	MOV	A, R0	
88	2-2	MOV	adr 8, R0	
F8	1-1	MOV	R0, A	
A8	2-2	MOV	R0, adr 8	

(Continued)

Opcode	Bytes/Cycles	Instruction		Flags affected
78	2-1	MOV	R0, #dat 8	
48	1-1	ORL	A, R0	
98	1-1	SUBB	A, R0	C OV AC
C8	1-1	XCH	A, R0	
68	1-1	XRL	A, R0	

A.4.6 | Instructions with Register Addressing Through R1

Opcode	Bytes/Cycles	Instruction		Flags affected
29	1-1	ADD	A, R1	C OV AC
39	1-1	ADDC	A, R1	C OV AC
59	1-1	ANL	A, R1	
B9	3-2	CJNE	R1, #dat 8, rel	C
19	1-1	DEC	R1	
D9	2-2	DJNZ	R1, rel	
09	1-1	INC	R1	
E9	1-1	MOV	A, R1	
89	2-2	MOV	adr 8, R1	
F9	1-1	MOV	R1, A	
A9	2-2	MOV	R1, adr 8	
79	2-1	MOV	R1, #dat 8	
49	1-1	ORL	A, R1	
99	1-1	SUBB	A, R1	C OV AC
C9	1-1	XCH	A, R1	
69	1-1	XRL	A, R1	

A.4.7 | Instructions with Register Addressing Through R2

Opcode	Bytes/Cycles	Instruction		Flags affected
2A	1-1	ADD	A, R2	C OV AC
3A	1-1	ADDC	A, R2	C OV AC
5A	1-1	ANL	A, R2	
BA	3-2	CJNE	R2, #dat 8, rel	C
1A	1-1	DEC	R2	
DA	2-2	DJNZ	R2, rel	
0A	1-1	INC	R2	
EA	1-1	MOV	A, R2	
8A	2-2	MOV	adr 8, R2	
FA	1-1	MOV	R2, A	
AA	2-2	MOV	R2, adr 8	
7A	2-1	MOV	R2, #dat 8	
4A	1-1	ORL	A, R2	
9A	1-1	SUBB	A, R2	C OV AC
CA	1-1	XCH	A, R2	
6A	1-1	XRL	A, R2	

A.4.8 | Instructions with Register Addressing Through R3

Opcode	Bytes/Cycles	Instruction		Flags affected
2B	1-1	ADD	A, R3	C OV AC
3B	1-1	ADDC	A, R3	C OV AC
5B	1-1	ANL	A, R3	
BB	3-2	CJNE	R3, #dat 8, rel	C
1B	1-1	DEC	R3	
DB	2-2	DJNZ	R3, rel	
0B	1-1	INC	R3	
EB	1-1	MOV	A, R3	
8B	2-2	MOV	adr 8, R3	
FB	1-1	MOV	R3, A	
AB	2-2	MOV	R3, adr 8	
7B	2-1	MOV	R3, #dat 8	
4B	1-1	ORL	A, R3	
9B	1-1	SUBB	A, R3	C OV AC
CB	1-1	XCH	A, R3	
6B	1-1	XRL	A, R3	

A.4.9 | Instructions with Register Addressing Through R4

Opcode	Bytes/Cycles	Instruction		Flags affected
2C	1-1	ADD	A, R4	C OV AC
3C	1-1	ADDC	A, R4	C OV AC
5C	1-1	ANL	A, R4	
BC	3-2	CJNE	R4, #dat 8, rel	C
1C	1-1	DEC	R4	
DC	2-2	DJNZ	R4, rel	
0C	1-1	INC	R4	
EC	1-1	MOV	A, R4	
8C	2-2	MOV	adr 8, R4	
FC	1-1	MOV	R4, A	
AC	2-2	MOV	R4, adr 8	
7C	2-1	MOV	R4, #dat 8	
4C	1-1	ORL	A, R4	
9C	1-1	SUBB	A, R4	C OV AC
CC	1-1	XCH	A, R4	
6C	1-1	XRL	A, R4	

A.4.10 | Instructions with Register Addressing Through R5

Opcode	Bytes/Cycles	Instruction		Flags affected
2D	1-1	ADD	A, R5	C OV AC
3D	1-1	ADDC	A, R5	C OV AC
5D	1-1	ANL	A, R5	
BD	3-2	CJNE	R5, #dat 8, rel	C
1D	1-1	DEC	R5	
DD	2-2	DJNZ	R5, rel	
0D	1-1	INC	R5	

(Continued)

Opcode	Bytes/Cycles	Instruction		Flags affected
ED	1-1	MOV	A, R5	
8D	2-2	MOV	adr 8, R5	
FD	1-1	MOV	R5, A	
AD	2-2	MOV	R5, adr 8	
7D	2-1	MOV	R5, #dat 8	
4D	1-1	ORL	A, R5	
9D	1-1	SUBB	A, R5	C OV AC
CD	1-1	XCH	A, R5	
6D	1-1	XRL	A, R5	

A.4.11 | Instructions with Register Addressing Through R6

Opcode	Bytes/Cycles	Instruction		Flags affected
2E	1-1	ADD	A, R6	C OV AC
3E	1-1	ADDC	A, R6	C OV AC
5E	1-1	ANL	A, R6	
BE	3-2	CJNE	R6, #dat 8, rel	C
1E	1-1	DEC	R6	
DE	2-2	DJNZ	R6, rel	
0E	1-1	INC	R6	
EE	1-1	MOV	A, R6	
8E	2-2	MOV	adr 8, R6	
FE	1-1	MOV	R6, A	
AE	2-2	MOV	R6, adr 8	
7E	2-1	MOV	R6, #dat 8	
4E	1-1	ORL	A, R6	
9E	1-1	SUBB	A, R6	C OV AC
CE	1-1	XCH	A, R6	
6E	1-1	XRL	A, R6	

A.4.12 | Instructions with Register Addressing Through R7

Opcode	Bytes/Cycles	Instruction		Flags affected
2F	1-1	ADD	A, R7	C OV AC
3F	1-1	ADDC	A, R7	C OV AC
5F	1-1	ANL	A, R7	
BF	3-2	CJNE	R7, #dat 8, rel	C
1F	1-1	DEC	R7	
DF	2-2	DJNZ	R7, rel	
0F	1-1	INC	R7	
EF	1-1	MOV	A, R7	
8F	2-2	MOV	adr 8, R7	
FF	1-1	MOV	R7, A	
AF	2-2	MOV	R7, adr 8	
7F	2-1	MOV	R7, #dat 8	
4F	1-1	ORL	A, R7	

(Continued)

Opcode	Bytes/Cycles	Instruction		Flags affected
9F	1-1	SUBB	A, R7	C OV AC
CF	1-1	XCH	A, R7	
6F	1-1	XRL	A, R7	

A.4.13 | Instructions with Register Addressing Through the Accumulator

Opcode	Bytes/Cycles	Instruction		Flags affected
25	2-1	ADD	A, adr 8	C OV AC
26	1-1	ADD	A, @R0	C OV AC
27	1-1	ADD	A, @R1	C OV AC
24	2-1	ADD	A, #dat 8	C OV AC
28	1-1	ADD	A, R0	C OV AC
29	1-1	ADD	A, R1	C OV AC
2A	1-1	ADD	A, R2	C OV AC
2B	1-1	ADD	A, R3	C OV AC
2C	1-1	ADD	A, R4	C OV AC
2D	1-1	ADD	A, R5	C OV AC
2E	1-1	ADD	A, R6	C OV AC
2F	1-1	ADD	A, R7	C OV AC
35	2-1	ADDC	A, adr 8	C OV AC
36	1-1	ADDC	A, @R0]	C OV AC
37	1-1	ADDC	A, @R1	C OV AC
34	2-1	ADDC	A, #dat 8	C OV AC
55	2-1	ANL	A, adr 8	
56	1-1	ANL	A, @R0	
57	1-1	ANL	A, @R1	
54	2-1	ANL	A, #dat 8	
58	1-1	ANL	A, R0	
59	1-1	ANL	A, R1	
5A	1-1	ANL	A, R2	
5B	1-1	ANL	A, R3	
5C	1-1	ANL	A, R4	
5D	1-1	ANL	A, R5	
5E	1-1	ANL	A, R6	
5F	1-1	ANL	A, R7	
52	2-1	ANL	adr 8, A	
B5	3-2	CJNE	A, adr 8,	C
B4	3-2	CJNE	A, #dat 8, rel	C
E4	1-1	CLR	A	
F4	1-1	CPL	A	
D4	1-1	DA	A	C
14	1-1	DEC	A	
84	1-4	DIV	AB	C OV
04	1-1	INC	A	
73	1-2	JMP	@A + DPTR	
70	2-2	JNZ	rel	

(Continued)

Opcode	Bytes/Cycles	Instruction	Flags affected
60	2-2	JZ rel	
E5	2-1	MOV A, adr 8	
E6	1-1	MOV A, @R0	
E7	1-1	MOV A, @R1	
74	2-1	MOV A, #dat 8	
E8	1-1	MOV A, R0	
E9	1-1	MOV A, R1	
EA	1-1	MOV A, R2	
EB	1-1	MOV A, R3	
EC	1-1	MOV A, R4	
ED	1-1	MOV A, R5	
EE	1-1	MOV A, R6	
EF	1-1	MOV A, R7	
F5	2-1	MOV adr 8, A	
F6	1-1	MOV @R0, A	
F7	1-1	MOV @R1, A	
F8	1-1	MOV R0, A	
F9	1-1	MOV R1, A	
FA	1-1	MOV R2, A	
FB	1-1	MOV R3, A	
FC	1-1	MOV R4, A	
FD	1-1	MOV R5, A	
FE	1-1	MOV R6, A	
FF	1-1	MOV R7, A	
93	1-2	MOVC A, @A + DPTR	
83	1-2	MOVC A, @A + PC	
E0	1-2	MOVX A, @DPTR	
E2	1-2	MOVX A, @R0	
E3	1-2	MOVX A, @R1	
F0	1-2	MOVX @DPTR, A	
F2	1-2	MOVX @R0, A	
F3	1-2	MOVX @R1, A	
A4	1-4	MUL AB	C OV
45	2-1	ORL A, adr 8	
46	1-1	ORL A, @R0	
47	1-1	ORL A, @R1	
44	2-1	ORL A, #dat 8	
48	1-1	ORL A, R0	
49	1-1	ORL A, R1	
4A	1-1	ORL A, R2	
4B	1-1	ORL A, R3	
4C	1-1	ORL A, R4	
4D	1-1	ORL A, R5	
4E	1-1	ORL A, R6	
4F	1-1	ORL A, R7	
42	2-1	ORL adr 8, A	

(Continued)

Opcode	Bytes/Cycles	Instruction		Flags affected
23	1-1	RL	A	
33	1-1	RLC	A	C
03	1-1	RR	A	
13	1-1	RRC	A	C
95	2-1	SUBB	A, adr 8	C OV AC
96	1-1	SUBB	A, @R0	C OV AC
97	1-1	SUBB	A, @R1	C OV AC
94	2-1	SUBB	A, #dat 8	C OV AC
98	1-1	SUBB	A, R0	C OV AC
99	1-1	SUBB	A, R1	C OV AC
9A	1-1	SUBB	A, R2	C OV AC
9B	1-1	SUBB	A, R3	C OV AC
9C	1-1	SUBB	A, R4	C OV AC
9D	1-1	SUBB	A, R5	C OV AC
9E	1-1	SUBB	A, R6	C OV AC
9F	1-1	SUBB	A, R7	C OV AC
C4	1-1	SWAP	A	
C5	2-1	XCH	A, adr 8	
C6	1-1	XCH	A, @R0	
C7	1-1	XCH	A, @R1	
C8	1-1	XCH	A, R0	
C9	1-1	XCH	A, R1	
CA	1-1	XCH	A, R2	
CB	1-1	XCH	A, R3	
CC	1-1	XCH	A, R4	
CD	1-1	XCH	A, R5	
CE	1-1	XCH	A, R6	
CF	1-1	XCH	A, R7	
D6	1-1	XCHD	A, @R0	
D7	1-1	XCHD	A, @R1	
65	2-1	XRL	A, adr 8	
66	1-1	XRL	A, @R0	
67	1-1	XRL	A, @R1	
64	2-1	XRL	A, #dat 8	
68	1-1	XRL	A, R0	
69	1-1	XRL	A, R1	
6A	1-1	XRL	A, R2	
6B	1-1	XRL	A, R3	
6C	1-1	XRL	A, R4	
6D	1-1	XRL	A, R5	
6E	1-1	XRL	A, R6	
6F	1-1	XRL	A, R7	
62	2-1	XRL	adr 8, A	

APPENDIX B

(Reprinted with Permission from Intel Corporation)

INSTRUCTION DEFINITIONS

ACALL   addr11

Function:	Absolute Call
Description:	ACALL unconditionally calls a subroutine located at the indicated address. The instruction increments the PC twice to obtain the address of the following instruction, then pushes the 16-bit result onto the stack (low-order byte first) and increments the Stack Pointer twice. The destination address is obtained by successively concatenating the five high-order bits of the incremented PC, opcode bits 7-5, and the second byte of the instruction. The subroutine called must therefore start within the same 2K block of the program memory as the first byte of the instruction following ACALL. No flags are affected.
Example:	Initially SP equals 07H. The label "SUBRTN" is at program memory location 0345 H. After executing the instruction,

ACALL   SUBRTN

at location 0123H, SP will contain 09H, internal RAM locations 08H and 09H will contain 25H and 01H, respectively, and the PC will contain 0345H.

Bytes:	2
Cycles:	2
Encoding:	a10 a9 a8 1 | 0 0 0 1 a7 a6 a5 a4 | a3 a2 a1 a0
Operation:	ACALL

$(PC) \leftarrow (PC) + 2$
$(SP) \leftarrow (SP) + 1$
$((SP)) \leftarrow (PC_{7-0})$
$(SP) \leftarrow (SP) + 1$
$((SP)) \leftarrow (PC_{15-8})$
$(PC_{10-0}) \leftarrow$ page address

ADD   A,<src-byte>

Function:	Add
Description:	ADD adds the byte variable indicated to the Accumulator, leaving the result in the Accumulator. The carry and auxiliary-carry flags are set, respectively, if there is a carry-out from bit 7 or bit 3, and cleared otherwise. When adding unsigned integers, the carry flag indicates an overflow occurred.
	OV is set if there is a carry-out of bit 6 but not out of bit 7, or a carry-out of bit 7 but not bit 6; otherwise OV is cleared. When adding signed integers, OV indicates a negative number produced as the sum of two positive operands, or a positive sum from two negative operands.
	Four source operand addressing modes are allowed: register, direct, register-indirect, or immediate.
Example:	The Accumulator holds 0C3H (11000011B) and register 0 holds 0AAH (10101010B). The instruction,

ADD   A,R0

will leave 6DH (01101101B) in the Accumulator with the AC flag cleared and both the carry flag and OV set to 1.

ADD   A,Rn

Bytes:	1
Cycles:	1
Encoding:	`0 0 1 0` `1 r r r`
Operation:	ADD
	(A) ← (A) + (Rn)

ADD   A,direct

Bytes:	2
Cycles:	1
Encoding:	`0 0 1 0` `0 1 0 1` `direct address`
Operation:	ADD
	(A) ← (A) + (direct)

ADD   A,@Ri

| Bytes: | 1 |
| Cycles: | 1 |

Encoding:

| 0 0 1 0 | 0 1 1 i |

Operation: ADD
$(A) \leftarrow (A) + ((Ri))$

ADD   A,#data

| Bytes: | 2 |
| Cycles: | 1 |

Encoding:

| 0 0 1 0 | 0 1 0 0 | | immediate data |

Operation: ADD
$(A) \leftarrow (A) + \#data$

ADDC   A,<src-byte>

Function: Add with Carry

Description: ADDC simultaneously adds the byte variable indicated, the carry flag and the Accumulator contents, leaving the result in the Accumulator. The carry and auxiliary-carry flags are set, respectively, if there is a carry-out from bit 7 or bit 3, and cleared otherwise. When adding unsigned integers, the carry flag indicates an overflow occurred.

OV is set if there is a carry-out of bit 6 but not out of bit 7, or a carry-out of bit 7 but not out of bit 6; otherwise OV is cleared. When adding signed integers, OV indicates a negative number produced as the sum of two positive operands or a positive sum from two negative operands.

Four source operand addressing modes are allowed: register, direct, register-indirect, or immediate.

Example: The Accumulator holds 0C3H (11000011B) and register 0 holds 0AAH (10101010B) with the carry flag set. The instruction,

ADDC   A,R0

will leave 6EH (01101110B) in the Accumulator with AC cleared and both the Carry flag and OV set to 1.

ADDC   A,Rn

Bytes:	1
Cycles:	1
Encoding:	0 0 1 1 | 1 r r r
Operation:	ADDC
	$(A) \leftarrow (A) + (C) + (R_n)$

ADDC   A,direct

Bytes:	2
Cycles:	1
Encoding:	0 0 1 1 | 0 1 0 1 direct address
Operation:	ADDC
	$(A) \leftarrow (A) + (C) + (\text{direct})$

ADDC   A,@Ri

Bytes:	1
Cycles:	1
Encoding:	0 0 1 1 | 0 1 1 i
Operation:	ADDC
	$(A) \leftarrow (A) + (C) + ((R_i))$

ADDC   A,#data

Bytes:	2
Cycles:	1
Encoding:	0 0 1 1 | 0 1 0 0 immediate data
Operation:	ADDC
	$(A) \leftarrow (A) + (C) + \#\text{data}$

AJMP   addr11

Function:	Absolute Jump
Description:	AJMP transfers program execution to the indicated address, which is formed at run-time by concatenating the high-order five bits of the PC (*after* incrementing the PC twice), opcode bits 7-5, and the second byte of the instruction. The destination must therefore be within the same 2K block of program memory as the first byte of the instruction following AJMP.
Example:	The label "JMPADR" is at program memory location 0123H. The instruction,

AJMP   JMPADR

is at location 0345H and will load the PC with 0123H.

Bytes:	2
Cycles:	2
Encoding:	a10 a9 a8 0 | 0 0 0 1 a7 a6 a5 a4 | a3 a2 a1 a0
Operation:	AJMP

$(PC) \leftarrow (PC) + 2$
$(PC_{10-0}) \leftarrow$ page address

ANL   <dest-byte>, <src-byte>

Function:	Logical-AND for byte variables
Description:	ANL performs the bitwise logical-AND operation between the variables indicated and stores the results in the destination variable. No flags are affected.

The two operands allow six addressing mode combinations. When the destination is the Accumulator, the source can use register, direct, register-indirect, or immediate addressing; when the destination is a direct address, the source can be the Accumulator or immediate data.

Note: When this instruction is used to modify an output port, the value used as the original port data will be read from the output data latch, *not* the input pins.

Example:	If the Accumulator holds 0C3H (11000011B) and register 0 holds 55H (01010101B) then the instruction,

ANL   A,R0

will leave 41H (01000001B) in the Accumulator.

When the destination is a directly addressed byte, this instruction will clear combinations of bits in any RAM location or hardware register. The mask byte determining the pattern of bits to be cleared would either be a constant contained in the instruction or a value computed in the Accumulator at run-time. The instruction,

ANL   P1, #01110011B

will clear bits 7, 3, and 2 of output port 1.

ANL   A,Rn

Bytes:	1
Cycles:	1
Encoding:	0 1 0 1 1 r r r
Operation:	ANL
	(A) ← (A) ∧ (Rn)

ANL   A,direct

Bytes:	2
Cycles:	1
Encoding:	0 1 0 1 0 1 0 1 direct address
Operation:	ANL
	(A) ← (A) ∧ (direct)

ANL   A,@Ri

Bytes:	1
Cycles:	1
Encoding:	0 1 0 1 0 1 1 i
Operation:	ANL
	(A) ← (A) ∧ ((Ri))

ANL   A,#data

Bytes:	2
Cycles:	1
Encoding:	0 1 0 1 0 1 0 0 immediate data
Operation:	ANL
	(A) ← (A) ∧ #data

ANL   direct,A

Bytes:	2
Cycles:	1
Encoding:	0 1 0 1 0 0 1 0 direct address
Operation:	ANL
	(direct) ← (direct) ∧ (A)

ANL   direct,#data

Bytes:	3
Cycles:	2

Encoding: | 0 1 0 1 | 0 0 1 1 | | direct address | | immediate data |

Operation: ANL
(direct) ← (direct) ∧ #data

ANL   C,<src-bit>

Function: Logical-AND for bit variables

Description: If the Boolean value of the source bit is a logical 0 then clear the carry flag; otherwise leave the carry flag in its current state. A slash ("/") preceding the operand in the assembly language indicates that the logical complement of the addressed bit is used as the source value, *but the source bit itself is not affected*. No other flags are affected.

Only direct addressing is allowed for the source operand.

Example: Set the carry flag if, and only if, P1.0 = 1, ACC. 7 = 1, and OV = 0:

MOV   C,P1.0   ; LOAD CARRY WITH INPUT PIN STATE

ANL   C,ACC.7   ; AND CARRY WITH ACCUM. BIT 7

ANL   C,/OV   ; AND WITH INVERSE OF OVERFLOW FLAG

ANL   C,bit

Bytes:	2
Cycles:	2

Encoding: | 1 0 0 0 | 0 0 1 0 | | bit address |

Operation: ANL
(C) ← (C) ∧ (bit)

ANL   C,/bit

Bytes:	2
Cycles:	2

Encoding: | 1 0 1 1 | 0 0 0 0 | | bit address |

Operation: ANL
(C) ← (C) ∧ ¬   (bit)

CJNE <dest-byte>, <src-byte>, rel

Function:	Compare and Jump if Not Equal.
Description:	CJNE compares the magnitudes of the first two operands, and branches if their values are not equal. The branch destination is computed by adding the signed relative-displacement in the last instruction byte to the PC, after incrementing the PC to the start of the next instruction. The carry flag is set if the unsigned integer value of <dest-byte> is less than the unsigned integer value of <src-byte>; otherwise, the carry is cleared. Neither operand is affected.

The first two operands allow four addressing mode combinations: the Accumulator may be compared with any directly addressed byte or immediate data, and any indirect RAM location or working register can be compared with an immediate constant.

Example: The Accumulator contains 34H. Register 7 contains 56H. The first instruction in the sequence,

```
            CJNE    R7, # 60H, NOT_EQ
;            ...      .......                    ; R7 = 60H.
NOT_EQ:     JC       REQ_LOW                     ; IF R7 < 60H.
;            ...      .......                    ; R7 > 60H.
```

sets the carry flag and branches to the instruction at label NOT_EQ. By testing the carry flag, this instruction determines whether R7 is greater or less than 60H.

If the data being presented to Port 1 is also 34H, then the instruction,

WAIT: CJNE A,P1,WAIT

clears the carry flag and continues with the next instruction in sequence, since the Accumulator does equal the data read from P1. (If some other value was being input on P1, the program will loop at this point until the P1 data changes to 34H.)

CJNE A,direct, rel

Bytes:	3
Cycles:	2
Encoding:	`1 0 1 1 0 1 0 1` `direct address` `rel. address`
Operation:	(PC) ← (PC) + 3
	IF (A) < > (*direct*)
	THEN
	(PC) ← (PC) + *relative offset*
	IF (A) < (*direct*)
	THEN
	(C) ← 1
	ELSE
	(C) ← 0

CJNE   A,#data,rel

         Bytes: 3

         Cycles: 2

         Encoding: | 1 0 1 1 | 0 1 0 0 | | immediate data | | rel. address |

         Operation: $(PC) \leftarrow (PC) + 3$
                       IF $(A) < > data$
                       THEN
                               $(PC) \leftarrow (PC) + relative\ offset$

                       IF $(A) < data$
                       THEN
                               $(C) \leftarrow 1$
                       ELSE
                               $(C) \leftarrow 0$

CJNE   Rn,#data,rel

         Bytes: 3

         Cycles: 2

         Encoding: | 1 0 1 1 | 1 r r r | | immediate data | | rel. address |

         Operation: $(PC) \leftarrow (PC) + 3$
                       IF $(Rn) < > data$
                       THEN
                               $(PC) \leftarrow (PC) + relative\ offset$

                       IF $(Rn) < data$
                       THEN
                               $(C) \leftarrow 1$
                       ELSE
                               $(C) \leftarrow 0$

CJNE   @Ri,#data,rel

         Bytes: 3

         Cycles: 2

         Encoding: | 1 0 1 1 | 0 1 1 i | | immediate data | | rel. address |

         Operation: $(PC) \leftarrow (PC) + 3$
                       IF $((Ri)) < > data$
                       THEN
                               $(PC) \leftarrow (PC) + relative\ offset$

                       IF $((Ri)) < data$

                       THEN
                               $(C) \leftarrow 1$
                       ELSE
                               $(C) \leftarrow 0$

intel₍ᵣ₎   **MCS®-51 PROGRAMMER'S GUIDE AND INSTRUCTION SET**

CLR A

Function:	Clear Accumulator
Description:	The Accumulator is cleared (all bits set on zero). No flags are affected.
Example:	The Accumulator contains 5CH (01011100B). The instruction,

CLR A

will leave the Accumulator set to 00H (00000000B).

Bytes:	1
Cycles:	1
Encoding:	1 1 1 0 0 1 0 0
Operation:	CLR (A) ←0

CLR bit

Function:	Clear bit
Description:	The indicated bit is cleared (reset to zero). No other flags are affected. CLR can operate on the carry flag or any directly addressable bit.
Example:	Port 1 has previously been written with 5DH (01011101B). The instruction,

CLR P1.2

will leave the port set to 59H (01011001B).

CLR C

Bytes:	1
Cycles:	1
Encoding:	1 1 0 0 0 0 1 1
Operation:	CLR (C) ← 0

CLR bit

Bytes:	2
Cycles:	1
Encoding:	1 1 0 0 0 0 1 0 bit address
Operation:	CLR (bit) ← 0

intel®

MCS®-51 PROGRAMMER'S GUIDE AND INSTRUCTION SET

CPL   A

Function:	Complement Accumulator
Description:	Each bit of the Accumulator is logically complemented (one's complement). Bits which previously contained a one are changed to a zero and vice-versa. No flags are affected.
Example:	The Accumulator contains 5CH (01011100B). The instruction,

CPL   A

will leave the Accumulator set to 0A3H (10100011B).

Bytes:	1
Cycles:	1
Encoding:	1 1 1 1 | 0 1 0 0
Operation:	CPL
	$(A) \leftarrow \neg (A)$

CPL   bit

Function:	Complement bit
Description:	The bit variable specified is complemented. A bit which had been a one is changed to zero and vice-versa. No other flags are affected. CLR can operate on the carry or any directly addressable bit.

Note: When this instruction is used to modify an output pin, the value used as the original data will be read from the output data latch, *not* the input pin.

Example:	Port 1 has previously been written with 5BH (01011101B). The instruction sequence,

CPL   P1.1

CPL   P1.2

will leave the port set to 5BH (01011011B).

CPL   C

Bytes:	1
Cycles:	1
Encoding:	1 0 1 1 | 0 0 1 1
Operation:	CPL
	$(C) \leftarrow \neg (C)$

CPL   bit

Bytes:	2
Cycles:	1
Encoding:	`1 0 1 1  0 0 1 0` `bit address`
Operation:	CPL
	(bit) ← ¬ (bit)

DA   A

Function: Decimal-adjust Accumulator for Addition

Description: DA A adjusts the eight-bit value in the Accumulator resulting from the earlier addition of two variables (each in packed-BCD format), producing two four-bit digits. Any ADD or ADDC instruction may have been used to perform the addition.

If Accumulator bits 3-0 are greater than nine (xxxx1010-xxxx1111), or if the AC flag is one, six is added to the Accumulator producing the proper BCD digit in the low-order nibble. This internal addition would set the carry flag if a carry-out of the low-order four-bit field propagated through all high-order bits, but it would not clear the carry flag otherwise.

If the carry flag is now set, or if the four high-order bits now exceed nine (1010xxxx-111xxxx), these high-order bits are incremented by six, producing the proper BCD digit in the high-order nibble. Again, this would set the carry flag if there was a carry-out of the high-order bits, but wouldn't clear the carry. The carry flag thus indicates if the sum of the original two BCD variables is greater than 100, allowing multiple precision decimal addition. OV is not affected.

All of this occurs during the one instruction cycle. Essentially, this instruction performs the decimal conversion by adding 00H, 06H, 60H, or 66H to the Accumulator, depending on initial Accumulator and PSW conditions.

Note: DA A *cannot* simply convert a hexadecimal number in the Accumulator to BCD notation, nor does DA A apply to decimal subtraction.

Example: The Accumulator holds the value 56H (01010110B) representing the packed BCD digits of the decimal number 56. Register 3 contains the value 67H (01100111B) representing the packed BCD digits of the decimal number 67. The carry flag is set. The instruction sequence.

ADDC   A,R3
DA   A

will first perform a standard twos-complement binary addition, resulting in the value 0BEH (10111110) in the Accumulator. The carry and auxiliary carry flags will be cleared.

The Decimal Adjust instruction will then alter the Accumulator to the value 24H (00100100B), indicating the packed BCD digits of the decimal number 24, the low-order two digits of the decimal sum of 56, 67, and the carry-in. The carry flag will be set by the Decimal Adjust instruction, indicating that a decimal overflow occurred. The true sum 56, 67, and 1 is 124.

BCD variables can be incremented or decremented by adding 01H or 99H. If the Accumulator initially holds 30H (representing the digits of 30 decimal), then the instruction sequence,

ADD   A,#99H

DA   A

will leave the carry set and 29H in the Accumulator, since 30 + 99 = 129. The low-order byte of the sum can be interpreted to mean 30 - 1 = 29.

Bytes: 1

Cycles: 1

Encoding: | 1 1 0 1 | 0 1 0 0 |

Operation: DA
                -contents of Accumulator are BCD
                IF $[[(A_{3-0}) > 9] \lor [(AC) = 1]]$
                        THEN$(A_{3-0}) \leftarrow (A_{3-0}) + 6$
                        AND

                IF $[[(A_{7-4}) > 9] \lor [(C) = 1]]$
                        THEN $(A_{7-4}) \leftarrow (A_{7-4}) + 6$

intel.   **MCS®-51 PROGRAMMER'S GUIDE AND INSTRUCTION SET**

DEC byte

Function:	Decrement
Description:	The variable indicated is decremented by 1. An original value of 00H will underflow to 0FFH. No flags are affected. Four operand addressing modes are allowed: accumulator, register, direct, or register-indirect.
	Note: When this instruction is used to modify an output port, the value used as the original port data will be read from the output data latch, *not* the input pins.
Example:	Register 0 contains 7FH (01111111B). Internal RAM locations 7EH and 7FH contain 00H and 40H, respectively. The instruction sequence,

DEC   @R0

DEC   R0

DEC   @R0

will leave register 0 set to 7EH and internal RAM locations 7EH and 7FH set to 0FFH and 3FH.

DEC A

Bytes:	1
Cycles:	1
Encoding:	0 0 0 1 | 0 1 0 0
Operation:	DEC (A) ← (A) – 1

DEC Rn

Bytes:	1
Cycles:	1
Encoding:	0 0 0 1 | 1 r r r
Operation:	DEC (Rn) ← (Rn) – 1

DEC   direct

Bytes:	2
Cycles:	1
Encoding:	`0 0 0 1	0 1 0 1`     `direct address`
Operation:	DEC
	(direct) ← (direct) − 1

DEC   @Ri

Bytes:	1
Cycles:	1
Encoding:	`0 0 0 1	0 1 1 i`
Operation:	DEC
	((Ri)) ← ((Ri)) − 1

DIV   AB

Function: Divide

Description: DIV AB divides the unsigned eight-bit integer in the Accumulator by the unsigned eight-bit integer in register B. The Accumulator receives the integer part of the quotient; register B receives the integer remainder. The carry and OV flags will be cleared.

Exception: if B had originally contained 00H, the values returned in the Accumulator and B-register will be undefined and the overflow flag will be set. The carry flag is cleared in any case.

Example: The Accumulator contains 251 (0FBH or 11111011B) and B contains 18 (12H or 00010010B). The instruction,

DIV   AB

will leave 13 in the Accumulator (0DH or 00001101B) and the value 17 (11H or 00010001B) in B, since 251 = (13 X 18) + 17. Carry and OV will both be cleared.

Bytes:	1
Cycles:	4
Encoding:	`1 0 0 0	0 1 0 0`
Operation:	DIV
	$(A)_{15-8} \leftarrow (A)/(B)$
	$(B)_{7-0}$

DJNZ   <byte>,<rel-addr>

Function:	Decrement and Jump if Not Zero
Description:	DJNZ decrements the location indicated by 1, and branches to the address indicated by the second operand if the resulting value is not zero. An original value of 00H will underflow to 0FFH. No flags are affected. The branch destination would be computed by adding the signed relative-displacement value in the last instruction byte to the PC, after incrementing the PC to the first byte of the following instruction.

The location decremented may be a register or directly addressed byte.

Note: When this instruction is used to modify an output port, the value used as the original port data will be read from the output data latch, *not* the input pins.

Example:	Internal RAM locations 40H, 50H, and 60H contain the values 01H, 70H, and 15H, respectively. The instruction sequence,

```
DJNZ    40H,LABEL__1
DJNZ    50H,LABEL__2
DJNZ    60H,LABEL__3
```

will cause a jump to the instruction at label LABEL__2 with the values 00H, 6FH, and 15H in the three RAM locations. The first jump was *not* taken because the result was zero.

This instruction provides a simple way of executing a program loop a given number of times, or for adding a moderate time delay (from 2 to 512 machine cycles) with a single instruction. The instruction sequence,

```
            MOV     R2,#8
TOGGLE:     CPL     P1.7
            DJNZ    R2,TOGGLE
```

will toggle P1.7 eight times, causing four output pulses to appear at bit 7 of output Port 1. Each pulse will last three machine cycles; two for DJNZ and one to alter the pin.

DJNZ   Rn,rel

Bytes:	2
Cycles:	2
Encoding:	`1 1 0 1` `1 r r r` `rel. address`
Operation:	DJNZ

$$(PC) \leftarrow (PC) + 2$$
$$(Rn) \leftarrow (Rn) - 1$$
IF   (Rn) > 0 or (Rn) < 0
       THEN
$$(PC) \leftarrow (PC) + rel$$

intel®   MCS®-51 PROGRAMMER'S GUIDE AND INSTRUCTION SET

DJNZ   direct,rel

Bytes:	3
Cycles:	2
Encoding:	`1 1 0 1	0 1 0 1`   `direct address`   `rel. address`
Operation:	DJNZ

$$(PC) \leftarrow (PC) + 2$$
$$(direct) \leftarrow (direct) - 1$$
IF (direct) > 0 or (direct) < 0
     THEN
            $$(PC) \leftarrow (PC) + rel$$

INC   <byte>

Function: Increment

Description: INC increments the indicated variable by 1. An original value of 0FFH will overflow to 00H. No flags are affected. Three addressing modes are allowed: register, direct, or register-indirect.

Note: When this instruction is used to modify an output port, the value used as the original port data will be read from the output data latch, *not* the input pins.

Example: Register 0 contains 7EH (011111110B). Internal RAM locations 7EH and 7FH contain 0FFH and 40H, respectively. The instruction sequence,

```
INC   @R0
INC   R0
INC   @R0
```

will leave register 0 set to 7FH and internal RAM locations 7EH and 7FH holding (respectively) 00H and 41H.

INC A

Bytes:	1
Cycles:	1
Encoding:	`0 0 0 0	0 1 0 0`
Operation:	INC

$$(A) \leftarrow (A) + 1$$

INC Rn

Bytes:	1
Cycles:	1
Encoding:	`0 0 0 0  1 r r r`
Operation:	INC (Rn) ← (Rn) + 1

INC   direct

Bytes:	2
Cycles:	1
Encoding:	`0 0 0 0 0 1 0 1` `direct address`
Operation:	INC (direct) ← (direct) + 1

INC @Ri

Bytes:	1
Cycles:	1
Encoding:	`0 0 0 0 0 1 1 i`
Operation:	INC ((Ri)) ← ((Ri)) + 1

INC   DPTR

Function: Increment Data Pointer

Description: Increment the 16-bit data pointer by 1. A 16-bit increment (modulo 2^{16}) is performed; an overflow of the low-order byte of the data pointer (DPL) from 0FFH to 00H will increment the high-order byte (DPH). No flags are affected.

This is the only 16-bit register which can be incremented.

Example: Registers DPH and DPL contain 12H and 0FEH, respectively. The instruction sequence,

```
INC   DPTR
INC   DPTR
INC   DPTR
```

will change DPH and DPL to 13H and 01H.

Bytes:	1
Cycles:	2
Encoding:	`1 0 1 0    0 0 1 1`
Operation:	INC (DPTR) ← (DPTR) + 1

JB   bit,rel

Function:	Jump if Bit set
Description:	If the indicated bit is a one, jump to the address indicated; otherwise proceed with the next instruction. The branch destination is computed by adding the signed relative-displacement in the third instruction byte to the PC, after incrementing the PC to the first byte of the next instruction. *The bit tested is not modified.* No flags are affected.
Example:	The data present at input port 1 is 11001010B. The Accumulator holds 56 (01010110B). The instruction sequence,

JB   P1.2,LABEL1

JB   ACC.2,LABEL2

will cause program execution to branch to the instruction at label LABEL2.

Bytes:	3
Cycles:	2
Encoding:	`0 0 1 0  0 0 0 0` `bit address` `rel. address`
Operation:	JB

$$(PC) \leftarrow (PC) + 3$$
IF   (bit) = 1
    THEN
$$(PC) \leftarrow (PC) + rel$$

JBC   bit,rel

Function:	Jump if Bit is set and Clear bit
Description:	If the indicated bit is one, branch to the address indicated; otherwise proceed with the next instruction. *The bit will not be cleared if it is already a zero.* The branch destination is computed by adding the signed relative-displacement in the third instruction byte to the PC, after incrementing the PC to the first byte of the next instruction. No flags are affected.

        Note: When this instruction is used to test an output pin, the value used as the original data will be read from the output data latch, *not* the input pin.

Example:	The Accumulator holds 56H (01010110B). The instruction sequence,

JBC   ACC.3,LABEL1
JBC   ACC.2,LABEL2

will cause program execution to continue at the instruction identified by the label LABEL2, with the Accumulator modified to 52H (01010010B).

Bytes:	3
Cycles:	2
Encoding:	`0 0 0 1 0 0 0 0` `bit address` `rel. address`
Operation:	JBC

$$(PC) \leftarrow (PC) + 3$$
$$\text{IF} \quad (\text{bit}) = 1$$
$$\qquad \text{THEN}$$
$$\qquad\qquad (\text{bit}) \leftarrow 0$$
$$\qquad\qquad (PC) \leftarrow (PC) + \text{rel}$$

JC   rel

Function:	Jump if Carry is set
Description:	If the carry flag is set, branch to the address indicated; otherwise proceed with the next instruction. The branch destination is computed by adding the signed relative-displacement in the second instruction byte to the PC, after incrementing the PC twice. No flags are affected.
Example:	The carry flag is cleared. The instruction sequence,

```
JC      LABEL1
CPL   C
JC      LABEL 2
```

will set the carry and cause program execution to continue at the instruction identified by the label LABEL2.

Bytes:	2
Cycles:	2
Encoding:	`0 1 0 0 0 0 0 0` `rel. address`
Operation:	JC

$$(PC) \leftarrow (PC) + 2$$
$$\text{IF} \quad (C) = 1$$
$$\qquad \text{THEN}$$
$$\qquad\qquad (PC) \leftarrow (PC) + \text{rel}$$

JMP   @A + DPTR

Function:	Jump indirect
Description:	Add the eight-bit unsigned contents of the Accumulator with the sixteen-bit data pointer, and load the resulting sum to the program counter. This will be the address for subsequent instruction fetches. Sixteen-bit addition is performed (modulo 2^{16}): a carry-out from the low-order eight bits propagates through the higher-order bits. Neither the Accumulator nor the Data Pointer is altered. No flags are affected.
Example:	An even number from 0 to 6 is in the Accumulator. The following sequence of instructions will branch to one of four AJMP instructions in a jump table starting at JMP_TBL:

	MOV	DPTR,# JMP_TBL
	JMP	@A + DPTR
JMP_TBL:	AJMP	LABEL0
	AJMP	LABEL1
	AJMP	LABEL2
	AJMP	LABEL3

If the Accumulator equals 04H when starting this sequence, execution will jump to label LABEL2. Remember that AJMP is a two-byte instruction, so the jump instructions start at every other address.

Bytes:	1
Cycles:	2
Encoding:	0 1 1 1 0 0 1 1
Operation:	MP (PC) ← (A) + (DPTR)

JNB   bit,rel

Function: Jump if Bit Not set

Description: If the indicated bit is a zero, branch to the indicated address; otherwise proceed with the next instruction. The branch destination is computed by adding the signed relative-displacement in the third instruction byte to the PC, after incrementing the PC to the first byte of the next instruction. *The bit tested is not modified*. No flags are affected.

Example: The data present at input port 1 is 11001010B. The Accumulator holds 56H (01010110B). The instruction sequence,

JNB   P1.3,LABEL1
JNB   ACC.3,LABEL2

will cause program execution to continue at the instruction at label LABEL2.

Bytes: 3

Cycles: 2

Encoding: | 0 0 1 1 | 0 0 0 0 | | bit address | | rel. address |

Operation: JNB
(PC) ← (PC) + 3
IF   (bit) = 0
         THEN (PC) ←(PC) + rel.

JNC rel

Function: Jump if Carry not set

Description: If the carry flag is a zero, branch to the address indicated; otherwise proceed with the next instruction. The branch destination is computed by adding the signed relative-displacement in the second instruction byte to the PC, after incrementing the PC twice to point to the next instruction. The carry flag is not modified.

Example: The carry flag is set. The instruction sequence,

JNC   LABEL1
CPL   C
JNC   LABEL2

will clear the carry and cause program execution to continue at the instruction identified by the label LABEL2.

Bytes: 2

Cycles: 2

Encoding: | 0 1 0 1 | 0 0 0 0 | | rel. address |

Operation: JNC
(PC) ← (PC) + 2
IF   (C) = 0
         THEN   (PC) ← (PC) + rel

JNZ rel

Function:	Jump if Accumulator Not Zero
Description:	If any bit of the Accumulator is a one, branch to the indicated address; otherwise proceed with the next instruction. The branch destination is computed by adding the signed relative-displacement in the second instruction byte to the PC, after incrementing the PC twice. The Accumulator is not modified. No flags are affected.
Example:	The Accumulator originally holds 00H. The instruction sequence,

```
JNZ   LABEL1
INC   A
JNZ   LABEL2
```

will set the Accumulator to 01H and continue at label LABEL2.

Bytes:	2
Cycles:	2
Encoding:	`0 1 1 1	0 0 0 0`   `rel. address`
Operation:	JNZ

$$(PC) \leftarrow (PC) + 2$$
$$IF \quad (A) \neq 0$$
$$\qquad THEN \quad (PC) \leftarrow (PC) + rel$$

JZ rel

Function:	Jump if Accumulator Zero
Description:	If all bits of the Accumulator are zero, branch to the address indicated; otherwise proceed with the next instruction. The branch destination is computed by adding the signed relative-displacement in the second instruction byte to the PC, after incrementing the PC twice. The Accumulator is not modified. No flags are affected.
Example:	The Accumulator originally contains 01H. The instruction sequence,

```
JZ    LABEL1
DEC   A
JZ    LABEL2
```

will change the Accumulator to 00H and cause program execution to continue at the instruction identified by the label LABEL2.

Bytes:	2
Cycles:	2
Encoding:	`0 1 1 0	0 0 0 0`   `rel. address`
Operation:	JZ

$$(PC) \leftarrow (PC) + 2$$
$$IF \quad (A) = 0$$
$$\qquad THEN \quad (PC) \leftarrow (PC) + rel$$

intel.   **MCS®-51 PROGRAMMER'S GUIDE AND INSTRUCTION SET**

LCALL   addr16

Function:	Long call
Description:	LCALL calls a subroutine located at the indicated address. The instruction adds three to the program counter to generate the address of the next instruction and then pushes the 16-bit result onto the stack (low byte first), incrementing the Stack Pointer by two. The high-order and low-order bytes of the PC are then loaded, respectively, with the second and third bytes of the LCALL instruction. Program execution continues with the instruction at this address. The subroutine may therefore begin anywhere in the full 64K-byte program memory address space. No flags are affected.
Example:	Initially the Stack Pointer equals 07H. The label "SUBRTN" is assigned to program memory location 1234H. After executing the instruction,

LCALL   SUBRTN

at location 0123H, the Stack Pointer will contain 09H, internal RAM locations 08H and 09H will contain 26H and 01H, and the PC will contain 1234H.

Bytes:	3
Cycles:	2
Encoding:	$\boxed{0\ 0\ 0\ 1 \mid 0\ 0\ 1\ 0}$ $\boxed{\text{addr15-addr8}}$ $\boxed{\text{addr7-addr0}}$
Operation:	LCALL

$(PC) \leftarrow (PC) + 3$
$(SP) \leftarrow (SP) + 1$
$((SP)) \leftarrow (PC_{7-0})$
$(SP) \leftarrow (SP) + 1$
$((SP)) \leftarrow (PC_{15-8})$
$(PC) \leftarrow addr_{15-0}$

LJMP   addr16

Function:	Long Jump
Description:	LJMP causes an unconditional branch to the indicated address, by loading the high-order and low-order bytes of the PC (respectively) with the second and third instruction bytes. The destination may therefore be anywhere in the full 64K program memory address space. No flags are affected.
Example:	The label "JMPADR" is assigned to the instruction at program memory location 1234H. The instruction,

LJMP   JMPADR

at location 0123H will load the program counter with 1234H.

Bytes:	3
Cycles:	2
Encoding:	$\boxed{0\ 0\ 0\ 0 \mid 0\ 0\ 1\ 0}$ $\boxed{\text{addr15-addr8}}$ $\boxed{\text{addr7-addr0}}$
Operation:	LJMP

$(PC) \leftarrow addr_{15-0}$

MOV   **<dest-byte>,<src-byte>**

        Function: Move byte variable

        Description: The byte variable indicated by the second operand is copied into the location speci-fied by the first operand. The source byte is not affected. No other register or flag is affected.

                This is by far the most flexible operation. Fifteen combinations of source and destina-tion addressing modes are allowed.

        Example: Internal RAM location 30H holds 40H. The value of RAM location 40H is 10H. The data present at input port 1 is 11001010B (0CAH).

MOV	R0,#30H	; R0 < = 30H
MOV	A,@R0	; A < = 40H
MOV	R1,A	; R1 < = 40H
MOV	B,@R1	; B < = 10H
MOV	@R1,P1	; RAM (40H) < = 0CAH
MOV	P2,P1	; P2 #0CAH

                leaves the value 30H in register 0, 40H in both the Accumulator and register 1, 10H in register B, and 0CAH (11001010B) both in RAM location 40H and output on port 2.

MOV   A,Rn

        Bytes: 1

        Cycles: 1

        Encoding: | 1 1 1 0 | 1 r r r |

        Operation: MOV

                $(A) \leftarrow (Rn)$

*MOV   A,direct

        Bytes: 2

        Cycles: 1

        Encoding: | 1 1 1 0 | 0 1 0 1 | | direct address |

        Operation: MOV

                $(A) \leftarrow (direct)$

MOV A,ACC is not a valid instruction.

MOV A,@Ri

Bytes:	1
Cycles:	1
Encoding:	`1 1 1 0` `0 1 1 i`
Operation:	MOV
	(A) ← ((Ri))

MOV A, #data

Bytes:	2
Cycles:	1
Encoding:	`0 1 1 1` `0 1 0 0` `immediate data`
Operation:	MOV
	(A) ← #data

MOV Rn,A

Bytes:	1
Cycles:	1
Encoding:	`1 1 1 1` `1 r r r`
Operation:	MOV
	(Rn) ← (A)

MOV Rn,direct

Bytes:	2
Cycles:	2
Encoding:	`1 0 1 0` `1 r r r` `direct addr.`
Operation:	MOV
	(Rn) ← (direct)

MOV Rn,#data

Bytes:	2
Cycles:	1
Encoding:	`0 1 1 1` `1 r r r` `immediate data`
Operation:	MOV
	(Rn) ← #data

int_el.   **MCS®-51 PROGRAMMER'S GUIDE AND INSTRUCTION SET**

MOV   direct,A

Bytes:	2
Cycles:	1

Encoding: | 1 1 1 1 | 0 1 0 1 | | direct address |

Operation: MOV
(direct) ← (A)

MOV   direct,Rn

Bytes:	2
Cycles:	2

Encoding: | 1 0 0 0 | 1 r r r | | direct address |

Operation: MOV
(direct) ← (Rn)

MOV   direct,direct

Bytes:	3
Cycles:	2

Encoding: | 1 0 0 0 | 0 1 0 1 | | dir. addr. (src) | | dir. addr. (dest) |

Operation: MOV
(direct) ← (direct)

MOV   direct,@Ri

Bytes:	2
Cycles:	2

Encoding: | 1 0 0 0 | 0 1 1 i | | direct addr. |

Operation: MOV
(direct) ← ((Ri))

MOV   direct,#data

Bytes:	3
Cycles:	2

Encoding: | 0 1 1 1 | 0 1 0 1 | | direct address | | immediate data |

Operation: MOV
(direct) ← #data

MOV @Ri,A

Bytes:	1
Cycles:	1
Encoding:	1 1 1 1 0 1 1 i
Operation:	MOV ((Ri)) ← (A)

MOV @Ri,direct

Bytes:	2
Cycles:	2
Encoding:	1 0 1 0 0 1 1 i direct addr.
Operation:	MOV ((Ri))← (direct)

MOV @Ri,#data

Bytes:	2
Cycles:	1
Encoding:	0 1 1 1 0 1 1 i immediate data
Operation:	MOV ((RI)) ← #data

MOV <dest-bit>,<src-bit>

Function:	Move bit data
Description:	The Boolean variable indicated by the second operand is copied into the location specified by the first operand. One of the operands must be the carry flag; the other may be any directly addressable bit. No other register or flag is affected.
Example:	The carry flag is originally set. The data present at input Port 3 is 11000101B. The data previously written to output Port 1 is 35H (00110101B).

```
MOV   P1.3,C
MOV   C,P3.3
MOV   P1.2,C
```

will leave the carry cleared and change Port 1 to 39H (00111001B).

MOV   C,bit

Bytes:	2
Cycles:	1
Encoding:		1 0 1 0	0 0 1 0		bit address	
Operation:	MOV
	(C) ← (bit)

MOV   bit,C

Bytes:	2
Cycles:	2
Encoding:		1 0 0 1	0 0 1 0		bit address	
Operation:	MOV
	(bit) ← (C)

MOV   DPTR,#data16

Function: Load Data Pointer with a 16-bit constant

Description: The Data Pointer is loaded with the 16-bit constant indicated. The 16-bit constant is loaded into the second and third bytes of the instruction. The second byte (DPH) is the high-order byte, while the third byte (DPL) holds the low-order byte. No flags are affected.

This is the only instruction which moves 16 bits of data at once.

Example: The instruction,

MOV   DPTR,#1234H

will load the value 1234H into the Data Pointer: DPH will hold 12H and DPL will hold 34H.

Bytes:	3
Cycles:	2
Encoding:		1 0 0 1	0 0 0 0		immed. data 15-8		immed. data 7-0	
Operation:	MOV
	$(DPTR) \leftarrow \#data_{15-0}$
	$DPH \,\square\, DPL \leftarrow \#data_{15-8} \,\square\, \#data_{7-0}$

MOVC A,@A+ <base-reg>

Function:	Move Code byte
Description:	The MOVC instructions load the Accumulator with a code byte, or constant from program memory. The address of the byte fetched is the sum of the original unsigned eight-bit Accumulator contents and the contents of a sixteen-bit base register, which may be either the Data Pointer or the PC. In the latter case, the PC is incremented to the address of the following instruction before being added with the Accumulator; otherwise the base register is not altered. Sixteen-bit addition is performed so a carry-out from the low-order eight bits may propagate through higher-order bits. No flags are affected.
Example:	A value between 0 and 3 is in the Accumulator. The following instructions will translate the value in the Accumulator to one of four values defined by the DB (define byte) directive.

```
REL_PC:  INC    A
         MOVC   A,@A+PC
         RET
         DB     66H
         DB     77H
         DB     88H
         DB     99H
```

If the subroutine is called with the Accumulator equal to 01H, it will return with 77H in the Accumulator. The INC A before the MOVC instruction is needed to "get around" the RET instruction above the table. If several bytes of code separated the MOVC from the table, the corresponding number would be added to the Accumulator instead.

MOVC A,@A + DPTR

Bytes:	1
Cycles:	2
Encoding:	`1 0 0 1	0 0 1 1`
Operation:	MOVC $(A) \leftarrow ((A) + (DPTR))$

MOVC A,@A + PC

Bytes:	1
Cycles:	2
Encoding:	`1 0 0 0	0 0 1 1`
Operation:	MOVC $(PC) \leftarrow (PC) + 1$ $(A) \leftarrow ((A) + (PC))$

MOVX   \<dest-byte\>,\<src-byte\>

Function: Move External

Description: The MOVX instructions transfer data between the Accumulator and a byte of external data memory, hence the "X" appended to MOV. There are two types of instructions, differing in whether they provide an eight-bit or sixteen-bit indirect address to the external data RAM.

In the first type, the contents of R0 or R1 in the current register bank provide an eight-bit address multiplexed with data on P0. Eight bits are sufficient for external I/O expansion decoding or for a relatively small RAM array. For somewhat larger arrays, any output port pins can be used to output higher-order address bits. These pins would be controlled by an output instruction preceding the MOVX.

In the second type of MOVX instruction, the Data Pointer generates a sixteen-bit address. P2 outputs the high-order eight address bits (the contents of DPH) while P0 multiplexes the low-order eight bits (DPL) with data. The P2 Special Function Register retains its previous contents while the P2 output buffers are emitting the contents of DPH. This form is faster and more efficient when accessing very large data arrays (up to 64K bytes), since no additional instructions are needed to set up the output ports.

It is possible in some situations to mix the two MOVX types. A large RAM array with its high-order address lines driven by P2 can be addressed via the Data Pointer, or with code to output high-order address bits to P2 followed by a MOVX instruction using R0 or R1.

Example: An external 256 byte RAM using multiplexed address/data lines (e.g., an Intel 8155 RAM/I/O/Timer) is connected to the 8051 Port 0. Port 3 provides control lines for the external RAM. Ports 1 and 2 are used for normal I/O. Registers 0 and 1 contain 12H and 34H. Location 34H of the external RAM holds the value 56H. The instruction sequence,

MOVX   A,@R1

MOVX   @R0,A

copies the value 56H into both the Accumulator and external RAM location 12H.

MOVX   A,@Ri

Bytes:	1
Cycles:	2
Encoding:	`1 1 1 0` `0 0 1 i`
Operation:	MOVX
	(A) ← ((Ri))

MOVX   A,@DPTR

Bytes:	1
Cycles:	2
Encoding:	`1 1 1 0` `0 0 0 0`
Operation:	MOVX
	(A) ← ((DPTR))

MOVX   @Ri,A

Bytes:	1
Cycles:	2
Encoding:	`1 1 1 1` `0 0 1 i`
Operation:	MOVX
	((Ri)) ← (A)

MOVX   @DPTR,A

Bytes:	1
Cycles:	2
Encoding:	`1 1 1 1` `0 0 0 0`
Operation:	MOVX
	(DPTR) ← (A)

MUL   AB

Function:	Multiply
Description:	MUL AB multiplies the unsigned eight-bit integers in the Accumulator and register B. The low-order byte of the sixteen-bit product is left in the Accumulator, and the high-order byte in B. If the product is greater than 255 (0FFH) the overflow flag is set; otherwise it is cleared. The carry flag is always cleared.
Example:	Originally the Accumulator holds the value 80 (50H). Register B holds the value 160 (0A0H). The instruction,

MUL   AB

will give the product 12,800 (3200H), so B is changed to 32H (00110010B) and the Accumulator is cleared. The overflow flag is set, carry is cleared.

Bytes:	1
Cycles:	4
Encoding:	1 0 1 0 0 1 0 0
Operation:	MUL $(A)_{7-0} \leftarrow (A) \times (B)$ $(B)_{15-8}$

NOP

Function:	No Operation
Description:	Execution continues at the following instruction. Other than the PC, no registers or flags are affected.
Example:	It is desired to produce a low-going output pulse on bit 7 of Port 2 lasting exactly 5 cycles. A simple SETB/CLR sequence would generate a one-cycle pulse, so four additional cycles must be inserted. This may be done (assuming no interrupts are enabled) with the instruction sequence,

CLR   P2.7
NOP
NOP
NOP
NOP
SETB   P2.7

Bytes:	1
Cycles:	1
Encoding:	0 0 0 0 0 0 0 0
Operation:	NOP $(PC) \leftarrow (PC) + 1$

ORL   <dest-byte> <src-byte>

Function: Logical-OR for byte variables

Description: ORL performs the bitwise logical-OR operation between the indicated variables, storing the results in the destination byte. No flags are affected.

The two operands allow six addressing mode combinations. When the destination is the Accumulator, the source can use register, direct, register-indirect, or immediate addressing; when the destination is a direct address, the source can be the Accumulator or immediate data.

Note: When this instruction is used to modify an output port, the value used as the original port data will be read from the output data latch, *not* the input pins.

Example: If the Accumulator holds 0C3H (11000011B) and R0 holds 55H (01010101B) then the instruction,

ORL   A,R0

will leave the Accumulator holding the value 0D7H (11010111B).

When the destination is a directly addressed byte, the instruction can set combinations of bits in any RAM location or hardware register. The pattern of bits to be set is determined by a mask byte, which may be either a constant data value in the instruction or a variable computed in the Accumulator at run-time. The instruction,

ORL   P1,#00110010B

will set bits 5, 4, and 1 of output Port 1.

ORL   A,Rn

Bytes: 1

Cycles: 1

Encoding: | 0 1 0 0 | 1 r r r |

Operation: ORL
$(A) \leftarrow (A) \vee (Rn)$

ORL A,direct

Bytes:	2
Cycles:	1

Encoding: `0 1 0 0 | 0 1 0 1` `direct address`

Operation: ORL
$(A) \leftarrow (A) \lor (\text{direct})$

ORL A,@Ri

Bytes:	1
Cycles:	1

Encoding: `0 1 0 0 | 0 1 1 i`

Operation: ORL
$(A) \leftarrow (A) \lor ((Ri))$

ORL A, # data

Bytes:	2
Cycles:	1

Encoding: `0 1 0 0 | 0 1 0 0` `immediate data`

Operation: ORL
$(A) \leftarrow (A) \lor \#\text{data}$

ORL direct,A

Bytes:	2
Cycles:	1

Encoding: `0 1 0 0 | 0 0 1 0` `direct address`

Operation: ORL
$(\text{direct}) \leftarrow (\text{direct}) \lor (A)$

ORL direct,#data

Bytes:	3
Cycles:	2

Encoding: `0 1 0 0 | 0 0 1 1` `direct addr.` `immediate data`

Operation: ORL
$(\text{direct}) \leftarrow (\text{direct}) \lor \#\text{data}$

ORL   C,<src-bit>

Function:	Logical-OR for bit variables
Description:	Set the carry flag if the Boolean value is a logical 1; leave the carry in its current state otherwise. A slash ("/") preceding the operand in the assembly language indicates that the logical complement of the addressed bit is used as the source value, but the source bit itself is not affected. No other flags are affected.
Example:	Set the carry flag if and only if P1.0 = 1, ACC. 7 = 1, or OV = 0:

    MOV   C,P1.0   ; LOAD CARRY WITH INPUT PIN P10
    ORL   C,ACC.7   ; OR CARRY WITH THE ACC. BIT 7
    ORL   C,/OV   ; OR CARRY WITH THE INVERSE OF OV.

ORL   C,bit

Bytes:	2
Cycles:	2
Encoding:	`0 1 1 1  0 0 1 0` `bit address`
Operation:	ORL
	$(C) \leftarrow (C) \vee (bit)$

ORL   C,/bit

Bytes:	2
Cycles:	2
Encoding:	`1 0 1 0  0 0 0 0` `bit address`
Operation:	ORL
	$(C) \leftarrow (C) \vee (\overline{bit})$

intel®   MCS®-51 PROGRAMMER'S GUIDE AND INSTRUCTION SET

POP   direct

Function:	Pop from stack.
Description:	The contents of the internal RAM location addressed by the Stack Pointer is read, and the Stack Pointer is decremented by one. The value read is then transferred to the directly addressed byte indicated. No flags are affected.
Example:	The Stack Pointer originally contains the value 32H, and internal RAM locations 30H through 32H contain the values 20H, 23H, and 01H, respectively. The instruction sequence,

POP   DPH

POP   DPL

will leave the Stack Pointer equal to the value 30H and the Data Pointer set to 0123H. At this point the instruction,

POP   SP

will leave the Stack Pointer set to 20H. Note that in this special case the Stack Pointer was decremented to 2FH before being loaded with the value popped (20H).

Bytes:	2
Cycles:	2
Encoding:	`1 1 0 1  0 0 0 0` `direct address`
Operation:	POP
	$(direct) \leftarrow ((SP))$
	$(SP) \leftarrow (SP) - 1$

PUSH   direct

Function:	Push onto stack
Description:	The Stack Pointer is incremented by one. The contents of the indicated variable is then copied into the internal RAM location addressed by the Stack Pointer. Otherwise no flags are affected.
Example:	On entering an interrupt routine the Stack Pointer contains 09H. The Data Pointer holds the value 0123H. The instruction sequence,

PUSH   DPL

PUSH   DPH

will leave the Stack Pointer set to 0BH and store 23H and 01H in internal RAM locations

0AH and 0BH, respectively.

Bytes:	2
Cycles:	2
Encoding:	`1 1 0 0  0 0 0 0` `direct address`
Operation:	PUSH
	$(SP) \leftarrow (SP) + 1$
	$((SP)) \leftarrow (direct)$

intel®   **MCS®-51 PROGRAMMER'S GUIDE AND INSTRUCTION SET**

RET

Function:	Return from subroutine
Description:	RET pops the high- and low-order bytes of the PC successively from the stack, decrementing the Stack Pointer by two. Program execution continues at the resulting address, generally the instruction immediately following an ACALL or LCALL. No flags are affected.
Example:	The Stack Pointer originally contains the value 0BH. Internal RAM locations 0AH and 0BH contain the values 23H and 01H, respectively. The instruction,

RET

will leave the Stack Pointer equal to the value 09H. Program execution will continue at location 0123H.

Bytes:	1
Cycles:	2
Encoding:	0 0 1 0 0 0 1 0
Operation:	RET

$(PC_{15-8}) \leftarrow ((SP))$
$(SP) \leftarrow (SP) - 1$
$(PC_{7-0}) \leftarrow ((SP))$
$(SP) \leftarrow (SP) - 1$

RETI

Function:	Return from interrupt
Description:	RETI pops the high- and low-order bytes of the PC successively from the stack, and restores the interrupt logic to accept additional interrupts at the same priority level as the one just processed. The Stack Pointer is left decremented by two. No other registers are affected; the PSW is *not* automatically restored to its pre-interrupt status. Program execution continues at the resulting address, which is generally the instruction immediately after the point at which the interrupt request was detected. If a lower- or same-level interrupt had been pending when the RETI instruction is executed, that one instruction will be executed before the pending interrupt is processed.
Example:	The Stack Pointer originally contains the value 0BH. An interrupt was detected during the instruction ending at location 0122H. Internal RAM locations 0AH and 0BH contain the values 23H and 01H, respectively. The instruction,

RETI

will leave the Stack Pointer equal to 09H and return program execution to location 0123H.

Bytes:	1
Cycles:	2
Encoding:	0 0 1 1 0 0 1 0
Operation:	RETI

$(PC_{15-8}) \leftarrow ((SP))$
$(SP) \leftarrow (SP) - 1$
$(PC_{7-0}) \leftarrow ((SP))$
$(SP) \leftarrow (SP) - 1$

intel.   **MCS®-51 PROGRAMMER'S GUIDE AND INSTRUCTION SET**

RL   A

Function:	Rotate Accumulator Left
Description:	The eight bits in the Accumulator are rotated one bit to the left. Bit 7 is rotated into the bit 0 position. No flags are affected.
Example:	The Accumulator holds the value 0C5H (11000101B). The instruction,
	RL A
	leaves the Accumulator holding the value 8BH (10001011B) with the carry unaffected.
Bytes:	1
Cycles:	1
Encoding:	`0 0 1 0	0 0 1 1`
Operation:	RL
	$(A_n + 1) \leftarrow (An)$ n = 0 - 6
	$(A0) \leftarrow (A7)$

RLC   A

Function:	Rotate Accumulator Left through the Carry flag
Description:	The eight bits in the Accumulator and the carry flag are together rotated one bit to the left. Bit 7 moves into the carry flag; the original state of the carry flag moves into the bit 0 position. No other flags are affected.
Example:	The Accumulator holds the value 0C5H (11000101B), and the carry is zero. The instruction,
	RLC A
	leaves the Accumulator holding the value 8BH (10001010B) with the carry set.
Bytes:	1
Cycles:	1
Encoding:	`0 0 1 1	0 0 1 1`
Operation:	RLC
	$(An + 1) \leftarrow (An)$ n = 0 - 6
	$(A0) \leftarrow (C)$
	$(C) \leftarrow (A7)$

intel®

RR A

Function:	Rotate Accumulator Right
Description:	The eight bits in the Accumulator are rotated one bit to the right. Bit 0 is rotated into the bit 7 position. No flags are affected.
Example:	The Accumulator holds the value 0C5H (11000101B). The instruction,

RR A

leaves the Accumulator holding the value 0E2H (11100010B) with the carry unaffected.

Bytes:	1
Cycles:	1
Encoding:	0 0 0 0 | 0 0 1 1
Operation:	RR $(An) \leftarrow (A_n + 1)$ $n = 0 - 6$ $(A7) \leftarrow (A0)$

RRC A

Function:	Rotate Accumulator Right through Carry flag
Description:	The eight bits in the Accumulator and the carry flag are together rotated one bit to the right. Bit 0 moves into the carry flag; the original value of the carry flag moves into the bit 7 position. No other flags are affected.
Example:	The Accumulator holds the value 0C5H (11000101B), the carry is zero. The instruction,

RRC A

leaves the Accumulator holding the value 62 (01100010B) with the carry set.

Bytes:	1
Cycles:	1
Encoding:	0 0 0 1 | 0 0 1 1
Operation:	RRC $(An) \leftarrow (An + 1)$ $n = 0 - 6$ $(A7) \leftarrow (C)$ $(C) \leftarrow (A0)$

intel® **MCS®-51 PROGRAMMER'S GUIDE AND INSTRUCTION SET**

SETB   <bit>

Function:	Set Bit
Description:	SETB sets the indicated bit to one. SETB can operate on the carry flag or any directly addressable bit. No other flags are affected.
Example:	The carry flag is cleared. Output Port 1 has been written with the value 34H (00110100B). The instructions,

SETB   C

SETB   P1.0

will leave the carry flag set to 1 and change the data output on Port 1 to 35H (00110101B).

SETB   C

Bytes:	1
Cycles:	1
Encoding:	1 1 0 1 | 0 0 1 1
Operation:	SETB
	(C) ← 1

SETB   bit

Bytes:	2
Cycles:	1
Encoding:	1 1 0 1 | 0 0 1 0 bit address
Operation:	SETB
	(bit) ← 1

SJMP   rel

Function:	Short Jump
Description:	Program control branches unconditionally to the address indicated. The branch destination is computed by adding the signed displacement in the second instruction byte to the PC, after incrementing the PC twice. Therefore, the range of destinations allowed is from 128 bytes preceding this instruction to 127 bytes following it.
Example:	The label "RELADR" is assigned to an instruction at program memory location 0123H. The instruction,

SJMP   RELADR

will assemble into location 0100H. After the instruction is executed, the PC will contain the value 0123H.

(*Note:* Under the above conditions the instruction following SJMP will be at 102H. Therefore, the displacement byte of the instruction will be the relative offset (0123H-0102H) = 21H. Put another way, an SJMP with a displacement of 0FEH would be a one-instruction infinite loop.)

Bytes:	2
Cycles:	2
Encoding:	`1 0 0 0  0 0 0 0` `rel. address`
Operation:	SJMP

$(PC) \leftarrow (PC) + 2$
$(PC) \leftarrow (PC) + rel$

intel₀   MCS®-51 PROGRAMMER'S GUIDE AND INSTRUCTION SET

SUBB   A,<src-byte>

Function:	Subtract with borrow
Description:	SUBB subtracts the indicated variable and the carry flag together from the Accumulator, leaving the result in the Accumulator. SUBB sets the carry (borrow) flag if a borrow is needed for bit 7, and clears C otherwise. (If C was set *before* executing a SUBB instruction, this indicates that a borrow was needed for the previous step in a multiple precision subtraction, so the carry is subtracted from the Accumulator along with the source operand.) AC is set if a borrow is needed for bit 3, and cleared otherwise. OV is set if a borrow is needed into bit 6, but not into bit 7, or into bit 7, but not bit 6.

When subtracting signed integers OV indicates a negative number produced when a negative value is subtracted from a positive value, or a positive result when a positive number is subtracted from a negative number.

The source operand allows four addressing modes: register, direct, register-indirect, or immediate.

Example:	The Accumulator holds 0C9H (11001001B), register 2 holds 54H (01010100B), and the carry flag is set. The instruction,

SUBB   A,R2

will leave the value 74H (01110100B) in the accumulator, with the carry flag and AC cleared but OV set.

Notice that 0C9H minus 54H is 75H. The difference between this and the above result is due to the carry (borrow) flag being set before the operation. If the state of the carry is not known before starting a single or multiple-precision subtraction, it should be explicitly cleared by a CLR C instruction.

SUBB   A,Rn

Bytes:	1
Cycles:	1
Encoding:	`1 0 0 1` `1 r r r`
Operation:	SUBB
	(A) ← (A) – (C) – (Rn)

SUBB   A,direct

Bytes:	2
Cycles:	1
Encoding:	`1 0 0 1` `0 1 0 1` `direct address`
Operation:	SUBB
	$(A) \leftarrow (A) - (C) - (direct)$

SUBB   A,@Ri

Bytes:	1
Cycles:	1
Encoding:	`1 0 0 1` `0 1 1 i`
Operation:	SUBB
	$(A) \leftarrow (A) - (C) - ((Ri))$

SUBB   A,#data

Bytes:	2
Cycles:	1
Encoding:	`1 0 0 1` `0 1 0 0` `immediate data`
Operation:	SUBB
	$(A) \leftarrow (A) - (C) - \#data$

SWAP   A

Function:	Swap nibbles within the Accumulator
Description:	SWAP A interchanges the low- and high-order nibbles (four-bit fields) of the Accumulator (bits 3-0 and bits 7-4). The operation can also be thought of as a four-bit rotate instruction. No flags are affected.
Example:	The Accumulator holds the value 0C5H (11000101B). The instruction,
	SWAP A
	leaves the Accumulator holding the value 5CH (01011100B).
Bytes:	1
Cycles:	1
Encoding:	`1 1 0 0` `0 1 0 0`
Operation:	SWAP
	$(A_{3-0}) \rightleftarrows (A_{7-4})$

int͢el.   **MCS®-51 PROGRAMMER'S GUIDE AND INSTRUCTION SET**

XCH   A,<byte>

Function:	Exchange Accumulator with byte variable
Description:	XCH loads the Accumulator with the contents of the indicated variable, at the same time writing the original Accumulator contents to the indicated variable. The source/destination operand can use register, direct, or register-indirect addressing.
Example:	R0 contains the address 20H. The Accumulator holds the value 3FH (00111111B). Internal RAM location 20H holds the value 75H (01110101B). The instruction,

XCH   A,@R0

will leave RAM location 20H holding the values 3FH (00111111B) and 75H (01110101B) in the accumulator.

XCH   A,Rn

Bytes:	1
Cycles:	1
Encoding:	`1 1 0 0` `1 r r r`
Operation:	XCH
	(A) ⇄ (Rn)

XCH   A,direct

Bytes:	2
Cycles:	1
Encoding:	`1 1 0 0` `0 1 0 1` `direct address`
Operation:	XCH
	(A) ⇄ (direct)

XCH   A,@Ri

Bytes:	1
Cycles:	1
Encoding:	`1 1 0 0` `0 1 1 i`
Operation:	XCH
	(A) ⇄ ((Ri))

intel. MCS®-51 PROGRAMMER'S GUIDE AND INSTRUCTION SET

XCHD A,@Ri

Function:	Exchange Digit
Description:	XCHD exchanges the low-order nibble of the Accumulator (bits 3-0), generally representing a hexadecimal or BCD digit, with that of the internal RAM location indirectly addressed by the specified register. The high-order nibbles (bits 7-4) of each register are not affected. No flags are affected.
Example:	R0 contains the address 20H. The Accumulator holds the value 36H (00110110B). Internal RAM location 20H holds the value 75H (01110101B). The instruction,

XCHD A,@R0

will leave RAM location 20H holding the value 76H (01110110B) and 35H (00110101B) in the Accumulator.

Bytes:	1
Cycles:	1
Encoding:	1 1 0 1 | 0 1 1 i
Operation:	XCHD $(A_{3-0}) \rightleftarrows ((Ri_{3-0}))$

XRL <dest-byte>,<src-byte>

Function:	Logical Exclusive-OR for byte variables
Description:	XRL performs the bitwise logical Exclusive-OR operation between the indicated variables, storing the results in the destination. No flags are affected.

The two operands allow six addressing mode combinations. When the destination is the Accumulator, the source can use register, direct, register-indirect, or immediate addressing; when the destination is a direct address, the source can be the Accumulator or immediate data.

(*Note:* When this instruction is used to modify an output port, the value used as the original port data will be read from the output data latch, *not* the input pins.)

Example:	If the Accumulator holds 0C3H (11000011B) and register 0 holds 0AAH (10101010B) then the instruction,

XRL A,R0

will leave the Accumulator holding the value 69H (01101001B).

When the destination is a directly addressed byte, this instruction can complement combinations of bits in any RAM location or hardware register. The pattern of bits to be complemented is then determined by a mask byte, either a constant contained in the instruction or a variable computed in the Accumulator at run-time. The instruction,

XRL P1,# 00110001B

will complement bits 5, 4, and 0 of output Port 1.

XRL　A,Rn

Bytes:	1
Cycles:	1
Encoding:	0 1 1 0	1 r r r
Operation:	XRL
	$(A) \leftarrow (A) \lor (Rn)$

$\lor$

XRL　A,direct

Bytes:	2
Cycles:	1
Encoding:	0 1 1 0	0 1 0 1 　　direct address
Operation:	XRL
	$(A) \leftarrow (A) \lor (direct)$

XRL　A,@Ri

Bytes:	1
Cycles:	1
Encoding:	0 1 1 0	0 1 1 i
Operation:	XRL
	$(A) \leftarrow (A) \lor ((Ri))$

XRL　A,#data

Bytes:	2
Cycles:	1
Encoding:	0 1 1 0	0 1 0 0 　　immediate data
Operation:	XRL
	$(A) \leftarrow (A) \lor \#data$

XRL　direct,A

Bytes:	2
Cycles:	1
Encoding:	0 1 1 0	0 0 1 0 　　direct address
Operation:	XRL
	$(direct) \leftarrow (direct) \lor (A)$

XRL   direct,#data

    Bytes: 3

    Cycles: 2

    Encoding: | 0 1 1 0 | 0 0 1 1 | | direct address | | immediate data |

    Operation: XRL

          (direct) ← (direct) ∀ #data

BIBLIOGRAPHY

Ayala, Kenneth J. *The 8051 Microcontroller – Architecture, Programming and Applications*, 3rd edition, 2008, Cengage Learning India Pvt. Ltd.

Deshmukh, Ajay V. *Microcontrollers [Theory and Applications]*, 1st edition, 2006, Tata McGraw-Hill

Gadre, Dhananjay V. *Programming and Customizing the AVR Microcontroller*, 1st edition (10th reprint), 2008, Tata McGraw-Hill

Ghoshal, Subrata. *Embedded Systems and Robots: Projects using the 8051 Microcontroller*, 1st edition, 2009, Cengage Learning India Pvt. Ltd.

Hall, Douglas V. *Microprocessors and Interfacing: Programming and Hardware*, 2nd edition, 2000, Tata McGraw-Hill

Kamal, Raj. *Microcontroller Architecture, Programming, Interfacing and System Design*, 1st edition, 2007, Pearson Education

Mazidi, Muhammad Ali, Mazidi, Janice Gillispie and McKinlay, Rolin D. *The 8051 Microcontroller and Embedded Systems (using assembly and C)*, 2nd edition, 2008, Pearson Education

Predko, Myke. *Programming and Customizing the 8051 Microcontroller*, 2003, Tata McGraw-Hill

http://www.8051projects.net/

http://www.fullandfree.info/software/kiel-c51-compiler/

http://www.win.tue.nl/~aeb/comp/8051/set8051.html

INDEX